STEAM-H: Science, Technology, Engineering, Agriculture, Mathematics & Health

Series Editor

Bourama Toni, Department of Mathematics, Howard University, Washington, USA

This interdisciplinary series highlights the wealth of recent advances in the pure and applied sciences made by researchers collaborating between fields where mathematics is a core focus. As we continue to make fundamental advances in various scientific disciplines, the most powerful applications will increasingly be revealed by an interdisciplinary approach. This series serves as a catalyst for these researchers to develop novel applications of, and approaches to, the mathematical sciences. As such, we expect this series to become a national and international reference in STEAM-H education and research.

Interdisciplinary by design, the series focuses largely on scientists and mathematicians developing novel methodologies and research techniques that have benefits beyond a single community. This approach seeks to connect researchers from across the globe, united in the common language of the mathematical sciences. Thus, volumes in this series are suitable for both students and researchers in a variety of interdisciplinary fields, such as: mathematics as it applies to engineering; physical chemistry and material sciences; environmental, health, behavioral and life sciences; nanotechnology and robotics; computational and data sciences; signal/image processing and machine learning; finance, economics, operations research, and game theory. The series originated from the weekly yearlong STEAM-H Lecture series at Virginia State University featuring world-class experts in a dynamic forum. Contributions reflected the most recent advances in scientific knowledge and were delivered in a standardized, self-contained and pedagogically-oriented manner to a multidisciplinary audience of faculty and students with the objective of fostering student interest and participation in the STEAM-H disciplines as well as fostering interdisciplinary collaborative research. The series strongly advocates multidisciplinary collaboration with the goal to generate new interdisciplinary holistic approaches, instruments and models, including new knowledge, and to transcend scientific boundaries.

Peer reviewing

All monographs and works selected for contributed volumes within the STEAM-H series undergo peer review. The STEAM-H series follows a single-blind review process. A minimum of two reports are asked for each submitted manuscript. The Volume Editors act in cooperation with the Series Editor for a final decision. The Series Editors agrees with and follows the guidelines published by the Committee on Publication Ethics.

Titles from this series are indexed by Scopus, Mathematical Reviews, and zbMATH.

Vladimir Anashin

Causality: The *p*-adic Theory

Springer

Vladimir Anashin
Moscow State University
Moscow, Russia

ISSN 2520-193X ISSN 2520-1948 (electronic)
STEAM-H: Science, Technology, Engineering, Agriculture, Mathematics & Health
ISBN 978-3-031-85817-8 ISBN 978-3-031-85818-5 (eBook)
https://doi.org/10.1007/978-3-031-85818-5

This Springer imprint is published by the registered company Springer Nature Switzerland AG
The registered company address is: Gewerbestrasse 11, 6330 Cham, Switzerland

To Alla and Masha, with gratitude for their love and care

Foreword

The book *Causality: The p-adic Theory* by Vladimir Anashin is devoted to the study of the question posed by Albert Einstein: "Does God play dice ?" Or in other words, is there randomness inherent in our world at a fundamental level?

According to modern quantum mechanics, there is an irreducible randomness of the fundamental physical laws. However, Einstein did not agree with this statement and in the discussion with Niels Bohr presented arguments that such a situation is impossible.

Although Einstein failed to convince Bohr that he was right, as a result of this discussion, the famous article of Einstein, Podolsky, and Rozen appeared, which became the initial stage in the development of modern quantum information. These studies were continued by Erwin Schrodinger, John Bell, and many others and now forms the basis of quantum technology.

However, the question of whether Nature is deterministic at a fundamental level or whether it has an element of chance continues to be debated in the literature. In particular, the Nobel laureate Gerald 't Hooft believes that there is no chance of randomness on the fundamental level. This point of view is shared by the author of this book, Vladimir Anashin, but with a caveat: He presents strong mathematical arguments that if causality is inherent to Nature, then on the base of observational data it is impossible to answer that question "yes" or "no."

Although the current dominant view is that there is irreducible randomness at the fundamental level, for the future development of the theory of quantum cosmology, black hole physics, chaos, big data theory, and artificial intelligence, it may be useful and even necessary to consider an alternative point of view when at the fundamental level the laws of nature are deterministic, and chaotic and random behavior arises as a consequence of an inevitable error in determining the initial data. The author of the book, Vladimir Anashin, is a professor of computer science at Moscow State University, and is a major specialist in computer science, so based on his expertise in that area he argues that a deterministic point of view is not less rationalized than a non-deterministic one and obtains remarkable results along this path in this book.

The book consists of three parts, it includes the necessary information on p-adic analysis, automata theory, as well as physical applications. The author managed to obtain a beautiful description of causal functions based on the following assumptions. On the one hand, he takes as a basis 't Hooft's deterministic approach to the concept of causality based on automata theory. On the other hand, he sets out the basic principles for describing fundamental laws based on postulates such as the observability of only rational numbers, the inability to observe and measure distances smaller than Planck distances, as well as the invariance of the laws of Nature under change of the number field. On this basis the author was able to obtain a complete description of causal functions in terms of special polynomials. There is no doubt that the Anashin's approach will play an important role in the further development of the theory.

This book may be useful to readers interested in the foundations of quantum theory, computer science, p-adic analysis, and their applications.

Moscow
June, 2024

Igor Volovich

Preface

This book is about causality, yet not of causality as a philosophy concept but rather of causality as a mathematical notion. That is, *the main content of the book is mathematical theory of causal functions over discrete time*. The mathematical theory is not a pure mathematical one: It is motivated by applied problems, and a number of mathematical results that are obtained within this theory already have (or may have in future) applications within other mathematical disciplines like automata theory and combinatorics as well as outside mathematics, namely, in applied computer science and maybe (or so I hope) in physics.

By this reason, the book is divided into three parts: The main content of Parts I and II is the said mathematical theory (which basically is the theory of p-adic 1-Lipschitz functions); Part III is devoted to applications of this theory. Though certain mathematical maturity of the readers is mandatory (MS degree in Mathematics, or in Computer Science, or in Physics is desirable), the readers who are interested mostly in applications may use Part II for references only and do not read proofs; however, for these readers it is necessary at least to look through Part I to get acquainted with basics of p-adic theory which is used throughout the book. Chapter 1 which belongs to neither of Parts I–III contains a variety of mathematical notions and facts which are used in the book; this chapter is mostly a collection of reminders to ease referencing. The most important of these reminders, the rigorous definition of causal function over discrete time, is given in this chapter, in Sect. 1.6.2.

Causal function is the main object examined in the book; the latter function may be thought of as a black box which accepts "causes" and produces "effects." The causes/effects are sequences of "elementary causes/effects" which follow one by one, each elementary cause/effect in each instant of time. Therefore, time is assumed to be discrete, and at each discrete instant of time only one of all elementary causes/effects is accepted/produced by the black box. The total number of elementary causes/effects is assumed to be finite; this *finiteness assumption* is assumed since no physical objects the total number of whose is infinite are known: Even the number of all elementary particles in our Universe is finite.

When the black box accepts an elementary cause, the box produces an elementary effect; but *if the next elementary cause is the same as the previous one, the next elementary effect may be different from the one which was produced just before.* This means that *when the black box accepts elementary cause, something is changed inside the box*; therefore, the elementary effects that produces the black box depend not only on elementary causes the box accepts but also on this "something" which is constantly changing inside the black box in dependence on elementary causes the black box accepts. These "something(s)" are called "states" of the black box. The *states are epistemic*; i.e., nobody can observe them directly: An observer may only make guesses about the states observing sequences of elementary causes and respective elementary effects. The guesses show that *generally the next state depends not only on the elementary cause the black box accepted but also on the state the black box was just before*, at the previous time instant. For mathematicians and computer scientists it should be clear now that the black box is *sequential machine*, a *letter-to-letter transducer*, a *synchronous automaton*: All these terms mean the same thing that is also known as *Mealy sequential machine* or *Moore sequential machine*. Yes, this is true, but with an adjustment that the total number of states (whatever could these be) of a black box need not be a finite. This does not contradict the finiteness assumptions since the "states" are epistemic, and not ontic, so one may think of infinity and use infinity in calculations despite no infinity exists in physical world: Infinity is a mathematical concept, not a physical entity; but the concept has proven useful over the centuries because it allows us to estimate the values of quantities that depend on variables that are too small or too large compared to others (i.e., when the values of these variables "tend to their limits").

A physicist may think of the black box as a model of open physical system, but with a caution: The physicist should not assume that the whole long strings of elementary causes and elementary effects, i.e., causes and effects as a whole, are completely observable. Very rarely causes and effects are the things that are available to observations with no "distortion." To put it in other words, the observables are just approximations of the causes and effects.

The question that should be asked immediately after the word "approximation" is that: With respect to what metric the approximation is? It is necessary to state rigorously which causes/effects are close to one another and which are far. This metric is defined in the book; actually *three* different metrics are defined.

1. The first one of the three metrics is standard real *Archimedean metric* which can be defined for the causes/effects as follows: Let there are only $p > 1$ elementary causes and elementary effects enumerated by $\{0, 1, \dots, p-1\}$; then, under the finiteness assumption, the causes/effects may be thought of as strings of some length ℓ. To each string $\alpha_{\ell-1}\alpha_{\ell-2}\dots\alpha_0$, where indices stand for time instances the elementary cause/effect happens, put into the correspondence the number $0.\alpha_{\ell-1}\alpha_{\ell-2}\dots\alpha_0 = \alpha_{\ell-1}p^{-1} + \alpha_{\ell-2}p^{-2} + \cdots \alpha_0 p^{-\ell}$ from the unit real segment $[0, 1]$. That is, the most significant digits of the base-p expansion of these numbers correspond to elementary causes/effects which happen later: The more significant is the digit in the base-p expansion of the number which corresponds to cause/effect, the later respective elementary cause/effect happen. The distance between the causes/effects is, by the definition, the ordinary distance between

real numbers $a, b \in [0, 1]$, the real absolute $|a - b|$. That is, the metric characterizes temporal evolution of the system represented by the black box: *Elementary causes/effects that happen later are more significant than the ones that have happened earlier; but the earlier ones may have a crucial influence on future behavior of the system* since each of these early elementary causes/effects change current states of the system and thus may lead to a sort of "dead butterfly effect" from R. Bradbury's science fiction short story *The Sound of Thunder*.

2. The second metric is defined similarly, but the digits of the numbers are taken in the opposite order compared to the first metric: To the sequence $\alpha_{\ell-1}\alpha_{\ell-2}\ldots\alpha_0$ there corresponds the number $0.\alpha_0\alpha_1\ldots\alpha_{\ell-1} = \alpha_0 p^{-1} + \alpha_1 p^{-2} + \cdots \alpha_{\ell-1} p^{-\ell}$, i.e., the earlier the elementary cause/effect happens, the more significant the corresponding digit is. The metric in this case is again the metric on reals; *this metric which reveals the "immediate response" of the system since the later the elementary causes happen, the less significant is the total effect; only the earliest elementary causes matter.*

   Metrics 1–2 have some deficiencies: *The first metric can be well-defined for finite sequences of elementary causes/effects only; therefore, it is not suitable for "predictions" of future behavior of the system or to "reveal the history" of the system.* Though the second metric can be defined for infinite sequences of elementary causes/effects, for a pair of different causes/effects there may correspond one real number, and there are infinitely many such pairs of causes/effects. *The second metric is therefore a pseudo-metric; moreover, it is "incapable to predict the future,"* e.g., to predict results of experiments.

3. To avoid these deficiencies, a yet one more metric is needed further; actually this metric is the main metric that is used throughout the book. The metric is a natural metric on one-side infinite sequences: Given two left-infinite sequences $\mathbf{a} = \cdots\alpha_3\alpha_2\alpha_1\alpha_0$ and $\mathbf{a}' = \cdots\alpha'_3\alpha'_2\alpha'_1\alpha'_0$ over $\{0, 1, \ldots, p-1\}$, put $d_p(\mathbf{a}, \mathbf{a}') = p^{-n}$, where $n = \max\{k : \alpha_k = \alpha'_k\}$ if that n exists, otherwise put $d_p(\mathbf{a}, \mathbf{a}') = 0$ if $\alpha_i = \alpha'_i$ for all $k = 0, 1, 2\ldots$. This metric is a metric rather than a pseudo-metric: $d_p(\mathbf{a}, \mathbf{a}') = 0$ if and only if $\mathbf{a} = \mathbf{a}'$. Moreover, the metric d_p is capable "to predict the future and to reveal the past" since to every black box $\mathfrak{B}$ there corresponds a unique well-defined map $f_{\mathfrak{B}}$ of "causes" to "effects" which is continuous with respect to this metric since $d_p(f_{\mathfrak{B}}(\mathbf{a}), f_{\mathfrak{B}}(\mathbf{a}')) \leqslant d_p(\mathbf{a}, \mathbf{a}')$. *The metric d_p is an ultrametric, the non-Archimedean metric* which satisfies strong triangle inequality $d_p(\mathbf{a}, \mathbf{c}) \leqslant \max\{d_p(\mathbf{a}, \mathbf{b}), d_p(\mathbf{b}, \mathbf{c})\}$; therefore, the mapping $f_{\mathfrak{B}}$ is 1-Lipschitz with respect to the metric d_p, i.e., satisfies Lipschitz condition with the constant 1. The most important is that, given any 1-Lipschitz map f of "causes" to "effects" there exists a "black box" $\mathfrak{B}$, the sequential machine, such that $f = f_{\mathfrak{B}}$. *This fact immediately gives a possibility to examine mappings of "causes" to "effects" that the black box (which a model of open system) performs, in order to understand and predict behavior of the system rather than trying unsuccessfully to "open the box in order to look what is inside the box." Facilities for that examination do exist: These are p-adic analysis and p-adic dynamics.*

By this reason, the claim that *causality is essentially non-Archimedean* might be taken as a motto of the book, but not as a motto only: *The non-Archimedean approach to causality results in a sort of a toolbox full of mathematical tools which can be (and some already are) effectively used in concrete applied areas* like the design of pseudo-random number generators (see Chap. 14) for computer simulations, data protection, etc.; experiment design (see Chap. 15); control of automata behavior, design of automata having prescribed properties, smart contracts design (see Chap. 13); straight-line computer programs (SLP) design (see Sect. 9.5). The underlying reason why the "non-Archimedean tools" may work (and already are working) in the said applied areas is that *main computer instructions, both arithmetical (addition and multiplication of integers) and bitwise logical (like bitwise negation* NOT, *bitwise logical conjunction* AND, *bitwise logical disjunction* OR, *etc.) as well as their compositions, are continuous with respect to non-Archimedean 2-adic metric*.

Another aspect of the non-Archimedean view at causality, the dominating view in the book, is related to contemporary *superdeterministic interpretations of quantum mechanics*, see Chap. 16. The reader should be warned that the author of this book is a mathematician rather than a physicist and therefore does not feel himself confident enough to develop a yet one more interpretation of quantum mechanics; but nonetheless the author takes the liberty to use the non-Archimedean approach to some fundamental questions of quantum theory in order to shed some mathematical light onto these questions. These fundamental questions are: Whether Nature on the smallest of the scales is "discrete" or "continuous"? Whether randomness is intrinsic property of Nature at the smallest of scales, or maybe Nature is strictly deterministic? And what is "free will" in case strict determinism rather than randomness is immanent? The answers are given in the form of *interpretations* rather than mathematical theorems; but all the interpretations are based on mathematical assertions which are rigorously proved in the book.

As a whole, these interpretations (supported by mathematical assertions) constitute an evidence for the claim that superdeterministic models of quantum mechanics may have same predictability power as "classical" models. *The very appearance of superdeterministic models of quantum mechanics is in some sense inevitable in view of model-dependent realism concept* (the term was coined by Stephen Hawking and Leonard Mlodinow in their 2010 book, *The Grand Design*). However, this is not the only reason why the author touches the questions related to very foundations of quantum mechanics: *The questions of that kind inevitably arise in connection with concrete applications* like semiconductor devices based on quantum processes that are assumed to be "chaotic." The devices are claimed to produce sequences of "true random" numbers or perform "true random transformations" of bitwords; these devices are aimed to be used in industry as parts of sensitive appliances where the randomness is a must. These practical questions are also mentioned (though briefly) in Chap. 16.

Moscow, Russia
May, 2024

Vladimir Anashin

Acknowledgments

Me, the author of this book, is indebted to many people thanks to whom the book appears; and especially for the questions the people asked me. Good questions matter! A good question is always a good starting point of interesting research.

First of all, I am indebted to my late Teacher and Supervisor Professor *Mikhail Mikhailovich Glukhov* not only for teaching and supervising but also for asking me a question which became a starting point of all my future research: How to produce the recurrence sequence of the longest period over a non-Abelian group using only operations of the group? This problem turned out to be fruitful and related to other interesting problems like, e.g., identities in groups, polynomial and affine completeness of groups, etc. Finally, my attempts to solve this problem resulted in ergodic theory for profinite groups with operators [13, 24], and papers on mixed identities in groups [6, 8].

The next good question was due to my colleague and friend *Mikhail Larin* who asked me once: "Why only the groups? What about recurrence sequences of the longest period produced only with the use of operations of other algebraic structures?". To that moment, Mikhail had already proved his criterion of transitivity of polynomials over residue rings modulo N which stayed unpublished until 2002, [132]. I got many ideas from his theorem (and published paper [7] developing these ideas); but the most important for me was the next question by Mikhail: "Do you see that the crucial property you and me are using in our works is that the mapping is compatible with all congruences of algebraic system? What are these mapping then, how to express them in a closed form, for instance, for residue rings?" This question turned out to be very fruitful because when I was trying to find the right answer I understood that the answer should be given in terms of p-adic 1-Lipschitz functions since p-adic integers constitute a profinite ring, the structure which is similar to profinite groups; and to that moment I already had developed the ergodic theory for polynomial transformations of profinite groups with operators. Thus, quite naturally my studies resulted in the ergodic theory of p-adic 1-Lipschitz functions which is the main mathematical theme of the current book. Thank you, Misha, for that question which you asked me in due time!

The third question was asked much later by my colleague and friend *Prof. Andrei Khrennikov*; to that moment our collaboration in algebraic dynamics had already resulted in monograph [24] and a number of papers, see References. Andrei did a lot of research in mathematical physics which was completely out of my area of expertise, not speaking of quantum mechanics; but Andrei's question was about quantum mechanics. Once during my stay in Sweden at Linnaeus University which Andrei was affiliated with, Andrei and me were walking around a beautiful lake and, of course, talking about mathematics. Andrei suddenly asked me: "Why mathematical formalism of quantum mechanics, of essentially nonlinear quantum phenomena, is the theory of *linear* operators? Why the linearity, anyway?". My reaction was unexpected even for me: "Because of causality and finiteness: Finite casual systems can 'compute' nothing more complex than linear functions." "Can you prove this?"—asked me Andrei. "Sure!"—I overconfidently replied. Well, finally I have found the proof that satisfies me [17, 18]. This happened about five years later the moment when the question was asked: That good was the question!

This way, step by step, I start getting involved in mathematical physics, mostly in connection with applications of the p-adic theory to quantum mechanics. Finally I have found myself in a good company which gathered around the Department of Mathematical Physics of Steklov Mathematical Institute, Russian Academy of Sciences. *Corresponding member of the Russian Academy of Sciences I. V. Volovich* who headed the Department at that time organized a number of my talks at the Department and at the Institute introduced me to Academician of the Russian Academy of Sciences V. S. Vladimirov and Dr. E. I. Zelenov, his co-authors of the book *p-adic Analysis and Mathematical Physics* [182], the Bible of the p-adic mathematical physics. During these talks I got a lot of questions; to some of these I gave answers immediately, to some after a while, and I am still thinking over the rest ones hoping to answer them sometimes. But the most important question which inspired me a lot was this one: What are the functions which satisfy Volovich postulates (see Sect. 16.1 about these)? My answer is Theorem 16.7 and, actually, the whole Chap. 16 which incorporates papers [22, 23].

My works related to interplay between p-adic dynamics, computer science, and quantum theory were unexpectedly noticed by people in the applied area of which I never heard before: The semiconductor superlattices, quite complex quantum devices. Once *Prof. Yaohui Zhang* and *Dr. Wei Liu* from the Institute of Nano-Tech and Nano-Bionics Chinese Academy of Sciences found me during one of my stays in Beijing, where I was a visiting professor of the Chinese Academy of Sciences, and asked me a question: "How to *prove* that the superlattice-based device which performs random mappings of bitstrings to bitstrings does perform the mappings which are *truly random*?" Wow, that was a challenge! "To prove" in that context means to convince a customer which is not supposed to believe in this "true randomness" just because a number of physicists assume that the randomness is inherent in Nature at quantum level, or because the devices pass standard statistical tests. Finally, I published work [23] results which are included in Chap. 16. I am not sure whether these results are convincing enough, but my collaboration with Prof. Yaohui Zhang and Dr. Wei Liu still continues. By the way, some results of the said paper were motivated by a question of *Prof. Franco Vivaldi* from Queen Mary University of London who asked me at some conference what is the entropy of the p-adic 1-Lipschitz maps, and I replied that the entropy is zero since all these maps are conjugate to the odometer, and odometer has zero entropy. But after that I thought that the entropy of the logistic map $L(x) = 4x(1-x)$ is positive, but the logistic map is a polynomial with integer coefficients, whence it is a p-adic 1-Lipschitz function for every p, hence ... it is conjugate to odometer. This is not a contradiction, of course; but that very question led me to the idea of free choice of metric by an observer with respect to which the observer makes his decision whether the dynamics is chaotic or not, see Sect. 16.4.2 of the current book.

And of course I ought to say thanks to my colleagues from the Institute of Information Engineering Chinese Academy of Sciences, especially to *Prof. Dongdai Lin* for hospitality during all my visiting professorship at the Institute but also for questions; for instance for the question about linear relations among coordinate sequences produced by T-functions, my results from Sect. 14.3 show that there are no such relations in a general T-function, but what about large important subclasses of T-functions? The answer was our joint work [169]: These relations exist if T-function satisfies some (very mild) smoothness conditions. I have not included the result into the current book since it is a somewhat specific one; but the proof of it demonstrates the usefulness of p-adic analysis to handle very specific problems which arise in applied computer science.

Last but not least: I must say lots of thanks to my colleagues from the Faculty of Computational Mathematics and Cybernetics of Lomonosov Moscow State University which I am affiliated with for a long time, for stimulating questions and discussions. My special thanks to *Academician of the Russian Academy of Sciences I. A. Sokolov*, the Dean of our faculty, who constantly supports me in research and teaching, who communicates my papers to "Doklady of the Russian Academy of Sciences," and who asked me questions, too! It was he who initiated the research on digital economy within which I have understood that smart contracts are causal functions whose interactions within digital economy in physical time can be examined by the methods of p-adic theory [20, 21]. Maybe, this examination will lead to some future results of practical value.

To list all the stimulating questions that my friends and colleagues throughout space-time asked me is to write a long paper; therefore, I have to stop and only mention names of some people to whom I am grateful: Dr. rer. nat. Patrick Erik Bradley, Prof. Ivan Chizhov, Prof. Branko Dragovich, Prof. Fabien Durand, Prof. Aihua Fan, Prof. Rostislav Grigorchuk, Prof. Alexander Grusho, Prof. Tor Helleseth, Prof. Sangtae Jeong, Dr. Vladimir Osipov, Prof. Igor Shparlinski, Prof. Livat Tyapaev, Prof. Yuefei Wang, Prof. Jia-Yan Yao, Dr. Igor Yurov, Prof. Michael Zieve, Prof. Wilson Zúñiga-Galindo, and many others to whom I am always thankful but haven't mentioned here, sorry.

Contents

Notation

$\mu \perp \nu$ (Mutual) singularity of the measures μ, ν
$\nu \ll \mu$ Continuity of the measure ν with respect to the measure μ
$\#S$ Cardinality of the set S; the number of elements in S if S is a finite set
$f\colon A \to B$ A map f of the set A to the set B
$f\colon A \twoheadrightarrow B$ A surjective map f of the set A onto the set B
$f\colon A \rightarrowtail B$ An injective map f of the set A into the set B
$f\colon A \rightleftarrows B$ A bijective map f of the set A onto the set B
$f|_S$ A restriction of the map f to the subset $S \subset A$
$f(S) = \{f(a)\colon a \in S\} \subset B$ An f-image of the set $S \subset A$
$f^{-1}(T) = \{a \in A\colon f(a) \in T \subset B\}$ An f-pre-image of the subset $T \subset B$
$f^n(\cdot) = \underbrace{f(f \dots (f}_{n}(\cdot)\dots)$ The n-th iterate of the map f
$f^{(n)}$ The n-th derivative of the function f
$A \times B$ Cartesian (direct) product of A and B
$S^n = \underbrace{S \times \cdots \times S}_{n}$ The n-th Cartesian (direct) power of S
$(a_i)_{i=0}^{\infty} = a_0, a_1, a_2, \dots$ (Infinite) sequence (sub- and superscripts may be omitted)
$\delta_i^p(z) = \alpha_i$ The i-th digit in the canonical expansion of p-adic number $z \in \mathbb{Q}_p$ (superscript may be omitted when it is clear what is p)
$\mathrm{wt}_p\, n = \sum_{k=0}^{m} \alpha_k$ The weight of the number $n = \alpha_0 + \alpha_1 p + \cdots + \alpha_m p^m$ represented by its base-p expansion
$\mathbb{N}$ The set of all natural numbers, $\mathbb{N} = \{1, 2, 3, \dots\}$
$\mathbb{N}_0$ The set of all natural numbers and 0, $\mathbb{N}_0 = \{0, 1, 2, 3, \dots\}$
$\mathbb{Z}$ The ring of (rational) integers, $\mathbb{Z} = \{0, \pm 1, \pm 2, \pm 3, \dots\}$
$\mathbb{Q}$ The field of rational numbers
$\mathbb{R}$ The field of real numbers
$\mathbb{R}_{\geqslant 0}$ The set of all real non-negative numbers
$\mathbb{C}$ The field of complex numbers
$\mathbb{I}$ Unit real segment, $\mathbb{I} = [0, 1] \subset \mathbb{R}$

$\mathbb{S}$ Unit circle; normally the length of the circle is assumed to be 1
$\mathbb{F}_{p^n}$ A finite field of p^n elements
$\mathbb{T}^n = \mathbb{S}^n$ An n-dimensional torus (the subscript n may be omitted when $n = 2$)
T_q Infinite routed tree; q is the out-degree of each vertice
$a \bmod n$ The least non-negative residue of a modulo $n \in \mathbb{N}$, $n \geqslant 2$
$\mathsf{mult}_b\, p$ The multiplicative order of p modulo $b \in \mathbb{N}$ if $b > 1$ is co-prime to p, or $\mathsf{mult}_b\, p = 1$ if $b = 1$
$m \mid n$ m is a factor of n, where $m, n \in \mathbb{N}$; $m \mid 0$ by the definition
$m \nmid n$ m is not a factor of n, where $m, n \in \mathbb{N}$
$\mathbb{Z}/n\mathbb{Z}$ The residue ring modulo n
R^+ The additive group of the ring R
R^* The multiplicative subgroup of the ring R; the group of units of R
$\Delta f(x) = f(x+1) - f(x)$ The difference operator
$\binom{x}{i} = \frac{x(x-1)\cdots(x-i+1)}{i!}$ Here $i = 1, 2, \ldots$; $\binom{x}{0} = 1$ by the definition
$|x|$ Real absolute value of $x \in \mathbb{R}$
$\mathrm{ord}_p\, n$ p-adic valuation of $n \in \mathbb{N}$; the largest $k \in \mathbb{N}_0$ such that p^k is a factor of n
$\mathbb{Q}_p$ The field of p-adic numbers
$|z|_p$ p-adic absolute value of $z \in \mathbb{Q}_p$
$[z]_p$ The integral part of the p-adic number z (subscript may be omitted if it is clear what is p)
$]z[_p$ The fractional part of the p-adic number z (subscript may be omitted if it is clear what is p)
$\lfloor a \rfloor$ The integral part of a, where $a \in \mathbb{R}_{\geqslant 0}$; the largest integer from $\mathbb{N}_0$ not exceeding a
$\lceil a \rceil$ The smallest integer from $\mathbb{N}_0$ which is not smaller than $a \in \mathbb{R}_{\geqslant 0}$
$y \bmod 1$ By the definition, $y \bmod 1 = y - \lfloor y \rfloor \in [0, 1) \subset \mathbb{R}$, for $y \in \mathbb{R}$
$\ln_p$ The p-adic logarithm; e.g., $\ln_p(1+x) = \sum_{i=1}^{\infty}(-1)^{i+1}\frac{x^i}{i}$
$\overset{p}{\lim}$ A limit in $\mathbb{Q}_p$, a limit with respect to the p-adic metric (superscript may be omitted if this leads to no misunderstanding)
$\mathbb{Z}_p$ The ring of p-adic integers
$\mathbf{B}_r^-(a)$ A p-adic open ball of radius r centered at a
$\mathbf{B}_r(a)$ A p-adic closed ball of radius r centered at a
$\mathbf{S}_r(a)$ A p-adic sphere of radius r centered at a
$\mathcal{W}(\mathcal{A})$ The set of all finite non-empty words over the alphabet$\mathcal{A}$
$\mathcal{W}_\oslash(\mathcal{A})$ The set of all finite words over the alphabet$\mathcal{A}$ including the empty word $\oslash$
$\mathcal{W}_n(\mathcal{A})$ The set of all finite words of length n over the alphabet$\mathcal{A}$
$\mathcal{W}_\infty(\mathcal{A})$ The set of all one-side infinite words of over the alphabet$\mathcal{A}$

Chapter 1
Background

Abstract This preliminary chapter is just a collection of reminders of some basic notions and results that are used throughout the book; this chapter is for further references. These facts are fundamental in the respective areas of mathematics; that is why they are mostly given with no proofs and with no references to corresponding textbooks and monographs.

1.1 Facts from Number Theory

In this section, we remind some important facts and useful formulas from number theory. We assume that the reader has basic knowledge of residues modulo N and their elementary properties.

1.1.1 Chinese Reminder Theorem

The following theorem holds true:

Theorem 1.1 *Let $N \in \mathbb{N}$ be a natural number, $N > 1$. Represent $N = p_1^{e_1} p_2^{e_2} \cdots p_r^{e_r}$, where p_j, $1 \leqslant j \leqslant r$, are prime numbers and r is the number of different primes in decomposition of N. Then, given arbitrary integers $a_j \in \{0, 1, \ldots, p_j^{e_j-1} - 1\}$, $1 \leqslant j \leqslant r$, there exists an integer $a \in \{0, 1, \ldots, N - 1\}$ such that $a \equiv a_j \pmod{p_j^{e_j}}$ for all $1 \leqslant j \leqslant r$.*

Note that the proof of Theorem 1.1 is constructive, that is, gives an algorithm to find this a explicitly (see any relevant book on number theory).

V. Anashin, *Causality: The* p*-adic Theory*, STEAM-H: Science, Technology, Engineering, Agriculture, Mathematics & Health,
https://doi.org/10.1007/978-3-031-85818-5_1

1.1.2 Binomial Coefficients

For $i \in \mathbb{N}_0$, $n \in \mathbb{N}$, the binomial coefficient is

$$\binom{n}{i} = \frac{n(n-1)\cdots(n-i+1)}{i!};$$

note that

$$\binom{n}{0} = 1,$$

by definition. Note also that $\binom{n}{i} = 0$ for $i > n$. The following result is known as *Lucas' Theorem* (see, e.g., [78, Section 5, Exercise 61] for a proof).

Theorem 1.2 *Let p be a prime number, $r = \sum_{i=0}^{N} r_i p^i$, and $n = \sum_{i=0}^{N} n_i p^i$ be the base-p expansions of $r, n \in \mathbb{N}_0$: $r_i, n_i \in \{0, 1, \ldots, p-1\}$ $(i = 0, 1, 2, \ldots)$. Then the following congruence for binomial coefficients holds:*

$$\binom{r}{n} \equiv \binom{r_0}{n_0}\binom{r_1}{n_1}\cdots\binom{r_N}{n_N} \pmod{p}.$$

Corollary 1.3 *Under conditions of Theorem 1.2, let $\ell \geqslant 1$, $k \geqslant 1$, $n \leqslant p^k - 1$. Then*

$$\binom{p^k\ell - 1}{n} \equiv (-1)^n \pmod{p}.$$

Proof of Corollary 1.3 Take $r = p^k\ell - 1$ in the statement of Theorem 1.2, and then $r_i = p - 1$ for $i = 0, 1, \ldots, k-1$. Now Theorem 1.2 implies that

$$\binom{p^k\ell - 1}{n} \equiv \binom{p-1}{n_0}\binom{p-1}{n_1}\cdots\binom{p-1}{n_{k-1}} \equiv (-1)^n \pmod{p}$$

as obviously $\binom{p-1}{j} = \frac{(p-1)(p-2)\cdots(p-j)}{j!} \equiv (-1)^j \pmod{p}$ for all $j \in \{0, 1, \ldots, p-1\}$. □

1.1.3 Stirling Numbers, Falling (Factorial) Powers

Remind that by definition, *Stirling number of the first kind* $\begin{bmatrix} n \\ i \end{bmatrix}$ is the number of ways to arrange n elements into i cycles, and *Stirling number of the second kind*

$\left\{ {n \atop i} \right\}$ is the number of ways to partition a set of n elements into i nonempty subsets. Note that by definition,

$$\left[{0 \atop 0} \right] = \left\{ {0 \atop 0} \right\} = 1.$$

In the sequel, we will use Stirling numbers to express powers via falling powers and vice versa. Remind that by definition, the n-th *falling power* (also known as *falling factorial power*) of x is

$$x^{\underline{n}} = n!\binom{x}{n} = x(x-1)\cdots(x-n+1), \quad x^{\underline{0}} = 1.$$

The falling factorial powers and (ordinary) powers are related one to another in the following way (see, e.g., [78, Section 6.1]):

$$x^n = \sum_{i=0}^{n} \left\{ {n \atop i} \right\} x^{\underline{i}}; \quad x^{\underline{n}} = \sum_{i=0}^{n} (-1)^{n-i} \left[{n \atop i} \right] x^i \tag{1.1}$$

1.1.4 Finite Differences

Definition 1.4 A (forward) *difference* (with respect to variable x_i) of the function $f(x_1, \ldots, x_n)$ is

$$\Delta_i f(x_1, \ldots, x_n) = f(x_1, \ldots, x_{i-1}, x_i + 1, x_{i+1}, \ldots, x_n) - f(x_1, \ldots, x_n),$$

and the s-th difference (with respect to the variable x_i) of the function f is

$$\Delta_i^s f = \Delta_i^{s-1}(\Delta_i f), \qquad (s = 1, 2, \ldots),$$

where $\Delta_i^0 f = f$ by definition. We write $\Delta f(x)$ rather than $\Delta_1 f(x)$ for a univariate function f.

One verifies directly that

$$\Delta\binom{n}{i} = \binom{n+1}{i} - \binom{n}{i} = \binom{n}{i-1} \tag{1.2}$$

Theorem 1.5 *The following identity holds for all $n \in \mathbb{N}$ and all functions g:*

$$g(y+n) = \sum_{i=0}^{\infty} \binom{n}{i} \Delta^i g(y),$$

The identity from the above theorem is known under the name of *Gregory-Newton formula*.

Theorem 1.6 *The following is true:*

$$\alpha_m = \sum_{k=0}^{\infty} \binom{m}{k} \beta_k$$

if and only if

$$\beta_k = \sum_{m=0}^{\infty} (-1)^{m+k} \binom{k}{m} \alpha_m$$

The statement of the above theorem is also known as *(binomial) inversion formula*. See, e.g., [78, Section 5.3] for Theorems 1.5 and 1.6

1.1.5 Euler's Totient Function

We recall the definition of Euler's totient function and Euler's Theorem.

Definition 1.7 Let $n \in \mathbb{N}$; denote by $\varphi(n)$ the number of all natural numbers less than n, which are relatively prime to n. The function φ is called *Euler's totient function*.

If p is a prime number, then $\varphi(p^l) = p^{l-1}(p-1)$. The following theorem is known as *Euler's Theorem*.

Theorem 1.8 *If a is an integer relatively prime to b, then $a^{\varphi(b)} \equiv 1 \pmod{b}$.*

1.2 Facts from Algebra

In this section, we recall some basic facts from algebra with a special focus on finite rings and finite fields.

1.2.1 Universal Algebras

We recall some basic concepts of universal algebra following mainly [133]. A *universal algebra* (or, briefly, an algebra, if this makes no confusion) is a non-empty set A endowed with a set of operations Ω (the latter set is often called a *signature* of the universal algebra). Every operation $\omega \in \Omega$ is a map from the n-th Cartesian power A^n into A; the number n is called the *arity* of the operation ω. Two algebras A and B with operations Ω and Ψ, respectively, are said to *have the same type* (or to be *similar*) if there exists a one-to-one correspondence ι between Ω and Ψ that preserves arities; that is, the arity of $\omega \in \Omega$ is equal to the arity of $\iota(\omega) \in \Psi$. If A and B are algebras of the same type, we do not differ further ω from $\iota(\omega)$, if it makes no confusion. A subset $S \subset A$ is called a *subalgebra* if it is closed with respect to all operations from Ω: $\omega(a_1, \ldots, a_n) \in S$ for all $a_1, \ldots, a_n \in S$ and every (n-ary) operation $\omega \in \Omega$. A binary relation $\sim$ on A is called a *congruence* of the algebra A whenever $\sim$ agrees with all operations from Ω; that is, given an n-ary operation $\omega \in \Omega$ and elements $a_1, \ldots, a_n, b_1, \ldots, b_n \in A$ such that $a_i \sim b_i$ for all $i = 1, 2, \ldots, n$, then necessarily $\omega(a_1, \ldots, a_n) \sim \omega(b_1, \ldots, b_n)$. It is clear that every congruence of A is an equivalence on A; so congruences are equivalence relations that agree with all operations of the algebra. If a class of equivalent elements contains more than one element, but not of all elements of the algebra, a congruence is said to be proper. Algebra that has no proper congruences is sometimes called *simple*.

If A and B are algebras of the same type, the map $\varphi\colon A \to B$ is called a homomorphism whenever it agrees with all operations; that is, for every (n-ary) operation $\omega \in \Omega$ and every $a_1, \ldots, a_n \in A$, we have that $\varphi(\omega(a_1, \ldots, a_n)) = \omega(\varphi(a_1), \ldots, \varphi(a_n))$. If φ is surjective (injective), it is called a epimorphism (monomorphism). If φ is simultaneously an epimorphism and a monomorphism, it is called an isomorphism. If $A = B$, the homomorphism φ is called an endomorphism, and if φ is an isomorphism, then φ is called an automorphism. Note that every homomorphism φ defines a congruence: $a \sim b$ if and only if $\varphi(a) = \varphi(b)$. Vice versa, every congruence defines an epimorphism of A onto algebra of classes of equivalent elements of A, the *factor algebra* of A with respect to the congruence $\sim$. The epimorphism ψ is said to be *natural* in this case. The congruence $\sim$ is sometimes called a *kernel* of φ.

Now we formulate one of the most important algebraic notions of the book, the compatibility. Loosely speaking, a mapping $F\colon A^k \to A^m$ is said to be compatible if it agrees with all congruences of A. Here is a formal definition:

Definition 1.9 (Compatibility) Let $F = (f_1 \ldots, f_m)\colon A^k \to A^m$ be a map of the k-th Cartesian power of the algebra A into its m-th Cartesian power; that is, $f_j\, A^k\, A$, for all $j = 1, 2, \ldots, m$. The map F is said to be *compatible*, if for every congruence $\sim$ of A and every elements $a_1, \ldots, a_n, b_1, \ldots, b_n \in A$ such that $a_i \sim b_i$ for all $i = 1, 2, \ldots, n$, we have that $f_j(a_1, \ldots, a_n) \sim f_j(b_1, \ldots, b_n)$, for all $j = 1, 2, \ldots, m$.

It is clear that every operation $\omega \in \Omega$ is compatible; whence, all compositions of operations from Ω are compatible as well.

1.2.2 Rings

In this subsection, we recall some notions and facts from ring theory, mainly following [149]. A *ring* R is a universal algebra with two operations $+$ (addition) and $\times$ multiplication, such that R with respect to $+$ is a commutative group (which is denoted as R^+) with neutral 0, which is called zero, and inverse $-$ (that is $-a$ is an additive inverse for $a \in R$, $a + (-a) = 0$), R is a semigroup with respect to $\times$, and $(a+b)\times c = (a\times c)+(b\times c)$, $c\times(a+b) = (c\times a)+(c\times b)$, for all $a, b, c \in R$. We mainly consider *commutative* rings in this book, that is, $a \times b = b \times a$, for all $a, b \in R$. As usual, we omit the sign of multiplication $\times$ in expressions, and we omit parentheses according to the common rule: $a + bc = a + (b \times c)$. Whenever the ring R has an *unity*, that is, a multiplicative neutral element, we denote it as 1: $a \times 1 = 1 \times a$, for all $a \in R$. A ring having the unity is called a *ring with unity*. Further within this subsection , "ring" stands for "commutative ring with unity." The additive order of 1, that is, the smallest $n \in \mathbb{N}$ such that $n \times 1 = 0$, if such n exists, called a *characteristic* of R, is denoted via $\mathrm{char}(R)$. A ring is said to be of *zero characteristic* if no such n exists.

If an element $a \in R$ has a multiplicative inverse, it is denoted via a^{-1}: $a \times a^{-1} = a^{-1} \times a = 1$. All *invertible* elements (those having multiplicative inverses) are called *units*. They form a group R^* with respect to ring multiplication; this group is called a *unit group*, or a *group of units*, or *multiplicative (sub)group* of the ring R. If $R^* = R \setminus \{0\}$, the ring R is called a *field*. Note that in ring theory, they often use the term *identity* rather than unity; we avoid using identity in this meaning in the book.

A non-zero element $a \in R$ is called a *zero divisor* whenever there exists an element $b \in R\setminus\{0\}$ such that $ab = 0$. An non-zero element $a \in R$ is called *nilpotent* whenever $a^n = 0$ for some $n \in \mathbb{N}$; the smallest such n is called the *nilpotency index of* a.

A ring R without zero divisors is called an *integral domain*. Every integral domain can be embedded into a field; the smallest one is called a *quotient field* of R and denoted as $Q(R)$. For instance, the ring $\mathbb{Z} = \{0, \pm1, \pm2, \ldots\}$ of all rational integers is an integral domain; its quotient field is $\mathbb{Q}$, the field of all rational numbers. An *integer-valued function* is a map $F\colon Q(R)^n \to Q(R)^m$ such that $F(R^n) \subset R^m$. We remind that any *integer-valued polynomial* f over $\mathbb{Q}$ in variable x can be expressed as

$$f(x) = \sum_{i=0}^{d} a_i \binom{x}{i},$$

where $a_i \in \mathbb{Z}$, $i = 0, 1, \ldots, d$ and vice versa (see monograph [34] on various aspects of integer-valued polynomials).

A *module* over a ring R is a commutative group M with respect to operation $\oplus$, endowed with an "external" operation of multiplication by elements of R: Given $r, s \in R$, $h, g \in M$, one defines this multiplication $r \times h \in M$ so that $(rs) \times h = r \times (s \times h)$ and $r \times (h \oplus g) = (r \times h) \oplus (r \times g)$. Vector spaces over fields are important example of modules; the other important example are ideals.

A non-empty subset $I \subset R$ is called an *ideal* whenever I is a subgroup with respect to ring addition $+$, and $ra \in I$ for all $r \in R$, $a \in I$. An ideal I is called *proper* whenever $I \neq R$ and $I \neq \{0\}$. Given $a \in R$, the set $a + I = \{a + r : r \in \mathrm{i}\}$ is called a *coset* of I with respect to a.

Ideals are kernels of ring homomorphisms, and vice versa. The *index* $k = |R : I|$ *of the ideal* I is the order of the factor ring R/I of the ring R by the ideal I. Given a ring R, a map $f : R \to R$ is called *compatible* if $f(a + I) \subset f(a) + I$ for every ideal I of R and every $a \in R$. To put it in other words, f is a compatible map if it induces well-defined maps on all factor rings of R. If $f(a + I) \subset f(a) + I$ for a concrete ideal I, the map f is called *compatible with the ring epimorphism whose kernel is* I. In this case, f induces a well-defined map on the factor ring R/I.

It is clear that given $a_1, \ldots, a_n \in R$, the set $a_1R + \cdots + a_nR$, which is a set of all sums $a_1r_1 + \cdots + a_nr_n$, $r_1, \ldots, r_n \in R$, is an ideal of R, the smallest ideal that contains $a_1, \ldots, a_n$. This ideal is called an *ideal generated by elements* $a_1, \ldots, a_n$. An ideal that is generated by a single element is called *principal*. A ring all whose ideals are principal is called a *principal ideal ring*. It is clear that factor rings of principal ideal rings are again principal ideal rings.

Theorem 1.10 *A ring* $R[x]$ *of all polynomials in a variable* x *over a field* R *is a principal ideal ring.*

In the sequel, we will also need a more special type of rings, the rings of formal power series: Given a ring R and a variable x, consider all formal expressions $\sum_{i=0}^{\infty} a_i x^i$, $a_i \in R$, $i = 0, 1, 2, \ldots$. We can define addition and multiplication of these sums by common rules for infinite series; as every coefficient of a sum or product is then a finite expression of finite number of coefficients of summands (respectively, factors), these operations are well defined. Thus we obtain a *ring* $R[[x]]$ *of formal power series*; its elements are called *formal power series* over R.

Now we remind some facts about finite rings. The following is true:

Proposition 1.11 *Every non-zero element of a finite ring is either a unit or a zero divisor.*

Throughout the book, we often deal with factor rings $\mathbb{Z}/N\mathbb{Z}$ of the ring $\mathbb{Z}$ of integers by principal ideals $N\mathbb{Z}$, where $N \geqslant 2$. For short, we call these rings the *residue rings modulo* N. Due to the independence of a choice of representatives in cosets modulo the ideal $N\mathbb{Z}$, we usually identify elements of the rings $\mathbb{Z}/N\mathbb{Z}$ with all non-negative residues $\{0, 1, \ldots, N-1\}$ modulo N of integers from $\mathbb{Z}$ rather than with the cosets. For the residue rings, the following restatement of Chinese Reminder Theorem is true:

Theorem 1.12 *Let $N \in \mathbb{N}$ be a natural number, $N > 1$. Represent $N = p_1^{e_1} p_2^{e_2} \cdots p_r^{e_r}$, where p_j, $1 \leqslant j \leqslant r$ are prime numbers and r is the number of different primes in decomposition of N. Then the residue ring $\mathbb{Z}/N\mathbb{Z}$ is a Cartesian product of residue rings $\mathbb{Z}/p_j^{e_j}\mathbb{Z}$, $1 \leqslant j \leqslant r$.*

Note that in the theory of finite rings, it is common to use a term *direct sum* rather than terms "direct product" or "Cartesian product."

For residue rings $\mathbb{Z}/N\mathbb{Z}$, there exists a simple way to determine whether a given element is invertible or a zero divisor (cf. Proposition 1.11):

Proposition 1.13 *Let N be a natural number, $N > 1$. Represent $N = p_1^{e_1} p_2^{e_2} \cdots p_r^{e_r}$, where p_j, $1 \leqslant j \leqslant r$, are prime numbers and r is the number of different primes in decomposition of N. Then the element a of the residue ring $\mathbb{Z}/N\mathbb{Z}$ is invertible if and only if $a \not\equiv 0 \pmod{p_j}$ for all $1 \leqslant j \leqslant r$.*

With the use of these results in combination with Proposition 1.14 that follows, it is easy to characterize multiplicative subgroups of residue rings.

Proposition 1.14 *Let p be a prime, let $k \in \mathbb{N}$. The group $(\mathbb{Z}/p^k\mathbb{Z})^*$ of all invertible elements of the residue ring $\mathbb{Z}/p^k\mathbb{Z}$ is a cyclic group of order $(p-1)p^{k-1}$ whenever p is odd. If $p = 2$ and $k > 2$, then $(\mathbb{Z}/2^k\mathbb{Z})^*$ is a direct product of a group of order 2 by a cyclic group of order 2^{k-2}. The group $(\mathbb{Z}/4\mathbb{Z})^*$ is a cyclic group of order 2, and the group $(\mathbb{Z}/2\mathbb{Z})^*$ is trivial.*

In case when the multiplicative subgroup $(\mathbb{Z}/p^k\mathbb{Z})^*$ of the residue ring $\mathbb{Z}/p^k\mathbb{Z}$ is cyclic, the number $a \in \mathbb{Z}$ is said to be *primitive modulo p^k* if the residue of a modulo p^k is a generator of $(\mathbb{Z}/p^k\mathbb{Z})^*$.

1.2.3 Finite Fields

A characteristic $\operatorname{char}(\mathbb{F})$ of a finite field $\mathbb{F}$ is a prime number p, and $\#\mathbb{F} = p^n$ for a suitable $n \in \mathbb{N}$. Given a prime p and a positive rational integer n, there exists (up to ring isomorphism) a unique field of order $q = p^n$. We denote this unique field of p^n elements via $\mathbb{F}_q$. In particular, if $n = 1$, then $\mathbb{F}_p$ is isomorphic to the residue ring $\mathbb{Z}/p\mathbb{Z}$ modulo p.

A multiplicative subgroup $\mathbb{F}_q^*$ is a cyclic group of order $q-1$; generators of this group are called *primitive elements* of the field $\mathbb{F}_q$. Thus, there are exactly $\varphi(q-1)$ different primitive elements in $\mathbb{F}_q$, where φ is Euler's totient function.

Finite fields are *polynomially complete*: Given a map $\omega\colon \mathbb{F}_q \to \mathbb{F}_q$, there exists a polynomial $f_\omega(x) \in \mathbb{F}_q[x]$ such that $f_\omega(z) = \omega(z)$, for all $z \in \mathbb{F}_q$:

$$f_\omega(x) = \sum_{z \in \mathbb{F}_q} \omega(z) \frac{x^q - x}{z - x}. \tag{1.3}$$

We note that $f_\omega(x)$ is indeed a polynomial over $\mathbb{F}_q$ as $x^q - x = \prod_{z\in\mathbb{F}_q}(x-z)$. Formula (1.3) holds since

$$\frac{x^q - x}{z - x} = \begin{cases} 1, & \text{whenever } x = z; \\ 0, & \text{otherwise.} \end{cases}$$

Using this method, which goes back to Lagrange, we can construct interpolation polynomial for arbitrary n-variate mapping from $\mathbb{F}_q^n$ to $\mathbb{F}_q^m$, as, e.g.,

$$\frac{x^q - x}{a - x} \times \frac{y^q - y}{b - y} = \begin{cases} 1, & \text{whenever } x = a \text{ and } y = b; \\ 0, & \text{otherwise,} \end{cases}$$

and henceforth. Moreover, we can interpolate simultaneously a polynomial mapping and its derivative (which is also a polynomial then), in the following way:

Proposition 1.15 *Given two mappings $\omega\colon \mathbb{F}_q \to \mathbb{F}_q$ and $\psi : \mathbb{F}_q \to \mathbb{F}_q$, there exists a polynomial $f_{\omega,\psi}(x) \in \mathbb{F}_p[x]$ such that:*

- *$f_{\omega,\psi}$ induces on $\mathbb{F}_q$ the mapping ω:*

$$f_{\omega,\psi}(z) = \omega(z) \text{ for all } z \in \mathbb{F}_q,$$

- *the derivative $f'_{\omega,\psi}(x)$ induces on $\mathbb{F}_q$ the mapping ψ:*

$$f'_{\omega,\psi}(z) = \psi(z) \text{ for all } z \in \mathbb{F}_q.$$

Proof Given mappings ω and ψ, construct interpolation polynomials f_ω and f_ψ according to formula (1.3). Then

$$f_{\omega,\psi}(x) = f_\omega(x) + (x^q - x)(f'_\omega(x) - f_\psi(x)).$$

Note that $z^q - z = 0$ for all $z \in \mathbb{F}_q$, that $(x^q - x)' = qx^{q-1} - 1$ is identically -1 on $\mathbb{F}_q$, and that $f'_\omega(x)$ is a polynomial over $\mathbb{F}_q$ (as $f_\omega(x)$ is a polynomial over $\mathbb{F}_q$). □

Note 1.16 This proposition can also be generalized for the case of arbitrary mappings $\mathbb{F}_q^n$ to $\mathbb{F}_q^m$ with the use of interpolation formulas for n-variate mappings we mentioned above, as well as for higher-order derivatives.

1.3 Facts from Dynamics

By definition, *dynamical system on a phase space* (or, in another terminology, on a *configuration space*) $\mathbb{X}$ is a triple $\langle \mathbb{X}; \mu; f\rangle$, where $\mathbb{X}$ is a *measure space* endowed with a measure μ and $f : \mathbb{X} \to \mathbb{X}$ is a *measurable transformation* $f : \mathbb{X} \to \mathbb{X}$; that is, f is such that an f-pre-image $f^{-1}(S)$ of any μ-measurable subset $S \subset \mathbb{X}$ is a μ-measurable subset of $\mathbb{S}$.

A *trajectory* (or, in another terminology, an *orbit*) of the dynamical system is a sequence

$$x_0, x_1 = f(x_0), \dots, x_i = f(x_{i-1}) = f^i(x_0), \dots$$

of points of the space $\mathbb{X}$; the point x_0 is called an *initial* point of the trajectory.

If $F: \mathbb{X} \to \mathbb{Y}$ is a *measurable map* of $\mathbb{X}$ to some other measure space $\mathbb{Y}$ with a measure ν (i.e., if an F-pre-image of any ν-measurable subset of $\mathbb{Y}$ is a μ-measurable subset of $\mathbb{S}$), the sequence

$$F(x_0), F(x_1), F(x_2), \dots$$

is sometimes called an *observable* (in physical models) or an *output* (in computer science). The *image measure* (or, in another terminology, a *pushforward measure*) on $\mathbb{Y}$ is the measure ν_F such that $\nu_F(M) = \mu(F^{-1}(M))$ for every ν-measurable set $M \subset \mathbb{Y}$.

A measurable map $F: \mathbb{X} \to \mathbb{Y}$ from a measure space $\mathbb{X}$ into a measure space $\mathbb{Y}$ endowed with measures μ and ν, respectively, is said to be *measure-preserving* if $\mu(F^{-1}(S)) = \nu(S)$ for each measurable subset $S \subset \mathbb{Y}$. In the case when $\mathbb{X} = \mathbb{Y}$ and $\mu = \nu$ are probability measures, a measure-preserving map F is said to be *ergodic* if, given a measurable subset S such that $F^{-1}(S) = S$, either $\mu(S) = 1$ or $\mu(S) = 0$.

The following example is trivial and, however, important. Let $\mathbb{X}$ be a finite set N-element set: $\#\mathbb{X} = N$; let $\mathbb{X}$ be endowed with a uniform probability measure μ: Given $A \subset \mathbb{X}$ and $\#A = M$, we put $\mu(A) = \frac{M}{N}$. A map $f: \mathbb{X} \to \mathbb{X}$ is measure-preserving if and only if f is bijective, i.e., if and only if f is a permutation on $\mathbb{X}$. The map f is ergodic if and only if this permutation consists of a single cycle, i.e., if and only if f is *transitive* on $\mathbb{X}$.

Definition 1.17 (Unique Ergodicity) A mapping $f: \mathbb{X} \to \mathbb{X}$ is called *uniquely ergodic* if there exists a unique f-invariant probability measure μ on $\mathbb{X}$; i.e., such that f is ergodic with respect to μ.

In many cases considered in the book further, the space $\mathbb{X}$ will be also a metric space; that is why we need more facts (and respective notions) about dynamics on spaces that are endowed not only with a measure but with a topology as well.

Definition 1.18 (Topological Transitivity) Given a topological space $\mathbb{X}$ and a continuous mapping $f: \mathbb{X} \to \mathbb{X}$, the mapping f (respectively, the dynamical system $\langle \mathbb{X}; f \rangle$) is called *topologically transitive* if there exists a dense orbit of f, that is, if there exists $x \in \mathbb{X}$ such that the set of iterations $\{f^i(x): i \in \mathbb{N}_0\}$ is dense in $\mathbb{X}$. The dynamical system $\langle \mathbb{X}; f \rangle$ is called *minimal* if every orbit is dense.

There is another one (generally, non-equivalent to the above) definition of topological transitivity: The map f is called topologically transitive if for every pair of non-empty open sets $U, V \subset \mathbb{X}$, there exists a non-negative integer $\ell \in \mathbb{N}_0$ such that $f^{\ell}(U) \cap V \neq \emptyset$. Note, however, that since in the sequel we mostly deal with n-dimensional p-adic space $\mathbb{X} = \mathbb{Z}_p^n$ and $n \in \mathbb{N}$, the two definitions are equivalent since that space has no isolated points and is separable and of second category.

Moreover, in the book, we mostly will deal with *non-expansive dynamics*, that is, the dynamics $\langle f, \mathbb{X}\rangle$ where $\mathbb{X}$ is a metric space equipped with a metric d and $d(f(x), f(x')) \leqslant d(x, x')$ for all $x, x' \in \mathbb{X}$. Recall also that a non-empty set $\mathbb{X}$ is a *metric space* with respect to the *metric* $d: \mathbb{X}\times\mathbb{X} \to \mathbb{R}_{\geqslant 0}$ if the following conditions are satisfied:

- $d(x, y) = 0$ if and only if $x = y$.
- $d(x, y) = d(y, x)$, for all $x, y \in \mathbb{X}$.
- For all $x, y, z \in \mathbb{X}$, the following *triangle inequality* holds:

$$d(x, z) \leqslant d(x, y) + d(y, z).$$

Recall also that a map $f: \mathbb{X} \to \mathbb{Y}$ of metric spaces $\mathbb{X}, \mathbb{Y}$ equipped with metrics d_1, d_2, respectively, is called an *isometry* if $d_2(f(x), f(x')) = d_1(x, x')$ for all $x, x' \in \mathbb{X}$. It is well-known (see, e.g., [112]) that any isometry $f: \mathbb{X} \to \mathbb{Y}$ of metric spaces $\mathbb{X}$ and $\mathbb{Y}$ is continuous and injective—moreover, that any isometry of a compact metric space into itself is surjective. The space $\mathbb{X} = \mathbb{Z}_p^n$ is a compact metric space; for compact metric spaces, the following theorem holds (see, e.g., [95, Corollary 4.3.6]):

Theorem 1.19 *A minimal isometry of a compact metric space is uniquely ergodic.*

1.4 Facts from Coding Theory

In this section, we recall necessary terminology and some facts from the coding theory, actually, from a very small part that deals with so-called *prefix codes* (or, in other terminology, *instantaneous codes*. The codes are some specific subsets of words over finite alphabets, so we begin with the respective notions and facts about the words.

1.4.1 A Few Words About Words

An *alphabet* is just a finite non-empty set $\mathcal{A}$; further in the book, usually, $\mathcal{A} = \{0, 1, \dots, p-1\}$, where $p > 1$ is an integer that is mostly (but not always) a prime. In the latter case, $0, 1, \dots, p-1$ will mostly be thought of as elements of a finite

p-element field $\mathbb{F}_p$, the residue ring $\mathbb{Z}/p\mathbb{Z}$ modulo p. Elements of $\mathcal{A}$ are called *symbols* or *letters*. By definition, a *word of length n over alphabet* $\mathcal{A}$ is a finite sequence (stretching from right to left) $\alpha_{n-1}\cdots\alpha_1\alpha_0$, where $\alpha_{n-1},\ldots,\alpha_1,\alpha_0\in\mathcal{A}$. The number n is called the *length* of the word $w=\alpha_{n-1}\cdots\alpha_1\alpha_0$ and is denoted via $\Lambda(w)$. The *empty word* $\oslash$ is a sequence of length 0, that is, the one that contains no symbols. Given a word $w=\alpha_{n-1}\cdots\alpha_1\alpha_0$, any word $v=\alpha_{k-1}\cdots\alpha_1\alpha_0$, $k\leqslant n$, is called a *prefix* of the word w, whereas any word $u=\alpha_{n-1}\cdots\alpha_{i+1}\alpha_i$, $0\leqslant i\leqslant n-1$ is called a *suffix* of the word w. Every word $\alpha_j\cdots\alpha_{i+1}\alpha_i$ where $n-1\geqslant j\geqslant i\geqslant 0$ is called a *subword* of the word $w=\alpha_{n-1}\cdots\alpha_1\alpha_0$. Given words $a=\alpha_{n-1}\cdots\alpha_1\alpha_0$ and $b=\beta_{k-1}\cdots\beta_1\beta_0$, the *concatenation ab* is the following word (of length $n+k$):

$$ab=\alpha_{n-1}\cdots\alpha_1\alpha_0\beta_{k-1}\cdots\beta_1\beta_0.$$

Given a word w, its k-times concatenation is denoted via $(w)^k$:

$$(w)^k=\underbrace{ww\ldots w}_{k\text{ times}}.$$

We denote via $\mathcal{W}=\mathcal{W}(\mathcal{A})$ the set of all non-empty words over $\mathcal{A}=\{0,1,\ldots,p-1\}$ and via $\mathcal{W}_\varnothing$ the set of all words including the empty word $\varnothing$. In the sequel, the set of all n-letter words over the alphabet $\mathcal{A}$ we denote as $\mathcal{W}_n$; so $\mathcal{W}=\cup_{n=1}^{\infty}\mathcal{W}_n$. To every word $w=\alpha_{n-1}\cdots\alpha_1\alpha_0$, we put into the correspondence a non-negative integer $\mathsf{num}(w)=\alpha_0+\alpha_1p+\cdots+\alpha_{n-1}p^{n-1}$. Thus, num maps the set $\mathcal{W}$ of all non-empty finite words over the alphabet $\mathcal{A}$ onto the set $\mathbb{N}_0=\{0,1,2,\ldots\}$ of all non-negative integers. If n is fixed, the number $\mathsf{num}(w)$ can be considered as an element of the residue ring $\mathbb{Z}/p^n\mathbb{Z}$ modulo p^n; that is, $\mathsf{num}\colon\mathcal{W}_n\twoheadrightarrow\mathbb{Z}/p^n\mathbb{Z}$. We denote via wrd_n an inverse mapping to num. The mapping wrd_n is a bijection of the set $\{0,1\ldots,p^n-1\}\subset\mathbb{N}_0$ onto the set $\mathcal{W}_n$ of all n-letter words over $\mathbb{F}_p$.

We will also need a map ρ of the set $\mathcal{W}$ into the real unit half-open interval $[0,1)$; the map ρ is defined as follows: Given $w=\beta_{r-1}\ldots\beta_0\in\mathcal{W}$, put

$$\rho(w)=\mathsf{num}(w)p^{-\Lambda(w)}=\frac{\beta_0+\beta_1p+\cdots+\beta_{r-1}p^{r-1}}{p^r}=0.\beta_{r-1}\ldots\beta_0\in[0,1). \tag{1.4}$$

We also use the notation $0.w$ for $0.\beta_{r-1}\ldots\beta_0$.

Along with finite words, we also consider (left-)infinite words over the alphabet $\mathcal{A}$; the ones are the infinite sequences of the form $\ldots\alpha_2\alpha_1\alpha_0$ where $\alpha_i\in\mathcal{A}$ and $i\in\mathbb{N}_0$. The set of all these words will be denoted via $\mathcal{W}_\infty(\mathcal{A})$.

In the book, we may write one-side infinite words either stretching from left to right or from right to left when convenient, i.e., both $\alpha_0\alpha_1\alpha_2\ldots$ and $\ldots\alpha_2\alpha_1\alpha_0$ denote the same word. For finite words, we may also use both notations, left and right; the order of indices of letters in the word shows which of the two notations is used.

For infinite words, notions of prefix, suffix, and subword are defined in the same way as for finite words; note that suffix is always an infinite word, while prefix and subwords are always finite words. Given a word w, by $\mathsf{pref}_n(w)$, we denote the prefix of length n of the word w; if $\Lambda(w) < n$, we have that $\mathsf{pref}_n(w) = w$. We denote $\mathsf{suff}_n(w)$ the subword of w such that $w = \mathsf{suff}_n(w)\,\mathsf{pref}_n(w)$ is a concatenation of the words $\mathsf{suff}_n(w)$ and $\mathsf{pref}_n(w)$. Note that $\Lambda(\mathsf{suff}_n(w)) = \Lambda(w) - n$ and that $\mathsf{suff}_n(w) = \oslash$, the empty word, whenever $\Lambda(w) \leqslant n$.

Let an infinite word w be *eventually periodic*; that is, let

$$w = \ldots\beta_{t-1}\beta_{t-2}\ldots\beta_0\beta_{t-1}\beta_{t-2}\ldots\beta_0\alpha_{r-1}\alpha_{r-2}\ldots\alpha_0$$

for $\alpha_i\beta_j \in \mathcal{A}$; then the subword $\beta_{t-1}\beta_{t-2}\ldots\beta_0$ is called a *period* of the word w, and the suffix $\alpha_{r-2}\ldots\alpha_0$ is called the *pre-period* of the word w. Note that a pre-period may be an empty word (in that case, the word is called *periodic*), while a period cannot. We write the eventually periodic word w as $w = (\beta_{t-1}\beta_{t-2}\ldots\beta_0)^\infty\alpha_{r-1}\alpha_{r-2}\ldots\alpha_0$.

1.4.2 Prefix (Instantaneous) Codes

Here we recall some basic properties of prefix codes, which may be found, e.g., in the book [75].

Definition 1.20 (Prefix Code) A non-empty set $\mathbf{C} \subset \mathcal{W}(\mathcal{A})$ of finite non-empty words over a finite alphabet $\mathcal{A}$ that consists of $p > 1$ letters is called a *prefix code* if each word from $\mathbf{C}$ is a prefix of no other word from $\mathbf{C}$.

Let words from the non-empty set $\mathbf{G}$ of finite non-empty words over $\mathcal{A}$ be ordered with respect to non-decreasing order of their lengths; let ℓ_i be the length of the i-th word (so $\ell_1 \leqslant \ell_2 \leqslant \cdots$). *The set* $\mathbf{G}$ *is a prefix code if and only if the following Kraft-McMillan inequality holds*:

$$\sum_{i=1}^{\infty} p^{-\ell_i} \leqslant 1. \tag{1.5}$$

Let X be a random variable on the prefix code $\mathbf{G} = (w_i)_{i=1}^{\infty}$; denote by $q_i = \mathrm{Prob}(X = w_j)$ the probability that X is equal to the word w_i. By definition, the *entropy* $\mathbf{H}(X)$ *of the random variable* X is $\mathbf{H}(X) = -\sum_{i=1}^{\infty} q_i \log_p q_i$, whereas the *mean* length of the codeword is $\mathbf{E}(X) = \sum_{i=0}^{\infty} q_i\ell_i$. If $\mathbf{H}(X), \mathbf{E}(X) < +\infty$, then

$$\mathbf{H}(X) \leqslant \mathbf{E}(X) < \mathbf{H}(X) + 1, \tag{1.6}$$

with left-hand-side inequality becoming an equality if and only if $q_i = p^{-\ell_i}$; there exists a prefix code such that $\ell_i = \lceil \log_p(1/q_i) \rceil$ for which the right-hand-side inequality (which is not true in general) holds [75, Theorem 4.3].

1.5 Facts from Graph Theory

The terminology in graph theory is somewhat diverse; the terminology we are using in the book is mostly taken from [85], with minor modifications. The graphs we are dealing with are mostly *directed graphs (digraphs)* whose (directed) edges in the sequel will also be called *arrows*. Given a vertex v, the number of arrows whose head is v we call *in-degree of the vertex*; accordingly, the number of arrows whose tail is v we call *out-degree of the vertex.*

Given digraphs G, G', the mapping f that maps the set of all vertices of the digraph G into the set of all vertices of the digraph G' is called a *homomorphism* of the digraph G into the digraph G' if, for every pair of vertices v_1, v_2 of the digraph G such that there is an arrow in G whose tail is v_1 and head is v_2, there is an arrow in the graph G' whose tail is $f(v_1)$ and whose head is $f(v_2)$. If $G = G'$, the homomorphism f is called the *endomorphism* of the digraph G; and if, moreover, the endomorphism f is bijective, then it is called the *automorphism* of the digraph G.

A *walk* from vertex v_0 to vertex v_n in a digraph is an alternating sequence

$$v_0;\, e_0;\, v_1;\, e_1;\, \cdots;\, v_{n-1};\, e_{n-1};\, v_n \tag{1.7}$$

of vertices v_i and edges (arrows) e_i such that tail of e_i is v_i and head of e_i is v_{i+1}. The *length of the walk* is the number of edges. We call a walk *simple* if it has no repeated edges and no repeated vertices except possibly the initial and final vertices v_0 and v_n; the sequence of edges in a simple walk we call a (finite) *path*; the length of the path is equal to the length of the walk by definition. If the initial and final vertices of the walk coincide, the path is called a *cycle* (a cycle of length 1 is called a *loop*). Given vertices v'_0, v'_1 ($v'_0 \neq v'_1$), the vertex v'_1 is called *reachable from the vertex* v'_0 if there is a walk (1.7) in the digraph such that $v'_1 = v_0$ and $v'_2 = v_n$.

A *rooted tree* T is a digraph with a designated vertex v called the *root* such that:

- v is a head of no arrow of T.
- There is a path from the root to any other vertex of T.
- There are no cycles in T.

In the book, all trees are mostly assumed to be rooted trees; necessary definitions are given in respective places otherwise.

Given a vertex v in a rooted tree, all vertices that are tails of arrows whose head is v are called *parents (of the vertex v)*; accordingly, all vertices that are heads of arrows whose tail is v are called *children (of the vertex v)*. Given a vertex v' in a rooted tree, all vertices that pass all paths that go from the root to the vertex v', except the vertex v' itself, are called *ancestors (of the vertex v')*; accordingly, all vertices whose ancestor the vertex v' is are called *descendants (of the vertex v')*. In a tree, a vertex that has no children is called a *leaf*; a path from the root to a leaf is called a *branch*.

A *ray (from the vertex v_i)*, $i \in \mathbb{N}_0$, or, which is the same, an *infinite path in a tree* that starts at v_i is infinite sequence

$$e_i; e_{i+1}, e_{i+2}; \cdots ; e_{i+n}; \cdots \tag{1.8}$$

of arrows e_{i+k}, $k \in \mathbb{N}_0$, such that the tail of e_{i+k} is a vertex v_{i+k} and the head of e_{i+k} is a vertex v_{i+k+1}. A tree is called *locally finite* if out-degree of every vertex is finite (and may be zero). A tree is called *infinite* if there is an infinite number of vertices in the tree.

The König's Infinity Lemma (which was independently discovered by Beth about a quarter of a century later; see, e.g., [187]) implies that the following proposition is true:

Proposition 1.21 *In every locally finite infinite tree, there is a ray.*

1.6 Facts from Automata Theory

In this section, we recall some basic notions and facts from automata theory, which will be needed further in the book.

1.6.1 Systems, Transducers, Automata, Sequential Machines

Terminology in automata theory is also diverse; in order to avoid misunderstanding of basic notions, we state them below.

Definition 1.22 (System, Transducer, Automaton, Sequential Machine) A (discrete) *system* (or, a system with discrete time $\mathbb{N}_0 = \{0, 1, 2, \ldots\}$) is a 5-tuple $\mathfrak{A} = \langle \mathcal{I}, \mathcal{S}, \mathcal{O}, S, O \rangle$ where:

- $\mathcal{I}$ is a non-empty finite set, the input alphabet
- $\mathcal{O}$ is a non-empty finite set, the output alphabet
- $\mathcal{S}$ is a non-empty (possibly, infinite) set of states
- $S\colon \mathcal{I} \times \mathcal{S} \to \mathcal{S}$ is a state transition function
- $O\colon \mathcal{I} \times \mathcal{S} \to \mathcal{O}$ is an output function

The system is called *autonomous* if neither S nor O depends on input letters (i.e., if $S\colon \mathcal{S} \to \mathcal{S}$, $O\colon \mathcal{S} \to \mathcal{O}$); otherwise the system is called non-autonomous. A *subsystem* $\mathfrak{A}'$ of $\mathfrak{A}$ is a system $\langle \mathcal{I}, \mathcal{S}', \mathcal{O}, S, O \rangle$ such that $\emptyset \neq \mathcal{S}' \subset \mathcal{S}$ and $S(\chi, s') \in$

$\mathcal{S}'$ for all $\chi \in \mathcal{I}$, $s' \in \mathcal{S}'$. A subsystem is called *minimal* if it has no subsystems other than itself.

An *initial automaton* (or, in other terminology, a *letter to letter transducer* [143], a *Mealy sequential machine* [167], an *initial synchronous automaton* [84]) $\mathfrak{A}(s_0)$ is a system where one of the states, $s_0 \in \mathcal{S}$, is fixed; s_0 is called the *initial state*. We stress that the definition of the initial automaton $\mathfrak{A}(s_0)$ is nearly the same as the one of *Mealy automaton* (see, e.g., [31]) with the only important difference: the set of states $\mathcal{S}$ of $\mathfrak{A}(s_0)$ is *not necessarily finite*.

In what follows, the term *automaton* mostly stands for an initial automaton such that its every non-initial state $s \in \mathcal{S}$ is *reachable* (or, *accessible*), i.e., there exists a finite sequence $\chi_0, \chi_1, \ldots, \chi_{N-1} \in \mathcal{I}$ such that $S(\chi_{N-1}, s_{N-1}) = s$, where $s_i = S(\chi_{i-1}, s_{i-1})$, $i = 1, 2, \ldots, N-1$, s_0 is the initial state (therefore the set $\mathcal{S}$ of all states of the automaton is countable). Automata whose state transition functions do not depend on input letters, i.e., such that $S(\chi, s)$ does not depend on $\chi \in \mathcal{I}$, are also known as *generators*; their initial states are called *seeds*. Automata whose output function does not depend on input letters, i.e., such that $O(\chi, s)$ does not depend on $\chi \in \mathcal{I}$, are also known as *Moore automata*, or *Moore (sequential) machines*, provided their state set $\mathcal{S}$ is finite. In the book, however, speaking of Moore automata, we do not assume the finiteness of $\mathcal{S}$.

An automaton $\mathfrak{A}$ determines a unique map $f_{\mathfrak{A}}\colon \ldots \chi_2\chi_1\chi_0 \mapsto \ldots \xi_2\xi_1\xi_0$ from the set $\mathcal{W}_\infty(\mathcal{I})$ of all (one-side) infinite words over the alphabet $\mathcal{I}$ to the set $\mathcal{W}_\infty(\mathcal{O})$ of all (one-side) infinite words over the alphabet $\mathcal{O}$ (cf. Fig. 1.1): At time instant $i = 0$, the automaton being in the state s_0 accepts the first input letter χ_0, updates its state to a newer state $s_1 = S(\chi_0, s_0)$, and produces an output letter $\xi_0 = O(\chi_0, s_0)$; at the next time instant $i = 1$, the automaton accepts χ_1, updates its state to $s_2 = S(\chi_1, s_1)$, and produces an output letter $\xi_1 = O(\chi_1, s_1)$. Therefore, $\xi_i = \varphi_i(\chi_0, \ldots, \chi_i)$, where $\varphi_i : \mathcal{I}^i \to \mathcal{O}$ is a uniquely determined sequence of maps.

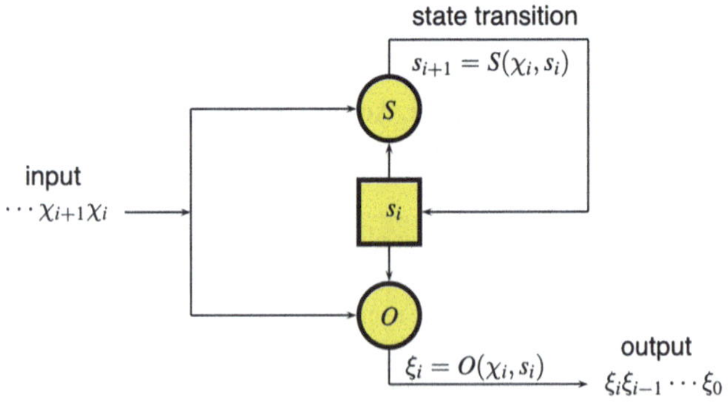

Fig. 1.1 Initial transducer, schematically

Definition 1.23 (Automaton Function) The mapping $f_{\mathfrak{A}}\colon \mathcal{W}_\infty(\mathcal{I}) \to \mathcal{W}_\infty(\mathcal{O})$ defined above is called *automaton function* of the automaton $\mathfrak{A}$.

Figure 1.1 is a schematic representation (block diagram) of an automaton; that representation is most applicable when it is needed to represent an automaton as a computer program. In automata theory, however, it is convenient to represent "internal structure" of an automaton via its *state transition diagram*, a directed graph (*digraph*) whose vertices are states of the automaton and whose edges (which we also call *arrows* for better clarity) are labelled by symbols $a|b$, where a (resp., b) is a letter of input (resp., output) alphabet, and, if being fed by input letter a, the automaton changes i-th state to j-th state, the tail of respective arrow is i-th vertex, and the arrowhead is j-th vertex. To put it in simple words, the arrow labelled by $a|b$ "flies" from the i-th state to the next j-th state of the automaton if the automaton at the i-th state, being fed by the letter a, changes its state to j-th and produces output letter b. Figure 1.2 depicts state transition diagram of the automaton $\mathfrak{A}(s_0)$ whose input/output alphabets are $\mathbb{F}_2 = \{0, 1\}$.

Given an automaton function f, the automaton $\mathfrak{A}_f$ whose automaton function is f is not unique in the meaning of Definition 1.22: There are infinitely many different automata (i.e., the ones whose sets of states are different, whose state transition functions are different, and whose output functions are different but whose automaton function is f). However, given f, there exists a unique automaton whose automaton function is f and whose set of states $\mathcal{S}$ is the "smallest." Namely, call the two states $s_i, s_j \in \mathcal{S}$ of the automaton $\mathfrak{A}$ *equivalent* if whenever s_i, s_j were taken as initial states, the word mappings performed by either of the two initial automata are equal to one another; i.e., if input words are equal to one another, then corresponding output words are also equal to one another. Factorizing the state set of the automaton $\mathfrak{A}$ by the equivalence relation, we get an automaton $\mathfrak{A}'$ that has no equivalent non-equal states and whose automaton function is $f_{\mathfrak{A}}$. The state transition diagram of the automaton $\mathfrak{A}'$ is called *reduced*. An automaton function $f_{\mathfrak{A}}$

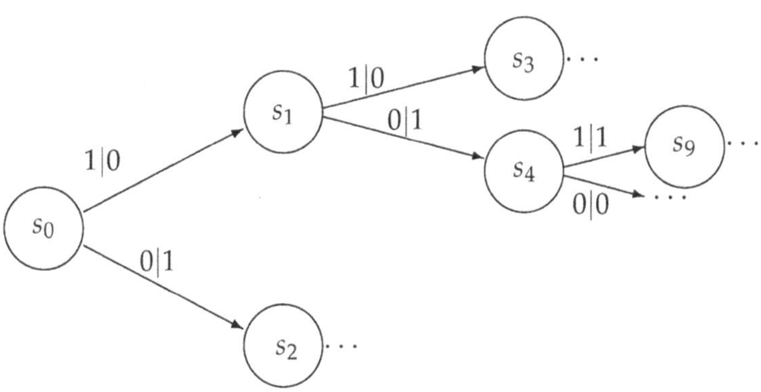

Fig. 1.2 Example state transition diagram of automaton. Both input and output alphabets are binary; their letters are $\{0, 1\} = \mathbb{F}_2$

(as well as the automaton $\mathfrak{A}$) is called *finite* if it can be produced by an automaton $\mathfrak{A}'$ whose set of states is finite; that is, $f_{\mathfrak{A}} = f_{\mathfrak{A}'}$, the factor set of the set of states of the automaton $\mathfrak{A}$ by the equivalence relation is finite, and thus the reduced state transition diagram of $\mathfrak{A}'$ is a finite digraph. To put it in other words, in most cases, we will assume that state transition diagram of automaton is reduced and, if the diagram is finite, call both the automaton and its automaton function a finite.

To the initial automaton $\mathfrak{A}(s_0)$, we put into a correspondence a family $\mathcal{F}(\mathfrak{A})$ of all *sub-automata* $\mathfrak{A}(s) = \langle \mathcal{I}, \tilde{\mathcal{S}}, \mathcal{O}, \tilde{S}, \tilde{O}, s \rangle$, $s \in \mathcal{S}$, where $\tilde{\mathcal{S}} = \tilde{\mathcal{S}}(s) \subset \mathcal{S}$ is the set of all states that are reachable from the state s and $\tilde{S}$, $\tilde{O}$ are respective restrictions of the state transition and output functions S, O on $\mathcal{I} \times \tilde{\mathcal{S}}$. A sub-automaton $\mathfrak{A}(s)$ is called *proper* if the set $\tilde{\mathcal{S}}$ of all its states is a proper subset of $\mathcal{S}$. A sub-automaton $\mathfrak{A}(s)$ is called *minimal* if it contains no proper sub-automata. It is obvious that a finite sub-automaton is minimal if and only if its every state is reachable from any other state. The set of all states of a minimal sub-automaton of the automaton $\mathfrak{A}$ is called an *ergodic component* of the (set of all states) of the automaton $\mathfrak{A}$. It is clear that once the automaton is in a state that belongs to an ergodic component, all its further states will also be in the same ergodic component. Therefore, all states of a finite automaton are of two types only: the *transient states*, which belong to no ergodic component, and *ergodic states*, which belong to ergodic components. It is clear that the set of all ergodic states is a disjoint union of ergodic components. Note that we use the term "minimal automaton" in a different meaning compared to the one used in automata theory: our terminology here is from the theory of Markov chains (see, e.g., [115]; since to the graph of state transitions of every automaton, there corresponds a Markov chain).

It is worth noting that *Moore sequential machines are equivalent to Mealy machines in the following meaning: Under assumption that an output of a Moore machine at initial state is an empty symbol* (i.e., there is no output)*, then the classes of functions represented by Mealy machines and by Moore machines coincide* (see, e.g., [167, Theorem 5.1]). However, to represent an automaton function via state transition diagram of a Moore machine, one needs more states compared to the diagram of respective Mealy machine. That is why in this book, the example state transition diagrams are given for Mealy machines though from a "physical" point of view; it might be more natural to deal with Moore machines since they are looking as defined on Markov chains, in a contrast to Mealy machines, as output of Moore machines formally depends only on states rather than on arrows reaching the states; but that view is misleading since Mealy machines do exactly what Moore machines do.

1.6.2 Discrete-Time Causality

The notion of causal function over discrete time, in terms of general system theory, reads as follows (cf., e.g., [74, 109]):

Definition 1.24 (Causality over Discrete Time) *Causal functions over discrete time* $\mathbb{N}_0 = \{0, 1, 2, \ldots\}$ are exactly functions f that satisfy the following conditions:

1. The domain (the "causes") and range (the "effects") of f are, accordingly, all sequences $\mathbf{c} = (c_i)_{i \in \mathbb{N}_0}$ and $\mathbf{e} = (e_i)_{i \in \mathbb{N}_0}$ over respective sets C, the "elementary causes," and $\mathcal{E}$, the "elementary effects."
2. If $f(\mathbf{c}) = (e_i)_{i \in \mathbb{N}_0}$, then e_i does not depend on $c_{i+1}, c_{i+2}, \ldots$, for all $i \in \mathbb{N}_0$.

In other words, the function f is causal if and only if there exists a sequence $(\varphi_i^f)_{i=0}^{\infty}$ of maps $\varphi_i^f : C^{i+1} \to \mathcal{E}$, $(i \in \mathbb{N}_0)$, such that

$$f(\mathbf{c}) = (\varphi_i^f(c_0, \ldots, c_i))_{i \in \mathbb{N}_0} \tag{1.9}$$

It is reasonable to assume that either of the sets C of "elementary causes" and $\mathcal{E}$ of "elementary effects" contain at least two elements and that both sets are *finite*. The first assumption is natural since otherwise, there is only one "reason" and/or only one "effect." The second assumption is more of a physical nature since no physical objects are known, which are proved to be infinite in some natural meaning: infinity is a mathematical rather than a physical notion, which is used in mathematical calculations in order to find good estimates of physical values since the values can be measured with a non zero error only.

Under that *finiteness assumption*, the following theorem is true (cf., e.g., [167, Chapter IV, Theorem 8.2]):

Theorem 1.25 *The class of all causal functions is exactly the class of all automata functions of automata whose input alphabet is C and output alphabet is* $\mathcal{E}$ (cf., Definitions 1.22 and 1.23).

Note that *Condition 2 of Definition 1.24 is just a Lipschitz condition with a constant 1* (1-Lipschitz, for brevity) with respect to the natural metric d on sequences.

The metric d can be defined as follows: Given two sequences $\mathbf{a} = (a_i)_{i \in \mathbb{N}_0}$ and $\mathbf{a}' = (a_i')_{i \in \mathbb{N}_0}$ over a finite non-empty set $\mathcal{A}$, put $d(\mathbf{a}, \mathbf{a}') = p^{-n}$ where $n = \max\{i \in \mathbb{N}: c_i = c_i'\}$ if such n exists and $d(\mathbf{a}, \mathbf{a}') = 0$ if $a_i = a_i'$ for all $i \in \mathbb{N}_0$ (here $p > 1$ is arbitrary real number). To put it in other words, $d(\mathbf{a}, \mathbf{a}')$ is distance between two sequences $\mathbf{a}$ and $\mathbf{a}'$, which is defined as follows: $d(\mathbf{a}, \mathbf{a}') = p^{-n}$ where n is the length of the longest common prefix of the two (left-)infinite words

$w_{\mathbf{a}} = \cdots \alpha_3\alpha_2\alpha_1\alpha_0$ and $w_{\mathbf{a}'} = \cdots \alpha'_3\alpha'_2\alpha'_1\alpha'_0$ over alphabet $\mathcal{A}$ (and $d(\mathbf{a}, \mathbf{a}') = 0$ if $\mathbf{a} = \mathbf{a}'$).

This is a very important observation since it implies that all *causal functions over discrete time are continuous* with respect to the metric d. The observation implies that it is possible to mathematically investigate causality over discrete time by using powerful tools of p-adic analysis and p-adic dynamics. This approach is the leading methodology of the book. The metric is defined as $d(\mathbf{a}, \mathbf{a}') = p^{-n}$, where n is the length of the longest common prefix of the two (left-)infinite words $w_{\mathbf{a}} = \cdots \alpha_3\alpha_2\alpha_1\alpha_0$ and $w_{\mathbf{a}'} = \cdots \alpha'_3\alpha'_2\alpha'_1\alpha'_0$ over alphabet $\mathcal{A}$ (and $d(\mathbf{a}, \mathbf{a}') = 0$ if $\mathbf{a} = \mathbf{a}'$). Of course, even more general functions that may be also judged as "causal" in some natural meaning are considered in literature (see, e.g., [29] about genetic code evolution or [171, 181, 195] for neural network(-like) models of Universe); in this book, we however restrict ourselves with the meaning introduced by Definition 1.24.

Part I
Basics of p-Adic Analysis

In this part, we develop p-adic analysis for 1-Lipschitz functions whose domain and co-domain are (modules over) p-adic integers. These functions are exactly the automata functions of automata which perform transformations of words over the alphabet $\mathcal{A} = \{0, 1, \ldots, p - 1\}$; or, in other terms, these functions are exactly the causal functions over discrete time.

Chapter 2
Rings and Fields of *p*-Adic Numbers

Abstract In this chapter, we recall basic concepts of p-adic theory: p-adic valuation, p-adic metrics, p-adic integers, and the likes. Most results are given without proofs since proofs can be found in a number of books on p-adic analysis (see, e.g., Mahler (p-adic Numbers and their Functions, 2nd edn., Cambridge Univ. Press (1981)), Schikhof (Ultrametric Calculus: An Introduction to p-adic Analysis, Cambridge Univ. Press (1984)), Koblitz (p-adic Numbers, p-adic Analysis, and Zeta-functions, Graduate texts in math., vol. 58, 2nd edn., Springer (1984)), Khrennikov, Non-Archimedean Analysis: Quantum Paradoxes, Dynamical Systems and Biological Models, Kluwer Academic Publishers, Dordrecht (1997), van Rooij, Non-Archimedean Functional Analysis, Marcel Dekker, New York (1978), Gouvêa, p-adic Numbers. An Introduction, 2nd edn., Springer, Berlin–Heidelberg–New York (1997), Robert, A Course in p-adic Analysis, Springer, Berlin–New York–Heidelberg (2000)), and, especially, Katok (p-adic Analysis in Comparison with Real, Mass. Selecta., American Mathematical Society (2003)).

2.1 *p*-Adic Valuation, *p*-Adic Absolute Value, *p*-Adic Metric

Let p be a fixed prime number. By the fundamental theorem of arithmetics, each non-zero integer n can be written uniquely as

$$n = p^{\operatorname{ord}_p n}\hat{n},$$

where $\hat{n}$ is a non-zero integer, $p \nmid \hat{n}$, and $\operatorname{ord}_p n$ is a unique non-negative integer. The function $\operatorname{ord}_p \colon \mathbb{Z} \setminus \{0\} \to \mathbb{N}_0$ is called the *p-adic valuation*. In the sequel, we will need the following classical result of Legendre (for the proof, see, e.g., [168, Lemma 25.5], [3, Corollary 3.2.2], [124, Ch.1, Section 2, Exercise 14]):

Lemma 2.1 (Valuation of Factorial) *Write a natural number $n \in \mathbb{N}$ in the base-p expansion: $n = a_0 + a_1 p + \cdots + a_m p^m$. Here $a_i = \delta_i(n)$ are uniquely defined numbers from $\{0, 1, \ldots, p-1\}$, $i = 0, 1, 2, \ldots$, the p-adic digits of n. Denote*

V. Anashin, *Causality: The p-adic Theory*, STEAM-H: Science, Technology, Engineering, Agriculture, Mathematics & Health,
https://doi.org/10.1007/978-3-031-85818-5_2

$\mathrm{wt}_p\, n = \sum_{k=0}^{m} a_k$, *the p-adic weight of n. Then*

$$\operatorname{ord}_p n! = \frac{n - \mathrm{wt}_p\, n}{p-1}.$$

Corollary 2.2 (Valuation of Binomial Coefficient) *For all* $i, k \in \mathbb{N}_0$,

$$\operatorname{ord}_p \binom{i+k}{i} = \frac{1}{p-1}(\mathrm{wt}_p\, i + \mathrm{wt}_p\, k - \mathrm{wt}_p(i+k)).$$

Let $\mathbb{K}$ be a commutative ring with a unity 1 and without zero divisors, i.e., let $\mathbb{K}$ be an integral domain.

Definition 2.3 (Absolute Value) An *absolute value* on $\mathbb{K}$ is a function $\|\cdot\|\colon \mathbb{K} \to \mathbb{R}_{\geqslant 0}$ such that

1. $\|x\| = 0$ if and only if $x = 0$
2. $\|xy\| = \|x\|\|y\|$, for all $x, y \in \mathbb{K}$
3. $\|x + y\| \leqslant \|x\| + \|y\|$, for all $x, y \in \mathbb{K}$ (*the triangle inequality*)

When inequality 3 is replaced by a stronger one, the *strong triangle inequality*,

$$\|x + y\| \leqslant \max(\|x\|, \|y\|), \text{ for all } x, y \in \mathbb{K},$$

the absolute value $\|\cdot\|$ is called *non-Archimedean*. If $\|x\| = 1$ for all non-zero $x \in \mathbb{K}$, then the absolute value $\|\cdot\|$ is called *trivial*.

Note that often some authors call absolute value from Definition 2.3 a *norm*; fields and rings endowed with an absolute value are called *normed fields* or *normed rings*, respectively. Further in the book, the notion "norm" is used in another meaning; that is why when speaking of fields and rings, we use the term "absolute value" rather than "norm."

The trivial absolute value is a trivial example of non-Archimedean absolute value. A non-trivial example of non-Archimedean absolute value is *p-adic absolute value* on the field $\mathbb{Q}$ of rational numbers, which is defined as follows.

Given $r \in \mathbb{Q} \setminus \{0\}$, represent r as a (simple) fraction $r = a/b$, where $a, b \in \mathbb{Z}$; put

$$\operatorname{ord}_p r = \operatorname{ord}_p a - \operatorname{ord}_p b;$$

then for $r \in \mathbb{Q}$ define

$$|r|_p = \begin{cases} p^{-\operatorname{ord}_p r}, & \text{if } r \neq 0; \\ 0, & \text{otherwise.} \end{cases} \tag{2.1}$$

This *p-adic absolute value* $|r|_p$ is well defined on $\mathbb{Q}$; i.e., does not depend on concrete representation of $r \in \mathbb{Q}$ as a simple fraction.

The field $\mathbb{Q}$ endowed with a metric d_p is a *non-Archimedean-valued field.* Recall that a field $\mathbb{K}$ endowed with the absolute value $\|\cdot\|$ is *Archimedean-valued field* if $\mathbb{K}$ satisfies the *Archimedean Axiom,* which in that case reads the following: *For every non-zero $x \in \mathbb{K}$ there exists $n \in \mathbb{N}$ such that*

$$\| \underbrace{x + \cdots + x}_{n \text{ terms}} \| > 1.$$

It is clear that the Archimedean Axiom is violated when $\mathbb{Q}$ is endowed with the absolute value $|\cdot|_p$: Take $x = p \in \mathbb{Q}$ then $|np|_p = |n|_p|p|_p = |n|_p/p < 1$ since $|n|_p \leqslant 1$ for every $n \in \mathbb{N}$.

Now we can define *p-adic metric* d_p on $\mathbb{Q}$ in a standard way assuming that $d_p(x, y) = |x - y|_p$, for every pair $x, y \in \mathbb{Q}$ thus endowing the field $\mathbb{Q}$ with a structure of metric space. For this metric, the *strong triangle inequality* holds. Namely, given $x, y, z \in \mathbb{Q}$,

$$d_p(x, y) \leqslant \max\{d_p(x, z), d_p(z, y)\}.$$

Metric that satisfies strong triangle inequality is called an *ultrametric*; metric spaces whose metrics satisfy strong triangle inequality are called *ultrametric spaces.*

Recall that a sequence $(a_i)_{i=0}^{\infty}$ of elements of a metric space $\mathbb{X}$ equipped with a metric d is called a *Cauchy sequence* if for any real $\varepsilon > 0$, there exists an $N \in \mathbb{N}_0$ such that $d(a_m, a_n) < \varepsilon$ whenever both $m > N$ and $n > N$. Two metrics d_1 and d_2 on a set X are called *equivalent* if a sequence $(a_i)_{i=0}^{\infty}$ of elements of X is Cauchy with respect to d_1 if and only if it is Cauchy with respect to d_2. Two absolute values are called *equivalent* if they induce equivalent metrics.

The following theorem was proved by Alexander Markowich Ostrowski in 1916 and now is known as *Ostrowski's theorem.*

Theorem 2.4 (Ostrowski's Theorem) *Every non-trivial absolute value on $\mathbb{Q}$ is either equivalent to the real absolute value $|\cdot|$ or to one of the p-adic absolute values $|\cdot|_p$, for some prime $p \in \{2, 3, 5, 7, 11, \dots\}$.*

For a proof of Ostrowski's theorem, see, e.g., [77, 124, 168]. In the book, we mostly deal with p-adic absolute value; in this case, the notion of Cauchy sequence can be refined:

Theorem 2.5 *A sequence $(x_i)_{i=0}^{\infty}$ of rational numbers is a Cauchy sequence with respect to the p-adic metric $d_p(x, y) = |x - y|_p$ if and only if*

$$\lim_{i\to\infty} |x_{i+1} - x_i|_p = 0. \tag{2.2}$$

See, e.g., [77, Lemma 3.2.2] for a proof.

2.2 The Field $\mathbb{Q}_p$ and the Ring $\mathbb{Z}_p$, Ostrowski's Theorem

It is an exercise to prove that if a sequence of elements of a metric space $\mathbb{X}$ (endowed with a metric d) converges to some element of $\mathbb{X}$, then the sequence is a Cauchy. It is well-known that converse is generally not true: for instance, the sequence of rational approximations of $\sqrt{2}$ is Cauchy with respect to the real metric $d_{\mathbb{R}}$, which is induced by the absolute value $|\cdot|$, i.e., $d_{\mathbb{R}}(x, y) = |x - y|$ for $x, y \in \mathbb{Q}$. Remind that a metric space $\mathbb{X}$ endowed with a metric d is called *complete* (with respect to the metric d) if every Cauchy sequence of elements of $\mathbb{X}$ converges to some element of $\mathbb{X}$. It is well-known that $\mathbb{R}$ is complete with respect to real metric $d_{\mathbb{R}}$. However, the field $\mathbb{Q}$ is complete neither with respect to real metric $d_{\mathbb{R}}$ nor to p-adic metric d_p, whatever prime p is taken (see, e.g., [77]). It turns out that there exist a number of fields that are completions of the field $\mathbb{Q}$ with respect to metrics d_p.

Recall that a complete metric space $\mathbb{X}^\star$ endowed with a metric $d^\star$ is said to be the *completion* of the space $\mathbb{X}$ if:

- $\mathbb{X}$ is a subspace of the space $\mathbb{X}^\star$ (i.e., $\mathbb{X} \subset \mathbb{X}^\star$ and the restriction $d^\star|_{\mathbb{X}}$ of the metric $d^\star$ to $\mathbb{X}$ is d)
- $\mathbb{X}$ is dense in $\mathbb{X}^\star$

Theorem 2.6 (Completion Theorem) *Every metric space has a completion, and all of its completions are isometric.*

For a proof of this theorem, see, e.g., [126, Ch. II, Sect. 13, Theorem 2].

Theorem 2.6 implies that, given a metric d_p for some prime p, there exists a completion $\mathbb{Q}^\star$ of the metric space $\mathbb{Q}$ with respect to the metric d_p. Addition and multiplication of elements of the space $\mathbb{Q}^\star$ can be defined in a standard manner by using addition and multiplication of Cauchy sequences over $\mathbb{Q}$; this way, $\mathbb{Q}^*$ becomes a field, the *field* $\mathbb{Q}_p$ *of* p*-adic numbers*. In what follows, we use notation d_p rather than $d_p^\star$ for the metric on $\mathbb{Q}_p$. For a detailed construction of $\mathbb{Q}_p$ as a completion of $\mathbb{Q}$, see [112].

> By Ostrowski's theorem 2.4, a complete list of all possible completions of $\mathbb{Q}$ are $\mathbb{R}$ and $\mathbb{Q}_p$, p a prime.

2.2.1 *Topology of* $\mathbb{Q}_p$ *and* $\mathbb{Z}_p$

Now we discuss some topological properties of fields (and rings) of p-adic numbers. Recall that a topological space is called *locally compact* if every point has a compact neighborhood. Recall also that a ring K (respectively, field) endowed with a topology is said to be a *topological ring* (respectively, *topological field*) if the operations of addition and multiplication are continuous with respect to that topology.

With respect to the topology induced by the p-adic metric d_p, the field $\mathbb{Q}_p$ is a locally compact topological field.

Recall that an *open* (resp., *closed*) *ball* of radius r centered at the point $a \in \mathbb{X}$ in a metric space $\mathbb{X}$ endowed with a metric d is a set $\mathbf{B}_r^-(a) = \{z \in \mathbb{X}\colon d(z, a) < r\}$ (respectively, $\mathbf{B}_r(a) = \{z \in M\colon d(z, a) \leq r\}$). As the absolute value $|\cdot|_p$ may be only p^ℓ for some $\ell \in \mathbb{Z}$, for p-adic balls (i.e., for balls in $\mathbb{Q}_p$), we have that $\mathbf{B}_{p^\ell}^-(a) = \mathbf{B}_{p^{\ell-1}}(a)$; thus:

p-adic balls (of non-zero radii) are open and closed simultaneously.

Sets that are open and closed simultaneously are called *clopen*; so $\mathbf{B}_{p^\ell}(a)$ is a clopen ball of radius p^ℓ centered at $a \in \mathbb{Q}_p$. Balls are *compact*; the set of all balls (of non-zero radii) forms a *topological base* of a topology of a metric space. Thus,

$\mathbb{Q}_p$ is a *totally disconnected* topological space.

Recall that a topological space $\mathbb{X}$ is said to be *totally disconnected* if for each pair $a, b \in \mathbb{X}$ there exist open subsets A, B of $\mathbb{X}$ such that $a \in A$, $b \in B$, $A \cap B = \emptyset$, and $A \cup B = \mathbb{X}$.

Every point z of a ball $\mathbf{B}_{p^\ell}(a)$ is a center of this ball, i.e.,

$$\mathbf{B}_{p^\ell}(a) = \mathbf{B}_{p^\ell}(z), \text{ for every } z \in \mathbf{B}_{p^\ell}(a).$$

Indeed, $d_p(z, a) \leqslant p^\ell$; so if $x \in \mathbb{Q}_p$ is such that $d_p(z, x) \leq p^\ell$, then, as $d_p(z, x) \leq \max\{d_p(z, a), d_p(a, x)\}$ by the strong triangle inequality, any x that lies in $\mathbf{B}_{p^\ell}(a)$ (i.e., satisfies $d_p(x, a) \leq p^\ell$) also satisfies $d_p(z, x) \leq p^\ell$; that is, lies in $\mathbf{B}_{p^\ell}(z)$.

A similar result holds for p-adic spheres as well: namely, every point that lies inside a p-adic sphere of non-zero radius is a center of this sphere.

By definition, the *p-adic sphere* of radius p^ℓ (where $\ell \in \mathbb{Z}$) around $a \in \mathbb{Q}_p$ is $\mathbf{S}_{p^\ell}(a) = \{z \in \mathbb{Q}_p : d_p(z, a) = p^\ell\}$; therefore, since $d_p(x, y) = p^k$ for some $k \in \mathbb{Z}$ if $x, y \in \mathbb{Q}_p, x \neq y$,

$$\mathbf{S}_{p^\ell}(a) = \mathbf{B}_{p^\ell}(a) \setminus \mathbf{B}_{p^{\ell-1}}(a),$$

so a point inside the sphere $\mathbf{S}_{p^\ell}(a)$ must lie in the ball $\mathbf{B}_{p^{\ell-1}}(a)$; but every point of a p-adic ball can be taken as a center of that ball. Note that from the definition, it immediately follows that the sphere is a disjoint union of $p-1$ balls of radii $p^{\ell-1}$; so a p-adic sphere is a compact clopen set.

The ball $\mathbf{B}_1(0) \subset \mathbb{Q}_p$ of radius 1 centered at 0 is of special interest further in the book. Since the absolute value $|\cdot|_p$ is non-Archimedean, from Definition 2.3, it immediately follows that $\mathbf{B}_1(0)$ is a ring. This ring is called *the ring of p-adic integers* and is denoted via $\mathbb{Z}_p$. It is clear that $\mathbb{Z}_p$ is integral domain (i.e., has no zero divisors) since it is a sub-ring of a field. Below we summarize the main properties of $\mathbb{Z}_p$ without proofs since the proofs can be found in, e.g., [112, 124, 145].

The ring $\mathbb{Z}_p$ of p-adic integers is a compact topological ring with respect to the topology induced by the metric d_p. The set $\mathbb{N}_0 = \{0, 1, 2, \ldots\}$ is a dense subset of $\mathbb{Z}_p$.

Recall that a family of sets $\mathcal{F}$ is said to have the *finite intersection property* if every finite intersection of sets of $\mathcal{F}$ is nonempty. It is well-known that a metric space is compact if and only if every family of closed sets that satisfies the finite intersection property has a nonempty intersection. This implies the following theorem:

Theorem 2.7 *Given a nested sequence of non-empty closed subsets*

$$M_0 \supset M_1 \supset \cdots \supset M_i \supset \cdots$$

in $\mathbb{Z}_p$, *the intersection* $\cap_{i=0}^{\infty} M_i$ *is not empty.*

An ultrametric space is said to be *spherically complete* if each nested sequence of balls has a nonempty intersection (cf., e.g., [168, Definition 20.1]). The space $\mathbb{Q}_p$ is spherically complete; that is, Theorem 2.7 (which is equivalent to the compactness) remains true for nested sequences of balls from $\mathbb{Q}_p$. However, the theorem does not hold for nested sequences of arbitrary closed subsets from $\mathbb{Q}_p$ since $\mathbb{Q}_p$ is locally compact, and not compact.

2.2.2 Algebraic Properties of the Ring $\mathbb{Z}_p$

In the book, we mostly deal with the compact space $\mathbb{Z}_p$ rather than with $\mathbb{Q}_p$ at whole; so the structure of $\mathbb{Z}_p$ is of the most interest in the sequel. The space $\mathbb{Z}_p$ is a ring, and it turns out that balls of that space are cosets of ideals of the ring $\mathbb{Z}_p$, and vice versa, cosets of ideals of the ring $\mathbb{Z}_p$ are balls of the space $\mathbb{Z}_p$. All proper ideals of the ring $\mathbb{Z}_p$ are principal ideals (cf. Sect. 1.2.2); namely, these are $p^n\mathbb{Z}_p$, and $n \in \mathbb{N}$. The ideals are kernels of the ring epimorphisms $\mathsf{mod}\, p^n : \mathbb{Z}_p \to \mathbb{Z}/p^n\mathbb{Z}$ of the ring $\mathbb{Z}_p$ onto the ring $\mathbb{Z}/p^n\mathbb{Z}$ of all non-negative residues $\{0, 1, \dots, p^n - 1\}$ of the ring $\mathbb{Z}$ modulo p^n, the *residue ring modulo* p^n. In what follows, we call integers from $\mathbb{Z}$ *rational integers* to distinguish them from p-adic integers since in the ring $\mathbb{Z}_p$ of p-adic integers, there are rational numbers from $\mathbb{Q}$ that are not in $\mathbb{Z}$; these are p-adic integers but not rational integers. Note that the map $\mathsf{mod}\ p^n$ is continuous provided $\mathbb{Z}/p^n\mathbb{Z}$ is endowed with standard discrete topology on a finite set. If $\mathsf{mod}\ p^n$-images of $x, y \in \mathbb{Z}_p$ are equal one to another, we write $x \equiv y \pmod{p^n}$.

Moreover, further for $x, y \in \mathbb{Q}_p$ and $n \in \mathbb{N}$, we use notation $x \equiv y \pmod{p^n}$ along with $|x - y|_p \leqslant p^{-n}$ and $d_p(x, y) \leqslant p^{-n}$.

This notation is justified by the following observation: From the definition of p-adic absolute value, it immediately follows that $|x - y|_p < p^{-n}$ is an equivalence relation that agrees with addition of p-adic numbers, and if $x, y \in \mathbb{Z}_p$, then this equivalence relation agrees also with multiplication of p-adic integers and thus is a congruence on $\mathbb{Z}_p$, which whence determines a canonical epimorphism $\mathsf{mod}\ p^n$ of the ring $\mathbb{Z}_p$ to its factor-ring modulo the ideal $p^n\mathbb{Z}_p$.

We summarize:

Let $n \in \mathbb{N}$, $a \in \mathbb{Q}_p$; then:

$$\mathbf{B}_{p^{-n}}(a) = a + p^n\mathbb{Z}_p = \{z \in \mathbb{Q}_p : z \equiv a \pmod{p^n}\}. \tag{2.3}$$

For $a \in \mathbb{Z}_p$, the ball $\mathbf{B}_{p^{-n}}(a)$ is a co-set of the ideal $p^n\mathbb{Z}_p$ of the ring $\mathbb{Z}_p$; the ball is a full pre-image of $a\ \mathsf{mod}\ p^n \in \mathbb{Z}/p^n\mathbb{Z}$ of the epimorphism $\mathsf{mod}\ p^n : \mathbb{Z}_p \to \mathbb{Z}/p^n\mathbb{Z}$ of the ring $\mathbb{Z}_p$ onto the ring $\mathbb{Z}/p^n\mathbb{Z}$ of all non-negative residues of rational integers modulo p^n.

Note that the ring $\mathbb{Z}_p$ can be viewed as a *projective limit* $\mathbb{Z}_p = \varprojlim_{n\to\infty} \mathbb{Z}/p^n\mathbb{Z}$ (also called the *inverse limit*) of the following *projective system* (respectively, the *inverse system*) of residue rings $\mathbb{Z}/p^n\mathbb{Z}$, $n = 1, 2, \ldots$ (cf., [99, Chapter I, Section "Projective Limits"] and [131, Chapter V, Section 1]):

$$\cdots \overset{\mathsf{mod}_n}{\longrightarrow} \mathbb{Z}/p^n\mathbb{Z} \overset{\mathsf{mod}_{n-1}}{\longrightarrow} \mathbb{Z}/p^{n-1}\mathbb{Z} \overset{\mathsf{mod}_{n-2}}{\longrightarrow} \cdots \overset{\mathsf{mod}_1}{\longrightarrow} \mathbb{Z}/p\mathbb{Z} \tag{2.4}$$

Here $\mathsf{mod}_k \colon \mathbb{Z}/p^{k+1}\mathbb{Z} \to \mathbb{Z}/p^k\mathbb{Z}$ is a natural epimorphism whose kernel is an ideal generated by p^k of the residue ring $\mathbb{Z}/p^k\mathbb{Z}_p$; that is, in plain words, mod_k is just a reduction modulo p^k of the residues $\{0, 1, \ldots, p^{k+1} - 1\}$, which constitutes the ring $\mathbb{Z}/p^{k+1}\mathbb{Z}$.

Finally, we note that a p-adic integer $z \in \mathbb{Z}_p$ is a unit of the ring $\mathbb{Z}_p$ (i.e., has a multiplicative inverse in $\mathbb{Z}_p$) if and only if $|z|_p = 1$ or, which is equivalent, if and only if $z \bmod p \neq 0$.

2.3 Representations of p-Adic Numbers

From algebraic point of view, the field $\mathbb{Q}_p$ is a quotient field of the ring $\mathbb{Z}_p$; therefore, every $z \in \mathbb{Q}_p$ can be represented as a fraction a/b for suitable $a, b \in \mathbb{Z}_p$, $b \neq 0$. However, this representation is not very convenient for practical calculations, which we need further in the book, so in this section, we discuss other representations of p-adic numbers.

2.3.1 *p-Adic Canonical Expansion*

In the book, the most important representations of p-adic numbers are *(p-adic) canonical expansions*. These are infinite series of special kind. Note that due to the strong triangle inequality, the following theorem holds (cf., e.g., [112, Proposition 3.3]):

Theorem 2.8 *Infinite series* $\sum_{j=0}^{\infty} z_i$ *of p-adic numbers z_j converges in* $\mathbb{Q}_p$ (i.e., with respect to the p-adic metric) *if and only if* $\lim_{j\to\infty} |z_j|_p = 0$.

The canonical p-adic expansion is the following series, which converges in $\mathbb{Q}_p$ since the series satisfies Theorem 2.8.

Given $z \in \mathbb{Q}_p$, there exists $k \in \mathbb{Z}$ and $\alpha_k, \alpha_{k+1}, \ldots \in \{0, 1, \ldots, p-1\}$ such that

$$z = \sum_{i=k}^{\infty} \alpha_i p^i. \tag{2.5}$$

(continued)

For $z \neq 0$, there exists the smallest $k \in \mathbb{Z}$ such that $\alpha_k \neq 0$; namely, this k is such that $|z|_p = p^{-k}$. For that k, the above representation of z is unique; the respective series is called a *canonical expansion* of the p-adic number z. In the book, these α_i are called *p-adic digits of $z \in \mathbb{Q}_p$* and denoted via $\delta_i^p(z)$ (note that $\delta_i^p(z) = 0$ when $i < k$); we usually omit the superscript when it is clear what p is meant. The p-adic number $z \neq 0$ is a p-adic integer if and only if $k \geqslant 0$. The canonical expansion of 0 is the one where k and all $\alpha_k, \alpha_{k+1}, \ldots$ are 0. Therefore, for $z \in \mathbb{Z}_p$, there exists a unique sequence $\alpha_0, \alpha_1, \alpha_2, \ldots \in \{0, 1, \ldots, p-1\}$ such that (2.5) is true.

Therefore, a p-adic number $z \in \mathbb{Q}_p$ whose p-adic absolute value is p^k for some $k \geqslant 1$ has a unique representation

$$z = \frac{m}{p^k} + [z]_p, \tag{2.6}$$

where $m \in \{1, 2 \ldots, p^k - 1\}$ and $[z]_p \in \mathbb{Z}_p$. The p-adic integer $[z]_p$ is called *integral part of the p-adic number* $z \in \mathbb{Q}_p$, whereas $\frac{m}{p^k}$ is called *fractional part* and denoted via $]z[_p$ ($]z[_p = 0$ for $z \in \mathbb{Z}_p$; subscript may be omitted when it is clear from the context what is p).

We stress once again that the ring $\mathbb{Z}_p$ is the main arena of the play the book is about; that is why we now focus on p-adic integers.

Note 2.9 Let $z \in \mathbb{Z}_p$; then the following is true:

- If $|z|_p = p^{-n}$, then $z = p^n \check{z}$, where $\check{z} \in \mathbb{Z}_p$ is a unit or the ring $\mathbb{Z}_p$, i.e., the multiplicative inverse $\check{z}^{-1}$ is also a p-adic integer from $\mathbb{Z}_p$. Therefore, $z \in \mathbb{Z}_p$ is a unit of $\mathbb{Z}_p$ if and only if $\delta_0(z) \neq 0$.
- $z \in \mathbb{N}_0$ if and only if all but a finite number of coefficients $\alpha_i = \delta_i(z)$ in the canonical expansion (2.5) of z are 0, while $z \in \{-1, -2, -3, \ldots\}$ if and only if all but a finite number of $\delta_i(z)$ are $p - 1$.

Note also that if

$$x = \delta_0(x) + \delta_1(x)p + \delta_2(x)p^2 + \ldots + \delta_j(x)p^j + \ldots$$

is the canonical expansion of the p-adic integer $x \in \mathbb{Z}_p$, then the reduction map $\bmod p^k \colon \mathbb{Z}_p \to \mathbb{Z}/p^k\mathbb{Z}$ (cf., Sect. 2.2.2) can be written as

$$\bmod p^k \colon x = \sum_{i=0}^{\infty} \delta_i(x)p^k \mapsto x \bmod p^k = \sum_{i=0}^{k-1} \delta_i(x)p^k \in \mathbb{Z}/p^k\mathbb{Z}. \tag{2.7}$$

We often omit the word "map" when speaking of the image of p-adic integer with respect to the reduction map.

2.3.2 Representations of Rational p-Adic Integers

In the book, rational p-adic integers are of our special attention since these numbers constitute a dense sunset both of $\mathbb{Z}_p$ with respect to p-adic metric and of $\mathbb{R}$ with respect to real metric since the *rational p-adic integers* are rational numbers z, which at the same time are p-adic integers, i.e., such that $z \in \mathbb{Z}_p \cap \mathbb{Q}$. In this subsection, we consider in detail various representations of rational p-adic integers. Note that $z \in \mathbb{Z}_p \cap \mathbb{Q}$ if and only if z can be represented by an irreducible fraction $z = a/b$, $a \in \mathbb{Z}$, $b \in \mathbb{N}$, where b is co-prime to p. The following proposition is well-known (see, e.g., [112]).

Proposition 2.10 *A p-adic integer z is rational* (i.e., $z \in \mathbb{Z}_p \cap \mathbb{Q}$) *if and only if the sequence of coefficients of its canonical form is eventually periodic:*

$$z = \alpha_0 + \alpha_1 p + \cdots + \alpha_{r-1} p^{r-1} + (\beta_0 + \beta_1 p + \cdots + \beta_{t-1} p^{t-1}) p^r + \\ + (\beta_0 + \beta_1 p + \cdots + \beta_{t-1} p^{t-1}) p^{r+t} + (\beta_0 + \beta_1 p + \cdots + \beta_{t-1} p^{t-1}) p^{r+2t} + \cdots \tag{2.8}$$

for suitable $\alpha_j, \beta_i \in \{0, 1, \ldots, p-1\}$, $r \in \mathbb{N}_0$, $t \in \mathbb{N}$ (the sum $\alpha_0 + \alpha_1 p + \cdots + \alpha_{r-1} p^{r-1}$ is absent in the above expression once $r = 0$).

To put this in other words: Once a p-adic integer z is represented via canonical expansion $z = \sum_{i=0}^{\infty} \gamma_i p^i$, then corresponding infinite word $\ldots \gamma_1 \gamma_0$ is eventually periodic: $\ldots \gamma_1 \gamma_0 = (\beta_{t-1} \ldots \beta_0)^{\infty} \alpha_{r-1} \ldots \alpha_0$. It is clear that given $z \in \mathbb{Z}_p \cap \mathbb{Q}$, both r and t are not unique: For instance,

$$(\beta_{t-1} \ldots \beta_0)^{\infty} \alpha_{r-1} \ldots \alpha_0 = (\beta_0 \beta_{t-1} \ldots \beta_1 \beta_0 \beta_{t-1} \ldots \beta_1)^{\infty} \alpha_r \alpha_{r-1} \ldots \alpha_0,$$

where $\alpha_r = \beta_0$. But once both pre-periodic and periodic parts (the prefix $\alpha_{r-1} \ldots \alpha_0$ and the word $\beta_{t-1} \ldots \beta_0$) are taken the shortest possible, both the *pre-period length* r and the *period length* t are unique for a given p-adic rational integer $z \in \mathbb{Z}_p \cap \mathbb{Q}$; we refer to $\alpha_{r-1} \ldots \alpha_0$ and to $\beta_{t-1} \beta_{t-2} \ldots \beta_1 \beta_0$ as to *pre-period* of z and *period* of z accordingly (cf., Sect. 1.4.1).

Given $z \in \mathbb{Z}_p \cap \mathbb{Q}$, we mostly assume further that in the representation $z = \alpha_0 + \cdots + \alpha_{r-1} p^{r-1} + (\beta_0 + \cdots + \beta_{t-1} p^{t-1}) \sum_{j=0}^{\infty} p^{r+tj}$ (respectively, in eventually periodic infinite word $\mathsf{wrd}(z) = (\beta_{t-1} \ldots \beta_0)^{\infty} \alpha_{r-1} \ldots \alpha_0$ that corresponds to z), r is a pre-period length and t is a period length. Note that a pre-period may be an empty word (i.e., of length 0), while a period cannot.

Rational p-adic integers can also be represented as fractions of special kind.

Proposition 2.11 *A p-adic integer $z \in \mathbb{Z}_p$ is rational if and only if there exist $t \in \mathbb{N}$, $c \in \mathbb{Z}$, $d \in \{0, 1, \dots, p^t - 2\}$ such that*

$$z = c + \frac{d}{p^t - 1} \tag{2.9}$$

Proof Indeed, $z \in \mathbb{Z}_p \cap \mathbb{Q}$ if and only if z is of the form (2.8); therefore,

$$\begin{aligned} z &= (\alpha_0+\alpha_1 p+\cdots+\alpha_{r-1}p^{r-1}-p^r)+p^r\left(1-\frac{\beta_0+\beta_1 p+\cdots+\beta_{t-1}p^{t-1}}{p^t-1}\right) = \\ &= (\alpha_0+\alpha_1 p+\cdots+\alpha_{r-1}p^{r-1}-p^r+q)+\frac{\zeta_0+\zeta_1 p+\cdots+\zeta_{t-1}p^{t-1}}{p^t-1} \end{aligned} \tag{2.10}$$

where $\zeta_0 + \zeta_1 p + \cdots + \zeta_{t-1} p^{t-1}$ is the base-p expansion of the least non-negative residue s of $p^r(p^t - 1 - (\beta_0 + \beta_1 p + \cdots + \beta_{t-1} p^{t-1})) = (p^t - 1)q + s$ modulo $p^t - 1$. □

Note 2.12 For every $m \in \mathbb{N}$, the p-adic canonical expansion of the multiplicative inverse $(1 - p^m)^{-1}$ of the p-adic integer $1 - p^m$ is $(1 - p^m)^{-1} = \sum_{i=0}^{\infty} p^{mi} \in \mathbb{Z}_p$.

Note 2.13 Note that once in (2.10) r is a pre-period length and t is a period length of $z \in \mathbb{Z}_p \cap \mathbb{Q}$, the representation (2.9) is unique; that is, the choice of c and d in (2.9) is unique.

In the sequel, we often use base-p expansions of p-adic rational integers reduced modulo 1 (recall that if $y \in \mathbb{R}$ then by the definition $y \operatorname{mod} 1 = y - \lfloor y \rfloor \in [0, 1) \subset \mathbb{R}$) along with their p-adic canonical expansions. For the reader's convenience, we now summarize some facts on connections between these representations.

It is very well-known that the base-p expansion of a rational number is eventually periodic; that is, given $x \in \mathbb{Q} \cap [0, 1]$, the base-$p$ expansion for x is

$$\begin{aligned} x &= 0.\chi_0 \dots \chi_{k-1}(\xi_0 \dots \xi_{n-1})^\infty = \\ &= \chi_0 p^{-1} + \chi_1 p^{-2} + \cdots + \chi_{k-1} p^{-k} + \xi_0 p^{-k-1} + \xi_1 p^{-k-2} + \cdots + \xi_{n-1} p^{-k-n} + \\ &\qquad + \xi_0 p^{-k-1-n} + \xi_1 p^{-k-2-n} + \cdots + \xi_{n-1} p^{-k-2n} + \cdots = \\ &= \frac{1}{p^k}(\chi_0 p^{k-1} + \chi_1 p^{k-2} + \cdots + \chi_{k-1}) + \frac{1}{p^k} \times \frac{\xi_0 p^{n-1} + \xi_1 p^{n-2} + \cdots + \xi_{n-1}}{p^n - 1}, \end{aligned} \tag{2.11}$$

where $\chi_i, \xi_j \in \{0, 1, \dots, p - 1\}$. Note that in the base-$p$ expansions of rational integers from $[0, 1]$, we use *right*-infinite words rather than left-infinite ones that correspond to canonical expansions of p-adic integers.

Proposition 2.14 *Given $z \in \mathbb{Z}_p \cap \mathbb{Q}$, represent z in the form* (2.8)*; then*

$$z \bmod 1 = 0.(\hat{\beta}_{t-1-\bar{r}}\hat{\beta}_{t-2-\bar{r}} \dots \hat{\beta}_0\hat{\beta}_{t-1}\hat{\beta}_{t-2} \dots \hat{\beta}_{t-\bar{r}})^{\infty} \bmod 1,$$

where $\hat{\beta} = p - 1 - \beta$ for $\beta \in \{0, 1, \dots, p-1\}$ and $\bar{r}$ is the least non-negative residue of r modulo t if $t > 1$ or $\bar{r} = 0$ if otherwise.

Proof Indeed, by Note 2.12, $\sum_{j=0}^{\infty} p^{r+tj} = -p^r(p^t - 1)^{-1}$ in $\mathbb{Z}_p$; so $z = u - vp^r(p^t - 1)^{-1}$ where $u = \alpha_0 + \alpha_1 p + \dots + \alpha_{r-1}p^{r-1}$ and $v = \beta_0 + \beta_1 p + \dots + \beta_{t-1}p^{t-1}$. Therefore,

$$z \bmod 1 = \left(-\frac{vp^r}{p^t - 1}\right) \bmod 1.$$

But $(p^t - 1)^{-1} = p^{-t} + p^{-2t} + p^{-3t} + \cdots$ in $\mathbb{R}$; so

$$(p^t - 1)^{-1} = 0.(\underbrace{00\dots0}_{t-1}1)^{\infty}$$

and thus $-v(p^t - 1)^{-1} = -0.(\beta_{t-1}\beta_{t-2}\dots\beta_0)^{\infty}$.

Now just note that

$$\begin{aligned}(p - 1 - \gamma_0) + (p - 1 - \gamma_1)p + \dots + (p - 1 - \gamma_{s-1})p^{s-1} \\ = p^s - 1 - (\gamma_0 + \gamma_1 p + \dots + \gamma_{s-1}p^{s-1})\end{aligned}$$

for $\gamma_0, \gamma_1, \ldots \in \{0, 1, \dots, p-1\}$, $s \in \mathbb{N}$; so

$$\begin{aligned}\frac{(p - 1 - \gamma_0) + (p - 1 - \gamma_1)p + \dots + (p - 1 - \gamma_{s-1})p^{s-1}}{p^s - 1} = \\ = 1 - \frac{\gamma_0 + \gamma_1 p + \dots + \gamma_{s-1}p^{s-1}}{p^s - 1}\end{aligned}$$

and therefore

$$(-0.(\gamma_{s-1}\gamma_{s-2}\dots\gamma_0)^{\infty}) \bmod 1 = (0.(\hat{\gamma}_{s-1}\hat{\gamma}_{s-2}\dots\hat{\gamma}_0)^{\infty}) \bmod 1 \tag{2.12}$$

where $\hat{\gamma} = p - 1 - \gamma$ for $\gamma \in \{0, 1, \dots, p-1\}$. □

Combining (2.11) with Proposition 2.11, we see that all real numbers whose base-p expansions are purely periodic must lie in $\mathbb{Z}_p \cap \mathbb{Q}$; therefore, the following criterion is true:

Corollary 2.15 *A real number x is in $\mathbb{Z}_p \cap \mathbb{Q}$ if and only if the base-p expansion of $x \bmod 1$ is purely periodic: $x \bmod 1 = 0.(\chi_0 \dots \chi_{n-1})^{\infty}$ for suitable $\chi_0, \dots, \chi_{n-1} \in \mathbb{F}_p$.*

The following corollary expresses the base-p expansion of a p-adic rational integer via its representation in the form given by Proposition 2.11:

Corollary 2.16 *Once a p-adic rational integer $z \in \mathbb{Z}_p \cap \mathbb{Q}$ is represented in the form as of Proposition 2.11, then $z \bmod 1 = 0.(\zeta_{t-1}\zeta_{t-2}\dots\zeta_0)^\infty$, where $d = \zeta_0 + \zeta_1 p + \cdots + \zeta_{t-1}p^{t-1}$.*

Proof Indeed, under notation of Proposition 2.11, $z \bmod 1 = (d(p^t - 1)^{-1}) \bmod 1$, and the result follows since $(p^t - 1)^{-1} = p^{-t} + p^{-2t} + p^{-3t} + \cdots$ in $\mathbb{R}$. □

Now we can find a period length of $z \in \mathbb{Z}_p \cap \mathbb{Q}$ provided z is represented as an irreducible fraction $z = a/b$, where $a \in \mathbb{Z}$, $b \in \mathbb{N}$.

Proposition 2.17 *Once a p-adic rational integer $z \neq 0$ is represented as an irreducible fraction $z = a/b$, and if $b > 1$, then the period length t of z is equal to the multiplicative order of p modulo b* (i.e., to the smallest $\ell \in \mathbb{N}$ such that $p^\ell \equiv 1 \pmod b$).

Proof Note that the multiplicative order ℓ of p modulo b is the smallest positive integer such that $p^\ell(a/b) = a/b \pmod 1$. Indeed, $p^\ell = eb + 1$ for a suitable $e \in \mathbb{Z}$; so $p^\ell(a/b) = ea + (a/b)$. On the other hand, if $p^s(a/b) = m + (a/b)$ for some $m \in \mathbb{Z}$, then $a(p^s - 1) = mb$ and thus $p^s - 1 \equiv 0 \pmod b$ since a is co-prime to b (as the fraction a/b is supposed to be irreducible).

Now, from Corollary 2.16, it immediately follows that $(p^t z) \bmod 1 = z \bmod 1$ once t is a period length of z and that t is the smallest positive integer with that property. Finally, we conclude that $\ell = t$. □

Given $b \in \mathbb{N}$, b co-prime to p, denote via $\mathsf{mult}_b\, p$ the multiplicative order of p modulo b if $b > 1$ or put $\mathsf{mult}_b\, p = 1$ if $b = 1$. Then $\mathsf{mult}_b\, p$ is the period length of $z \in \mathbb{Z}_p \cap \mathbb{Q}$ once z is represented as an irreducible fraction $z = a/b$ where $a \in \mathbb{Z}$ and $b \in \mathbb{N}$. Note that we consider here only infinite words that correspond to p-adic rational integers; thus, to, e.g., 0, there corresponds a word $(0)^\infty$ (so a period of 0 is 0 and a pre-period is empty), and the respective base-p expansion of 0 is $0.(0)^\infty$. Also, $1 = 1 + 0 \times p + 0 \times p^2 + \cdots$, so the corresponding infinite word is $(0)^\infty 1$; therefore, 1 is a pre-period of 1, 0 is a period of 1, and the representation of 1 in the form (2.9) is $1 = 1 + (0/p - 1)$.

Example 2.18 Let $p = 2$; then $1/3 = 1 \times 1 + 1 \times 2 + 0 \times 4 + 1 \times 8 + 0 \times 16 + \cdots = 1 - 2 \times 3^{-1}$ is a canonical 2-adic expansion of $1/3$; so the corresponding infinite binary word is $(01)^\infty 1$. Therefore, the period length of $1/3$ is 2 (and note that the multiplicative order of 2 modulo 3 is indeed 2), the period is 01, and the pre-period is 1. Also, $c = 0$ and $d = 1$ once $1/3$ is represented in the form of Proposition 2.11; $1/3 = 0.(01)^\infty$ is the base-2 expansion of $1/3$ (cf. Proposition 2.14 and Corollary 2.16.

As every real number can be approximated by fractions of the form u/v, where $u, v \in \mathbb{Z}$ and $p \nmid v$, all rational p-adic integers are dense in $\mathbb{R}$, i.e., $\mathbb{Z}_p \cap \mathbb{Q}$ is a dense subset of $\mathbb{R}$. Now we prove a stronger result of this kind.

Theorem 2.19 *Rational p-adic integers from every p-adic ball of non-zero radius constitute a dense subset of* $\mathbb{R}$.

Of course, Theorem 2.19 can be proved in a few lines, by using known results on discrete subgroups of $\mathbb{R}$ (see, e.g., [164, Chapter 1, Section 3.4]): As every discrete additive subgroup of $\mathbb{R}$ is $a\mathbb{Z}$ for $a \in \mathbb{R}_{\geqslant 0}$, rational p-adic integers from any ball $p^{\ell}\mathbb{Z}_p$ obviously constitute a non-discrete (thus, dense) subgroup of $\mathbb{R}$. Therefore, rational p-adic integers from any ball $n + p^{\ell}\mathbb{Z}_p$ also constitute a dense subset of $\mathbb{R}$ since multiplication by $n \in \mathbb{N}$ is a continuous automorphism of the additive group $\mathbb{R}$ onto itself. However, below we give a longer proof in order to illustrate rigorously a fact, which, in very simple terms, can be stated as follows:

In loose terms, higher-order digits of canonical p-adic representation of a rational p-adic integer h are low-order digits of the base-p expansion of h as of a real number and vice versa. This fact in many cases is an underlying source of many properties of p-adic 1-Lipschitz functions and respective models, which we study further in the book (see, e.g., Sects. 16.5, 16.6, and 16.7).

As a first step, we prove a restricted version of the theorem:

Proposition 2.20 *Given a ball* $\mathbf{B}_{p^{-r}}(c) \subset \mathbb{Z}_p$ *of non-zero radius, the set* $\mathbf{B}_{p^{-r}}(c) \cap [0, 1]$ *is dense in* $[0, 1]$, *whatever* $c \in \mathbb{Z}_p$ *and* $r \in \mathbb{N}_0$ *are taken.*

Proof For $r = 0$, the ball $\mathbf{B}_{p^{-r}}(c)$ is the whole space $\mathbb{Z}_p$, and we already know that $\mathbb{Z}_p \cap \mathbb{Q}$ is dense in $\mathbb{R}$; whence the assertion is trivial, so let $r > 0$.

We have that $\mathbf{B}_{p^{-r}}(c) = \{n + p^r z \colon z \in \mathbb{Z}_p\}$ for a suitable $n \in \mathbb{N}_0$ such that $\lfloor \log_p n \rfloor < r$ and $n \equiv c \pmod{p^r}$. If $n > 0$, let $n - 1 = \nu_0 + \nu_1 p + \cdots + \nu_{r-1} p^{r-1}$ be the base-p expansion of $n - 1$ (some higher-order digits $\nu_j \in \{0, 1 \ldots, p - 1\}$ may be 0). For every $t \in \mathbb{N}$ and every $\gamma_r, \ldots, \gamma_{r+t-1} \in \{0, 1, \ldots, p - 1\}$, consider p-adic integer $u = \frac{v}{1 - p^{r+t}}$, where

$$v = \nu_0 + \nu_1 p + \cdots + \nu_{r-1} p^{r-1} + \gamma_r p^r + \cdots + \gamma_{r+t-1} p^{r+t-1}.$$

As the canonical p-adic expansion of u is

$$\begin{aligned} u = {} & \nu_0 + \nu_1 p + \cdots + \nu_{r-1} p^{r-1} + \gamma_r p^r + \cdots + \gamma_{r+t-1} p^{r+t-1} + \\ & + \nu_0 p^{r+t} + \nu_1 p^{r+t+1} + \cdots + \nu_{r-1} p^{2r+t-1} + \gamma_r p^{2r+t} + \cdots + \gamma_{r+t-1} p^{2r+2t-1} + \cdots, \end{aligned} \tag{2.13}$$

the number $1 + u$ is of the form $1 + (n - 1) + p^r d = n + p^r d$ for respective $d \in \mathbb{Z}_p$; whence, $1 + u \in \mathbf{B}_{p^{-r}}(c)$. On the other hand, $1 + u = 1 + \frac{v}{1 - p^{r+t}} \in [0, 1]$ as $0 \leqslant v \leqslant p^{r+t} - 1$ by the construction.

Further, as

$$\sum_{j=0}^{\infty} \alpha_j p^j + \sum_{j=0}^{\infty} (p - 1 - \alpha_j) p^j = -1$$

for all $\alpha_0, \alpha_1, \ldots \in \{0, 1, \ldots, p-1\}$, we see that

$$1 + u = \frac{p^{r+t} - (v+1)}{p^{r+t} + 1} = 0.(\hat{\gamma}_{r+t-1}\hat{\gamma}_{r+t-2} \ldots \hat{\gamma}_r \hat{\eta}_{r-1}\hat{\eta}_{r-2} \ldots \hat{\eta}_0)^{\infty}, \tag{2.14}$$

where $\hat{\alpha} = p - 1 - \alpha$ for $\alpha \in \{0, 1, \ldots, p-1\}$ and $n = \eta_0 + \eta_1 p + \cdots + \eta_{r-1} p^{r-1}$ is the base-p expansion of n (cf., (2.8) and Proposition 2.14). As (2.14) is true for all $t \in \mathbb{N}$ and all $\gamma_j \in \{0, 1, \ldots, p-1\}$, we conclude that any real number from $[0, 1]$ can be approximated by numbers of the form (2.14) with arbitrary accuracy, for suitable choice of $t \in \mathbb{N}$ and γ_j; that is, $\mathbf{B}_{p^{-r}}(c) \cap [0, 1]$ is dense in $[0, 1]$ provided $n > 0$, i.e., if $0 \notin \mathbf{B}_{p^{-r}}(c)$.

If $0 \in \mathbf{B}_{p^{-r}}(c)$, then $\mathbf{B}_{p^{-r}}(c) = p^r \mathbb{Z}_p$; as $r > 0$, the claim of Proposition 2.20 immediately follows from Proposition 2.14 in view of (2.8). □

Now we prove a full version of the theorem:

Proof of Theorem 2.19 Given a ball $\mathbf{B} = \mathbf{B}_{p^{-r}}(c) = \{n + p^r z \colon z \in \mathbb{Z}_p\}$, where $n \in \mathbb{N}_0$ is such that $\lfloor \log_p n \rfloor < r$ and $n \equiv c \pmod{p^r}$, given any $m \in \mathbb{N}_0$, take any integer $\ell \geqslant r$ so that $n + p^\ell > m$. Then ball $\mathbf{B}' = \mathbf{B}_{p^{-\ell-1}}(c') = \{n' + p^{\ell+1} z \colon z \in \mathbb{Z}_p\}$, where $n' = n + p^\ell$ and $c' \equiv n' \pmod{p^{\ell+1}}$, lies in $\mathbf{B}$. Let $\sum_{j=0}^{\ell} \nu_j p^j$ be the base-p expansion of $k = n' - m$. Construct numbers u as in the proof of Proposition 2.20, taking $\ell + 1$ instead of r. Then the numbers $k + u$ for all so constructed $u = \frac{v}{1 - p^{\ell+t+1}}$ constitute a dense subset of the closed segment $[m, m-1] \subset \mathbb{R}$, and all these numbers are rational p-adic integers from $\mathbf{B}' \subset \mathbf{B}$ since $m + k + u = n + p^\ell + p^{\ell+1} d$ for suitable $d \in \mathbb{Z}_p$. We have shown therefore that all rational p-adic integers from the ball $\mathbf{B}$ are dense in $\mathbb{R}_{\geqslant 0}$.

Given $m \in \mathbb{N}_0$, take ℓ large enough so that $0 < n + p^\ell + m < p^{\ell+1}$, construct u as above for $k = n + p^\ell + m$; then the numbers $-m + k + u$ are all in $\mathbf{B}$ for every m and are dense in $[-m-1, -m]$. This finishes the proof. □

2.3.3 Representations via Rays of a p-Adic Rooted Tree

Let T_p be an infinite rooted tree with a countable number of vertices such that every vertex has exactly p children (cf., Sect. 1.5). In simple words, from each vertex there, fly exactly p arrows, which hit p different vertices. Then by labelling each arrow with numbers from $\{0, 1, \ldots, p-1\}$ so that two different arrows having common tails are labelled by different numbers, one establishes a one-to-one correspondence between the set of all rays whose tail is the root vertex (which further are called *rooted rays*) and the set $\mathcal{W}_\infty(\mathcal{A})$ of all one-side infinite words over the alphabet $\mathcal{A}_p = \{0, 1, \dots, p-1\}$ (cf., Sect. 1.4.1). The labelling defines also a one-to-one correspondence between finite paths that start at the root vertex and finite words over the alphabet $\mathcal{A}_p$; so we denote when convenient both finite and infinite paths that start at the root vertex via respective sequences (finite or infinite) of labels or/and words (finite and infinite) over the alphabet $\mathcal{A}_p$.

Between the set $\mathcal{W}_\infty(\mathcal{A}_p)$ and p-adic integers from $\mathbb{Z}_p$, there is a natural one-to-one correspondence: Given $\alpha_0\alpha_1\alpha_2\cdots \in \mathcal{W}_\infty(\mathcal{A}_p)$, the corresponding p-adic integer is the one whose canonical p-adic expansion is $\sum_{i=0}^{\infty}\alpha_i p^i$ (cf., Fig. 2.1), which depicts the tree T_2. Under that bijection between T_p and $\mathbb{Z}_p$, to every ball $\mathbf{B}_{p^{-n}}(z)$ of radius p^{-n}, $n \in \mathbb{N}$, centered at $z = \sum_{i=0}^{\infty}\alpha_i p^i$, there corresponds a rooted sub-tree whose root is the vertex to which the path that corresponds to the finite word $\alpha_0\alpha_1\cdots\alpha_{n-1}$ leads. Note that under that said bijection, to the map $\bmod p^n : \mathbb{Z}_p \to \mathbb{Z}/p^n\mathbb{Z}$, there corresponds a map of the tree T_p onto a finite tree having exactly p^n leaves: The map just "cut off tails of rays," thus truncating the rays to finite paths of length n.

Example 2.21 In the 2-adic tree T_2 (cf., Fig. 2.1) take all finite paths with labels $0, 10, 110, 1110, \dots$; the paths define a sequence of balls (subtrees)

$$\mathbf{B}_{1/2}(0), \mathbf{B}_{1/4}(1), \mathbf{B}_{1/8}(3), \dots, \mathbf{B}_{2^{-n}}(2^{n-1}-1), \dots$$

The balls are pairwise disjoint, their union is open in p-adic topology, and its complement (in the 2-adic tree) is the ray $1111\dots$, which corresponds to the number -1 as the 2-adic canonical expansion of -1 is $\sum_{i=0}^{\infty}2^i$. The probability measure μ_2 of the union of these balls is 1.

Note 2.22 A q-adic tree T_q can be defined for any integer $q > 1$, not necessarily for $q = p$ a prime. The tree T_q in that case is still a non-Archimedean compact space whose elements are infinite words $a = \alpha_0\alpha_1\cdots$ over the alphabet $\mathcal{A}_q = \{0, 1\dots, q-1\}$, which are in a one-to-one correspondence with rays starting from

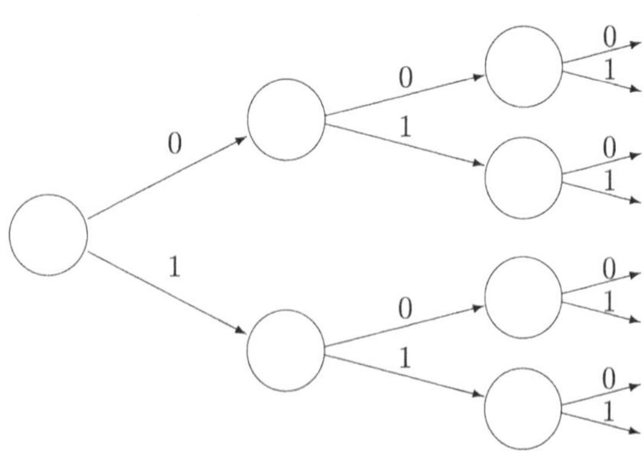

Fig. 2.1 Tree representation of $\mathbb{Z}_2$. Every infinite path that starts from the root vertex represents a 2-adic integer. The set of infinite paths that start from a fixed vertex represent a ball

the root of the tree; the distance $d_q(a, b)$ between the words a, b is $q^{-\ell}$, where ℓ is the length of the longest common prefix of the words a, b if $a \neq b$ and $d(a, b)$=0 if $a = b$. In the definition of the metric, the constant q may be chosen arbitrarily; usually q is the cardinality of the alphabet, but any real $q > 1$, which is not necessarily an integer, can be taken if convenient.

The so-defined metric is an ultrametric, a ball $\mathbf{B}_{q^{-\ell}}(u)$ of radius $q^{-\ell}$ centered at the ray u is a set of all rays whose prefixes of length ℓ are the same as the one of the ray u; the balls of non-zero radii are clopen; given a pair of such balls, they are either disjoint or one of them is a subset of the other one. The metric space T_q is compact; it is a ball of radius 1.

In the case $q = p^n$, letters of the alphabet $\mathcal{A}_q$ can be viewed either as elements of the residue ring $\mathbb{Z}/p^n\mathbb{Z}$ or as n-dimensional vectors over a p-element field $\mathbb{F}_p$. Therefore, to each ray $a = \alpha_0\alpha_1\cdots$, there corresponds either an infinite sequence $(\alpha_i)_{i=0}^{\infty}$ of residues modulo p^n or an infinite sequence of n-dimensional vectors $((\chi_{1i}, \ldots, \chi_{ni}))_{i=0}^{\infty}$ over $\mathbb{F}_p$. The latter sequence can be considered as an element $\mathbf{x} = (x_1, \ldots, x_n)$ of $\mathbb{Z}_p^n$, the n-th Cartesian power of $\mathbb{Z}_p$: just put $x_j = \sum_{i=0}^{\infty} \chi_{ji} p^i \in \mathbb{Z}_p$ for all $i = 0, 1, 2, \ldots,$ $j = 1, 2, \ldots, n$. The equalities $\chi_{ji} = \delta_{j-1}(\alpha_i)$, where $\delta_k(\alpha)$ is the k-th digit in the base-p expansion of α, $k = 0, 1, 2, \ldots, n-1$ (cf., Sect. 2.3.1) define a bijection between rays, which start at the root of T_q and elements of $\mathbb{Z}_p^n$.

2.4 Probability Measure on $\mathbb{Z}_p^n$

In the book, we study dynamical systems whose phase space is $\mathbb{Z}_p^n$, the n-th Cartesian power of the space $\mathbb{Z}_p$, $(n \in \mathbb{N})$; the space $\mathbb{Z}_p^n$ can also be viewed as a tree T_q where $q = p^n$. Throughout the book, we mostly assume that the phase space is also a probability space that is endowed with a *natural probability measure* $\mu_q = \mu_p^n$.

The probability measure μ_q on T_q can be defined by a σ-algebra generated by all the balls $\mathbf{B}_{q^{-\ell}}(u) \subset \mathrm{T}_q$ of non-zero radii (cf. Note 2.22): Namely, $\mu_q(\mathbf{B}_{q^{-\ell}}(u)) = q^{-\ell}$; the balls are elementary μ_q-measurable subsets; all μ_q-measurable subsets of T_q constitute the σ-algebra. Actually, the measure μ_q on T_q is a *Hausdorff net measure* with respect to the net, which is a collection of all balls of non-zero radii; this measure is a regular metric measure (see, e.g., [165, Chapter 2, §7]). To put it in simple words, a probability of an event when a randomly chosen ray that starts at the root of the tree T_q and meets a given ball $\mathbf{B} \subset \mathrm{T}_q$ is exactly $\mu_q(\mathbf{B})$.

The space $\mathbb{Z}_p^n$ can be endowed with a metric d_p^n induced by the metric d_p on $\mathbb{Z}_p$; namely, given $\mathbf{x} = (x_1, \ldots, x_n), \mathbf{y} = (y_1, \ldots, y_n) \in \mathbb{Z}_p^n$, let

$$d_p^n(\mathbf{x}, \mathbf{y}) = \max\{d_p(x_i, y_i) \colon i = 1, \ldots, n\}.$$

The so-defined metric is an ultrametric on $\mathbb{Z}_p^n$; the (clopen) balls $\mathbf{B}_{p^{-k}}(\mathbf{x})$ of radius p^{-k} ($k = 0, 1, 2, \ldots$) centered at $\mathbf{x} \in \mathbb{Z}_p^n$ of that ultrametric space are the cosets $\mathbf{x}+p^k\mathbb{Z}_p^n$. Note that for $q = p^n$, the metric d_p^n on $\mathbb{Z}_p^n$ coincides with the metric d_q on T_q due to the one-to-one correspondence between $\mathbf{x} = (x_1, \ldots, x_n) \in \mathbb{Z}_p^n$ and rooted rays (the ones that start at the root) of the tree T_q (cf., Note 2.22).

In plain words, rather than to consider the metric space $\mathbb{Z}_p^n$ as a Cartesian product of n copies of the metric space $\mathbb{Z}_p$, one may consider it as a ring $\mathbb{Z}_q$ of q-adic integers, for $q = p^n$, with the respective metric: Note that the notion of p-adic integer is well defined not only for p a prime but for any integer $p > 1$ (see [145]): speaking informally, the elements of $\mathbb{Z}_p$ have infinite base-p expansions; the elements of $\mathbb{Z}_{p^n}$ have infinite base-p^n expansions. The one-to-one correspondence described in Note 2.22 between the rays that start at the root of the tree T_{p^n} and elements of $\mathbb{Z}_p^n$ defines a metric isomorphism between $\mathbb{Z}_p^n$ and T_{p^n}; actually the spaces $\mathbb{Z}_{p^n}$ and T_{p^n} are isomorphic one to another as metric spaces since *the metric on both spaces is just a one-side-infinite-word metric defined via lengths of common prefixes of infinite words*, so the metric d_p^n is "causal" (cf. Sect. 1.6.2).

The probability measure μ_p^n is the "basic uniform measure," which we use further in the book to describe how causal functions "transform probability distributions": In the course of the book, the measure μ_p^n plays a role similar to that of Lebesgue measure on the unit real hypercube $[0; 1]^n \subset \mathbb{R}^n$.

The probability measure μ_p^n on $\mathbb{Z}_p^n$ shares the following properties:

- The measure of every ball $\mathbf{B}_{p^{-k}}(\mathbf{x})$ of a non-zero radius p^{-k} is equal to the n-th power of the radius of that ball: $\mu_p^n(\mathbf{B}_{p^{-k}}(\mathbf{x})) = p^{-nk}$; the measure of every ball of zero radius (i.e., of a point) is 0.
- The measure μ_p^n is a *Borel measure*: That is, every open subset of $\mathbb{Z}_p^n$ is μ_p^n-measurable; whence, every closed subset is μ_p^n-measurable as well.
- The measure μ_p^n is *regular*: That is, for any μ_p^n-measurable subset $A \subset \mathbb{Z}_p^n$, one has

$$\mu_p^n(A) = \sup\{\mu_p^n(S) \colon S \subset A, S \text{ is closed in } \mathbb{Z}_p^n\} = \\ = \inf\{\mu_p^n(S) \colon S \supset A, S \text{ is open in } \mathbb{Z}_p^n\}.$$

The metric space $\mathbb{Z}_p^n$ is a compact Abelian group with respect to addition; therefore, $\mathbb{Z}_p^n$ can be endowed also with a *Haar measure*. To be a probability measure, the Haar measure should be normalized to make the measure of the whole space $\mathbb{Z}_p^n$ equal to 1. The so-normalized Haar measure on $\mathbb{Z}_p^n$ coincides with μ_p^n.

In the sequel, we will need more facts about $\mathbb{Z}_p^n$ as of a probability space; though we recall necessary definitions in respective places of the book, the interested reader is referred to, e.g., the books [49, 68, 130, 165, 172, 182], where all the necessary information, which we have mentioned only briefly in this subsection, without proofs, can be found.

Chapter 3
p-Adic Calculus

Abstract In this chapter, we explain p-adic calculus for functions whose arguments and values are p-adic numbers, especially when the domain of the functions is $\mathbb{Z}_p^n$, the n-th Cartesian power of $\mathbb{Z}_p$, and the functions take values in $\mathbb{Z}_p^m$, for some $n, m \in \mathbb{N} = \{1, 2, 3, \ldots\}$, (usually $m \leqslant n$). Though general p-adic calculus can be found in a number of respective books that are mentioned earlier (in Chap. 2), in the current chapter, we are mostly focused on the main object of our investigation, the p-adic 1-Lipschitz functions, since the latter functions can be treated as causal functions over discrete time. This is why we give only references where the proofs of general results of p-adic calculus can be found and prove only the results that are specific for the theme of the book. Therefore, this chapter should not be judged as a crash course of p-adic calculus; there are much better books for a novice (see, e.g., (Katok, p-adic Analysis in Comparison with Real, Mass. Selecta. American Mathematical Society (2003); Mahler, p-adic Numbers and their Functions, 2nd edn., Cambridge Univ. Press (1981); Koblitz, p-adic Numbers, p-adic Analysis, and Zeta-functions, Graduate Texts in Math., vol. 58, 2nd edn., Springer-Verlag (1984))).

3.1 Univariate Functions

For better convenience, we restate basic notions of p-adic analysis via congruences rather than via absolute values. According to conventions we made in Sect. 2.2.2, given $s \in \mathbb{N}$ and $a, b \in \mathbb{Q}_p$, we write $a \equiv b \pmod{p^s}$ if and only if $|a-b|_p \leqslant p^{-s}$ (or, which is the same, if and only if $a = b + cp^s$ for suitable $c \in \mathbb{Z}_p$). To put it in other words, we use $a \equiv b \pmod{p^s}$ rather than $|a - b|_p \leqslant p^{-s}$, meaning that both a and b lie in a ball of radius p^{-s} of the space $\mathbb{Q}_p$.

Therefore, we can restate definition of the p-adic limit in the following manner:

Definition 3.1 (p-Adic Limit) A p-adic number z is said to be a *limit* of the sequence $(z_i)_{i=0}^{\infty}$ of p-adic numbers $z_i \in \mathbb{Q}_p$,

$$z = \overset{p}{\lim_{i \to \infty}} z_i,$$

V. Anashin, *Causality: The* p*-adic Theory*, STEAM-H: Science, Technology, Engineering, Agriculture, Mathematics & Health,
https://doi.org/10.1007/978-3-031-85818-5_3

if and only if for every (sufficiently large) positive rational integer $K \in \mathbb{N}$, there exists $N \in \mathbb{N}_0$ such that $z_i \equiv z \pmod{p^K}$ for all $i \geqslant N$.

3.1.1 Continuous Functions, 1-Lipschitz Functions

With the above restatement of the p-adic limit, we restate the notion of (uniformly) continuous function:

Definition 3.2 (p-Adic (Uniform) Continuity) Let U be a non-empty subset of $\mathbb{Z}_p$. A function $f: U \to \mathbb{Q}_p$ is said to be *continuous* at the point $z \in \mathbb{Z}_p$ if and only if for every (sufficiently large) positive rational integer $M \in \mathbb{N}$, there exists a positive rational integer $L \in \mathbb{N}$ such that $f(x) \equiv f(z) \pmod{p^M}$ whenever $x \equiv z \pmod{p^L}$. The function f is said to be *uniformly continuous* on $U \subset \mathbb{Z}_p$ if and only if for each $K \in \mathbb{N}$, there exists $N \in \mathbb{N}$ such that for every $x, y \in U$, the congruence $x \equiv y \pmod{p^K}$ implies that $f(x) \equiv f(y) \pmod{p^N}$.

It is well-known that if a function whose domain is a compact metric space and which takes values in a metric space is continuous at every point of the domain, then it is uniformly continuous on the domain. Hence, the following proposition is true:

Proposition 3.3 *Let U be a compact subspace of $\mathbb{Z}_p$* (e.g., let U be a ball or a sphere of a non-zero radius). *Then any function $f: U \to \mathbb{Q}_p$ that is continuous at every point of U is uniformly continuous on U.*

The next theorem serves as a tool to expand functions that are uniformly continuous on a dense subset of $\mathbb{Z}_p$ to functions that are uniformly continuous on the whole $\mathbb{Z}_p$ (cf., [112, Theorem 4.15]):

Theorem 3.4 *Let E be a subset of $\mathbb{Z}_p$ and let $\bar{E}$ be its closure. Let $f: E \to \mathbb{Q}_p$ be a function uniformly continuous on E. Then there exists a unique function $\bar{f}: \bar{E} \to \mathbb{Q}_p$ uniformly continuous and bounded on E such that $\bar{f}(x) = f(x)$ if $x \in E$.*

An important example of functions that are uniformly continuous on $\mathbb{Z}_p$ is p-adic 1-Lipschitz functions. We are especially focused at these functions since these are automatic (whence, causal) functions (cf., Sect. 1.6.2); that is why we recall here some equivalent definitions of that class of functions.

By definition, 1-Lipschitz p-adic functions $f: \mathbb{Z}_p \to \mathbb{Z}_p$ are the ones that satisfy the p-adic Lipschitz condition with a constant 1

$$|f(x) - f(y)|_p \leqslant |x - y|_p,$$

(continued)

1-Lipschitz functions are called *non-expansive* since they are exactly those functions that do not increase the p-adic distance between points,

$$d_p(f(x), f(y) \leqslant d_p(x, y),$$

or, also, *compatible* since they are exactly those functions that agree with all epimorphisms of the ring $\mathbb{Z}_p$ (cf., Sect. 1.2.2):

$$\text{If } x \equiv y \pmod{p^n} \text{ then } f(x) \equiv f(y) \pmod{p^n}.$$

That is, given a 1-Lipschitz function $f\colon \mathbb{Z}_p \to \mathbb{Z}_p$ and $k \in \mathbb{N}$, the map $f \bmod p^k\colon \mathbb{Z}/p^k\mathbb{Z} \to \mathbb{Z}/p^k\mathbb{Z}$ is a well-defined map of the residue ring $\mathbb{Z}/p^k\mathbb{Z}$; we call the map $f \bmod p^k$ the *reduction of function f modulo p^k* or, when it is clear what the modulo is, simply the *reduced function*. In simple words, 1-Lipschitz functions $f\colon \mathbb{Z}_p \to \mathbb{Z}_p$ are exactly those functions that "commute" with all reduction maps: $f(x \bmod p^k) \equiv f(x) \pmod{p^k}$ for all $x \in \mathbb{Z}_p$ and all $k \in \mathbb{N}$, cf. (2.7).

The function $f\colon \mathbb{Z}_p \to \mathbb{Z}_p$ is 1-Lipschitz if and only if its canonical representation has a special triangular-like form:

$$\chi_0 + \chi_1 p + \chi_2 p^2 + \cdots \overset{f}{\mapsto} \psi_0(\chi_0) + \psi_1(\chi_0, \chi_1)p + \psi_2(\chi_0, \chi_1, \chi_2)p^2 + \cdots, \tag{3.1}$$

where $\chi_i = \delta_i(x) \in \mathbb{F}_p$, $\psi_i = \delta_i(f)\colon \mathbb{F}_p^{j+1} \to \mathbb{F}_p$, $i = 0, 1, 2, \ldots$, $\mathbb{F}_p = \{0, 1, \ldots, p-1\}$ is a p-element field.

By this reason, in the case $p = 2$, 1-Lipschitz functions are also known under the name of *T-functions*, "T" from "triangular." T-functions play an important role in computer science; we examine these functions in detail in Sect. 9.5.

From the compatibility, it immediately follows that a *function $f\colon \mathbb{Z}_p \to \mathbb{Z}_p$ is 1-Lipschitz if and only if it maps balls into balls*:

$$f(x + p^k\mathbb{Z}_p) \subset f(x) + p^k\mathbb{Z}_p,$$

for every $x \in \mathbb{Z}_p$ and every $k \in \mathbb{N}$.

(continued)

In turn, this property can be equivalently restated in terms of endomorphisms of the rooted tree T_p (cf., Sects. 1.5 and 2.3.3): *The function* $f: \mathbb{Z}_p \to \mathbb{Z}_p$ *is 1-Lipschitz if and only if* f *induces an endomorphism of the p-adic tree* T_p. From the definition, it immediately follows that *the function* f *induces an automorphism of the rooted tree* T_p *if and only if* f *is an isometry of the metric space* $\mathbb{Z}_p$.

Along with 1-Lipschitz functions whose domain and co-domain are $\mathbb{Z}_p$, we also will study 1-Lipschitz functions whose domain is a proper closed subset of $\mathbb{Z}_p$. An example of such functions is the function $f(x) = x^{-1}$ on the sphere $\mathbf{S}_{1/p}(0)$ of radius $1/p$ centered at 0. The sphere is a disjoint union of $p-1$ balls of radii $1/p$ centered at points $1, 2, \ldots, p-1$ (cf. Sect. 2.2.1); so all p-adic integers from the sphere are units of the ring $\mathbb{Z}_p$ (see Note 2.9).

3.1.2 *Differentiability*

The following definition is just a common definition of a derivative of a function in the case when metric is the p-adic metric restated in terms of congruences rather than of inequalities (cf., Sect. 2.2.2); thus, it is a direct analog of its classical counterpart from, e.g., real analysis:

Definition 3.5 (p-Adic Differentiability) A function $f: \mathbb{Z}_p \to \mathbb{Q}_p$ is said to be *differentiable* at the point $x \in \mathbb{Z}_p$ if there exists $f'(x) \in \mathbb{Q}_p$ (called the *derivative of* f *at the point* x) such that, given arbitrary (sufficiently large) $k \in \mathbb{N}$,

$$f(x+h) \equiv f(x) + f'(x)h \pmod{p^{\operatorname{ord}_p h+k}} \tag{3.2}$$

for all sufficiently small $h \in \mathbb{Z}_p$: in other words, if there exists $N = N_k \in \mathbb{N}$ such that congruence (3.2) holds once $|h|_p \leqslant p^{-N}$, that is, once $h \equiv 0 \pmod{p^N}$ (equivalently, once $\operatorname{ord}_p h \geqslant N$).

The function f is called *uniformly differentiable* (on $\mathbb{Z}_p$) if it is differentiable at every point $x \in \mathbb{Z}_p$ and N_k does not depend on x; that is, given sufficiently large $k \in \mathbb{N}$, there exists $N = N_k = N_k(f) \in \mathbb{N}$ such that congruence (3.2) holds simultaneously for all $x \in \mathbb{Z}_p$ and all $h \in \mathbb{Z}_2$ once $\operatorname{ord}_p h \geqslant N$.

Note 3.6 One can define the notion of uniform differentiability on a closed subset $U \subset \mathbb{Q}_p$ rather than on $\mathbb{Z}_p$ in an obvious way, following the lines of Definition 3.5 (cf., e.g., [112, Definition 3.18]).

Rules of derivation in the p-adic case are the same as in real case; that is, given (uniformly) differentiable functions $u, v\colon U \to \mathbb{Q}_p$, functions $u+v$ (sum), uv (product), and $u(v)$ (composition) are (uniformly) differentiable, and:

- $(u+v)' = u' + v'$ (derivative of a sum).
- $(uv)' = u'v + v'u$ (derivative of a product).
- $(\frac{u}{v})' = \frac{u'v - v'u}{v^2}$ (provided v vanishes nowhere on U).
- $(u(v))' = u'(v)v'$ (chain rule).

The proofs follow the same lines as in real case and thus are omitted. Derivative of a power function $f(x) = x^n$, $n \in \mathbb{N}$, has also the same form as in real case. We supply a proof in order to calculate $N_k(f)$ for that function.

Proposition 3.7 *For $m \in \mathbb{N}$, the function $f(x) = x^m$ is uniformly differentiable on $\mathbb{Z}_p$, $f'(x) = mx^{m-1}$, and $N_k(f) = k$.* (Note that f is 1-Lipschitz.)

Proof Indeed, by Newton's binomial formula, for all $n \in \mathbb{N}$ and all $z \in \mathbb{Z}_p$, one has

$$(x + p^n z)^m = \sum_{i=0}^{m} \binom{m}{i} x^{m-i} p^{ni} z^i;$$

so $(x+p^n z)^m \equiv x^m + mx^{m-1}p^n z + \pmod{p^{n+k}}$ for $n \geqslant k$. That is, $(x+h)^m \equiv x^m + mx^{m-1}h \pmod{p^{\mathrm{ord}_p h + k}}$ once $h \equiv 0 \pmod{p^k}$. In other words, we may take $N_k(f) = k$ in Definition 3.5. □

Corollary 3.8 *Every polynomial $f(x)$ in variable x over $\mathbb{Z}_p$ is uniformly differentiable on $\mathbb{Z}_p$, and $N_k(f) = k$.* (The polynomials are 1-Lipschitz functions $\mathbb{Z}_p \to \mathbb{Z}_p$; so they are automatic, whence causal, functions.)

Note 3.9 Due to the rules of derivation, Proposition 3.7 remains true if m is negative integer, $m \in \{-1, -2, \ldots\}$, and the domain is the sphere $\mathbf{S}_{1/p}(0) \subset \mathbb{Z}_p$ rather than the whole space $\mathbb{Z}_p$: For instance, $(1/x)' = -1/x^2$. Recall that the function $f(x) = 1/x$ is a 1-Lipschitz function $f\colon \mathbf{S}_{1/p}(0) \to \mathbb{Z}_p$, cf., the very end of Sect. 3.1.1.

In the p-adic case the derivative of a constant is 0, as in real case; the peculiarity is that the converse is not true in the p-adic case:

Example 3.10 The function $\delta_0(x)$ (the lowest-order digit in the canonical p-adic expansion of $x \in \mathbb{Z}_p$; cf., Sect. 2.3.1) is uniformly differentiable on $\mathbb{Z}_p$ and $\delta_0'(x) = 0$ for all $x \in \mathbb{Z}_p$.

Indeed, $\delta_0(x + p^n z) = \delta_0(x)$ for all $n \in \mathbb{N}$, $z \in \mathbb{Z}_p$; so $\delta_0'(x) = 0$ and $N_k(\delta_0) = 1$ for all $k \in \mathbb{N}$.

Note 3.11 The function $\delta_0(x)$ is a 1-Lipschitz function that maps $\mathbb{Z}_p$ to $\mathbb{Z}_p$, whence $\delta_0(x)$ is automatic, thus causal.

The function δ_0 serves as an example of the so-called *pseudo-constants*; the latter are non-constant functions that are differentiable everywhere on the domain and whose derivatives are identically zero. In the case of real analysis, there are no pseudo-constants due to Rolle's theorem, which, in turn, is based on the fact that $\mathbb{R}$ is a totally ordered field, in contrast to $\mathbb{Q}_p$, which cannot be ordered: That is, no relation of total order that agrees with both addition and multiplication in $\mathbb{Q}_p$ does exist (see, e.g., [112]). The ring $\mathbb{Z}_p$ cannot be ordered as well.

There are 1-Lipschitz functions that are infinitely many times differentiable everywhere on $\mathbb{Z}_p$ but a one point. The following example is of importance to applications in computer science (cf. Sect. 9.5.1 and Chap. 14).

Example 3.12 Given $x \in \mathbb{Z}_p$, the *generalized multiplicative inverse* $\operatorname{inv}_p(x)$ of x is defined as follows:

$$\operatorname{inv}_p(x) = \begin{cases} p^{\operatorname{ord}_p x}\left(\frac{x}{p^{\operatorname{ord}_p x}}\right)^{-1}, \text{ if } x \neq 0; \\ 0, \text{ if } x = 0. \end{cases} \tag{3.3}$$

The function inv_p shares the following properties:

1. The function inv_p is well defined on $\mathbb{Z}_p$.
2. The function inv_p is 1-Lipschitz.
3. The function inv_p is infinitely many times differentiable everywhere on $\mathbb{Z}_p \setminus \{0\}$, and the i-th derivative is

$$\operatorname{inv}_p^{(i)}(x) = \frac{(-1)^i i!}{p^{(i-1)\operatorname{ord}_p x}}\left(\frac{x}{p^{\operatorname{ord}_p x}}\right)^{-i-1} = \frac{(-1)^i i!}{p^{2i \operatorname{ord}_p x}}(\operatorname{inv}_p x)^{i+1}, \tag{3.4}$$

$i = 0, 1, 2, 3, \ldots$

4. The function inv_p is not differentiable at 0.

In simple words, the function inv_p acts on $\mathbb{Z}_p$ as follows: Given $x \in \mathbb{Z}_p \setminus \{0\}$, the first ord_p digits in the canonical p-adic expansion of x are all 0, whereas the (ord_p+1)-st digit is not zero, so the number $xp^{-\operatorname{ord}_p x}$ is a unity of $\mathbb{Z}_p$ and, thus, has a multiplicative inverse in $\mathbb{Z}_p$. So to calculate $\operatorname{inv}_p x$, one must just find the multiplicative inverse of $xp^{-\operatorname{ord}_p x}$ and then multiply it by $p^{\operatorname{ord}_p x}$. We leave proofs of the assertions of Example 3.12 as an exercise for the reader. It is worth mentioning that the function inv_p originates from [58] where the function was introduced for the case $p = 2$ only, as a part of construction of a pseudo-random number generator.

Now we prove an important property of differentiable 1-Lipschitz functions:

Proposition 3.13 *Provided the derivative $f'(x)$ of the 1-Lipschitz function $f: \mathbb{Z}_p \to \mathbb{Z}_p$ exists, $f'(x) \in \mathbb{Z}_p$.*

Proof Indeed, $f(x + p^n z) \equiv f(x) \pmod{p^n}$ for all $x, z \in \mathbb{Z}_p$ and all $n \in \mathbb{N}$ since f is 1-Lipschitz; so as $f(x + p^n z) \equiv f(x) + f'(x)p^n z \pmod{p^{n+k}}$ for all sufficiently large $k \in \mathbb{N}$ by the differentiability of f at x, one has

$$\mathbb{Z}_p \ni a = \frac{f(x + p^n) - f(x)}{p^n} \equiv f'(x) \pmod{p^{n+k}}.$$

Hence, $f'(x) = a + p^{n+k} b$ for suitable $b \in \mathbb{Z}_p$. □

The following Theorem is known under the name of Hensel's Lemma, a result which is fundamental for p-adic analysis. The proof of the Lemma is very important in further considerations since it exploits *Hensel lifting*, the approach we widely use further in the proofs of various results. The approach, in simple terms, can be stated as follows: if some property of a p-adic 1-Lipschitz function "holds modulo p^{n_0}" for some $n_0 \in \mathbb{N}$, and if under assumption that the property "holds modulo p^n" for $n \geqslant n_0$ it follows that the property "holds modulo p^{n+1}," then the property "holds globally." That is, respective proof is the "lifting" of the property from residue ring modulo p^{n_0} to the whole ring $\mathbb{Z}_p$.

Since 1-Lipschitz functions are exactly automata (i.e., causal) functions (cf. Theorem 1.25), the *Hensel lifting* may be restated in terms of automata properties: if some property holds for transformation that performs an automaton on words of length n_0, then this property holds for transformations of words of all lengths (including infinite words) that performs the automaton. This is a cornerstone of the methodology of the book.

Now we state and proof Hensel's Lemma in its original from:

Theorem 3.14 *If a polynomial $f(x) = a_0 + a_1x + a_2x^2 + \cdots$ over $\mathbb{Z}$ has a root z modulo p* (i.e., if $f(z) \equiv 0 \pmod{p}$) *such that $f'(z) \not\equiv 0 \pmod{p}$, then f has a root $\hat{z}$ in $\mathbb{Z}_p$, and $z \equiv \hat{z} \pmod{p}$.*

Proof Put $z_0 = z$; we will show that given $z_n \in \mathbb{Z}_p$ such that $f(z_n) \equiv 0 \pmod{p^n}$, there exists $z_{n+1} \in \mathbb{Z}_p$ such that

$$f(z_{n+1}) \equiv 0 \pmod{p^{n+1}}; \tag{3.5}$$

$$z_{n+1} \equiv z_n \pmod{p^n}. \tag{3.6}$$

Write $z_{n+1} = z_n + p^n t$; we will find $t \in \mathbb{Z}_p$ to satisfy these conditions.

By Proposition 3.7, for $n \geqslant 1$, we have that $f(z_n + p^n t) \equiv f(z_n) + f'(x)p^n t \pmod{p^{n+1}}$; so we must find t to satisfy

$$0 \equiv p^n s + p^n t f'(x) \pmod{p^{n+1}} \tag{3.7}$$

where $p^n s \equiv f(z_n) \pmod{p^{n+1}}$. Remind that $f(z_n) \equiv 0 \pmod{p^n}$ implies that $f(z_n) = p^n s$ for a suitable $s \in \mathbb{Z}_p$. However, congruence (3.7) is equivalent to the congruence $0 \equiv s + tf'(x) \pmod{p}$, which has a unique (modulo p) solution with respect to unknown t since $f'(x) \not\equiv 0 \pmod{p}$ and thus $f'(x)$ is a unit of $\mathbb{Z}_p$ (cf., Sect. 2.3.1).

In view of (3.6), the sequence (z_n) is a p-adic Cauchy sequence by Theorem 2.5; so it has a limit that is a p-adic integer $\hat{z}$ since $\mathbb{Z}_p$ is a complete metric space. However, $f(\hat{z}) = \overset{p}{\lim_{n\to\infty}} f(z_n) = 0$ by (3.5) as f is continuous. □

Corollary 3.15 *From the proof of Hensel's Lemma, it immediately follows that the Lemma is also true for polynomials over* $\mathbb{Z}_p$.

Let us highlight a very important moment: As a matter of fact, nowhere in the proof of Hensel's Lemma we have used an assumption that f is a polynomial over $\mathbb{Z}_p$; moreover, we assumed only that f is 1-Lipschitz and satisfy a condition that is much less restrictive that uniform differentiability. Namely, we assumed that congruence (3.2) holds for all $x \in \mathbb{Z}_p$ uniformly but only for $k = 1$, and not for every sufficiently large k (cf. Definition 3.5). We call this condition a *uniform differentiability modulo p* and discuss the latter (in a more general setting) in the next subsection. This condition is used further in the book in order to expand the Hensel lifting to a much wider class of functions than polynomials over $\mathbb{Z}_p$ and to apply that tool for investigation of dynamical properties of causal functions.

3.1.3 *Differentiability Modulo* p^k

Now we introduce, following [9], a notion of differentiability modulo p^k that is of high importance in the sequel; the notion seems to have no direct analogs in classical real Calculus.

Definition 3.16 (Differentiability Modulo p^k) Given $k \in \mathbb{N}$, a function $f\colon \mathbb{Z}_p \to \mathbb{Q}_p$ is said to be *differentiable modulo* p^k at the point $x \in \mathbb{Z}_p$ if there exists $f'_k(x) \in \mathbb{Q}_p$ (called the *derivative modulo* p^k *of the function* f *at the point* x) such that

$$f(x+h) \equiv f(x) + f'_k(x)h \pmod{p^{\operatorname{ord}_p h+k}} \tag{3.8}$$

for all sufficiently small $h \in \mathbb{Z}_p$ (i.e., if there exists $N_k \in \mathbb{N}$ such that the congruence (3.8) holds once $\operatorname{ord}_p h \geqslant N_k$).

The function f is called *uniformly differentiable modulo* p^k (on $U \subset \mathbb{Z}_p$) if it is differentiable modulo p^k at every point $x \in U \subset \mathbb{Z}_p$ and N_k does not depend on x; that is, there exists $N \in \mathbb{N}$ such that congruence (3.8) holds simultaneously for all $x \in \mathbb{Z}_p$ and all $h \in \mathbb{Z}_2$ once $\operatorname{ord}_p h \geqslant N$. The smallest N with this property is denoted via $N_k(f)$.

From Definition 3.16, it immediately follows that if f is differentiable modulo p^{k+1} at x, then f is differentiable modulo p^k at x and that $f'_{k+1}(x) \equiv f'_k(x) \pmod{p^k}$. Therefore, from Definitions 3.5 and 3.16, we easily conclude that the function f is differentiable at x if and only if it is differentiable modulo p^k for all $k \in \mathbb{N}$ and that $f'(x) \equiv f'_k(x) \pmod{p^k}$.

In simple words, the *differentiability* of f means that

$$\frac{f(x+h) - f(x)}{h} \approx f'_k(x)$$

with *arbitrarily high* accuracy, whereas for differentiability modulo p^k, the accuracy is only *not worse than* p^{-k}.

Note 3.17 Following the lines of the proof of Proposition 3.13, it can be easily shown that the proposition is true for derivative modulo p^k as well; that is, if the derivative $f'_k(x)$ modulo p^k of the 1-Lipschitz function $f: \mathbb{Z}_p \to \mathbb{Z}_p$ exists, then $f'(x) \in \mathbb{Z}_p$. But as derivative modulo p^k is defined up to the term that is 0 modulo p^k, for a 1-Lipschitz function f, we may assume when convenient that $f'_k(x) \in \mathbb{Z}/p^k\mathbb{Z} = \{0, 1, \dots, p^k - 1\}$; so *given a 1-Lipschitz function* $f: \mathbb{Z}_p \to \mathbb{Z}_p$ *which is differentiable modulo* p^k *at every point* $x \in \mathbb{Z}_p$*, the derivative* f'_k *maps* $\mathbb{Z}_p$ *into the residue ring* $\mathbb{Z}/p^k\mathbb{Z}$.

A 1-Lipschitz function $f: \mathbb{Z}_p \to \mathbb{Z}_p$ can be non-differentiable at $x \in \mathbb{Z}_p$ but differentiable modulo p^k for some k. An example of a 2-adic 1-Lipschitz function that is uniformly differentiable modulo 2 on $\mathbb{Z}_2$ and is differentiable modulo 4 at no point of $\mathbb{Z}_2$ (whence is differentiable nowhere on $\mathbb{Z}_2$) is as follows:

Example 3.18 Define the map $f: \mathbb{Z}_2 \to \mathbb{Z}_2$ as follows: Given $x \in \mathbb{Z}_p$, let $x = \sum_{i=0}^{\infty} \chi_i 2^i$ be its 2-adic canonical expansion; put $\delta_i(f(x)) = \chi_i$ if i is odd; put $\delta_i(f(x)) \equiv \chi_i + 1 \pmod 2$ if i is even. By (3.1), the function f is 1-Lipschitz with respect to the 2-adic metric. We claim:

1. The function f is uniformly differentiable modulo 2 on $\mathbb{Z}_2$ and its derivative modulo 2 is $f'_1(x) \equiv 1 \pmod 2$ everywhere on $\mathbb{Z}_2$.
2. The derivative $f'_2(x)$ modulo 4 of the function f exists for no $x \in \mathbb{Z}_2$.

Indeed, take $h \in \mathbb{Z}_2$; let $\ell = \operatorname{ord}_2 h \geqslant 1$. Then $h = 2^\ell z$ where $z \in \mathbb{Z}_2$ is a unit of $\mathbb{Z}_2$, i.e., $z \equiv 1 \pmod 2$ (cf., Sect. 2.3.1). Consider the congruence $f(x + h) \equiv f(x) + 2^\ell z u(x) \pmod{2^{\ell+1}}$, with unknown $u(x)$. Since f is 1-Lipschitz, $u(x)$ must be a 2-adic integer; whence, if solution exists, then there exists a solution $u(x) \in \{0, 1\}$ since $2^\ell z u(x) \equiv 2^\ell \gamma_0(x) \pmod{2^{\ell+1}}$, where $\gamma_0(x) = \delta_0(u(x))$. Then from the definition of the function f, we immediately deduce that the congruence

$\gamma_0(x) \equiv 1 \pmod 2$ must hold for every $x \in \mathbb{Z}_2$; therefore, $f_2'(x) \equiv \gamma_0(x) \equiv 1 \pmod 2$ for all $x \in \mathbb{Z}_p$. This proves the first claim.

Assume that $f_2'(x)$ exists for some $x \in \mathbb{Z}_2$. As $f_2'(x) \in \mathbb{Z}/4\mathbb{Z} = \{0, 1, 2, 3\}$, represent $f_2'(x) = \gamma_0(x) + 2\gamma_1(x)$ where $\gamma_0(x) = \delta_0(f_2'(x))$, $\gamma_1(x) = \delta_1(f_2'(x))$. We already know that $f_1'(x) \equiv 1 \pmod 2$, but $f_2'(x) \equiv f_1'(x) \pmod 2$; therefore, $f_2'(x) = 1 + 2\gamma_1(x)$.

By the differentiability modulo 4,

$$f(x+2^\ell(1+2\zeta_1)) \equiv f(x)+2^\ell(1+2\zeta_1)(1+2\gamma_1(x)) \pmod{p^{\ell+2}}, \text{ here } \zeta_1 = \delta_1(z)$$

This, in view of the definition of the function f and the claim we have already proved, implies that the following congruence must hold:

$$\begin{aligned} 2^\ell(\zeta_0 \oplus \chi_\ell \oplus \xi_\ell) + 2^{\ell+1}(\chi_{\ell+1} \oplus \zeta_1 \oplus \chi_\ell\zeta_0 \oplus \xi_{\ell+1}) &\equiv \\ \equiv 2^\ell(\chi_\ell \oplus \xi_\ell) + 2^{\ell+1}(\chi_{\ell+1} \oplus \xi_{\ell+1}) + 2^\ell\zeta_0\gamma_0(x) + \\ + 2^{\ell+1}(\zeta_1\gamma_0(x) + \zeta_0\gamma_1(x)) \pmod{2^{\ell+2}}, \end{aligned} \tag{3.9}$$

where $\zeta_i = \delta_i(z)$, $\chi_i = \delta_i(x)$, $\oplus$ stands for addition modulo 2, $\xi_i = 1$ if i is even, and $\xi_i = 0$ otherwise. Now, taking into the account that $\xi_{i+1} = \xi_i \oplus 1$ by the definition of the function f, that $\gamma_0(x) = 1$ by the claim that we have already proved and that $\zeta_0 = 1$ by the definition of ℓ, from (3.9), we conclude that

$$\begin{aligned} \chi_{\ell+1} + \zeta_1 + \chi_\ell + \xi_\ell + 1 &\equiv \\ \equiv \chi_{\ell+1} + \xi_\ell + 1 + \zeta_1 + \gamma_1(x) + \chi_\ell + \xi_\ell \pmod 2, \end{aligned}$$

which implies that $\gamma_1(x) + \xi_\ell \equiv 0 \pmod 2$. The latter congruence must hold for all sufficiently large $\ell \in \mathbb{N}$, by the differentiability modulo 4. However, no such $\gamma_1(x)$ exists since $\gamma_1(x)$ must not depend on ℓ; but the sequence $(\xi_i)_{i=0}^\infty$ is an alternating sequence of zeros and ones, by the definition of the function f. This proves claim 2.

Note 3.19 Actually the function f is constructed by using the computer instruction XOR which is defined as follows: $\delta_i(x \,\mathsf{XOR}\, y) \equiv \delta_i(x) + \delta_i(y) \pmod 2$ for all $x, y \in \mathbb{Z}_2$ and all $i \in \mathbb{N}_0$. Namely, $f(x) = x \,\mathsf{XOR}\, (-1/3)$ since the 2-adic canonical representation of $-1/3$ is $\sum_{i=0}^\infty 2^i\xi_i$ where $\xi_i = 1$ if i is even, and $\xi_i = 0$ otherwise. It is not difficult to show that nonetheless *if $c \in \mathbb{Z}$, then the function $f(x) = x \,\mathsf{XOR}\, c$ is uniformly differentiable on $\mathbb{Z}_2$; $f'(x) = 1$ if $c \geqslant 0$, and $f'(x) = -1$ if $c < 0$.* As a matter of fact, all standard computer instructions, both arithmetical and bitwise logical ones, are 1-Lipschitz 2-adic functions; this observation makes it possible to apply 2-adic analysis (and 2-adic ergodic theory) to investigate dynamics of computer programs. We discuss in detail applications of the 2-adic Calculus and 2-adic ergodic theory to computer science in respective places of the book (see Sect. 9.5, Chap. 14).

With the notion of uniform differentiability modulo p, it is possible to expand Hensel's Lemma for a much wider class of functions than polynomials over $\mathbb{Z}_p$:

Note 3.20 Nowhere in the proof of Hensel's Lemma (cf., Theorem 3.14) we have used that f is a polynomial over $\mathbb{Z}$ (or over $\mathbb{Z}_p$). During the proof, only the following has been assumed:

- The function f is 1-Lipschitz.
- The function f is uniformly differentiability modulo p on $\mathbb{Z}_p$.
- $N_1(f) = 1$.

Under these assumptions, if the function $f : \mathbb{Z}_p \to \mathbb{Z}_p$ has a zero z modulo p (i.e., if $f(z) \equiv 0 \pmod{p}$) such that $f_1'(z) \not\equiv 0 \pmod{p}$, then f has a zero $\hat{z} \in \mathbb{Z}_p$, and $z \equiv \hat{z} \pmod{p}$. Moreover, since the condition $N_1(f) = 1$ is used in the proof of Hensel's Lemma as a base of induction, if $N_1(f) = M > 1$, then under the assumption that z is a zero modulo p^M rather than modulo p, the conclusion remains true; that is, f has a zero $\hat{z} \in \mathbb{Z}_p$, and $z \equiv \hat{z} \pmod{p^M}$. This follows directly from the proof of Hensel's Lemma.

3.2 Multivariate Functions

To deal with multivariate p adic functions, we first define an (ultra)metric on $\mathbb{Q}_p^n$, the n-th Cartesian power of $\mathbb{Q}_p$, in a standard way: given $\mathbf{a} = (a_1, \ldots, a_n), \mathbf{b} = (b_1, \ldots, b_n) \in \mathbb{Q}_p^n$, put

$$|\mathbf{a} - \mathbf{b}|_p = \max\{|a_i - b_i|_p : i = 1, 2, \ldots, n\}.$$

Further, in most cases, we deal with metric spaces $\mathbb{Z}_p^n$ and $n \in \mathbb{N}$, which are compact since the metric space $\mathbb{Z}_p$ is compact. As in the case of univariate p-adic functions, it is more convenient sometimes to deal with congruences rather than with absolute values; thus, we need to define congruences for vectors from $\mathbb{Q}_p^n$.

Let $s \in \mathbb{N}$; let $\mathbf{a} = (a_1, \ldots, a_n), \mathbf{b} = (b_1, \ldots, b_n) \in \mathbb{Q}_p^n$. We write $\mathbf{a} \equiv \mathbf{b} \pmod{p^s}$ if and only if $|a_i - b_i|_p \leqslant p^{-s}$ or, which is the same, if and only if $a_i = b_i + c_i p^s$ for suitable $c_i \in \mathbb{Z}_p$, $i = 1, 2, \ldots, n$. In other words, further $\mathbf{a} \equiv \mathbf{b} \pmod{p^s}$ stands for $|\mathbf{a} - \mathbf{b}|_p \leqslant p^{-s}$, meaning that both $\mathbf{a}$ and $\mathbf{b}$ lie in a ball of radius p^{-s} of the space $\mathbb{Q}_p^n$.

The function $F = (f_1, \ldots, f_m) : \mathbb{Z}_p^n \to \mathbb{Z}_p^m$ is *1-Lipschitz* (i.e., $|F(\mathbf{a}) - F(\mathbf{b})|_p \leqslant |\mathbf{a} - \mathbf{b}|_p$ for all $\mathbf{a}, \mathbf{b} \in \mathbb{Z}_p^n$) if and only if

$$F(\mathbf{a}) \equiv F(\mathbf{b}) \pmod{p^\ell} \text{ once } \mathbf{a} \equiv \mathbf{b} \pmod{p^\ell}. \tag{3.10}$$

(continued)

The multivariate p-adic 1-Lipschitz functions

$$F = (f_1, \ldots, f_m)\colon \mathbb{Z}_p^n \to \mathbb{Z}_p^m$$

constitute the class of automata (i.e., causal) functions of automata having p^n-letter input alphabet and p^m-letter output alphabet (cf., Sect. 1.6.2). When convenient, the functions will be treated as *automata functions of automata having n input channels and m output channels over the p-letter alphabet* (see Chap. 13).

Note 3.21 Note that for all $s \in \mathbb{N}$, the binary relation $\cdot \equiv \cdot \pmod{p^s}$ is a congruence on $\mathbb{Q}_p^n$ whenever $\mathbb{Q}_p^n$ is considered as a module over the ring $\mathbb{Z}_p$. In other words, we can work with the relation $\cdot \equiv \cdot \pmod{p^s}$ in a usual manner; e.g., multiply both parts by a p-adic integer, and add congruences sidewise. This implies in particular that the *reduction map* $\operatorname{mod} p^s\colon \mathbb{Z}_p^n \to (\mathbb{Z}/p^s\mathbb{Z})^n$ is well defined (cf. (2.7)):

$$\mathsf{mod}\, p^s : \mathbf{a} = (a_1, \ldots, a_n) \mapsto \mathbf{a} \,\mathsf{mod}\, p^s = (a_1 \,\mathsf{mod}\, p^s, \ldots, a_n \,\mathsf{mod}\, p^s). \qquad (3.11)$$

3.2.1 Differentiability Modulo p^k of Multivariate Functions

Both the definition of differentiability and uniform differentiability for p-adic univariate functions (see Definition 3.5) can be restated for multivariate functions in a standard way, like in real calculus, so we omit these statements. The following definition is an analog of Definition 3.16 of differentiability modulo p^k for the case of multivariate functions.

Definition 3.22 (Differentiability Modulo p^k) Given $k \in \mathbb{N}$, a function $F = (f_1, \ldots, f_m)\colon \mathbb{Z}_p^n \to \mathbb{Q}_p^m$ is said to be *differentiable modulo p^k* at the point $\mathbf{u} = (u_1, \ldots, u_n) \in \mathbb{Z}_p^n$ if there exists a positive rational integer N and an $n \times m$ matrix $F_k'(\mathbf{u})$ over $\mathbb{Q}_p$ (called *the Jacobi matrix modulo p^k* of the function F at the point $\mathbf{u}$) such that for every positive rational integer $K \geqslant N$ and every $\mathbf{h} = (h_1, \ldots, h_n) \in \mathbb{Z}_p^n$, the congruence

$$F(\mathbf{u} + \mathbf{h}) \equiv F(\mathbf{u}) + \mathbf{h} F_k'(\mathbf{u}) \pmod{p^{k+K}} \qquad (3.12)$$

holds whenever $|\mathbf{h}|_p \leqslant p^{-K}$. In the case $m = 1$, the Jacobi matrix modulo p^k is called a *differential modulo p^k*. In the case $m = n$, a determinant of the Jacobi matrix modulo p^k is called a *Jacobian modulo p^k*. Entries of the Jacobi matrix modulo p^k are called *partial derivatives modulo p^k* of the function F at the point $\mathbf{u}$.

Similarly to univariate functions, the function F is called *differentiable at the point* $\mathbf{u}$ if it is differentiable modulo p^k at $\mathbf{u}$ for all $k \in \mathbb{N}$.

The function $F: \mathbb{Z}_p^n \to \mathbb{Q}_p^m$ is said to be *uniformly differentiable modulo* p^k *on* $\mathbb{Z}_p^n$ if there exists $K \in \mathbb{N}$ such that the congruence (3.12) holds simultaneously for all $\mathbf{u} \in \mathbb{Z}_p^n$ whenever $|\mathbf{h}|_p \leqslant p^{-K}$. The smallest of these K is denoted via $N_k(F)$.

Uniform differentiability (modulo p^k) on a closed subset $U \subset \mathbb{Q}_p^n$ rather than on $\mathbb{Z}_p^n$ can be defined by direct analogy with the univariate case (cf., Note 3.6). We denote partial derivative (respectively, differential) modulo p^k via $\frac{\partial_k f_i(\mathbf{u})}{\partial_k x_j}$ (respectively, via $\mathrm{d}_k F(\mathbf{u}) = \sum_{i=1}^{n} \frac{\partial_k F(\mathbf{u})}{\partial_k x_i} \mathrm{d}_k x_i$). In cases when all partial derivatives modulo p^k at all points of $\mathbb{Z}_p^n$ are p-adic integers, we say that the function F has *integer-valued derivatives modulo* p^k. In these cases, we can associate to each partial derivative modulo p^k a unique element of the ring $\mathbb{Z}/p^k\mathbb{Z}$; a Jacobi matrix modulo p^k at every point $\mathbf{u} \in \mathbb{Z}_p^n$ can then be considered as a *matrix over the ring* $\mathbb{Z}/p^k\mathbb{Z}$. It is easy to show that Proposition 3.13 holds true for partial derivatives modulo p^k:

If a 1-Lipschitz function $F = (f_1, \ldots, f_m): \mathbb{Z}_p^n \to \mathbb{Z}_p^m$ is differentiable modulo p^k, then all its partial derivatives modulo p^k are integer-valued. Note that integer-valued functions that have integer-valued derivatives are called *twice integer-valued*.

Example 3.23 The 2-variate function $F(x, y) = x \,\mathsf{XOR}\, y$ (cf., Note 3.19) delivers an example of a 1-Lipschitz function $\mathbb{Z}_2^2 \to \mathbb{Z}_2$, which is:

- Differentiable at some points of $\mathbb{Z}_2^2$ and not differentiable modulo 4 (thus, not differentiable) at some other points of $\mathbb{Z}_2^2$
- Uniformly differentiable modulo 2 on $\mathbb{Z}_2^2$

Indeed, for every fixed $c \in \mathbb{Z}$, the univariate function $f(x) = F(x, c) = x \,\mathsf{XOR}\, c$ is uniformly differentiable on $\mathbb{Z}_2$; and $f'(x) = 1$ for $c \geqslant 0$, $f'(x) = -1$ for $c < 0$ (cf., Note 3.19). Therefore, the function F is not differentiable as a bivariate function since a 2-adic integer can be simultaneously considered as a limit of a sequence of positive rational integers and as a limit of a sequence of negative rational integers as both $\mathbb{N}$ and $-\mathbb{N} = \{-1, -2, -3, \ldots\}$ are dense in $\mathbb{Z}_p$. However, at the same time, this implies that F is uniformly differentiable modulo 2 on $\mathbb{Z}_2$, as $-1 \equiv 1 \pmod 2$, and that $N_1(F) = 1$.

The rules of derivation modulo p^k of functions that have integer-valued derivatives modulo p^k are similar to the ones in the classical real calculus, with this difference only: These rules are congruences modulo p^k, and not equalities. The following proposition is true:

Proposition 3.24 *Let the functions* $G\colon \mathbb{Z}_p^s \to \mathbb{Z}_p^n$ *and* $F\colon \mathbb{Z}_p^n \to \mathbb{Z}_p^m$ *be differentiable modulo* p^k *at points* $\mathbf{v} = (v_1, \dots, v_s)$ *and* $\mathbf{u} = G(\mathbf{v})$, *, respectively, and let all partial derivatives modulo* p^k *of the functions G and F at points* $\mathbf{v}$ *and* $\mathbf{u}$, *respectively, be p-adic integers. Then the composition function* $F \circ G\colon \mathbb{Z}_p^s \to \mathbb{Z}_p^m$ *is differentiable modulo* p^k *at the point* $\mathbf{v}$, *all its partial derivatives modulo* p^k *at this point are p-adic integers, and*

$$(F \circ G)'_k(\mathbf{v}) \equiv G'_k(\mathbf{v})F'_k(\mathbf{u}) \pmod{p^k}.$$

In particular, if the functions $f, g\colon \mathbb{Z}_p \to \mathbb{Z}_p$ *are differentiable modulo* p^k *at the point* $u \in \mathbb{Z}_p$, *and if their derivatives modulo* p^k *at this point are integer-valued, then*

$$(f + g)'_k(u) \equiv f'_k(u) + g'_k(u) \pmod{p^k};$$

$$(fg)'_k(u) \equiv f'_k(u)g(u) + f(\mathbf{u})g'_k(u) \pmod{p^k}.$$

If, moreover, there exists a ball $U \ni u$ *of non-zero radius such that* $g(r) \not\equiv 0 \pmod{p}$ *for every point* $r \in U$, *then the function* $\frac{f}{g}\colon U \to \mathbb{Z}_p$ *is differentiable modulo* p^k *at the point u and has integer-valued derivative modulo* p^k *at this point, and*

$$\left(\frac{f}{g}\right)'_k(u) \equiv \frac{f'_k(u)g(u) - f(u)g'_k(u)}{g(u)^2} \pmod{p^k}.$$

If additionally functions F, G, f, and g are uniformly differentiable modulo p^k, *and if their derivatives modulo* p^k *are integer-valued everywhere on* $\mathbb{Z}_p$, *then the same is true for functions* $F \circ G$, $f + g$, *and* fg.

Finally, if $g(v) \not\equiv 0 \pmod{p}$ *for all* $v \in \mathbb{Z}_p$, *then the function* $\frac{f}{g}$ *is integer-valued and uniformly differentiable modulo* p^k *everywhere on* $\mathbb{Z}_p$, *and its partial derivative modulo* p^k *is integer-valued at all points of* $\mathbb{Z}_p$.

A proof of this proposition with minor changes due to the non-Archimedean metric follows (up to congruences modulo p^n rather than equations) the one of the classical real calculus. The argument is still valid since a congruence modulo p^n is a congruence relation on the ring $\mathbb{Z}_p$: We can, for instance, multiply both sides of some congruence modulo p^n by a p-adic unit without affecting the validity of this congruence. We omit further details.

Note 3.25 The rules of derivation modulo p^k generally do not hold for functions whose derivatives modulo p^k are not integer-valued. However, both a sum of (uniformly) differentiable modulo p^k functions and a product of such a function by a p-adic integer are still (uniformly) differentiable modulo p^k since a congruence modulo p^n is a congruence relation on $\mathbb{Q}_p$ when $\mathbb{Q}_p$ is considered as module over the ring $\mathbb{Z}_p$.

3.2.2 Properties of Derivatives Modulo p^k

Recall that by definition, a locally constant function is a function such that, given a point in the domain, there exists a neighborhood where the function is a constant. Therefore, a function G whose domain is $\mathbb{Z}_p^n$ is called a *locally constant* if, given $\mathbf{x} = (x_1, \ldots, x_n) \in \mathbb{Z}_p^n$, there exists a ball of non-zero radius around $\mathbf{x}$ such that G is a constant on that ball. Following [112, 145, 168], we call a function G with domain $\mathbb{Z}_p^n$ a *step function* if there exists $N \in \mathbb{N}_0$ such that $G(\mathbf{x}) = G(\mathbf{y})$ once $|\mathbf{x} - \mathbf{y}|_p \leqslant p^{-N}$; the smallest N with this property is called the *order* of the step function G.

From the compactness of $\mathbb{Z}_p^n$, it follows that, given a locally constant function F on $\mathbb{Z}_p^n$, the space $\mathbb{Z}_p^n$ is a disjoint union of finite number of clopen balls of non-zero radii such that F is a constant on every that ball. Therefore, if p^{-N} is the smallest radius of the balls that constitute the union, the function F is periodic: $F(\mathbf{z} + p^N) = F(\mathbf{z})$ for every $\mathbf{z} \in \mathbb{Z}_p^n$. Therefore, F is a step function of order $\geqslant N$. Vice versa, from the strong triangle inequality, it easily follows that any step function G on $\mathbb{Z}_p^n$ of order N is periodic: $G(\mathbf{z} + p^N) = G(\mathbf{z})$ for every $\mathbf{z} \in \mathbb{Z}_p^n$.

Now, from the definition of differentiability, it immediately follows that locally constant functions are uniformly differentiable and their derivatives vanish at every point of $\mathbb{Z}_p^n$. The following proposition (cf., [9, Proposition 2.12]) therefore shows that if a function $F\colon \mathbb{Z}_p^n \to \mathbb{Z}_p^m$ is uniformly differentiable modulo p^k, then it is "smooth modulo p^k": All its partial derivatives modulo p^k are uniformly differentiable on $\mathbb{Z}_p^n$, and second derivatives modulo p^k are identically 0.

Proposition 3.26 *If the function $F = (f_1, \ldots, f_m)\colon \mathbb{Z}_p^n \to \mathbb{Z}_p^m$ is uniformly differentiable modulo p^k, then every derivative modulo p^k is a step function of order not greater than $N_k(F)$* (cf., Definition 3.22).

Proof The proof can obviously be restricted to the case $m = n = 1$. According to Definition 3.22, if $|h|_p \leqslant p^{-K}$, then for all $u \in \mathbb{Z}_p$ and $K \geqslant N_k(F)$, the following congruence holds:

$$F(u+h) - F(u) \equiv h\frac{\partial_k F(u)}{\partial_k x} \pmod{p^{k+K}}. \tag{3.13}$$

Taking $|h_1|_p \leqslant |h|_p$ and substituting $u = u_1 + h_1$ into (3.13), represent

$$F(u+h) - F(u) = F(u_1+h_1+h) - F(u_1) - (F(u_1+h_1) - F(u_1)). \tag{3.14}$$

Now, applying (3.13) to (3.14), we obtain that

$$F(u+h) - F(u) \equiv (h_1+h)\frac{\partial_k F(u_1)}{\partial_k x} - h_1\frac{\partial_k F(u_1)}{\partial_k x} \pmod{p^{k+K}},$$

and conclude that

$$F(u+h) - F(u) \equiv h\frac{\partial_k F(u_1)}{\partial_k x} \pmod{p^{k+K}} \tag{3.15}$$

since a congruence modulo p^{k+K} is a congruence relation of the module $\mathbb{Q}_p$ over the ring $\mathbb{Z}_p$ (see Note 3.25). Now, comparing (3.13) with (3.15) and putting $h = p^K$, we obtain that

$$\frac{\partial_k F(u_1)}{\partial_k x} \equiv \frac{\partial_k F(u)}{\partial_k x} \pmod{p^k}$$

whenever $|u_1 - u|_p \leqslant p^{-N_k(F)}$. □

Note that nowhere in the proof we assume that the derivatives modulo p^k must be integer-valued!

Note 3.27 Proposition 3.26 implies that *each partial derivative modulo p^k can be considered as a function defined on the residue ring $\mathbb{Z}/p^{N_k(F)}\mathbb{Z}$ and valuated in* $\mathbb{Z}/p^k\mathbb{Z}$. Moreover, if a continuation $\tilde{F}$ of the function $F = (f_1, \ldots, f_m)\colon \mathbb{N}_0^n \to \mathbb{N}_0^m$ to the space $\mathbb{Z}_p^n$ is uniformly differentiable modulo p^k on $\mathbb{Z}_p^n$, then one can simultaneously continue the function F together with all its (partial) derivatives modulo p^k to the whole space $\mathbb{Z}_p^n$. Consequently, we may study if necessary (partial) derivatives modulo p^k of the function $\tilde{F}$ rather than those of F, and vice versa. For example, a partial derivative $\frac{\partial_k f_i(\mathbf{u})}{\partial_k x_j}$ modulo p^k vanishes modulo p^k at no point of $\mathbb{Z}_p^n$ (i.e., $\frac{\partial_k f_i(\mathbf{u})}{\partial_k x_j} \not\equiv 0 \pmod{p^k}$ for all $u \in \mathbb{Z}_p^n$, or equivalently, $\left|\frac{\partial_k f_i(\mathbf{u})}{\partial_k x_j}\right|_p > p^{-k}$ everywhere on $\mathbb{Z}_p^n$) if and only if $\frac{\partial_k f_i(\mathbf{u})}{\partial_k x_j} \not\equiv 0 \pmod{p^k}$ for all $u \in \{0, 1, \ldots, p^{N_k(F)} - 1\}$.

Now we give yet one more example of Hensel lifting where the differentiability modulo p^k plays a crucial role:

Example 3.28 Let 1-Lipschitz functions $f_i\colon \mathbb{Z}_p^n \to \mathbb{Z}_p$ be uniformly differentiable modulo p on $\mathbb{Z}_p^n$, $i = 1, 2, \ldots, n$. If the system of equations $f_i(x_1, \ldots, x_n) = 0$ and $i = 1, 2, \ldots, n$ in unknowns $x_1, \ldots, x_n$ has a solution $\bar{\mathbf{u}} = (\bar{u}_1, \ldots, \bar{u}_n)$ modulo p^N, where $N \in \mathbb{N}$ is some (equivalently, any) number not less than $\max\{N_1(f_i)\colon i = 1, 2, \ldots, n\}$, (i.e., if $f_i(\bar{u}_1, \ldots, \bar{u}_n) \equiv 0 \pmod{p^N}$, $i = 1, 2, \ldots, n$) such that the Jacobian of the system does not vanish modulo p at $\bar{\mathbf{u}}$, then the system has a solution $\mathbf{u} = (u_1, \ldots, u_n)$ in $\mathbb{Z}_p^n$ such that $\mathbf{u} \equiv \bar{\mathbf{u}} \pmod{p^N}$.

As in the univariate case, the proof follows the lines of classical Hensel's Lemma (cf., Note 3.20) and thus is omitted.

Chapter 4
p-Adic Series

Abstract In this chapter we, mostly following (Mahler, p-adic Numbers and Their Functions, 2nd edn., Cambridge University Press (1981); Schikhof, Ultrametric Calculus: An Introduction to p-adic Analysis, Cambridge University Press (1984); Katok, p-adic Analysis in Comparison with Real, Mass. Selecta. American Mathematical Society (2003)), introduce some p-adic series, which are used as technical tools further in the book.

4.1 Taylor Series

The following definition is just a restatement of the respective definition from real analysis.

Definition 4.1 (Taylor Series) Let the function $f : \mathbb{Z}_p \to \mathbb{Z}_p$ be infinitely many times differentiable at the point $a \in \mathbb{Z}_p$. The power series

$$\sum_{i=0}^{\infty} \frac{f^{(i)}(a)}{i!}(x-a)^i \tag{4.1}$$

are called *Taylor series of the function f at the point a.*

As in real analysis, central questions about Taylor series are where the series (4.1) converges and, if converges for some x, whether the series converges to $f(x)$. These questions lead to the notion of analytic function.

4.1.1 Analytic Functions on Balls

Definition 4.2 (Analytic Function) The function $f : \mathbb{Z}_p \to \mathbb{Z}_p$ is said to be *analytic* on the ball $\mathbf{B}_{p^{-r}}(a) \subset \mathbb{Z}_p$ if it can be represented by power series that

V. Anashin, *Causality: The* p*-adic Theory*, STEAM-H: Science, Technology, Engineering, Agriculture, Mathematics & Health,
https://doi.org/10.1007/978-3-031-85818-5_4

converges everywhere on $\mathbf{B}_{p^{-r}}(a)$:

$$f(x) = \sum_{i=0}^{\infty} a_i (x - a)^i$$

for all $x \in \mathbf{B}_{p^{-r}}(a)$ (here $a_i \in \mathbb{Q}_p, i = 0, 1, 2, \ldots$).

If the function f is analytic on the ball $\mathbf{B}_{p^{-r}}(a)$, it can be represented by Taylor series everywhere on the ball; that is, $a_i = \frac{f^{(i)}(a)}{i!}$, $i = 0, 1, 2, \ldots$ (see, e.g., [168, Theorem 40.2]).

Example 4.3 The following functions are analytic on their respective balls:

- The *p-adic exponential* function

$$\exp_p x = \sum_{j=0}^{\infty} \frac{x^j}{j!}$$

  is analytic on $\mathbf{B}_{p^{-1}}(0) = p\mathbb{Z}_p$ if $p \geqslant 3$ and on $\mathbf{B}_{\frac{1}{4}}(0) = 4\mathbb{Z}_2$ if $p = 2$. Note that there are no "p-adic e" in $\mathbb{Q}_p$ since the series does not converge at $x = 1$.
- In the same way, i.e., by considering corresponding power series, we can introduce *p-adic trigonometric functions*:

$$\sin_p x = \sum_{j=0}^{\infty} \frac{(-1)^j x^{2j+1}}{(2j+1)!}, \quad \cos_p x = \sum_{j=0}^{\infty} \frac{(-1)^j x^{2j}}{2j!}.$$

  They are analytic on the same balls as the exponential function.
- The *p-adic logarithm*

$$\ln_p x = \sum_{k=1}^{\infty} \frac{(-1)^{k+1}(x-1)^k}{k}$$

  is analytic on $\mathbf{B}_{p^{-1}}(1) = 1 + p\mathbb{Z}_p$.

In the sequel, we are mostly interested in the functions that are analytic on the whole space $\mathbb{Z}_p$, which is a ball of radius 1 centered at 0; these functions are represented by power series $\sum_{i=0}^{\infty} a_i x^i$.

Theorem 4.4 *Given $a_0, a_1, \ldots \in \mathbb{Q}_p$, the power series $\sum_{i=0}^{\infty} a_i x^i$ converges everywhere on $\mathbb{Z}_p$ if and only if $\lim_{i\to\infty}^{p} a_i = 0$; under the latter condition, the series defines a continuous function on $\mathbb{Z}_p$.*

Proof Follows immediately from Theorem 2.8 and [112, Proposition 3.17]. □

4.2 Mahler (Interpolation) Series

In this section, we first recall Mahler expansion, which is also known as interpolation series (see., e.g., [112, 145, 168]). We use this expansion in further chapters of the book as tool to investigate dynamics of 1-Lipschitz (i.e., causal) mappings. Then in this section, we also prove, by applying that tool, an auxiliary result on identities modulo p^k, which we need in the sequel.

Every function $f\colon \mathbb{N}_0 \to \mathbb{Z}_p$ (or, respectively, $f\colon \mathbb{N}_0 \to \mathbb{Z}$) has the only *Mahler expansion*, that is, has a unique representation via the so-called *Mahler (interpolation) series*

$$f(x) = \sum_{i=0}^{\infty} a_i \binom{x}{i}, \tag{4.2}$$

where $a_i \in \mathbb{Z}_p$ (respectively, $a_i \in \mathbb{Z}$), $i = 0, 1, 2, \ldots$, and

$$\binom{x}{i} = \frac{x(x-1)\cdots(x-i+1)}{i!}$$

for $i = 1, 2, \ldots$;

$$\binom{x}{0} = 1,$$

by definition.

If f is uniformly continuous on $\mathbb{N}_0$ with respect to the p-adic metrics, it can be uniquely expanded to a uniformly continuous function on $\mathbb{Z}_p$. Hence, the interpolation series for f converges uniformly on $\mathbb{Z}_p$. The following is true: The series (4.2) converges uniformly on $\mathbb{Z}_p$ if and only if

$$\overset{p}{\lim_{i\to\infty}} a_i = 0. \tag{4.3}$$

Hence, the uniformly convergent series defines a uniformly continuous function on $\mathbb{Z}_p$. The function f represented by the interpolation series (4.2) is (uniformly) differentiable everywhere on $\mathbb{Z}_p$ if and only if

$$\overset{p}{\lim_{i\to\infty}} \frac{a_{i+n}}{i} = 0 \tag{4.4}$$

for all $n \in \mathbb{N}_0$; in this case, the following formula for the derivative holds:

$$f'(x) = \sum_{i=1}^{\infty} (-1)^{i+1} \frac{\Delta^i f(x)}{i}. \tag{4.5}$$

The function f is analytic on $\mathbb{Z}_p$ if and only if

$$\overset{p}{\lim_{i\to\infty}} \frac{a_i}{i!} = 0. \tag{4.6}$$

To represent functions of several variables, we use interpolation series of the following form:

$$f(x_1, \dots, x_n) = \sum_{(i_1,\dots,i_n)\in\mathbb{N}_0^n} a_{i_1,\cdots,i_n} \binom{x_1}{i_1}\binom{x_2}{i_2}\cdots\binom{x_n}{i_n}; \tag{4.7}$$

here $a_{i_1,\dots,i_n} \in \mathbb{Z}_p$.

Below are some examples of how to apply the Mahler series to determine whether a function is analytic.

Example 4.5 Given the function $f(x) = (1 + pz)^x$ of variable x on $\mathbb{Z}_p$:

- If p is odd, then the function f is analytic on $\mathbb{Z}_p$, for every $z \in \mathbb{Z}_p$.
- If $p = 2$, then the function f is analytic on $\mathbb{Z}_2$ if and only if $z \in \mathbf{B}_{\frac{1}{2}}(0) = 2\mathbb{Z}_2$.

Indeed, $(1 + pz)^x = \sum_{i=0}^{\infty} p^i z^i \binom{x}{i}$. By Lemma 2.1, we see that

$$\operatorname{ord}_p \frac{p^i z^i}{i!} = i - \frac{1}{p-1}(i - \operatorname{wt}_p i) + \operatorname{ord}_p z^i \to \infty \text{ as } i \to \infty$$

Therefore, $\overset{p}{\lim_{i\to\infty}} \frac{p^i z^i}{i!} = 0$, and the first claim follows. The proof of the second claim is similar and is thus omitted.

4.2.1 *Identities Modulo p^k*

This an auxiliary subsection; we describe here a special class of functions introduced in [9], which are, loosely speaking, sufficiently small with respect to the p-adic metric, but not too small.

Definition 4.6 A function $F\colon \mathbb{Z}_p^n \to \mathbb{Z}_p^m$ is called *an identity modulo p^k* if for every $\mathbf{u} \in \mathbb{Z}_p^n$, the following congruence holds:

$$F(\mathbf{u}) \equiv (0, \dots, 0) \pmod{p^k}.$$

In other words, F is an identity modulo p^k if and only if $|F(\mathbf{u})|_p \leqslant p^{-k}$ for all $\mathbf{u} \in \mathbb{Z}_p^n$.

The following easy proposition (cf., [9, Proposition 4.2]) characterizes identities modulo p^k in terms of Mahler expansion.

Proposition 4.7 *A function $f: \mathbb{Z}_p^n \to \mathbb{Z}_p$ is an identity modulo p^k if and only if all coefficients of its Mahler expansion* (4.7) *are 0 modulo p^k:*

$$|a_{i_1,\ldots,i_n}|_p \leqslant p^{-k}$$

for all $(i_1, \ldots, i_n) \in \mathbb{N}_0^n$.

Proof Induction on n. Let $n = 1$. As f is a continuous function, and as $\mathbb{N}_0$ is a dense subset in $\mathbb{Z}_p$, f is an identity modulo p^k if and only if

$$\sum_{i=0}^{s} a_i \binom{s}{i} \equiv 0 \pmod{p^k} \tag{4.8}$$

for all $s = 0, 1, 2, \ldots$. However, a triangular system of congruences (4.8) has a unique solution

$$0 \equiv a_0 \equiv a_1 \equiv a_2 \equiv \ldots \pmod{p^k}; \tag{4.9}$$

hence, for $n = 1$, the proposition is true. As

$$f(x_1, \ldots, x_{n-1}, s) = \sum_{i=0}^{s} g_i(x_1, \ldots, x_{n-1}) \binom{x}{i}$$

for every $s \in \mathbb{N}_0$, then by similar argument, we conclude that $f(x_1, \ldots, x_n)$ is an identity modulo p^k if and only if

$$g_i(x_1, \ldots, x_{n-1}) \equiv 0 \pmod{p^k}$$

for all $x_1, \ldots, x_{n-1} \in \mathbb{Z}_p$ and all $i = 0, 1, 2, \ldots$. By the induction hypothesis, in view of (4.7), the latter condition holds if and only if the following congruences

$$a_{i_1,\ldots,i_{n-1},i} \equiv 0 \pmod{p^k}$$

hold for all $i = 0, 1, 2, \ldots$. □

4.3 van der Put Series

In this section, we, following [145, 168], briefly recall yet one more type of series, the van der Put series, which is also a very convenient tool to investigate various properties of p-adic functions. There is a slight difference between the definitions of van der Put series given by K. Mahler in [145] and by W. Schikhof in [168]. In the sequel, we, following [83], call these two versions of van der Put series the Mahler's version and the Schikhof's version, accordingly.

We start with Mahler's version. Given a continuous function $f: \mathbb{Z}_p \to \mathbb{Z}_p$, there exists a unique sequence $B_0, B_1, B_2, \ldots$ of p-adic integers such that

$$f(x) = \sum_{m=0}^{\infty} B_m \chi(m, x) \tag{4.10}$$

for all $x \in \mathbb{Z}_p$, where

$$\chi(m, x) = \begin{cases} 1, \text{ if } |x - m|_p \leqslant p^{-n} \\ 0, \text{ otherwise} \end{cases} \tag{4.11}$$

and $n = 1$ if $m = 0$; n is uniquely defined by the inequality $p^{n-1} \leqslant m \leqslant p^n - 1$ otherwise. The series (4.10) is called the *van der Put series* of the function f. Note that the sequence $B_0, B_1, \ldots, B_m, \ldots$ of *van der Put coefficients* of the function f tends p-adically to 0 as $m \to \infty$, and the series converges uniformly on $\mathbb{Z}_p$. Vice versa, if a sequence $B_0, B_1, \ldots, B_m, \ldots$ of p-adic integers tends p-adically to 0 as $m \to \infty$, then the series (4.10) converges uniformly on $\mathbb{Z}_p$ and thus defines a continuous function $f: \mathbb{Z}_p \to \mathbb{Z}_p$.

The number n in the definition of $\chi(m, x)$ has a very natural meaning; it is merely the number of digits in the base-p expansion of $m \in \mathbb{N}_0$:

$$\lfloor \log_p m \rfloor = (the\ number\ of\ digits\ in\ the\ base - p\ \ expansion\ \ for\ \ m) - 1 \tag{4.12}$$

Thus, $n = \lfloor \log_p m \rfloor + 1$ for all $m \in \mathbb{N}_0$; that is why, in view of (4.12), in the sequel everywhere in the book, we assume that $\lfloor \log_p 0 \rfloor = 0$. Therefore, the so-defined functions $\chi(x, m)$ and $m = 0, 1, 2, \ldots$ are characteristic functions of balls $\mathbf{B}_{p^{-\lfloor \log_p m \rfloor - 1}}(m)$ of radii $p^{-\lfloor \log_p m \rfloor - 1}$ centered at $m \in \mathbb{N}_0$. Coefficients B_m in (4.10) can be expressed via values of the function f as follows:

$$B_m = \begin{cases} f(m), \text{ if } m = 0, 1, \ldots, p-1; \\ f(m) - f(m \bmod p^{\lfloor \log_p m \rfloor}), \text{ otherwise.} \end{cases} \tag{4.13}$$

The difference between Schikhof's and Mahler's versions of van der Put series is the choice of balls whose characteristic functions are used as a base; actually, in Schikhof's version, the very first ball (the one for $m = 0$) is $\mathbb{Z}_p$, whereas in Mahler's version, the ball is $\mathbf{B}_{1/p}(0) = p\mathbb{Z}_p$. Therefore, in Schikhof's version, the coefficients in (4.10) are related to the values the function f takes in $0, 1, 2, \ldots$ (cf. (4.13)):

$$B_m = \begin{cases} f(m), \text{ if } m = 0; \\ f(m) - f(0), \text{ if } m = 1, 2, \ldots, p-1; \\ f(m) - f(m \bmod p^{\lfloor \log_p m \rfloor}), \text{ otherwise.} \end{cases} \tag{4.14}$$

Further in the book, we use Mahler's version of the van der Put series; we make respective remarks. Though the difference between the two versions of van der Put series usually does not change much (if any) in the proofs and statements, it must be taken into the account when considering representation via van der Put series of automata functions acting on infinite words (see discussion in [83]).

Chapter 5
1-Lipschitz Functions

Abstract This chapter is devoted to the class of functions the whole book is focused at: 1-Lipschitz p-adic functions. As we have explained before, these functions can also be called causal, automatic, non-expansive, or compatible since all these classes of functions coincide. That is why in what follows, we often use these terms as synonyms. We mostly (but not always) assume that 1-Lipschitz functions are *integer-valued* functions; that is, they map $\mathbb{Z}_p^n$ into $\mathbb{Z}_p^m$.

5.1 1-Lipschitzness and Local 1-Lipschitzness

From Sect. 3.2, we already know that an integer-valued function $F\colon \mathbb{Z}_p^n \to \mathbb{Z}_p^m$ is 1-Lipschitz if and only if

$$|F(\mathbf{a}) - F(\mathbf{b})|_p \leqslant |\mathbf{a} - \mathbf{b}|_p \tag{5.1}$$

for all $\mathbf{a}, \mathbf{b} \in \mathbb{Z}_p^n$ (note that we use the same symbol for absolute values on $\mathbb{Z}_p^n$ and on $\mathbb{Z}_p^m$). Often, 1-Lipschitz maps from $\mathbb{Z}_p^n$ into $\mathbb{Z}_p^n$ are also called *non-expansive* since the maps do not increase distances between points.

The function F is 1-Lipschitz if and only if it is *compatible* (cf. (3.10)):

$$F(\mathbf{a}) \equiv F(\mathbf{b}) \pmod{p^\ell} \text{ once } \mathbf{a} \equiv \mathbf{b} \pmod{p^\ell}.$$

The *compatibility* is an algebraic notion: A function defined on a Cartesian power of a universal algebra and valuated in a Cartesian power of the universal algebra is called compatible once the function agrees with all congruences of the algebra (cf. Definition 1.9). All congruences of the ring $\mathbb{Z}_p$ (and of $\mathbb{Z}_p^n$) are the ones that correspond to epimorphisms $\operatorname{mod} p^k$ for a suitable $k \in \mathbb{N}$ (cf. (2.7) and (3.11)). Thus, in the book, 1-Lipschitz maps and compatible maps are just synonyms.

Along with compatible functions, we also consider *locally compatible functions* (or, *locally 1-Lipschitz functions*). The latter are functions that satisfy (5.1) locally; that is, given $\mathbf{a} \in \mathbb{Z}_p^n$, there exists an open neighborhood $O_{\mathbf{a}}$ of $\mathbf{a}$ such that (5.1)

V. Anashin, *Causality: The p-adic Theory*, STEAM-H: Science, Technology, Engineering, Agriculture, Mathematics & Health,
https://doi.org/10.1007/978-3-031-85818-5_5

holds for all $\mathbf{b} \in O_{\mathbf{a}}$. As $\mathbb{Z}_p^n$ is compact, the function $F\colon \mathbb{Z}_p^n \to \mathbb{Z}_p^m$ is locally compatible if and only if (5.1) holds for all $\mathbf{a}, \mathbf{b} \in \mathbb{Z}_p^n$, which are sufficiently close to one another, that is, there exists $r \in \mathbb{N}_0$ such that (5.1) is true once $|\mathbf{a}-\mathbf{b}|_p \leqslant p^{-r}$. Thus, we conclude:

The function $F\colon \mathbb{Z}_p^n \to \mathbb{Z}_p^m$ is locally 1-Lipschitz if and only if there exists a finite number of pairwise disjoint balls in $\mathbb{Z}_p^n$ such that their union is $\mathbb{Z}_p^n$ and the function F satisfy a p-adic 1-Lipschitz condition on every ball of these.

5.1.1 Coordinate Functions

Now we characterize multivariate 1-Lipschitz functions in terms of the so-called *coordinate functions*; the latter are functions $\delta_i(f(x_1, \dots, x_n))$, whose domain is $\mathbb{Z}_p^n$ and co-domain is $\{0, 1, \dots, p-1\}$: The i-th coordinate function is merely a value of coefficient of the i-th term in a canonical p-adic expansion of $f(x_1, \dots, x_n)$ (cf., Sect. 2.3.1). The following proposition is an obvious generalization of respective claim about univariate functions from Sect. 3.1 to a multivariate case.

Proposition 5.1 *The function $f\colon \mathbb{Z}_p^n \to \mathbb{Z}_p$ is 1-Lipschitz if and only if for every $i = 1, 2, \dots$, the i-th coordinate function $\delta_i(f(x_1, \dots, x_n))$ does not depend on $\delta_{i+k}(x_s)$, for all $s = 1, 2, \dots, n$ and $k = 1, 2, \dots$.*

Proof Let the function $\delta_i(f(x_1, \dots, x_n))$ depend on $\delta_{i+k}(x_s)$ for some i, s, k; i.e., let there exist $(u_1, \dots, u_n)$ and $(v_1, \dots, v_n)$ in $\mathbb{Z}_p^n$ such that $u_j = v_j$ for $j = 1, 2, \dots, n$; $j \neq s$, and

$$\delta_{i+k}(u_s) \neq \delta_{i+k}(v_s), \quad \delta_r(u_s) = \delta_r(v_s)$$

for all $r = 0, 1, 2, \dots$; $r \neq i+k$, and

$$\delta_i(f(u_1, \dots, u_n)) \neq \delta_i(f(v_1, \dots, v_n)). \tag{5.2}$$

This means that $(u_1, \dots, u_n) \equiv (v_1, \dots, v_n) \pmod{p^{i+k}}$, i.e., in particular

$$(u_1, \dots, u_n) \equiv (v_1, \dots, v_n) \pmod{p^{i+1}},$$

whereas in view of (5.2),

$$f(u_1, \dots, u_n) \not\equiv f(v_1, \dots, v_n) \pmod{p^{i+1}}.$$

This is a contradiction with the compatibility of f. □

Note 5.2 From the proof of Proposition 5.1, it immediately follows that the *function $f: \mathbb{Z}_p^n \to \mathbb{Z}_p$ is locally 1-Lipschitz if and only if there exists $N \in \mathbb{N}_0$ such that for every $i = N, N+1, N+2, \ldots$ the i-th coordinate function $\delta_i(f(x_1, \ldots, x_n))$ does not depend on $\delta_{i+k}(x_s)$, for all $s = 1, 2, \ldots, n$ and $k = 1, 2, \ldots$.*

Proposition 5.1 shows that a 1-Lipschitz function $F: \mathbb{Z}_p^n \to \mathbb{Z}_p^m$ is a *determined function*, in terminology of [190], which is also known as a *triangular function of p-valued logic*; vice versa, every triangular function defines a 1-Lipschitz function $F: \mathbb{Z}_p^n \to \mathbb{Z}_p^m$. In cryptology, triangular functions are usually considered only for $p = 2$ and are called *T-functions*.

Definition 5.3 An n-variate *triangular function* is a mapping

$$\Phi: (\alpha_0^\downarrow, \alpha_1^\downarrow, \alpha_2^\downarrow, \ldots) \mapsto (\Phi_0^\downarrow(\alpha_0^\downarrow), \Phi_1^\downarrow(\alpha_0^\downarrow, \alpha_1^\downarrow), \Phi_2^\downarrow(\alpha_0^\downarrow, \alpha_1^\downarrow, \alpha_2^\downarrow), \ldots),$$

where $\alpha_i^\downarrow \in \mathbb{F}_p^n$ is an n-dimensional columnar vector; $\mathbb{F}_p = \{0, 1, \ldots, p-1\}$, and the mapping $\Phi_i^\downarrow: (\mathbb{F}_p^n)^{i+1} \to \mathbb{F}_p^m$ maps n-dimensional vectors $\alpha_0^\downarrow, \ldots, \alpha_i^\downarrow$ to an m-dimensional vector $\Phi_i^\downarrow(\alpha_0^\downarrow, \ldots, \alpha_i^\downarrow) \in \mathbb{F}_p^m$. Accordingly, a univariate triangular function f is a mapping

$$(\chi_0, \chi_1, \chi_2, \ldots) \overset{f}{\mapsto} (\psi_0(\chi_0); \psi_1(\chi_0, \chi_1); \psi_2(\chi_0, \chi_1, \chi_2); \ldots),$$

where $\chi_j \in \{0, 1, \ldots, p-1\}$, and each $\psi_j(\chi_0, \ldots, \chi_j) \in \{0, 1, \ldots, p-1\}$ is a function in variables $\chi_0, \ldots, \chi_j$ of p-valued logic (cf. (3.1)).

Due to the polynomial completeness of finite fields (cf. Sect. 1.2.3), if p is a prime, then all coordinate functions are polynomials of respective number of variables over a finite field $\mathbb{F}_p = \mathbb{Z}/p\mathbb{Z}$, the residue ring modulo p.

5.1.2 *Differences of 1-Lipschitz Functions*

In this subsection, we give a yet one more characterization of p-adic 1-Lipschitz functions following [9]. Recall that given an n-variate function $f(x_1, \ldots, x_n)$, its (first) difference with respect to j-th variable is $\Delta_j(f(x_1, \ldots, x_n)) = f(x_1, \ldots, x_j + 1, \ldots, x_n) - f(x_1, \ldots, x_j, \ldots, x_n)$ by definition.

Proposition 5.4 *A continuous function $f: \mathbb{Z}_p^n \to \mathbb{Z}_p$ is 1-Lipschitz if and only if all functions $\frac{1}{i}\Delta_j^i f$ (where $j = 1, 2, \ldots, n$; $i = 1, 2, \ldots$) are integer-valued on $\mathbb{Z}_p$* (i.e., all the values that take these functions on $\mathbb{Z}_p$ are p-adic integers).

Proof By (5.1), we see that $f\colon \mathbb{Z}_p^n \to \mathbb{Z}_p$ is 1-Lipschitz if and only if

$$|f(x_1, \ldots, x_{i-1}, x_i + h, x_{i+1}, \ldots, x_n) - f(x_1, \ldots, x_n)|_p \leqslant |h|_p \tag{5.3}$$

for all $x_1, \ldots, x_n, h \in \mathbb{Z}_p$ and all $i = 1, 2, \ldots, n$ or, equivalently, if and only if the p-adic number

$$\alpha_h = \frac{1}{h}(f(x_1, \ldots, x_{i-1}, x_i + h, x_{i+1}, \ldots, x_n) - f(x_1, \ldots, x_n)) \tag{5.4}$$

is a p-adic integer for all $h \in \mathbb{Z}_p \setminus \{0\}$ and all $x_1, \ldots, x_n \in \mathbb{Z}_p$.

As $f(x_1, \ldots, x_n)$ is continuous, (5.3) holds for all $h \in \mathbb{Z}_p$ if and only if it holds for all $h \in \mathbb{N}$, since $\mathbb{N}$ is a dense subset in $\mathbb{Z}_p$. Thus, a continuous function f is 1-Lipschitz if and only if α_h is a p-adic integer for every $h \in \mathbb{N}$. Now applying by Theorem 1.5, we conclude that for $h \in \mathbb{N}$, the p-adic number α_h can be expressed as

$$\alpha_h = \frac{1}{h}\sum_{j=1}^{h}\binom{h}{j}\Delta_i^j f(x_1, \ldots, x_n) = \sum_{j=1}^{h}\binom{h-1}{j-1}\frac{1}{j}\Delta_i^j f(x_1, \ldots, x_n).$$

Thus, the function f is 1-Lipschitz if and only if each p-adic number

$$\alpha_m = \sum_{k=0}^{m-1}\binom{m}{k}\frac{1}{k+1}\Delta_i^{k+1} f(x_1, \ldots, x_n) \tag{5.5}$$

is a p-adic integer for $m = 1, 2, 3, \ldots$. Now applying combinatorial relations of Theorem 1.6, we express $\frac{1}{k+1}\Delta_i^{k+1} f(x_1, \ldots, x_n)$ from (5.5) via the numbers α_m:

$$\frac{1}{k+1}\Delta_i^{k+1} f(x_1, \ldots, x_n) = \sum_{m=0}^{k}(-1)^{m+k}\binom{k}{m}\alpha_{m+1}, \tag{5.6}$$

where $k = 0, 1, 2, \ldots$. Now (5.6) implies that all fractions $\frac{1}{k+1}\Delta_i^{k+1} f(x_1, \ldots, x_n)$ are p-adic integers whenever all α_n for $n = 0, 1, 2, 3, \ldots$ are p-adic integers; whereas (5.5) implies the converse. Whence, all α_m for $m = 0, 1, 2, \ldots$ are p-adic integers if and only if all fractions $\frac{1}{k+1}\Delta_i^{k+1} f(x_1, \ldots, x_n)$ for $k = 0, 1, 2, \ldots$ are p-adic integers. □

5.2 1-Lipschitzness and Differentiability

We already know that if a 1-Lipschitz function is uniformly differentiable modulo p^k, then all its partial derivatives modulo p^k are integer-valued (cf., Proposition 3.13, Note 3.17, Sect. 3.2.1). The following theorem (cf., [9, Theorem 2.10]) demonstrates that 1-Lipschitz functions are tightly related to functions that are uniformly differentiable (or at least are uniformly differentiable modulo p^k) and have integer-valued derivatives.

Theorem 5.5 *Let a function* $F = (f_1, \dots, f_m)\colon \mathbb{Z}_p^n \to \mathbb{Z}_p^m$ *be uniformly differentiable modulo p; let F have integer-valued derivatives modulo p at all points of* $\mathbb{Z}_p^n$. *Then*

$$F(x_1, \dots, x_n) = P(x_1, \dots, x_n) + C(x_1, \dots, x_n)$$

where P is a step function of order not greater than $N_1(F)$ *and C is 1-Lipschitz. Consequently, F is locally 1-Lipschitz, and C is uniformly differentiable modulo p.*

Proof Put

$$P(x_1, \dots, x_n) = (f_1(x_1, \dots, x_n) \bmod p^{N_1(F)}, \dots, f_m(x_1, \dots, x_n) \bmod p^{N_1(F)})$$

$$C(x_1, \dots, x_n) = F(x_1, \dots, x_n) - P(x_1, \dots, x_n).$$

For $\ell \geqslant N_1(F)$ and all $s_1, \dots, s_n \in \mathbb{Z}_p$, Definition 3.22 implies that

$$F(x_1 + s_1 p^\ell, \dots, x_n + s_n p^\ell) \equiv F(x_1, \dots, x_n) \pmod{p^\ell} \tag{5.7}$$

since $F_1'(x_1, \dots, x_n)$ is a matrix over $\mathbb{Z}/p\mathbb{Z}$, and consequently,

$$(s_1 p^\ell, \dots, s_n p^\ell) F_1'(x_1, \dots, x_n) \equiv (0, .., 0) \pmod{p^\ell}.$$

In particular, (5.7) implies that F is locally 1-Lipschitz. This in turn means that for $i \geqslant N_1(F)$, the function $\delta_i(f_j(x_1, \dots, x_n))$ depends only on

$$\delta_0(x_1), \dots, \delta_0(x_n), \dots, \delta_i(x_1), \dots, \delta_i(x_n);$$

i.e., the function $\delta_i(f_j(x_1, \dots, x_n))$ is a step function of order not greater than $i + 1$. Hence, C is 1-Lipschitz.

On the other hand, (5.7) implies that if $i < N_1(F)$, then $\delta_i(f_j(x_1, \dots, x_n))$ does not depend on $\delta_r(x_t)$ for $r = N_1(F), N_1(F) + 1, \dots$ and $t = 1, 2, \dots, n$; that is, for all $i = 0, 1, \dots, N_1(F) - 1$ and all $j = 1, 2, \dots, m$, the function $\delta_i(f_j(x_1, \dots, x_n))$ is a step function of order not greater than $N_1(F)$. Hence, the function $P(x_1, \dots, x_n)$ is a step function of order not greater than $N_1(F)$ since

$$f_j(x_1, \dots, x_n) \bmod p^{N_1(F)} = \sum_{i=0}^{N_1(F)-1} \delta_i(f_j(x_1, \dots, x_n)) p^i$$

for $j = 1, 2, \ldots, m$. Thus, $P(x_1, \ldots, x_n)$ is a pseudo-constant, whence has zero derivatives. We conclude finally that the function $C = F - P$ is uniformly differentiable modulo p and that the corresponding partial derivatives of C and F modulo p pairwise coincide. □

Note 5.6 The following can be easily deduced from the proof of Theorem 5.5:

Every locally 1-Lipschitz function is a sum of a 1-Lipschitz function and of a step function of order K, for some $K \in \mathbb{N}_0$, and vice versa, any such sum is locally 1-Lipschitz.

This is so since the congruence (5.7) from the proof of Theorem 5.5 is equivalent to the local 1-Lipschitzness of F. Moreover, K is equal to N from the statement of Note 5.2: Actually from the proof of Theorem 5.5, as well as from the proof of Proposition 5.1, it can be easily deduced that the following claim is true:

The function $f: \mathbb{Z}_p^n \to \mathbb{Z}_p$ is locally 1-Lipschitz if and only if there exists $N \in \mathbb{N}_0$ such that $f(x_1, \ldots, x_n) = g(x_1, \ldots, x_n) + c(x_1, \ldots, x_n)$, where $c: \mathbb{Z}_p^n \to \mathbb{Z}_p$ is a 1-Lipschitz function (which is an identity modulo p^N) and $g: \mathbb{Z}_p^n \to \{0, 1, \ldots, p^N - 1\}$ is a step function of order not greater than N.

Indeed, from the proof of Theorem 5.5, as well as from the proof of Proposition 5.1, it follows that $g(x_1, \ldots, x_n) = f(x_1, \ldots, x_n) \bmod p^N$ and $c(x_1, \ldots, x_n) = f(x_1, \ldots, x_n) - g(x_1, \ldots, x_n)$. Thus, the most essential component of any locally 1-Lipschitz function is a 1-Lipschitz function: for instance, the function f is differentiable if and only if its 1-Lipschitz summand c is differentiable since every step function is differentiable everywhere on $\mathbb{Z}_p$ and its derivative is 0. By the same reason, differentiable modulo p^k locally 1-Lipschitz functions have integer-valued derivatives modulo p^k; the derivatives are the ones of the 1-Lipschitz function c. So in the sequel, we focus on 1-Lipschitz functions and make remarks about locally 1-Lipschitz ones whenever it is reasonable.

5.2.1 Differentiability Modulo p

In this subsection, we state and prove a criterion of differentiability modulo p of a univariate 1-Lipschitz function and find a formula for its derivative modulo p following [9].

Theorem 5.7 *A 1-Lipschitz function $f : \mathbb{Z}_p \to \mathbb{Z}_p$ is differentiable modulo p at the point $u \in \mathbb{Z}_p$ if and only if*

$$\frac{\Delta^i f(u)}{i} \equiv 0 \pmod{p}$$

for all sufficiently large i. If the function f is differentiable at $u \in \mathbb{Z}_p$, then the derivative $f_1'(u)$ modulo p of the function f at the point u is

$$f_1'(u) \equiv \sum_{i=1}^{\infty}(-1)^{i-1}\frac{\Delta^i f(u)}{i} \equiv \sum_{t=0}^{\infty}\sum_{k=1}^{p-1}(-1)^{k-1}\frac{\Delta^{kp^t} f(u)}{kp^t} \pmod{p}. \tag{5.8}$$

Note that the series from the statement of the second claim of Theorem 5.7 is actually finite. To prove the theorem, we need some technical lemmas.

Lemma 5.8 *Let $f : \mathbb{Z}_p \to \mathbb{Z}_p$ be a 1-Lipschitz function, let $u \in \mathbb{Z}_p$, and let the base-p expansion of i contain more than one nonzero digit* (i.e., $i \neq p^\alpha l$ for $\alpha \in \{0, 1, 2, \ldots\}$, $l \in \{1, 2, \ldots, p-1\}$). *Then $\frac{1}{i}\Delta^i f(u) \equiv 0 \pmod{p}$.*

Proof of Lemma 5.8 Since

$$\frac{1}{i}\Delta^i f(u) = \sum_{j=1}^{i}(-1)^{i+j}\binom{i-1}{j-1}\frac{1}{j}(f(u+j) - f(u)),$$

(see (5.6) and (5.4) of Proposition 5.4) it is sufficient to show that

$$S(i) = \sum_{j=1}^{\infty}(-1)^j\binom{i-1}{j-1}\frac{1}{j}(f(u+j) - f(u)) \equiv 0 \pmod{p}$$

whenever $i \neq lp^\alpha$, where $l \in \{1, 2, \ldots, p-1\}$ and $\alpha \in \mathbb{N}_0$. Note that all fractions $\frac{1}{j}(f(u+j) - f(u))$ are p-adic integers since f is 1-Lipschitz (cf., Proposition 5.4).

For $j \in \mathbb{N}$, represent $j = p^r l + p^{r+1} t$ where $r = \operatorname{ord}_p j$, $l \in \{1, 2, \ldots, p-1\}$, $t \in \mathbb{N}_0$. We have then

$$S(i) = \sum_{r=0}^{\infty}\sum_{l=1}^{p-1}\sum_{t=0}^{\infty}(-1)^{p^r l + p^{r+1} t}\binom{i-1}{p^r l + p^{r+1} t - 1}\frac{f(u + p^r l + p^{r+1} t) - f(u)}{p^r l + p^{r+1} t}. \tag{5.9}$$

The 1-Lipschitzness of f implies that

$$f(u + p^r l + p^{r+1} t) = f(u + p^r l) + p^{r+1}\xi$$

for a suitable $\xi \in \mathbb{Z}_p$; hence,

$$\frac{f(u+p^r l+p^{r+1}t)-f(u)}{p^r l+p^{r+1}t} \equiv \frac{f(u+p^r l)-f(u)}{p^r l+p^{r+1}t} \pmod p \tag{5.10}$$

since $l+pt$ is a unit in $\mathbb{Z}_p$. Therefore,

$$\frac{f(u+p^r l)-f(u)}{p^r l+p^{r+1}t} \equiv \frac{f(u+p^r l)-f(u)}{p^r l} \pmod p \tag{5.11}$$

since $(l+pt)^{-1} \equiv l^{-1} \pmod p$. Now from (5.9) in view of (5.10) and of (5.11), we conclude that

$$S(i) \equiv \sum_{r=0}^{\infty}\sum_{l=1}^{p-1} \frac{f(u+p^r l)-f(u)}{p^r l} \sum_{t=0}^{\infty}(-1)^{l+t}\binom{i-1}{p^r l+p^{r+1}t-1} \pmod p, \tag{5.12}$$

since $(-1)^p \equiv -1 \pmod p$ for every prime p. Denote

$$\sigma_r(i) = \sum_{t=0}^{\infty}(-1)^{l+t}\binom{i-1}{p^r l+p^{r+1}t-1}. \tag{5.13}$$

Note that if s is a p-adic integer and $\operatorname{ord}_p s = k$, then the j-th digit $\delta_j(s-1)$ of the base-p expansion of $s-1$ is $p-1$ for $j<k$. With this in mind, we consider cases $\operatorname{ord}_p i < r$, $\operatorname{ord}_p i > r$, and $\operatorname{ord}_p i = r$ separately.

Case 1: $\operatorname{ord}_p i < r$. The above note in view of Lucas' Theorem 1.2 implies that

$$\binom{i-1}{p^r l+p^{r+1}t-1} \equiv 0 \pmod p$$

whenever $\operatorname{ord}_p i < r$ and, consequently, that $\sigma_r(i) \equiv 0 \pmod p$ in this case.

Case 2: $\operatorname{ord}_p i > r$. In this case, Lucas' Theorem 1.2 implies that

$$\binom{i-1}{p^r l+p^{r+1}t-1} \equiv \binom{p-1}{l-1}\binom{\nu(i,r)}{t} \pmod p, \tag{5.14}$$

where $\nu(i,r) = \lfloor \frac{i-1}{p^{r+1}} \rfloor$, the integral part of $\frac{i-1}{p^{r+1}}$. Now, since

$$\binom{p-1}{l-1} \equiv (-1)^{l-1} \pmod p,$$

combining (5.14) and(5.13), we conclude that

$$\sigma_r(i) \equiv -\sum_{t=0}^{\infty}(-1)^t\binom{\nu(i,r)}{t} \pmod{p}. \tag{5.15}$$

Further, since

$$\sum_{\ell=0}^{\infty}(-1)^\ell\binom{m}{\ell} = \begin{cases}1, & \text{if } m = 0;\\ 0, & \text{otherwise,}\end{cases}$$

the right-hand part of (5.15) is zero modulo p if $\nu(i,r) \neq 0$, that is, when $i > p^{r+1}$. However, we are considering the case $\operatorname{ord}_p i > r$; thus, since the conditions $\operatorname{ord}_p i > r$ and $i \leqslant p^{r+1}$ hold simultaneously only when $i = p^{r+1}$, the condition $\sigma_r(i) \not\equiv 0 \pmod{p}$ necessarily implies that $i = p^{r+1}$ in the case under consideration.

Case 3: $\operatorname{ord}_p i = r$. In a manner similar to the one of Case 2, we prove that

$$\sigma_r(i) \equiv \sum_{t=0}^{\infty}(-1)^{l+t}\binom{\delta_r(i)-1}{l-1}\binom{\nu(i,r)}{t} \pmod{p},$$

and that the sum in the right-hand side of this congruence may not vanish modulo p only if the two conditions $\delta_r(i) \geqslant l$ and $\nu(i,r) = 0$ hold simultaneously. But these two conditions hold simultaneously only when $i = p^r\delta_r(i)$. This in view of (5.12) and (5.13) finishes the proof of Lemma 5.8. □

Lemma 5.9 *Let $f\colon \mathbb{Z}_p \to \mathbb{Z}_p$ be a 1-Lipschitz function, and let $u, h \in \mathbb{Z}_p$. Then the following congruence holds:*

$$f(u+h) \equiv f(u) + h\left(\tilde{f}_m(u) + \sum_{i=2}^{p-1}\frac{\Delta^{ip^m}f(u)}{ip^m}\binom{h-1}{ip^m-1}\right) \pmod{p^{m+1}},$$

where $m = \operatorname{ord}_p h$ and

$$\tilde{f}_m(u) \equiv \sum_{t=0}^{m-1}\sum_{l=1}^{p-1}(-1)^{l-1}\frac{\Delta^{lp^t}f(u)}{lp^t} + \frac{\Delta^{p^m}f(u)}{p^m} \pmod{p}.$$

In particular, if $p = 2$, then

$$f(u+h) \equiv f(u) + h\sum_{i=0}^{m}\frac{\Delta^{2^i}f(u)}{2^i} \pmod{2^{m+1}}.$$

Proof of Lemma 5.9 In view of the 1-Lipschitzness of f, it is sufficient to prove the lemma under assumption that $h = p^m\chi$, $\chi \in \{1, 2, \dots, p-1\}$. By Theorem 1.5, we see that

$$f(u + p^m\chi) = \sum_{i=0}^{p^m\chi} \binom{p^m\chi}{i} \Delta^i f(u);$$

thus,

$$f(u + p^m\chi) = f(u) + p^m\chi \sum_{i=1}^{p^m\chi} \binom{p^m\chi - 1}{i-1} \frac{\Delta^i f(u)}{i}$$

since

$$\ell\binom{\ell-1}{j-1} = j\binom{\ell}{j}.$$

Now, Lemma 5.8 implies that

$$f(u + p^m\chi) \equiv f(u) + p^m\chi\Bigg(\sum_{t=0}^{m-1}\sum_{l=1}^{p-1} \binom{p^m\chi - 1}{p^t l - 1} \frac{\Delta^{lp^t} f(u)}{lp^t} + \\ + \sum_{j=1}^{\chi} \binom{p^m\chi - 1}{p^m j - 1} \frac{\Delta^{jp^m} f(u)}{jp^m}\Bigg) \pmod{p^{m+1}}.$$

From here, combining together the congruence

$$\binom{\chi - 1}{j-1} \equiv \binom{p^m\chi - 1}{p^m j - 1} \pmod{p}, \tag{5.16}$$

which follows immediately from Lucas' Theorem 1.2, with an obvious congruence

$$\binom{p-1}{k} \equiv (-1)^k \pmod{p},$$

we deduce that

$$f(u + p^m\chi) \equiv f(u) + p^m\chi\Bigg(\sum_{t=0}^{m-1}\sum_{l=1}^{p-1} (-1)^{l-1} \frac{\Delta^{lp^t} f(u)}{lp^t} + \\ + \sum_{j=1}^{\chi} \binom{\chi - 1}{j-1} \frac{\Delta^{jp^m} f(u)}{jp^m}\Bigg) \pmod{p^{m+1}}.$$

The latter congruence implies that

$$f(u+p^m\chi) \equiv f(u) + p^m\chi\left(\tilde{f}_m(u) + \sum_{j=2}^{p-1}\binom{\chi-1}{j-1}\frac{\Delta^{jp^m} f(u)}{jp^m}\right) \pmod{p^{m+1}},$$

since $\chi \in \{1, 2, \ldots, p-1\}$. This in view of (5.16) proves Lemma 5.9. □

Proof of Theorem 5.7 If $\frac{\Delta^i f(u)}{i} \equiv 0 \pmod p$ for all $i \geqslant N$, then in view of Lemma 5.9, the following congruences hold for all sufficiently small $h \in \mathbb{Z}_p$ (i.e., for all h with $|h|_p = p^{-m}$, where m is sufficiently large):

$$f(u+h) \equiv f(u) + h\tilde{f}_m(u) \pmod{p^{m+1}},$$
$$\tilde{f}_m(u) \equiv \tilde{f}_{m+1}(u) \equiv \ldots \pmod{p},$$

Consequently, f is differentiable modulo p at the point $u \in \mathbb{Z}_p$.

Vice versa, let the function f be differentiable modulo p at the point u, i.e., let there exist $N \in \mathbb{N}$ and $c \in \mathbb{Q}_p$ such that

$$f(u+h) \equiv f(u) + hc \pmod{p^{m+1}}, \tag{5.17}$$

where $|h|_p = p^{-m}$, $m \geqslant N$. From (5.17) in view of Lemma 5.9, we deduce that

$$\tilde{f}_m(u) + \sum_{j=2}^{p-1}\binom{h-1}{jp^m-1}\frac{\Delta^{jp^m} f(u)}{jp^m} \equiv c \pmod{p} \tag{5.18}$$

for all $m \geqslant N$. In the case $p = 2$, the sum in the left-hand side of congruence (5.18) vanishes, so suppose for a moment that $p \neq 2$. Then, according to Lucas' Theorem 1.2, we have that

$$\sum_{j=2}^{p-1}\binom{h-1}{jp^m-1}\frac{\Delta^{jp^m} f(u)}{jp^m} \equiv \sum_{j=2}^{p-1}\binom{h_m-1}{j-1}\frac{\Delta^{jp^m} f(u)}{jp^m} \pmod{p},$$

where $h_m = \delta_m(h)$, the m-th p-adic digit of h. So in view of (5.18), the function $\Psi_u(h_m)$ defined by the equation

$$\Psi_u(h_m) = \delta_0\left(\sum_{j=2}^{p-1}\binom{h_m-1}{j-1}\frac{\Delta^{jp^m} f(u)}{jp^m}\right)$$

is a constant whenever $|h|_p = p^{-m}, m \geqslant N$. In particular, $\Psi_u(h_m) = \Psi_u(1) = 0$, and this implies that for all $m \geqslant N$, the following system of congruences modulo p holds:

$$\sum_{j=2}^{p-1} \binom{k-1}{j-1} \frac{\Delta^{jp^m} f(u)}{jp^m} \equiv 0 \pmod{p}, \quad (k = 2, 3, \ldots, p-1). \tag{5.19}$$

System (5.19) of congruences is triangular, so necessarily

$$\frac{\Delta^{jp^m} f(u)}{jp^m} \equiv 0 \pmod{p}, \quad (j = 2, 3, \ldots, p-1) \tag{5.20}$$

for all $m \geqslant N$. Now from (5.18) in view of (5.20) and of Lemma 5.9, we deduce that for all prime p, the following congruence holds

$$\sum_{t=0}^{N-1} \sum_{l=1}^{p-1} (-1)^{l-1} \frac{\Delta^{lp^t} f(u)}{lp^t} + \sum_{s=N}^{m} \frac{\Delta^{p^s} f(u)}{p^s} \equiv c \pmod{p}, \tag{5.21}$$

where c does not depend on m. Hence,

$$\frac{\Delta^{p^s} f(u)}{p^s} \equiv 0 \pmod{p} \tag{5.22}$$

for all $s \geqslant N + 1$. Finally, combining (5.20) and (5.22) with Lemma 5.8, we obtain that

$$\frac{\Delta^i f(u)}{i} \equiv 0 \pmod{p}$$

for all $i \geqslant p^N + 1$. The second claim of Theorem 5.7 follows from (5.21) in view of Lemma 5.8 since $c \equiv f_1'(u) \pmod{p}$ (see (5.17)). □

It is worth comparing differentiability and differentiability modulo p^k yet one more time. The following theorem concerning differentiability of a function $f : \mathbb{Z}_p \to \mathbb{Q}_p$ at the point $u \in \mathbb{Z}_p$ is known (see [145, Chapter 13, Theorem 1])):

Theorem 5.10 *The function $f : \mathbb{Z}_p \to \mathbb{Q}_p$ is differentiable at the point $u \in \mathbb{Z}_p$ if and only if*

$$\overset{p}{\lim_{i \to \infty}} \frac{\Delta^i f(u)}{i} = 0.$$

If the function f is differentiable at the point u, then the following series converges to the derivative $f'(u)$:

$$f'(u) = \sum_{i=1}^{\infty} (-1)^{i-1} \frac{\Delta^i f(u)}{i}.$$

Comparing Theorems 5.7 and 5.10, it is natural to ask whether a similar criterion holds for differentiability modulo p^k if $k > 1$. We may conjecture that such a criterion is true, but currently, we are able to prove sufficiency only:

Proposition 5.11 *Given a 1-Lipschitz function $f: \mathbb{Z}_p \to \mathbb{Z}_p$, if $\frac{\Delta^i f(u)}{i} \equiv 0 \pmod{p^k}$ for all sufficiently large i, then the function f is differentiable modulo p^k at the point $u \in \mathbb{Z}_p$, and*

$$f_k'(u) \equiv \sum_{i=1}^{\infty} (-1)^{i-1} \frac{\Delta^i f(u)}{i} \pmod{p^k}$$

Proof Let $N \in \mathbb{N}_0$ be such that $\frac{\Delta^i f(u)}{i} \equiv 0 \pmod{p^k}$ for all $i > N$. Take $h \in \mathbb{N}$; then by Theorem 1.5, we see that

$$f(u+h) = \sum_{i=0}^{h} \binom{h}{i} \Delta^i f(u);$$

thus,

$$f(u+h) = f(u) + h \sum_{i=1}^{h} \binom{h-1}{i-1} \frac{\Delta^i f(u)}{i}$$

since

$$h \binom{h-1}{j-1} = j \binom{h}{j}.$$

Therefore,

$$\frac{f(u+h) - f(u)}{h} \equiv \sum_{i=1}^{N} \binom{h-1}{i-1} \frac{\Delta^i f(u)}{i} \pmod{p^k}. \tag{5.23}$$

As $\binom{z}{j} = \frac{z(z-1)\cdots(z-j+1)}{j!}$ is a continuous function of $z \in \mathbb{Z}_p$, we see that $\overset{p}{\lim_{z \to 0}} \binom{z-1}{j} = (-1)^j$, and thus,

$$\binom{h-1}{j} \equiv (-1)^j \pmod{p^k}$$

for all h such that $\mathrm{ord}_p h$ is sufficiently large, say, greater than $M \in \mathbb{N}$. Consequently, in view of (5.23), we have for all these h that

$$f(u+h) \equiv f(u) + h \sum_{i=1}^{\infty} (-1)^{i-1} \frac{\Delta^i f(u)}{i} \pmod{p^{k+\mathrm{ord}_p h}}. \tag{5.24}$$

The series from the right-hand side of (5.24) actually is finite; due to the 1-Lipschitzness of f, the sum is a p-adic integer by Proposition 5.4; the latter congruence holds for all $h \in \mathbb{Z}_p$ such that $\mathrm{ord}_p h > M$ since representing $h = p^{\mathrm{ord}_p h}\tilde{h} + p^{k+\mathrm{ord}_p h}\breve{h}$, where $\tilde{h} \in \{1, 2, \ldots, p^k - 1\}$, $\breve{h} \in \mathbb{Z}_p$, we have that $f(u+h) \equiv f(u+p^{\mathrm{ord}_p h}\tilde{h}) \pmod{p^{k+\mathrm{ord}_p h}}$. Therefore, (5.24) holds for all $h \in \mathbb{Z}_p$ such that $\mathrm{ord}_p h > M$. This proves Proposition 5.11 in view of Definition 3.22. □

5.3 Mahler Expansion of 1-Lipschitz Functions

In this section, we characterize 1-Lipschitz functions in terms of Mahler expansions following [9]. Recall that for $\alpha \in \mathbb{N}$, the number $\lfloor \log_p \alpha \rfloor$ is merely a reduced by 1 the number of digits in the base-p expansion of α (see (4.12)).

Theorem 5.12 *The function $f : \mathbb{Z}_p^n \to \mathbb{Z}_p$ represented by Mahler expansion* (4.7) *is 1-Lipschitz if and only if*

$$|a_{i_1,\ldots i_n}|_p \leqslant p^{-\mu(i_1,\ldots,i_n)},$$

where $\mu(i_1, \ldots, i_n) = \max\{\lfloor \log_p i_k \rfloor : k = 1, 2, \ldots, n\}$.

In particular, a univariate function $f : \mathbb{Z}_p \to \mathbb{Z}_p$ represented by Mahler expansion (4.2) *is 1-Lipschitz if and only if*

$$|a_i|_p \leqslant p^{-\lfloor \log_p i \rfloor}$$

for all $i = 1, 2, \ldots$. *In other words, the function* $f : \mathbb{Z}_p \to \mathbb{Z}_p$ *is 1-Lipschitz if and only if it can be represented as*

$$f(x) = \sum_{i=0}^{\infty} p^{\lfloor \log_p i \rfloor} c_i \binom{x}{i}, \tag{5.25}$$

for suitable $c_i \in \mathbb{Z}_p$; $i = 0, 1, 2, \ldots$.

Proof of Theorem 5.12 Induction on n. Let $n = 1$. Then

$$f(x) = \sum_{i=0}^{\infty} a_i \binom{x}{i},$$

where $a_i \in \mathbb{Z}_p$ (see (4.2)). As f is 1-Lipschitz, thus continuous, this series converges uniformly on $\mathbb{Z}_p$ to f (cf., Sect. 4.2).

According to Proposition 5.4, the function f is 1-Lipschitz if and only if $\frac{\Delta^i f(x)}{i}$ is a p-adic integer for all $x \in \mathbb{Z}_p$, $i = 1, 2, \ldots$. Consequently,

$$\frac{\Delta^i f(x)}{i} = \frac{1}{i} \sum_{j=i}^{\infty} a_j \binom{x}{j-i} \tag{5.26}$$

in view of (1.2). By (5.26), $\frac{\Delta^i f(x)}{i}$ is a p-adic integer if and only if

$$\sum_{j=i}^{\infty} a_j \binom{x}{j-i}$$

is an identity modulo $p^{\operatorname{ord}_p i}$. Proposition 4.7 implies now that $\frac{\Delta^i f(x)}{i}$ is a p-adic integer if and only if the following congruences hold simultaneously for all $j \geqslant i$:

$$a_j \equiv 0 \pmod{p^{\operatorname{ord}_p i}} \tag{5.27}$$

Thus, f is 1-Lipschitz if and only if (5.27) hold simultaneously for all $i = 1, 2, \ldots$ and all $j \geqslant i$. This means (since $\lfloor \log_p j \rfloor = \max\{\operatorname{ord}_p i : i = 1, 2, 3, \ldots, j\}$) that the following congruences hold simultaneously:

$$a_j \equiv 0 \pmod{p^{\lfloor \log_p j \rfloor}}, \ (j = 1, 2, 3, \ldots).$$

This proves Theorem 5.12 for $n = 1$.

Now let the statement of the theorem be true for all r-variate functions that satisfy conditions of the theorem, $r < n$. In view (4.7), we can write

$$f(x_1, \ldots, x_n) = \sum_{j=0}^{\infty} g_j(x_1, \ldots, x_{n-1}) \binom{x_n}{j},$$

where all functions g_j are:

$$g_j(x_1, \ldots, x_{n-1}) = \sum_{(i_1,\ldots,i_{n-1}) \in \mathbb{N}_0^{n-1}} a_{i_1,\ldots,i_{n-1},j} \binom{x_1}{i_1}\binom{x_2}{i_2} \cdots \binom{x_{n-1}}{i_{n-1}},$$

As f is 1-Lipschitz, thus uniformly continuous on $\mathbb{Z}_p^n$, all the functions g_j are uniformly continuous on $\mathbb{Z}_p^{n-1}$, for all $j = 1, 2, \ldots$ (cf. Sect. 4.2).

According to Proposition 5.4, the function $f(x_1, \ldots, x_n)$ is 1-Lipschitz if and only if all fractions $\frac{1}{i}\Delta_s^i f(x_1, \ldots, x_n)$ are p-adic integers, for all $i = 1, 2, \ldots$, all $s = 1, 2, \ldots, n$, and all $x_1, \ldots, x_n \in \mathbb{Z}_p$. Using an argument similar to that of the case $n = 1$, we conclude that

$$\frac{1}{i}\Delta_s^i f(x_1, \ldots, x_n) = \begin{cases} \sum_{j=i}^{\infty} \frac{1}{i} g_j(x_1, \ldots, x_{n-1})\binom{x_n}{j-i}, \text{ if} s = n; \\ \sum_{j=0}^{\infty} \frac{1}{i}\Delta_s^i g_j(x_1, \ldots, x_{n-1})\binom{x_n}{j}, \text{ otherwise.} \end{cases} \tag{5.28}$$

If $s \neq n$, all functions $\frac{1}{i}\Delta_s^i f(x_1, \ldots, x_n)$, $i = 1, 2, \ldots$, are simultaneously integer-valued if and only if all functions $\frac{1}{i}\Delta_s^i g_j(x_1, \ldots, x_{n-1})$ are simultaneously integer-valued, for all $j = 0, 1, 2, \ldots$ and all $i = 1, 2, \ldots$. This in force of Proposition 5.4 implies that every function $g_j(x_1, \ldots, x_n)$, $j = 0, 1, 2, \ldots$, is 1-Lipschitz. By induction hypothesis, the latter holds if and only if the following inequalities hold simultaneously:

$$|a_{i_1,\ldots,i_{n-1},j}|_p \leqslant p^{-\mu(i_1,\ldots,i_{n-1})}, \quad (j, i_1, \ldots, i_n \in \mathbb{N}_0). \tag{5.29}$$

If $s = n$, then by an argument similar to that of the case $n = 1$, we deduce from (5.28) that all the functions $\frac{1}{i}\Delta_n^i f(x_1, \ldots, x_n)$, $i = 1, 2, \ldots$, are integer-valued if and only if the following inequalities hold simultaneously for all $j = 1, 2, \ldots$ and all $x_1, \ldots, x_{n-1} \in \mathbb{Z}_p$:

$$|g_j(x_1, \ldots, x_{n-1})|_p \leqslant p^{-\lfloor \log_p j \rfloor}. \tag{5.30}$$

But these conditions imply that every function $g_j(x_1, \ldots, x_{n-1})$ is an identity modulo $p^{\lfloor \log_p j \rfloor}$; whence, in view of Proposition 4.7, the following conditions hold simultaneously for all $i_1, \ldots, i_{n-1} \in \mathbb{N}_0$ and all $j \in \mathbb{N}$:

$$|a_{i_1,\ldots i_{n-1},j}|_p \leqslant p^{-\lfloor \log_p j \rfloor}. \tag{5.31}$$

Now by combining (5.29) with (5.31), we finish the proof of Theorem 5.12. □

Note that from this theorem, by using Chinese Reminder Theorem 1.1, it can be deduced complete description of an important class of functions, which we consider later in Part II of the book, the so-called *universal functions* ([32]) or *pseudo-polynomials* ([90]). These functions are maps $g\colon \mathbb{N}_0 \to \mathbb{Z}$ such that $g(m) \equiv g(n) \pmod q$ once $m \equiv n \pmod q$, where $m, n, q \in \mathbb{N}_0$, $q > 1$. Obviously, these functions constitute the class of all maps $\mathbb{N}_0 \to \mathbb{Z}$, which are p-adic 1-Lipschitz for every prime p.

Corollary 5.13 *A map $g\colon \mathbb{N}_0 \to \mathbb{Z}$ is a universal function if and only if g can be represented as*

$$g(z) = c_0 + \sum_{i=1}^{\infty} c_i \operatorname{lcm}\{1, 2, \dots, i\} \binom{z}{i}$$

where $c_i \in \mathbb{Z}$, $\operatorname{lcm}\{1, 2, \dots, i\}$ is the least common multiple of the numbers $1, 2, \dots, i$.

It is worth noting also that $\operatorname{lcm}\{1, 2, \dots, i\} = e^{\psi(i)}$, where

$$\psi(i) = \sum_{q \leqslant i,\ q \text{ prime}} \lfloor \log_q i \rfloor \ln q$$

is the second Chebyshev function, $i = 1, 2, \dots$ (recall that $\psi(i) = i + o(i)$).

We omit the proof of Corollary 5.13 since other proofs (though somewhat different from the one based on Theorem 5.12) can be found in [32, 90] but note that there are pseudo-polynomials (i.e., universal functions), which are not polynomials.

The class of pseudo-polynomials is wide; it is a subject of study by number theorists mostly around Ruzsa's conjecture, which is about sufficient conditions when a pseudo-polynomial is a polynomial (see, e.g., [46]). Classical examples of pseudo-polynomials that are not polynomials are $\sum_{i=0}^{\infty} x^{\underline{i}}$ and $\sum_{i=0}^{\infty} (-1)^i x^{\underline{i}}$, where $x^{\underline{i}}$ is the i-th falling factorial power: $x^{\underline{i}} = x(x-1)\cdots(x-i+1)$ if $i > 0$ and $x^{\underline{0}} = 1$. The p-adic 1-Lipschitz functions, which can be represented via falling factorial power series, constitute an important class that will be investigated in detail further (see Sect. 6.3).

5.4 van der Put Expansion of 1-Lipschitz Functions

In this section, we prove, following [27], a criterion of 1-Lipschitzness of a map $\mathbb{Z}_p \to \mathbb{Z}_p$, which is defined by convergent van der Put series.

Theorem 5.14 *Given van der Put expansion of the function* $f: \mathbb{Z}_p \to \mathbb{Z}_p$ (either via Mahler base (4.13) or Schikhof base (4.14); cf., (4.10))

$$f(x) = \sum_{m=0}^{\infty} B_m \chi(m, x),$$

the function f *is 1-Lipschitz if and only if* $|B_m|_p \leqslant p^{-\lfloor \log_p m \rfloor}$ *for all* $m = 0, 1, 2, \ldots$.

To put this in other words, the function $f: \mathbb{Z}_p \to \mathbb{Z}_p$ *is 1-Lipschitz if and only if it can be represented as*

$$f(x) = \sum_{m=0}^{\infty} p^{\lfloor \log_p m \rfloor} b_m \chi(m, x), \tag{5.32}$$

for suitable $b_m \in \mathbb{Z}_p$; $m = 0, 1, 2, \ldots$.

Proof of Theorem 5.14 To prove necessity of the conditions, take $m \in \mathbb{N}_0$, and consider the base-p expansion $m = m_0 + \ldots + m_{n-1} p^{n-1}$. Here, $m_j \in \{0, \ldots, p-1\}$, $m_{n-1} \neq 0$, $n = \lfloor \log_p m \rfloor + 1$. As

$$m_0 + \ldots + m_{n-2} p^{n-2} \equiv m_0 + \ldots + m_{n-2} p^{n-2} + m_{n-1} p^{n-1} \pmod{p^{n-1}};$$

then

$$f(m_0 + \ldots + m_{n-2} p^{n-2}) \equiv f(m_0 + \ldots + m_{n-1} p^{n-1}) \pmod{p^{n-1}}$$

by the 1-Lipschitzness of f. The latter congruence, by either of (4.13)–(4.14), implies that $B_m \equiv 0 \pmod{p^{n-1}}$ for $m \geqslant p$; so $|B_m| \leqslant p^{-\lfloor \log_p m \rfloor}$.

To prove sufficiency, we first note that since $|B_m|_p \leqslant p^{-\lfloor \log_p m \rfloor}$, the sequence $B_0, B_1, \ldots$ tends p-adically to 0 as $m \to \infty$, and so the function f is continuous (cf., Sect. 4.3). Hence, while proving that $|f(x) - f(y)|_p \leqslant |x - y|_p$ for $x, y \in \mathbb{Z}_p$, we may assume that $x, y \in \mathbb{N}_0$ as $\mathbb{N}_0$ is dense in $\mathbb{Z}_p$. Moreover, by the same reason, in order to prove that f satisfies the p-adic Lipschitz condition with a constant 1, it suffices to prove only that given $x, y \in \mathbb{N}_0$ such that $|x - y|_p = p^{-n}$, then $f(x) \equiv f(y) \pmod{p^n}$. As $y = x + hp^n$, without loss of generality, we may assume that $h, n \in \mathbb{N}$ and $x \in \{0, 1, \ldots, p^n - 1\}$.

With this in mind, let $h = h_0 + h_1 p + \ldots + h_k p^k$ be the base-p expansion of $h \in \mathbb{N}$; let $n_0 < n_1 < n_2 < \ldots < n_k$ be all indices in the latter base-p expansion such that $h_{n_0}, h_{n_1}, \ldots, h_{n_k}$ are nonzero; so

$$h = h_{n_0} p^{n_0} + h_{n_1} p^{n_1} + \ldots + h_{n_k} p^{n_k}.$$

Now, since $n \geqslant 1$, by either of (4.13)–(4.14), we get that

$$f(x+hp^n) = f(x) + [f(x + p^n h_{n_0} p^{n_0}) - f(x)] +$$

$$+\sum_{j=1}^{k}[f(x+p^n(h_{n_0}p^{n_0}+\cdots+h_{n_j}p^{n_j}))-f(x+p^n(h_{n_0}p^{n_0}+\cdots+h_{n_{j-1}}p^{n_{j-1}}))] =$$

$$= f(x) + \sum_{j=0}^{k} B_{x+h_{n_0}p^{n+n_0}+\cdots+h_{n_j}p^{n+n_j}} \tag{5.33}$$

However, due to our assumption,

$$|B_{x+h_{n_0}p^{n+n_0}+\cdots+h_{n_j}p^{n+n_j}}|_p \leqslant p^{-(n+n_j)};$$

so (5.33) implies that $|f(x+hp^n) - f(x)|_p \leqslant p^{-n}$ by virtue of the strong triangle inequality.

□

Chapter 6
Special Classes of 1-Lipschitz Functions

Abstract In this chapter, we, following Anashin (Discrete Math Appl 12(6):527–590, 2002), introduce and examine three important special classes, $\mathcal{A}$, $\mathcal{B}$, and C, of p-adic 1-Lipschitz functions. These functions play an important role in the p-adic ergodic theory (see Part II). All these functions belong to a wider class, the class of locally analytic functions. The class C is the class of all power series over $\mathbb{Z}_p$, which converge everywhere in $\mathbb{Z}_p$; the class $\mathcal{B}$ is a completion of C with respect to sup-norm.

6.1 Locally Analytic Functions

Recall the definition of p-adic locally analytic functions given in [168, Definition 25.3]):

Definition 6.1 (Locally Analytic Function) A function $f : \mathbb{Z}_p \to \mathbb{Z}_p$ is said to be *locally analytic of order r* if and only if f is analytic on all balls of radii p^{-r}.

The definition can be restated in equivalent form:

Definition 6.2 (Locally Analytic Functions) The function $f : \mathbb{Z}_p \to \mathbb{Z}_p$ is said to be *locally analytic of order r* if and only if

$$f(a+h) = \sum_{i=0}^{\infty} \frac{f^{(i)}(a)}{i!} h^i$$

for all $a \in \mathbb{Z}_p$ whenever $|h|_p \leqslant p^{-r}$.

V. Anashin, *Causality: The p-adic Theory*, STEAM-H: Science, Technology, Engineering, Agriculture, Mathematics & Health,
https://doi.org/10.1007/978-3-031-85818-5_6

Here as usual $f^{(i)}(a)$ stands for the i-th derivative of the function f at the point $a \in \mathbb{Z}_p$. Note that functions that are analytic of order 0 are analytic on $\mathbb{Z}_p$, and vice versa. The following theorem by Y. Amice, (see [4, Ch. III, Sec. 10, Th. 3, Cor. 1(c)]) gives complete characterization of locally analytic functions represented via Mahler series:

Theorem 6.3 (Y. Amice) *The function*

$$f(x) = \sum_{i=0}^{\infty} a_i \binom{x}{i},$$

where $a_i \in \mathbb{Q}_p$, *is locally analytic of order* r *if and only if*

$$\lim_{i\to\infty} \left(\operatorname{ord}_p a_i + \frac{1}{p-1}\left(\operatorname{wt}_p i - \frac{i}{p^r}\right)\right) = +\infty$$

6.2 Class C

According to Theorem 4.4, the power series $\sum_{i=0}^{\infty} c_i x^i$, where $c_i \in \mathbb{Q}_p$ for $i = 0, 1, 2\ldots$, converges everywhere on $\mathbb{Z}_p$ if and only if $\overset{p}{\lim\limits_{i\to\infty}} c_i = 0$; under the latter condition, the series defines a continuous function on $\mathbb{Z}_p$. Of course, the function defined by these series may not be integer-valued, not speaking about being 1-Lipschitz. Consider, however, a special case when all coefficients c_i are p-adic integers. Namely, given the ring $\mathbb{Z}_p[[x]]$ of all formal power series in one variable x over the ring $\mathbb{Z}_p$, let $C(x)$ be the set of all series

$$s(x) = \sum_{i=0}^{\infty} c_i x^i \quad (c_i \in \mathbb{Z}_p,\ i = 0, 1, 2\ldots), \tag{6.1}$$

which converge everywhere on $\mathbb{Z}_p$. In other words, $s(x) \in C(x)$ if and only if $\overset{p}{\lim\limits_{i\to\infty}} c_i = 0$. Once these conditions are satisfied, the series $s(x) \in C(x)$ defines an integer-valued function $s\colon \mathbb{Z}_p \to \mathbb{Z}_p$, which is called a *C-function*; the class of all C-functions we denote via C. This is an exercise to show that s is a 1-Lipschitz function.

Proposition 6.4 *The C-function* $s\colon \mathbb{Z}_p \to \mathbb{Z}_p$ *defined by power series* (6.1) *is uniformly differentiable on* $\mathbb{Z}_p$*; its derivative* $s'(x)$*is integer-valued everywhere on*

$\mathbb{Z}_p$, *and*

$$s'(x) = \sum_{i=1}^{\infty} i c_i x^{i-1}$$

Corollary 6.5 *The class C is closed with respect to derivations; all C-functions are infinitely many times differentiable.*

Proofs of Proposition 6.4 and Corollary 6.5 See [112, Proposition 3.21]. □

Consider now Mahler expansions of C-functions. Let

$$s(x) = \sum_{i=0}^{\infty} s_i \binom{x}{i} \tag{6.2}$$

be Mahler expansion of the function $s \in C$ defined by convergent power series (6.1). The following proposition is true:

Proposition 6.6 *All fractions $\frac{s_i}{i!}$ are p-adic integers, for all $i = 0, 1, 2, \ldots$.*

Proof Indeed, using formulas for Stirling numbers and falling powers (cf., (1.1) from Sect. 1.1.3), we see that

$$s(x) = \sum_{k=0}^{\infty} c_k x^k = \sum_{k=0}^{\infty} c_k \sum_{i=0}^{k} \left\{ {k \atop i} \right\} i! \binom{x}{i} = \sum_{i=0}^{\infty} i! \binom{x}{i} \sum_{k=i}^{\infty} \left\{ {k \atop i} \right\} c_k. \tag{6.3}$$

Further, since all Stirling numbers $\left\{ {k \atop i} \right\}$ are rational integers,

$$\left\| \left\{ {k \atop i} \right\} \right\|_p \leqslant 1;$$

whence, as power series (6.1) is convergent, $\lim^p_{i\to\infty} c_i = 0$, and thus

$$\lim^p_{k\to\infty} \left\{ {k \atop i} \right\} c_k = 0.$$

Consequently, the series

$$\sum_{k=i}^{\infty} \left\{ {k \atop i} \right\} c_k$$

converges to a certain $A_i \in \mathbb{Z}_p$, for all $i = 0, 1, 2, \ldots$. This proves our assertion since

$$s_i = i! \sum_{k=i}^{\infty} \left\{ {k \atop i} \right\} c_k = i! A_i \quad (i = 0, 1, 2, \ldots) \tag{6.4}$$

(see (6.3)). □

We finally summarize:

The class C is the class of all 1-Lipschitz functions $\mathbb{Z}_p \to \mathbb{Z}_p$, which are analytic on the ball $\mathbb{Z}_p$.

Example 6.7 Here are 1-Lipschitz functions $\mathbb{Z}_p \to \mathbb{Z}_p$ that are or are not C-functions (cf., Sect. 4.1.1):

1. If $p \neq 2$, then functions $f(x) = \exp_p px$, $f(x) = \sin_p px$, and $f(x) = \cos_p px$ are C-functions (cf. Example 4.3).
2. If $p = 2$, then functions $f(x) = \exp_p px$, $f(x) = \sin_p px$, and $f(x) = \cos_p px$ are *not* C-functions.
3. Functions $f(x) = \exp_p p^2x$, $f(x) = \sin_p p^2x$, $f(x) = \cos_p p^2x$, and $f(x) = \ln_p(1 + px)$ are C-functions, for any prime p.
4. The function $f(x) = (1 + px)^{-1}$ is a C-function, for any prime p.

6.3 Class $\mathcal{B}$

Proposition 6.6 shows that any C-function can be represented by *falling power series* over $\mathbb{Z}_p$; i.e., via falling powers $x^{\underline{i}}$:

$$s(x) = \sum_{i=0}^{\infty} b_i x^{\underline{i}} \ \ (b_i = \frac{s_i}{i!} \in \mathbb{Z}_p; i = 0, 1, 2, \ldots).$$

Let us consider a wider class $\mathcal{B}(x)$ of falling factorial series with p-adic integer coefficients; that is, $f(x) \in \mathcal{B}(x)$ if and only if $f(x) = \sum_{i=0}^{\infty} b_i x^{\underline{i}}$, $(b_i \in \mathbb{Z}_p)$. In other words,

$$\mathcal{B}(x) = \left\{ \sum_{i=0}^{\infty} a_i \binom{x}{i} : \frac{a_i}{i!} \in \mathbb{Z}_p;\ i = 0, 1, 2, \ldots \right\}. \tag{6.5}$$

By the convergence criterion for Mahler interpolation series (see (4.3)), the series from $\mathcal{B}(x)$ are uniformly convergent on $\mathbb{Z}_p$ and thus define uniformly continuous functions on $\mathbb{Z}_p$; call these functions *$\mathcal{B}$-functions*. Note that as $a_i/i! \in \mathbb{Z}_p$, then $\overset{p}{\lim_{i\to\infty}} a_i = 0$ since $\overset{p}{\lim_{i\to\infty}} i! = 0$.

Proposition 6.8 *Every $\mathcal{B}$-function 1-Lipschitz. Every $\mathcal{B}$-function is uniformly differentiable on $\mathbb{Z}_p$.*

Proof The inequality $\mathrm{wt}_p\, i \leqslant (p-1)(\lfloor \log_p i \rfloor + 1)$ holds for all $i = 1, 2, \ldots$ and all prime p since the function $\mathrm{wt}_p\, n$ attains the maximum on numbers $n \in \{0, 1, \ldots, p^k - 1\}$ if and only if $n = (p-1) + (p-1)p + \cdots + (p-1)p^{k-1}$. Now the 1-Lipschitzness follows from Theorem 5.12 and Lemma 2.1. The uniform differentiability follows from (4.4). □

Denote via $\mathcal{B}$ the class of all functions that are defined by series from $\mathcal{B}(x)$. It turns out that distinct series define distinct functions, so in the sequel, we do not differ series from functions they define.

Proposition 6.9 *Any two distinct series from $\mathcal{B}(x)$ (respectively, from $C(x)$) define two distinct functions on $\mathbb{Z}_p$.*

Proof Indeed, for functions defined by series from $\mathcal{B}(x)$, the assertion follows from the definition of $\mathcal{B}$-functions in view of Proposition 4.7. As for functions defined by series from $C(x)$, we note that the above mentioned interpolation series (6.2) for $s(x) \in C(x)$ defines a function that is identically 0 on $\mathbb{Z}_p$ if and only if all coefficients s_i are 0. Whence, $A_i = 0$ for $i = 0, 1, 2, \ldots$ (see (6.4)). However, $A_i = \sum_{k=i}^{\infty} \left\{ {k \atop i} \right\} c_k$; thus, $c_i = \sum_{k=i}^{\infty} \left[{k \atop i} \right] A_k = 0$, and the assertion follows. □

6.3.1 Stone-Weierstrass Theorem for $\mathcal{B}$-Functions

The class of all 1-Lipschitz functions $\mathbb{Z}_p \to \mathbb{Z}_p$ can be endowed with a supremum norm $\|f\|_p$ in standard way: $\|f\|_p = \sup\{|f(z)|_p : z \in \mathbb{Z}_p\}$. The norm is non-Archimedean and thus defines the metric D_p on this class:

$$D_p(f, g) = \max\{|f(z) - g(z)|_p : z \in \mathbb{Z}_p\}.$$

In other words, we put the distance D_p between 1-Lipschitz functions f and g to be equal to p^{-N} where N is the largest natural integer $N \in \mathbb{N}$ such that the corresponding reduced maps $f \bmod p^N$ and $g \bmod p^N$ of the residue ring $\mathbb{Z}/p^N\mathbb{Z}$

coincide: $f(z) \bmod p^N = g(z) \bmod p^N$ for all $z \in \mathbb{Z}/p^N\mathbb{Z}$; we put $D_p(f, g) = 0$ if no such N exists ($f(z) = g(z)$ for all $z \in \mathbb{Z}_p$ in the latter case).

A $\mathcal{B}$-function is not necessarily analytic on $\mathbb{Z}_p$: Indeed, the function $\sum_{i=0}^{\infty} x^{\underline{i}}$ is a $\mathcal{B}$-function; however, conditions (4.6) are not satisfied (so C is a proper subclass of $\mathcal{B}$). On the other hand, $\mathcal{B}$-functions behave somewhat like real analytic functions: A composition of real analytic functions is again an analytic function; though generally this is not the case for p-adic analytic functions (see [168, Section 41] for details).

Fortunately, for $\mathcal{B}$-functions, the following analog of Stone-Weierstrass theorem holds:

Theorem 6.10 *The class $\mathcal{B}$ is a complete* (with respect to D_p) *metric space. The class $\mathcal{B}$ is closed with respect to additions, multiplications, derivations, and compositions of functions. Polynomials with non-negative rational integer coefficients constitute a dense subset of $\mathcal{B}$.*

Before the proof of Theorem 6.10, it should be noted that although Theorem 6.10 can be considered as an analog of Stone-Weierstrass theorem, it is *not* the p-adic Stone-Weierstrass theorem: The latter theorem yields conditions when a uniform closure of a class of functions defined on a compact set is a class of *all* continuous functions on the set (see, e.g., [168, Appendix A.4]). Furthermore, the p-adic Stone-Weierstrass theorem says nothing special about differentiability of the uniform closure; moreover, in contrast to Theorem 6.10, the uniform closure of polynomials over $\mathbb{Q}_p$ contains not only uniformly differentiable functions as in the case of uniform closure of polynomials over $\mathbb{N}_0$. Theorem 6.10 deals with a uniform closure of a smaller class of functions, the polynomials over $\mathbb{N}_0$ rather than over $\mathbb{Q}_p$; however, the closure is with respect to the same metric as in the p-adic Stone-Weierstrass theorem. From this view, Theorem 6.10 can be considered also as an analog of p-adic Kaplansky's theorem (see, e.g., [168, Theorem 43.3]), however, not for all polynomials over the *field* $\mathbb{Q}_p$ as the p-adic Kaplansky's theorem deals with, but only for polynomials over $\mathbb{N}_0$ or, equivalently, over the *ring* $\mathbb{Z}$. The p-adic Kaplansky's theorem yields that all continuous functions defined on a compact set can be uniformly approximated (with respect to D_p) by polynomials over $\mathbb{Q}_p$, whereas Theorem 6.10 deals with uniform closure of polynomials over $\mathbb{Z}$ (or, which is equivalent, with uniform closure of polynomials over $\mathbb{Z}_p$), and all functions of the closure turn out to be uniformly differentiable, and not only continuous. Finally, we note that classical Weierstrass approximation theorem also only yields that continuous functions defined on a real compact set and valuated in $\mathbb{R}$ can be uniformly approximated by polynomials over $\mathbb{R}$; that is, the limit of uniformly convergent sequence of polynomials over $\mathbb{R}$ (which are differentiable functions) need not be a differentiable function. However, Theorem 6.10 yields that in the case of polynomials over $\mathbb{Z}_p$, the limit is a uniformly differentiable function. All these remarks are to explain why Theorem 6.10 can be deduced neither from p-adic Stone-Weierstrass theorem nor from p-adic Kaplansky's theorem and needs special proof.

Proof of Theorem 6.10 First we prove that $\mathcal{B}$-functions are uniformly differentiable on $\mathbb{Z}_p$ and that $\mathcal{B}$ is closed with respect to derivations: If $f \in \mathcal{B}$, then $f' \in \mathcal{B}$. Recall that a uniformly continuous function $f: \mathbb{Z}_p \to \mathbb{Z}_p$, which is represented by the interpolation series (4.2), is uniformly differentiable on $\mathbb{Z}_p$ if and only if

$$\overset{p}{\lim_{i\to\infty}} \frac{a_{i+n}}{i} = 0$$

holds for all $n \in \mathbb{N}_0$ (cf., (4.4)).

Obviously, $\lim_{i\to\infty}(\mathrm{ord}_p(i+n)! - \mathrm{ord}_p\, i) = +\infty$; so by the above criterion, any $\mathcal{B}$-function f is uniformly differentiable on $\mathbb{Z}_p$, and by (4.5),

$$f'(x) = \sum_{i=1}^{\infty}(-1)^{i+1}\frac{\Delta^i f(x)}{i}.$$

Next we prove that the derivative f' is also a $\mathcal{B}$-function. As $f(x) = \sum_{i=0}^{\infty} a_i\binom{x}{i}$, then by applying the formula $\Delta^i\binom{x}{j} = \binom{x}{i-j}$ and changing the order of summing, we see that

$$f'(x) = \sum_{i=1}^{\infty}(-1)^{i+1}\frac{\Delta^i f(x)}{i} = \sum_{k=0}^{\infty}\binom{x}{k}\sum_{i=1}^{\infty}(-1)^{i+1}\frac{a_{k+i}}{i} \tag{6.6}$$

since

$$\frac{\Delta^i f(x)}{i} = \frac{1}{i}\sum_{j=i}^{\infty} a_j\binom{x}{j-i}.$$

Since (4.4) holds, the series $\sum_{i=1}^{\infty}(-1)^{i+1}\frac{a_{k+i}}{i}$ converges for every $k \in \mathbb{N}_0$ to some $S_k \in \mathbb{Q}_p$. Moreover, $\mathrm{ord}_p\frac{a_{k+i}}{i} = \mathrm{ord}_p\, a_{k+i} - \mathrm{ord}_p\, i \geqslant \mathrm{ord}_p(k+i)! - \lfloor\log_p i\rfloor = \frac{1}{p-1}(i+k-\mathrm{wt}_p(i+k)) - \lfloor\log_p i\rfloor = \frac{1}{p-1}(i-\mathrm{wt}_p\, i) - \lfloor\log_p i\rfloor + \frac{1}{p-1}(k-\mathrm{wt}_p\, k) + \frac{1}{p-1}(\mathrm{wt}_p\, k - \mathrm{wt}_p(i+k) + \mathrm{wt}_p\, i) \geqslant \frac{1}{p-1}(k-\mathrm{wt}_p\, k) = \mathrm{ord}_p\, k!$. The latter inequality holds since $\frac{1}{p-1}(i-\mathrm{wt}_p\, i) \geqslant \lfloor\log_p i\rfloor$ and $\frac{1}{p-1}(\mathrm{wt}_p\, k - \mathrm{wt}_p(i+k) + \mathrm{wt}_p\, i) = \mathrm{ord}_p\binom{i+k}{i} \geqslant 0$ (see Lemma 2.1 and Corollary 2.2). Thus, $\frac{S_k}{k!} \in \mathbb{Z}_p$ for all $k \in \mathbb{N}_0$; whence f' is a $\mathcal{B}$-function.

Now we prove that $\mathcal{B}$ is a closure (with respect to the metric D_p) of the class of all functions that are polynomials with non-negative rational integer coefficients. Since every polynomial from $\mathbb{Z}_p[x]$ is congruent modulo p^k to some polynomial all whose coefficients are non-negative rational integers from $\mathbb{N}_0$, it suffices to prove that $\mathcal{B}$ is a closure of $\mathbb{Z}_p[x]$ with respect to the metric D_p.

From the definition of the class $\mathcal{B}$, it easily follows that every function $f \in \mathcal{B}$ can be uniformly approximated by polynomials over $\mathbb{Z}_p$: For each $n \in \mathbb{N}$, there exists a polynomial $f_n(x) \in \mathbb{Z}_p[x]$ such that $f(z) \equiv f_n(z) \bmod p^n$ for all $z \in \mathbb{Z}_p$. The series $\sum_{j=0}^{\infty} r_j \binom{x}{j}$ defines a function that is identically 0 modulo p^n if and only if all $r_j \equiv 0 \pmod{p^n}$, by Proposition 4.7. Therefore, in view of Lemma 2.1, we may put $f_n(x) = \sum_{i=0}^{\omega(n)} a_i \binom{x}{i}$, where $\omega(n) = \max\{j \in \mathbb{N}_0 \colon \frac{1}{p-1}(j - \mathrm{wt}_p\, j) < n\}$.

The converse assertion is also true: If a function $f \colon \mathbb{Z}_p \to \mathbb{Z}_p$ can be uniformly approximated by polynomials over $\mathbb{Z}_p$ in the above mentioned meaning, then $f \in \mathcal{B}$. To prove the assertion, assume that $f(z) \equiv f_i(z) \pmod{p^i}$ for all $z \in \mathbb{Z}_p$, where $f_i(x) \in \mathbb{Z}_p[x]$, $i = 1, 2, \dots$. Every polynomial $f_i(x)$ of degree d_i has a unique Mahler expansion (4.2): $f_i(x) = \sum_{j=0}^{d_i} a_{ij} \binom{x}{j}$, where $a_{ij} \in \mathbb{Z}_p$ and $\mathrm{ord}_p\, a_{ij} \geqslant \mathrm{ord}_p(j!)$ in view of (6.4), since $f_i \in \mathcal{C} \subset \mathcal{B}$. Given a function f, every polynomial $f_i(x)$ is uniquely determined up to a summand that is 0 modulo p^i everywhere on $\mathbb{Z}_p$. So we may assume that $d_i = \omega(i)$; then coefficients of the polynomial $f_i(x)$ are determined uniquely up to summands whose p-adic absolute values do not exceed p^{-i}. This implies that $a_{i+1,j} \equiv a_{ij} \pmod{p^i}$; we assume $a_{ij} = 0$ for $j > \omega(i)$. Hence, $\overset{p}{\lim\limits_{i\to\infty}} a_{ij} = a_j \in \mathbb{Z}_p$, and $\frac{a_j}{j!} \in \mathbb{Z}_p$. Consequently, the series $\sum_{i=0}^{\infty} a_i \binom{x}{i}$ defines a function $\tilde{f} \in \mathcal{B}$, which is uniformly continuous on $\mathbb{Z}_p$. The function $\tilde{f}$ is equal to f since $f(z) \equiv f_i(z) \equiv \tilde{f}(z) \pmod{p^i}$ for all $z \in \mathbb{Z}_p$ and all $i = 1, 2, \dots$.

So we have proved that $\mathcal{B}$ is a complete metric space with respect to the metric D_p; from here, it follows that $\mathcal{B}$ is closed with respect to additions, multiplications, and compositions of functions: If $f, g \in \mathcal{B}$, then $f + g, fg, f(g) \in \mathcal{B}$. Indeed, let g be uniformly approximated by the sequence $(g_n)_{n\in\mathbb{N}}$ of polynomials over $\mathbb{Z}_p$; that is, $g_n(z) \equiv g(z) \pmod{p^n}$ for all $z \in \mathbb{Z}_p$. Now the 1-Lipschitzness of the function f implies that $D_p(f(g), f(g_n)) \leqslant p^{-n}$, i.e., that the sequence $(f(g_n))_{n\in\mathbb{N}}$ converges to $f(g)$ with respect to the metric D_p as $n \to \infty$. However, $f(g_n) \in \mathcal{B}$ for every $n = 1, 2, \dots$: If f is uniformly approximated by the sequence $(f_m)_{m\in\mathbb{N}}$ of polynomials over $\mathbb{Z}_p$, then $f_m(g_n(z)) \equiv f(g_n(z)) \pmod{p^m}$ for all $z \in \mathbb{Z}_p$. Hence, the sequence $(f_m(g_n(x)))_{m\in\mathbb{N}}$ of polynomials over $\mathbb{Z}_p$ converges to the function $f(g_n)$ with respect to the metric D_p, and $f_m(g_n) \in \mathcal{B}$ since $f_m(g_n)$ is a polynomial over $\mathbb{Z}_p$, for all $m, n \in \mathbb{N}$. Consequently, $f(g) \in \mathcal{B}$ in view of completeness of $\mathcal{B}$ with respect to D_p. □

6.3.2 *Taylor Theorem for $\mathcal{B}$-Functions*

Although a $\mathcal{B}$-function is not necessarily analytic on $\mathbb{Z}_p$, it turns out that every $\mathcal{B}$-function is analytic on all balls of radii less than 1; that is, all $\mathcal{B}$-functions are locally analytic of order 1 (cf., Definition 6.1). Namely, the following Taylor theorem for $\mathcal{B}$-functions holds:

Theorem 6.11 *For every $\mathcal{B}$-function f and all $a, h \in \mathbb{Z}_p$, $k = 1, 2, 3, \ldots$, the following equality holds:*

$$f(a + p^k h) = f(a) + f'(a)p^k h + \frac{f''(a)}{2!}p^{2k}h^2 + \frac{f'''(a)}{3!}p^{3k}h^3 + \cdots \qquad (6.7)$$

Moreover, all $\frac{f^{(j)}(a)}{j!}$ are p-adic integers, $j = 0, 1, 2, \ldots$.

Proof The first claim of Theorem 6.11 immediately follows from Theorem 6.3, which obviously holds with $r = 1$ for any $\mathcal{B}$-function f by virtue of the definition of the class $\mathcal{B}$ (see (6.5)).

To prove the second claim of the theorem, we note that in view of (4.5), the following equality, which can be easily proved by induction on n, holds for the n-th derivative $f^{(n)}$ of the function f:

$$f^{(n)}(x) = \sum_{k=0}^{\infty} \binom{x}{k} \sum_{i_1, i_2, \ldots, i_n \geqslant 1} \frac{a_{k+i_1+i_2+\cdots+i_n}}{i_1 i_2 \cdots i_n}(-1)^{n+i_1+i_2+\cdots+i_n}.$$

Further,

$$\sum_{i_1, i_2, \ldots, i_n \geqslant 1} \frac{a_{k+i_1+i_2+\cdots+i_n}}{i_1 i_2 \cdots i_n}(-1)^{n+i_1+i_2+\cdots+i_n} =$$

$$= \sum_{s=n}^{\infty} \sum_{\substack{i_1, i_2, \ldots, i_n \geqslant 1 \\ i_1+i_2+\ldots+i_n=s}} \frac{a_{k+s}}{i_1 i_2 \cdots i_n}(-1)^{n+s}, \qquad (6.8)$$

and $\frac{a_{k+s}}{i_1 i_2 \cdots i_n} = \frac{a_{k+s}}{s!}\frac{s!}{i_1 i_2 \cdots i_n} \in \mathbb{Z}_p$ since both $\frac{(i_1+i_2+\cdots+i_n)!}{i_1 i_2 \cdots i_n}$ and $\frac{a_{k+s}}{(k+s)!}$ are p-adic integers (see the definition of a $\mathcal{B}$-function (6.5) for the latter). Thus, the sum

$$\sigma_s = \sum_{\substack{i_1, i_2, \ldots, i_n \geqslant 1 \\ i_1+i_2+\ldots+i_n=s}} \frac{a_{k+s}}{i_1 i_2 \cdots i_n}(-1)^{n+s}$$

in the right-hand side of (6.8) is a p-adic integer. Moreover, as $\frac{a_{k+s}}{i_1 i_2 \cdots i_n} = \frac{a_{k+s}}{j_1 j_2 \cdots j_n}$ whenever $j_1, j_2, \ldots, j_n$ is a permutation of $i_1, i_2, \ldots, i_n$, the sum σ_s is a multiple of $n!$, i.e., $\frac{\sigma_s}{n!} \in \mathbb{Z}_p$. This proves the theorem. □

6.3.3 Important Subclasses of the Class $\mathcal{B}$

In this subsection, we consider some important types of $\mathcal{B}$-functions. We start with *exponential $\mathcal{B}$-functions*. Consider p-adic functions of the form $f(x) = u(x)^{v(x)}$

where $u, v\colon \mathbb{Z}_p \to \mathbb{Z}_p$ are 1-Lipschitz functions. The domain of the function f may be smaller than $\mathbb{Z}_p$ (and actually may be empty), not speaking of the 1-Lipschitzness. The following proposition states sufficient conditions for the function f to be 1-Lipschitz, and moreover to lie in the class $\mathcal{B}$:

Proposition 6.12 *Let $u, v\colon \mathbb{Z}_p \to \mathbb{Z}_p$ be 1-Lipschitz functions; let $u(z) \equiv 1 \pmod p$ for all $z \in \mathbb{Z}_p$. Then the function $f(z) = u(z)^{v(z)}$ is well defined for all $z \in \mathbb{Z}_p$, integer-valued and 1-Lipschitz. If w, v are $\mathcal{B}$-functions, and $u(z) = 1 + pw(z)$, then f is a $\mathcal{B}$-function.*

Proof As $u(z) \equiv 1 \pmod p$ for all $z \in \mathbb{Z}_p$, then necessarily $u(z) = 1 + pt(z)$ for a suitable function $t\colon \mathbb{Z}_p\, \mathbb{Z}_p$; so $f(z) = \sum_{i=0}^{\infty} p^i t(z)^i \binom{v(z)}{i}$ is a p-adic integer since $p^i t(z)^i \binom{v(z)}{i} \in \mathbb{Z}_p$ and $\overset{p}{\lim\limits_{i\to\infty}} p^i t(z)^i \binom{v(z)}{i} = 0$. Hence, the function f is well defined on $\mathbb{Z}_p$, and f is integer-valued.

To prove the 1-Lipschitzness of f, note that as basic properties of powers are of the same form, both in real and p-adic cases (see, e.g., [145, Ch. 14, Section 5]), one has for arbitrary $b, c, d \in \mathbb{Z}_p$ and $n = 1, 2, \ldots$ that $(a + p^n b)^{c+p^n d} = (a + p^n b)^c((a+p^n b)^{p^n})^d$. As both u and v are 1-Lipschitz functions, for arbitrary $z, r \in \mathbb{Z}_p$, there exist $s, t \in \mathbb{Z}_p$ such that $(u(z + p^n r))^{v(z+p^n r)} = (u(z) + p^n t)^{v(z)+p^n s}$; hence, $(u(z+p^n r))^{v(z+p^n r)} = (u(z)+p^n t)^{v(z)}((u(z)+p^n t)^{p^n})^s \equiv (u(z)+p^n t)^{v(z)} \pmod{p^n}$ since $(u(z) + p^n t)^{p^n} \equiv 1 \pmod{p^n}$.

Here is a proof of the latter congruence: As $u(z) \equiv 1 \pmod p$, for a suitable $k \in \mathbb{Z}_p$, we have $u(z) + p^n t = 1 + pk$; yet $(1 + pk)^{p^n} = \sum_{i=0}^{p^n} k^i p^i \binom{p^n}{i} = \sum_{i=0}^{p^n} k^i \frac{p^i}{i!}(p^n)^{\underline{i}} \equiv 1 \pmod{p^n}$ since $\frac{p^i}{i!} \in \mathbb{Z}_p$ in view of Lemma 2.1. Finally denoting by $\overline{v(z)} = v(z) \bmod p^n$ the least nonnegative residue of $v(z)$ modulo p^n, for a suitable $h \in \mathbb{Z}_p$, we see that $f(z + p^n r) \equiv (u(z) + p^n t)^{v(z)} = (u(z)+p^n t)^{\overline{v(z)}}(u(z)+p^n t)^{p^n h} \equiv (u(z)+p^n t)^{\overline{v(z)}} = \sum_{i=0}^{\overline{v(z)}} u(z)^{\overline{v(z)}-i} p^{ni} t^i \binom{\overline{v(z)}}{i} \equiv (u(z))^{\overline{v(z)}} \equiv (u(z))^{\overline{v(z)}}(u(z))^{p^n h} = (u(z))^{v(z)}$, where $\equiv$ stands for congruence modulo p^n. This proves that f is 1-Lipschitz.

To prove the rest of the claim of the proposition, note that for every $z \in \mathbb{Z}_p$ and every $n = 1, 2, \ldots$, the congruence $(u(z))^{v(z)} \equiv \sum_{i=0}^{n-1} (u(z) - 1)^i \binom{v(z)}{i} \pmod{p^n}$ holds since $|u(z) - 1|_p \leqslant \frac{1}{p}$. This implies that:

- By Theorem 6.10, all functions $f_n = \sum_{i=0}^{n-1} \frac{p^i}{i!} v^{\underline{i}} w^i$ are in $\mathcal{B}$ since all fractions $\frac{p^i}{i!}$ are p-adic integers (see Lemma 2.1).
- The sequence $(f_n)_{n\in\mathbb{N}}$ converges to f with respect to the metric D_p.

From here it follows that f is a $\mathcal{B}$-function by Theorem 6.10. □

Example 6.13 Given $a \in \mathbb{Z}_p$, $a \equiv 1 \pmod p$, the function $f(x) = a^x$ is a $\mathcal{B}$ function. Moreover, f is a C-function, with the only exception of the case when $p = 2$ and $a \not\equiv 1 \pmod 4$.

Yet one more important subclass of the class $\mathcal{B}$ constitute *rational $\mathcal{B}$-functions*. A rational function over $\mathbb{Z}_p$ is a function $f(x) = \frac{g(x)}{u(x)}$, where $g(x), u(x)$ are polynomials with p-adic integer coefficients. From Proposition 6.12, it immediately follows that f is a $\mathcal{B}$-function once $u(x) = 1 + pw(x)$, where w is a polynomial over $\mathbb{Z}_p$. However, the latter assertion can be slightly generalized:

Proposition 6.14 *The rational function $f(x) = \frac{g(x)}{u(x)}$ is a $\mathcal{B}$-function if the denominator $u(x)$ vanishes modulo p nowhere on $\mathbb{Z}_p$* (i.e., if $u(z) \not\equiv 0 \pmod p$ for all $z \in \mathbb{Z}_p$).

Proof Indeed, once $v(z) \not\equiv 0 \pmod p$ for every $z \in \mathbb{Z}_p$, there exists a multiplicative inverse for $v(z)$ in the residue ring $\mathbb{Z}/p^n\mathbb{Z}$, for every $n = 1, 2, \ldots$. Thus, $\frac{u(z)}{v(z)} \equiv u(z)v(z)^{\varphi(p^n)-1} \pmod{p^n}$, where φ is Euler's totient function. Hence, the function f can be uniformly approximated (with respect to the metric D_p) by polynomials $u(x)v(x)^{\varphi(p^n)-1}$ over $\mathbb{Z}_p$, $n = 1, 2, \ldots$; so f is a $\mathcal{B}$-function by Theorem 6.10. □

6.4 Class $\mathcal{A}$

Some important functions do not lie in $\mathcal{B}$. For instance, these are some 1-Lipschitz integer-valued polynomials over $\mathbb{Q}_p$; i.e., the polynomials that not necessarily have integer p-adic coefficients yet map $\mathbb{Z}_p$ into itself and satisfy a Lipschitz condition with a constant 1 everywhere on $\mathbb{Z}_p$. However, they lie in a wider class $\mathcal{A}$:

Definition 6.15 A function $f: \mathbb{Z}_p \to \mathbb{Z}_p$ lies in $\mathcal{A}$ (and is said to be an *$\mathcal{A}$-function*) if and only if f satisfies a Lipschitz condition with a constant 1 and $p^n f$ is a $\mathcal{B}$-functions for some non-negative rational integer $n \in \mathbb{N}_0$.

Now, since $f = \frac{1}{p^n} g$ for a suitable $\mathcal{B}$-function g and suitable non-negative rational integer n, by Theorem 6.11, we immediately conclude that Taylor theorem for every $\mathcal{A}$-function f holds in the following form:

Theorem 6.16 *For every $\mathcal{A}$-function f and every $a, h \in \mathbb{Z}_p$, $k = 1, 2, 3, \ldots$ the function $f(a + p^k h)$ in variable h can be represented via convergent power series*

$$f(a + p^k h) = f(a) + f'(a)p^k h + \frac{f''(a)}{2!} p^{2k} h^2 + \frac{f'''(a)}{3!} p^{3k} h^3 + \cdots . \tag{6.9}$$

By Definition 6.2, we immediately get the corollary:

Corollary 6.17 *Any $\mathcal{A}$-function is locally analytic of order 1.*

Note 6.18 Note that $\frac{f^{(j)}(a)}{j!}$ are *not* necessarily p-adic integers now; however, in view of the second claim of Theorem 6.11, $\left|\frac{f^{(j)}(a)}{j!}\right|_p \leqslant p^n$ for all $j = 1, 2, \ldots$. Moreover, $f'(a)$ is still a p-adic integer by Proposition 3.13.

The most important examples of $\mathcal{A}$-functions that are not necessarily $\mathcal{B}$-functions are 1-Lipschitz integer-valued polynomials over $\mathbb{Q}_p$, i.e., functions of the form $f(x) = \sum_{i=0}^{d} a_i p^{\lfloor \log_p i \rfloor} \binom{x}{i}$, where $a_i \in \mathbb{Z}_p$, $i = 0, 1, 2, \ldots$. This stresses the importance of $\mathcal{A}$-functions: In view of Theorem 5.12 and Proposition 4.7, every 1-Lipschitz function can be uniformly approximated on $\mathbb{Z}_p$ (with respect to the metric D_p) by $\mathcal{A}$-functions.

Example 6.19 The function $f(x) = \frac{(x^p - x)^2}{p}$ is an $\mathcal{A}$-function, but not a $\mathcal{B}$-function.

Part II
The p-Adic Ergodic Theory

In this part, we develop ergodic theory of p-adic 1-Lipschitz functions whose domain and co-domain are (modules over) p-adic integers. These functions are exactly the automata functions, thus causal functions over discrete time. Respective automata can serve as models of open systems in physics as well as models of computational processes in computer science. The ergodic theory of these functions thus describes important features of temporal dynamics of the said systems, i.e., how these systems evolve in time; so mathematical results of this part, which is purely mathematical, are intensively used in Part III of the book which is focused at various applications.

Chapter 7
Ergodic Theory: Preliminaries for the p-Adic Case

Abstract In this chapter, we mostly recall some basic concepts of ergodic theory (cf., the book (Silva Invitation to Ergodic Theory, 2008) and Sect. 1.3) and restate them for p-adic 1-Lipschitz dynamics. We also examine image probabilities, which are associated with the p-adic 1-Lipschitz maps.

7.1 Some Notions from General Ergodic Theory

Basic questions in general ergodic theory are: Given a dynamical system $\langle \mathbb{X}; \mu; f\rangle$ on a probability space $\langle \mathbb{X}; \mu\rangle$, how often does an orbit $(f^i(x))_{i=0}^{\infty}$ visit given region $S \subset \mathbb{X}$ (i.e., a μ-measurable subset $S \subset \mathbb{X}$)? What is the frequency with which the orbit of $x = x_0$ hits S, for instance, when the frequency is equal to the probability that a randomly chosen point of $\mathbb{X}$ lies in S? That is, if $\chi(S, x)$ is a characteristic function of S (i.e., $\chi(S, x) = 1$ if and only if $x \in S$; otherwise $\chi(S, x) = 0$), when does the following condition hold?

$$\lim_{N\to\infty} \frac{1}{N} \sum_{i=0}^{N-1} \chi(S, x_i) = \mu(S). \tag{7.1}$$

In the book, we call a sequence $(x_i)_{i=0}^{\infty}$ *uniformly distributed* if (7.1) holds for all measurable subsets S of $\mathbb{X}$. A word of caution: This definition must *not* be considered as a definition of uniform distribution on arbitrary measure space $\mathbb{X}$ with arbitrary probability measure μ! Some restrictions must be imposed both on measurable sets S and on the measure μ. However, all these restrictions are satisfied for all dynamical systems we consider in the book, so we are free to use the above definition of uniform distribution throughout the book. For further details on these restrictions, as well as in the theory of uniform distribution of sequences, interested readers are referred to [130]. One of the fundamental results of ergodic theory yields that *once the function f is ergodic, the orbit $(f^i(x_0))_{i=0}^{\infty}$ is uniformly distributed.* Again, to make this assertion true in general, some additional restrictions must be

V. Anashin, *Causality: The p-adic Theory*, STEAM-H: Science, Technology, Engineering, Agriculture, Mathematics & Health,
https://doi.org/10.1007/978-3-031-85818-5_7

imposed (see [130]). However, for dynamical systems, we consider further the ones on finite spaces, and on $\mathbb{Z}_p^n$, the restrictions are satisfied; so we use the just stated assertion without limitations.

In Part II, we mainly focus on the following questions: Given a map $f: \mathbb{X} \to \mathbb{Y}$, is it measure preserving? Is it ergodic? These questions are crucial to applied problems, which we address in Part III.

7.1.1 Finite Dynamics

Finite dynamical systems look somewhat trivial; however, they constitute an important class of dynamical systems, which we consider on the book. Of course, when dynamics is finite, all points are eventually periodic. Now let us look at what the notions of measure-preservation and ergodicity (cf., Sect. 1.3) mean for finite dynamical systems.

Let $\mathbb{X}$, $\mathbb{Y}$ be finite. For $S \subset \mathbb{X}$ and $T \subset \mathbb{Y}$, put

$$\mu(S) = \frac{\#S}{\#\mathbb{X}}, \quad \nu(T) = \frac{\#T}{\#\mathbb{Y}}.$$

That is, μ and ν are "standard" probability measures on $\mathbb{X}$ and $\mathbb{Y}$, respectively. Let $F: \mathbb{X} \twoheadrightarrow \mathbb{Y}$ be a surjective map. General definitions of measure-preservation/ergodicity from Sect. 1.3 imply:

1. The map F is measure-preserving if and only if f is *balanced*. i.e., if and only if the number $\#F^{-1}(t)$ of F-pre-images of the point $t \in \mathbb{T}$ does not depend on the point t:

$$\#F^{-1}(t) = \frac{\#\mathbb{X}}{\#\mathbb{Y}}$$

   for all $t \in \mathbb{Y}$.
2. In particular, in the case when $\mathbb{X} = \mathbb{Y}$ and $\mu = \nu$, the map F is measure-preserving if and only if F is bijective, i.e., if and only if F is a permutation on $\mathbb{S}$.
3. Finally, F is ergodic if and only if F is *transitive* on $\mathbb{X}$, i.e., if and only if F is a cycle of length $\#\mathbb{X}$ (i.e., F cyclically permutes points in $\mathbb{X}$).

It is clear that if a map $f: \mathbb{X} \to \mathbb{X}$ is transitive, then its every orbit is uniformly distributed since the orbit is a periodic sequence such that the length of its shortest period is $\#\mathbb{X}$, and every element from $\mathbb{X}$ appears at the period exactly once. Sequences having this property are called *strictly uniformly distributed.*

Balanced, bijective, and transitive maps may under certain conditions be "projected" onto smaller domains so that the "images" of these maps remain balanced, bijective, and transitive, respectively. We consider these conditions for the case of maps of finite rings.

Let A, B be rings; let $\varphi\colon A \twoheadrightarrow B$ be a ring epimorphism. According to the general definition of the compatibility with ring epimorphisms (cf., Sect. 1.2.2), a mapping $f\colon A \to A$ is called *compatible with the epimorphism* $\varphi\colon A \to B$ if there exists a commutative closure $\varphi(f)$ of the following diagram:

$$\begin{array}{ccc} A & \xrightarrow{f} & A \\ \downarrow{\scriptstyle\varphi} & & \downarrow{\scriptstyle\varphi} \\ B & \xrightarrow{\varphi(f)} & B \end{array}$$

Here by definition, $(\varphi(f))(b) = f(\varphi(a))$ for $b \in B$, where $a \in A$ is a φ-pre-image of b: $a \in \varphi^{-1}(b)$. In other words, the compatibility of f with respect to the epimorphism φ means that the map $\varphi(f)$ is well defined: given arbitrary $x, y \in A$ such that their φ-images coincide, $\varphi(x) = \varphi(y)$, and then $\varphi(f(x)) = \varphi(f(y))$.

So each compatible transformation on A defines a unique transformation on each epimorphic image of A. As each epimorphism of A defines a unique ideal I_φ of A, the kernel of φ, and vice versa (see Sect. 1.2.2), we also say that f is *compatible with respect to the ideal* I_φ. We say that f *has some property* $\mathfrak{P}$ *modulo the ideal* I_φ or, which is the same, *modulo epimorphism* φ if the mapping φf induced by f on the epimorphic image B of A has the property $\mathfrak{P}$.

The notion of compatibility with an epimorphism can be defined for multivariate maps as well: if $F\colon A^n \twoheadrightarrow A^m$, $m \leqslant n$ is a map from the n-th Cartesian power of A onto its m-th Cartesian power, and if I is an ideal in A, then I^n and I^m are ideals in A^n and A^m, respectively; so the map F is called compatible with respect to the ideal I of A if there exists a commutative closure $\varphi(F)$ of the following diagram:

$$\begin{array}{ccc} A^n & \xrightarrow{F} & A^m \\ \downarrow{\scriptstyle\varphi} & & \downarrow{\scriptstyle\varphi} \\ B^n & \xrightarrow{\varphi(F)} & B^m \end{array}$$

Note that we denote epimorphisms $A^n \twoheadrightarrow B^n$ and $A^m \twoheadrightarrow B^m$ with kernels I^n and I^m, respectively, by the same symbol φ just to avoid complicated notation. We will also say that F has some property $\mathfrak{P}$ modulo I (modulo φ) if the map $\varphi(F)$ has the property $\mathfrak{P}$. The following proposition holds:

Proposition 7.1 *Let A be a finite ring, let I be an ideal of A, and let $F\colon A^n \twoheadrightarrow A^m$* (where $m \leqslant n$) *be a balanced* (*respectively, bijective, transitive*) *compatible* (with respect to I) *mapping of the n-th Cartesian power A^n onto the m-th Cartesian power A^m of the ring A. Then F is balanced* (*respectively, bijective, transitive*) *modulo I.*

Further, if $k = |A : I|$ is the index of I in A, then the mapping $F: A^n \twoheadrightarrow A^n$ is transitive if and only if F is transitive modulo I and the iterated mapping $F^{kn}: I^n \to I^n$ is transitive on I^n.

Moreover, if A is a Cartesian product of rings B and C, $A = B \times C$, then F is balanced on A if and only if F is balanced both on B and C, i.e., modulo each projection onto a Cartesian factor. Finally, the mapping $F: A \to A$ is transitive if and only if it is transitive both on B and C and orders #B and #C are coprime.

Proof Choose arbitrary element $c \in A^m$, and consider the following inclusion:

$$F(x_1 + I, \ldots, x_n + I) \subseteq c + I^m \tag{7.2}$$

Let $\mathcal{S} \subseteq I^n$ be arbitrary complete system of representatives of cosets with respect to the ideal I^n; that is, $\mathcal{S}$ is a set that contains one and only one element of each coset $h + I^n$. Let t be a number of elements of $\mathcal{S}$ that satisfy (7.2). Consider the inclusion

$$F(a_1, \ldots, a_n) \in c + I^m. \tag{7.3}$$

If $\mathbf{x} = (x_1, \ldots, x_n) \in S$ and if $\mathbf{x}$ satisfies (7.2), then every $(a_1, \ldots, a_n)$ that lies in the coset $(x_1, \ldots, x_n) + I^n$ satisfies (7.3) since F is compatible. Thus, the number of elements of A^n that satisfy (7.3) is exactly $t\#I^n$.

On the other hand, let F be balanced. Then for every $d \in c + I^m$, the equation $F(a_1, \ldots, a_n) = d$ has exactly $\#A^{n-m}$ solutions in A^n, and consequently, there exist exactly $\#A^{n-m}\#I^m$ elements of A^n that satisfy (7.3). In view of the argument above, this implies that $\#A^{n-m}\#I^m = t\#I^n$. Hence, $t = \#(A/I)^{n-m}$. Thus, t does not depend on the choice of c and, consequently, F induces a balanced mapping of the factor-ring $(A/I)^n$ onto the factor-ring $(A/I)^m$. The reader is encouraged to prove other assertions of the proposition. □

7.2 1-Lipschitz Dynamics on $\mathbb{Z}_p^n$

Configuration spaces of dynamical systems we are focused at in the book are metric spaces $\mathbb{Z}_p^n$ endowed with a natural probability measure μ_p^n (cf., Sect. 2.4). In the sequel, we may omit either of sub- and super-scripts in μ_p^n when it is clear what measure space is considered. For brevity, we may also refer to this probability measure on $\mathbb{Z}_p^n$ as to a *p-adic measure*. The measure μ_p^n may be viewed as an analog of Lebesgue measure λ on the closed unit interval $[0; 1] \subset \mathbb{R}$; moreover, the measure spaces $\mathbb{Z}_p^n$ and $[0; 1] \subset \mathbb{R}$ are isomorphic as probability spaces endowed with the probability measures μ_p^n and λ, respectively. In the following subsection, we describe this analogy in detail.

7.2.1 The Monna Map and the Likes

Given $z \in \mathbb{Z}_p$, represent $z = \zeta_0 + \zeta_1 p + \zeta_2 p^2 + \cdots$ via canonical expansion; that is, $\zeta_i = \delta_i(z) \in \{0, 1, \ldots, p-1\}$, (cf., Sect. 2.3.1). The *Monna map* (cf., [154], [130, Chapter 5, Section 2, Exercise 2.4]) is the map $\mathsf{Mon}\colon \mathbb{Z}_p \twoheadrightarrow [0, 1] \subset \mathbb{R}$ of $\mathbb{Z}_p$ onto the real unit segment $[0, 1] \subset \mathbb{R}$ such that

$$\mathsf{Mon}(z) = \sum_{i=1}^{\infty} \zeta_{i-1} p^{-i} = 0.\zeta_0\zeta_1 \cdots \tag{7.4}$$

The Monna map is surjective but not bijective: It may happen that $\mathsf{Mon}(z) = \mathsf{Mon}(y)$ for some $z, y \in \mathbb{Z}_p, z \neq y$, since $p^{-k} = p^{-k-1}(p-1) + p^{-k-2}(p-1) + \cdots$ in $\mathbb{R}$, for every $k \in \mathbb{N}_0$.

Namely, a p-adic integer z is a negative rational integer, i.e., $z \in \{-1, -2, \ldots\}$ if and only if $\delta_i(z) = p - 1$ for all $i \geqslant \ell$ for some $\ell \in \mathbb{N}_0$. Denote via $\ell(z)$ the smallest $\ell \in \mathbb{N}_0$ having that property. If $\ell = 0$, then necessarily $z = (p-1) + (p-1)p + (p-1)p^2 + \cdots = -1$, and therefore, $\mathsf{Mon}(z) = 1$. If $\ell(z) > 0$, then denote $\tilde{z} = \zeta_0 + \zeta_1 p + \cdots + \zeta_{\ell(z)-1} p^{\ell(z)-1}$; thus, $z = \tilde{z} - p^{\ell(z)}$. Note that $0 \leqslant \zeta_{\ell(z)-1} < p - 1$ and that $-p^{\ell(z)} = (p-1)p^{\ell(z)} + (p-1)p^{\ell(z)+1} + \cdots$; therefore,

$$\begin{aligned} \mathsf{Mon}(z) = \mathsf{Mon}(\tilde{z} - p^{\ell(z)}) &= \zeta_0 p^{-1} + \cdots + \zeta_{\ell(z)-1} p^{-\ell(z)-2} + p^{-\ell(z)-2} = \\ &= 0.\zeta_0 \cdots \zeta_{\ell(z)-2}(\zeta_{\ell(z)-1} + 1) = \mathsf{Mon}(\tilde{z} + p^{\ell(z)-1}) \end{aligned} \tag{7.5}$$

This argument proves the following assertion:

Proposition 7.2 *Given $z, y \in \mathbb{Z}_p$, $z \neq y$, the equality $\mathsf{Mon}(z) = \mathsf{Mon}(y)$ holds if and only if $z = n - p^{k+1}$ and $y = n + p^k$, where $n, k \in \mathbb{N}_0$, $\lfloor \log_p n \rfloor \leqslant k$, and $\delta_k(n) \neq p - 1$.*

Note 7.3 The Monna map can be defined via (7.4) for any integer $p > 1$, which is not necessarily a prime: Actually, the Monna map just maps the tree T_p onto $[0, 1]$ by putting into a correspondence to each ray $z = \zeta_0\zeta_1 \cdots$ a real number whose base-p expansion is $\mathsf{Mon}(z)$. Proposition 7.2 remains true for a non-prime p as well.

It is well-known that Monna map is a measure-preserving map of the probability space $\mathbb{Z}_p$ equipped with Haar probability measure μ_p onto the probability space, which is a real unit segment $[0, 1]$ equipped with a probability measure that is Lebesgue measure λ (see, e.g., [129, 154]).

This assertion can be generalized as follows. Let $q > 1$ be an integer; let T_q be a q-adic tree (see Note 2.22). For every ray $z = \zeta_0\zeta_1 \cdots$ that starts at the root of the tree T_q, let $\mathsf{Mon}_q\colon z \mapsto \sum_{j=1}^{\infty} \zeta_{j-1} q^{-j} \in [0, 1]$. That is, in plain words, to every such ray z the mapping Mon_q put into the correspondence a real integer $\mathsf{Mon}_q(z) \in [0, 1]$ whose base-q expansion is that ray.

The metric space $\langle \mathrm{T}_q; d_q \rangle$ is equipped with a probability measure μ_q, which is a Hausdorff net-measure induced by the metric d_p (see Sect. 2.4). The following proposition is true.

Proposition 7.4 *Given any λ-measurable subset M of $[0, 1]$, a full pre-image $\mathsf{Mon}_q^{-1}(M) \subset \mathrm{T}_q$ of the tree T_q equipped with a measure μ_q is μ_q-measurable, and $\mu_q(\mathsf{Mon}_q^{-1}(M)) = \lambda(M)$.*

Proof From the construction of the mapping $\mathsf{Mon}_q \colon \mathrm{T}_q \to [0, 1]$, it follows that

$$|\mathsf{Mon}_q(z) - \mathsf{Mon}_q(y)| \leqslant d_q(z, y); \tag{7.6}$$

i.e., the mapping Mon_q is a 1-Lipschitz map of the metric space $\langle \mathrm{T}_q; d_q \rangle$ onto the metric space $[0, 1]$ equipped with a metric $|a - b|$, which is a distance between the points $a, b \in [0, 1]$ on the real line $\mathbb{R}$. To show this, we mimic the proof of [129, Lemma 3]. Let ℓ be the length of the longest common prefix of the rays z, y; that is,

$$\zeta_0 = \gamma_0, \ldots, \zeta_{\ell-1} = \gamma_{\ell-1}, \zeta_\ell \neq \gamma_\ell,$$

where $y = \gamma_0\gamma_1 \cdots$. Then

$$|\mathsf{Mon}_q(z) - \mathsf{Mon}_q(y)| = \left| \sum_{i=0}^{\infty} (\zeta_i - \gamma_i) q^{-i-1} \right| \leqslant$$

$$\leqslant (q-1) \sum_{i=\ell}^{\infty} q^{-i-1} = q^{-\ell} = d_q(z, y). \tag{7.7}$$

Hence, the map Mon_q is continuous.

Given an open interval $(a, b) \subset [0, 1]$, let $\mathcal{D}'$ be a collection all balls $\mathbf{B} \subset \mathrm{T}_q$ of non-zero radii such that $\mathsf{Mon}(\mathbf{B}) \subset (a, b)$. As every two balls are either disjoint or one of the pair is a subset of the other ball, let $\mathcal{D}$ be a set of all maximal with respect to the inclusion balls from the collection $\mathcal{D}'$; then $\mathcal{B}$ consists of not more than a countable number of balls, and every two of balls from $\mathcal{B}$ are disjoint.

Given a ball $\mathbf{B} = \mathbf{B}_{q^{-\ell}}(w) \subset \mathrm{T}_q$ centered at the root ray $w = \omega_0\omega_1 \cdots$, we have that

$$\mathsf{Mon}_q(\mathbf{B}) = [0.\omega_0\omega_1 \cdots \omega_{\ell-1}, 0.\omega_0\omega_1 \cdots \omega_{\ell-1} + q^{-\ell}]. \tag{7.8}$$

Therefore, for every $c \in (a, b)$, there is a ball $\mathbf{B}' \in \mathcal{D}'$ such that $\mathsf{Mon}_q(\mathbf{B}') \ni c$; whence there is a ball $\mathbf{B} \in \mathcal{D}$ such that $\mathsf{Mon}_q(\mathbf{B}) \ni c$.

As the intersection $\mathsf{Mon}_q(\mathbf{B}_1) \cap \mathsf{Mon}_q(\mathbf{B}_2)$ of two disjoint balls $\mathbf{B}_1, \mathbf{B}_2 \in \mathcal{D}$, is either empty or consists of a single point from an open real interval (a, b), we conclude by the regularity of measures μ_q and λ that:

- The set $C = \cup_{\mathbf{B}\in\mathcal{D}}\mathbf{B}$ is open in T_q; thus, μ_q-measurable.
- $\mathsf{Mon}_q^{-1}((a, b)) = C$.
- $\mu_q(C) = \lambda(\mathsf{Mon}_q(C)) = \lambda((a, b)) = b - a$.

This completes the proof since both measures μ_q and λ are Borel (see Sect. 2.4). □

Corollary 7.5 *Let $q = p^n$; then Proposition 7.4 holds for $\mathbb{Z}_p^n$ as well, rather than for T_q only.*

Indeed, when $q = p^n$, the probability spaces $\langle \mathrm{T}_q; \mu_q\rangle$ and $\langle \mathbb{Z}_p^n; \mu_p^n\rangle$ are isomorphic as metric spaces equipped with metrics d_q and d_p^n, respectively; whence the spaces are isomorphic as probability spaces since the probability measures are Hausdorff net measures induced by respective metrics (see Sect. 2.4). The map $\mathsf{Mon}_q : \mathbb{Z}_p^n \to [0, 1]$ can be defined by using the one-to-one correspondence between root rays of $\mathbf{T}_q$ and elements of $\mathbb{Z}_p^n$ described by Note 2.22. *That is why in further considerations in both cases, for T_q and $\mathbb{Z}_p^n$, we use the same notation* Mon *for the Monna map, without a subscript.* We conclude:

> The probability spaces T_q for every $q \in \mathbb{N}\setminus\{1\}$, and the probability spaces $\mathbb{Z}_p^n$ for every prime p and every $n \in \mathbb{N}$, are all isomorphic (via the Monna map) to the unit real segment $[0, 1]$ equipped with a probability measure, which is Lebesgue measure λ on the segment.

7.2.2 *Measure Properties of 1-Lipschitz Maps: Image Measures*

Though the main theme of Part II of the book is dynamics produced by surjective 1-Lipschitz maps $\mathbb{Z}_p^n \twoheadrightarrow \mathbb{Z}_p^n$, in this subsection, we discuss measure properties of 1-Lipschitz maps $F : \mathbb{Z}_p^n \to \mathbb{Z}_p^m$ in a general setting for $m \leqslant n$ without the assumption that F is surjective. These results will be needed later in the book.

Given a 1-Lipschitz map $F : \mathbb{Z}_p^n \to \mathbb{Z}_p^m$, consider $F(\mathbb{Z}_p^n)$, the range of F (or, in other terms, the F-image of $\mathbb{Z}_p^n$), and the subset $F(\mathbb{Z}_p^n) \subset \mathbb{Z}_p^m$. Since F is continuous, $F(\mathbb{Z}_p^n)$ is compact. This makes assertions of the following proposition nearly obvious, but we nevertheless give a proof just to stress some moments, which we will often use later.

Proposition 7.6 *If $F : \mathbb{Z}_p^n \to \mathbb{Z}_p^m$ is 1-Lipschitz, then the following assertions are true:*

1. *$F(\mathbb{Z}_p^n)$ is closed in $\mathbb{Z}_p^m$.*
2. *$F(\mathbb{Z}_p^n)$ is a μ_p^n-measurable subset of $\mathbb{Z}_p^m$.*
3. *F is a measurable map $F\colon \mathbb{Z}_p^n \twoheadrightarrow F(\mathbb{Z}_p^n) \subset \mathbb{Z}_p^m$ of measure spaces endowed with respective measures μ_p^n and μ_p^m.*

Proof We first prove that $F(\mathbb{Z}_p^n)$ is a closed subset of $\mathbb{Z}_p^m$. Indeed, let $(u_j \in \mathbb{Z}_p^m)_{j=0}^{\infty}$ be a Cauchy sequence all whose terms lie in $F(\mathbb{Z}_p^n)$. We may assume that all the terms u_j are pairwise distinct since otherwise $F(\mathbb{Z}_p^n)$ is a finite set, and the proof becomes trivial as points in $\mathbb{Z}_p^m$ are closed, and thus their F-pre-images in $\mathbb{Z}_p^n$ are closed in $\mathbb{Z}_p^n$ as F is continuous.

Let $z_j \in \mathbb{Z}_p^n$ be arbitrary F-pre-images of u_j, $j = 0, 1, 2, \ldots$. All terms of the sequence $(z_j)_{j=0}^{\infty}$ are pairwise distinct since all the terms of the sequence $(u_j = f(z_j))$ are pairwise distinct. As $\mathbb{Z}_p^n$ is compact, the sequence $(z_j)_{j=0}^{\infty}$ contains a convergent subsequence $(z_{j_k})_{k=0}^{\infty}$ whose limit is, say, z; then $z \in \mathbb{Z}_p^n$ as $\mathbb{Z}_p^n$ is compact. Thus, the subsequence $(u_{j_k} = F(z_{j_k}))_{k=0}^{\infty}$ converges to $F(z) \in F(\mathbb{Z}_p^n)$ as F is continuous; therefore, $F(\mathbb{Z}_p^n)$ is a closed subset of $\mathbb{Z}_p^m$. As $F(\mathbb{Z}_p^n)$ is a closed subset of $\mathbb{Z}_p^m$, then $F(\mathbb{Z}_p^n)$ is measurable with respect to the measure μ_p^m on $\mathbb{Z}_p^m$.

Finally, as every closed subset of $F(\mathbb{Z}_p^n)$ is $A \cap F(\mathbb{Z}_p^n)$ for a suitable closed subset $A \subset \mathbb{Z}_p^m$, and as an F-pre-image $F^{-1}(S)$ of every closed subset $S = A \cap F(\mathbb{Z}_p^n)$ is closed in $\mathbb{Z}_p^n$ by the continuity of F, all these F-pre-images are measurable. Since the measures μ_p^n and μ_p^m are Haar measures, the respective σ-algebras are completely determined by closed subsets. Therefore, an F-pre-image $F^{-1}(M)$ of every μ_p^m-measurable subset M from $F(\mathbb{Z}_p^n)$ is μ_p^n-measurable. □

Note that all μ_p^m-measurable subsets of $\mathbb{Z}_p^m$ constitute a σ-algebra, which, in turn, defines respective σ-algebra on $F(\mathbb{Z}_p^n)$ of μ_p^m-measurable subsets of $F(\mathbb{Z}_p^n)$: These are exactly all intersections of $F(\mathbb{Z}_p^n)$ with μ_p^m-measurable sunsets of $\mathbb{Z}_p^n$. The latter σ-algebra therefore defines a measure $\bar{\mu}_p^m$ on $F(\mathbb{Z}_p^n)$, the restriction of the measure μ_p^m to the closed subset $F(\mathbb{Z}_p^n)$, and the mapping $F\colon \mathbb{Z}_p^n \twoheadrightarrow F(\mathbb{Z}_p^n) \subset \mathbb{Z}_p^m$ is then a measurable mapping of the measure space $\langle \mathbb{Z}_p^n; \mu_p^n \rangle$ onto the measure space $\langle F(\mathbb{Z}_p^n); \bar{\mu}_p^m \rangle$ (cf., Sect. 1.3).

In order to define a "natural" probability measure on $F(\mathbb{Z}_p^n)$, one may proceed as follows. If $\mu_p^m(F(\mathbb{Z}_p^n)) \neq 0$, then one can define a probability measure $\tilde{\mu}_p^n$ on $F(\mathbb{Z}_p^n)$ by assuming $\tilde{\mu}_p^m(S) = \mu_p^m(S)/\mu_p^m(F(\mathbb{Z}_p^n))$ for every μ_p^m-measurable subset of $F(\mathbb{Z}_p^n)$. We call the measure $\tilde{\mu}_p^m$ an *induced probability measure* on $F(\mathbb{Z}_p^n)$. This method does not work if $\mu_p^m(F(\mathbb{Z}_p^n)) = 0$. In the latter case, one may define an *image measure* μ_F on $F(\mathbb{Z}_p^n)$ by assuming $\mu_F(M) = \mu_p^n(F^{-1}(M))$ for every μ_p^m-measurable set $M \subset F(\mathbb{Z}_p^n)$ (cf., Sect. 1.3).

Therefore, loosely stated, any 1-Lipschitz (thus, causal; thus, automaton) map $F: \mathbb{Z}_p^n \twoheadrightarrow F(\mathbb{Z}_p^n) \subset \mathbb{Z}_p^m$ transforms uniform probability distribution on $\mathbb{Z}_p^n$ (the distribution that corresponds to the probability measure μ_p^n) into some other probability distribution on $F(\mathbb{Z}_p^n)$ and, respectively on $\mathbb{Z}_p^m$, assuming the "event" $\mathbb{Z}_p^m \setminus F(\mathbb{Z}_p^n)$ is "impossible" (thus, of zero probability) since it is out of range of the map F. The latter distribution on $F(\mathbb{Z}_p^n)$ may or may be not uniform any longer; it may be very diverse depending on F. Moreover, it turns out that, in plain words, *the uniform probability distribution can be transformed to a probability distribution of any kind, by a suitable 1-Lipschitz map F.*

In the rest of the subsection, we are going to express this feature of 1-Lipschitz maps in precise mathematical terms.

To begin with, it turns out that *any nonempty closed subset $S \subset \mathbb{Z}_p^m$ is an F-image of a suitable 1-Lipschitz map $F: \mathbb{Z}_p^n \to \mathbb{Z}_p^m$.* Moreover, the subset S is an *F-retract*; i.e., there is a closed subset $\tilde{S} \subset \mathbb{Z}_p^n$ such that $S = F(\tilde{S}) = F(\mathbb{Z}_p^n)$, and the restriction $F|_{\tilde{S}}: \tilde{S} \to F(\tilde{S})$ is one-to-one. As a matter of fact, this result holds in a more general setting, for the so-called monotone maps of trees (see, e.g., [113, Sections 2B–2C]), but we give a proof in order to emphasize some important details, which will be needed further in the book. We start with the case $m = n$.

Proposition 7.7 *Every non-empty closed subset $S \subset \mathbb{Z}_p^n$ is an F-retract for a suitable 1-Lipschitz map $F: \mathbb{Z}_p^n \to \mathbb{Z}_p^n$.*

Proof There is a one-to-one correspondence between $\mathbb{Z}_p^n$ and one-side infinite words $a = \alpha_0\alpha_1\cdots$ over the alphabet $\mathcal{A} = \mathbb{F}_p^n$, which, in turn, are in a one-to-one correspondence with rooted rays of the tree $\mathrm{T} = \mathrm{T}_{p^n}$ (cf., Note 2.22). So in what follows, we do not distinguish elements from $\mathbb{Z}_p^n$, infinite words from $\mathcal{W}_\infty(\mathcal{A})$, and rooted rays from T. Denote $P(S)$ as a collection of all non-empty prefixes of all infinite words from S; that is, a finite word $z = \zeta_0\zeta_1\cdots\zeta_{\ell-1}$ over $\mathcal{A}$ belongs to $P(S)$ if and only if there is an infinite word in S whose prefix of length ℓ is z (cf. Sect. 1.4.1).

By virtue of the said correspondence, the set S is closed if and only if, given a word $w \in \mathcal{W}_\infty(\mathcal{A})$ whose prefixes $w_{\ell_1}, w_{\ell_2}, \ldots$ of strictly increasing lengths $\ell_1 < \ell_2 < \cdots$ are such that for every $j = 1, 2, \ldots$ there is a word $w_j \in S$ whose prefix $\mathsf{pref}_{\ell_j}(w_j)$ of length ℓ_j is w_{ℓ_j}, the word w necessarily belongs to S: It is clear that the sequence of respective elements of $\mathbb{Z}_p^n$ that correspond to w_j, $j = 1, 2, \ldots$, converges to the element that corresponds to w.

Now to finish the proof, we use Definition 1.23 to construct a causal map (i.e., an automaton function) $\Phi: \mathcal{W}_\infty(\mathcal{A}) \twoheadrightarrow S$ such that the restriction $\Phi|_S(w) = w$ for every $w \in S$. That is, in view of Definition 1.23, we need to construct for every $j = 0, 1, 2\ldots$ a map φ_j from the $(j+1)$th Cartesian power of $\mathcal{A}$ into $\mathcal{A}$

such that, for every infinite word $w = \omega_0\omega_1 \cdots \omega_{\ell-1} \cdots$ over $\mathcal{A}$, the infinite word $\varphi_0(\omega_0)\varphi_1(\omega_0, \omega_1) \cdots$ belongs to S and is equal to w whenever $w \in S$.

Given $w \in \mathcal{W}_\infty(\mathcal{A})$, put $\varphi_j(\omega_0, \ldots, \omega_j) = \omega_j$ for all $j = 0, 1, 2 \ldots$ if $w \in S$. If $w \notin S$, then, as S is closed, there is a prefix of w that is not in $P(S)$; let $w_{\ell_1} = \omega_0\omega_1 \cdots \omega_{\ell_1-1}$ be the shortest of these prefixes. Then put $\varphi_\ell(\omega_0, \ldots, \omega_\ell) = \omega_\ell$ for $\ell = 0, 1, \ldots, \ell_1 - 1$; then choose any $\xi \in \mathcal{A}$ so that $\omega_0 \cdots \omega_{\ell-1}\xi \in P(\mathcal{A})$, and put $\varphi_\ell(\omega_0, \ldots, \omega_\ell) = \xi$ where $\xi \in \mathcal{A}$. This ξ exists since S is closed.

If $\omega_0 \cdots \omega_{\ell_1-2}\xi\omega_{\ell+1} \cdots \in S$, then put $\varphi_\ell(\omega_0, \ldots, \omega_\ell) = \omega_\ell$ for all $\ell \geqslant \ell_1$; otherwise, take the smallest $\ell \geqslant \ell_1$ such that $\omega_0 \cdots \omega_\ell \notin P(S)$, choose any $\chi \in \mathcal{A}$ such that $\omega_0 \cdots \omega_{\ell-1}\chi \in P(S)$, and put $\varphi_\ell(\omega_0, \ldots, \omega_\ell) = \chi$. Proceed in the same way.

The so-defined map $\Phi\colon \mathcal{W}_\infty(\mathcal{A}) \twoheadrightarrow S$ is surjective, and $\Phi|_S(w) = w$ for every $w \in S$. The map Φ is 1-Lipschitz since if $u, v \in \mathcal{W}_\infty(\mathcal{A})$, then the construction of Φ implies that $\mathsf{pref}_\ell(\Phi(u)) = \mathsf{pref}_\ell(\Phi(v))$ once $\mathsf{pref}_\ell(u) = \mathsf{pref}_\ell(v)$, for every $\ell = 0, 1, 2, \ldots$ (cf., (1.9) from Sect. 1.6.2). □

From the proof of Proposition 7.7, it immediately follows that its statement remains true whenever $\mathbb{Z}_p^n$ is replaced by any closed subset $R \supset S$; that is:

Corollary 7.8 *Given nonempty closed subsets $R, S \subset \mathbb{Z}_p^n$, $R \supset S$, there is a 1-Lipschitz map $F\colon \mathbb{Z}_p^n \to \mathbb{Z}_p^n$ such that S is an F-retract of R.*

Consider the following instructive examples: while constructing the examples, we do not distinguish one-side infinite words from $\mathcal{W}_\infty(\{0, 1\})$, rooted rays from the tree T_2, and 2-adic integers from $\mathbb{Z}_2$ represented by their canonical 2-adic expansions.

Example 7.9 For $\ell = 1, 2, 3, \ldots$, let $\mathbf{B}(\ell) \subset \mathrm{T}_2$ be a ball of radius $1/2^{\ell+1}$ centered at the word $w_\ell = 1^\ell 01^\infty \cdots$, which is a concatenation of the prefix consisting of ℓ symbols 1 with a suffix $011 \cdots$ (cf. Sect. 1.4.1) (actually one may take any suffix of the form $0\alpha_\ell\beta_\ell\gamma_\ell \cdots$ where $\alpha_\ell, \beta_\ell, \gamma_\ell, \ldots \in \mathbb{F}_2 = \{0, 1\}$). It is worth noting here that under the one-to-one correspondence between T_2 and $\mathbb{Z}_2$, the balls $\mathbf{B}(\ell)$ correspond, respectively, to the balls $\mathbf{B}_{1/2^{\ell+1}}(2^\ell - 1) \subset \mathbb{Z}_2$ (cf. Sect. 2.2.1 and Example 2.21). It is evident that the ball $\mathbf{B}(0)$ is the ball $\mathbf{B}_{1/2}(0)$ of radius $1/2$ centered at 0 since the prefix 1^ℓ in this case is the empty word. We see that $\mathbb{Z}_2$ is a disjoint union of all balls $\mathbf{B}(\ell)$ for $\ell = 0, 1, 2, \ldots$ and of a one-point set $\{-1\} = \mathbf{B}_0(-1)$, which is a ball of radius 0 consisting of a single point -1 since its 2-adic canonical representation is $\sum_{j=0}^\infty 2^j$; i.e., $\mathbb{Z}_p^n = (\cup_{\ell=0}^\infty \mathbf{B}(\ell)) \cup \mathbf{B}_0(-1)$.

Take $0 < \mu < 1$ and consider the base-2 expansion of $\mu = \sum_{\ell=0}^\infty \mu_\ell 2^{-\ell-1}$, where $\mu_\ell \in \{0, 1\}$, $\ell \in \mathbb{N}$. For better illustration, assume that the base-2 of μ has infinitely many non-zero μ_ℓ. Recall that the 2-adic probability measure μ_2^1 of every ball is equal to the radius of that ball (see Sect. 2.4), so $\mu_2^1(\mathbf{B}(\ell)) = 2^{-\ell-1}$, $\ell \in \mathbb{N}_0$. Take all the balls $\mathbf{B}(\ell)$ such that $\mu_\ell \neq 0$; i.e., such that $\mu_\ell = 1$; then the set $\mathbf{B} = \cup_{\mu_\ell=1}\mathbf{B}(\ell)$ is an open set whose measure is $\mu = \mu_2^1(\mathbf{B})$.

Now define a 1-Lipschitz map $F\colon \mathbb{Z}_2 \to \mathbb{Z}_2$ as follows: Consider representations of 2-adic integers as right-infinite words over the alphabet $\mathbb{F}_2 = \{0, 1\}$; take a word

$w_0 = 0\alpha_1\alpha_2\cdots$; put $F(w_0) = w_0$ if $\mu_0 = 0$; put $F(w_0) = 111\cdots$ otherwise (i.e., if $\mu_0 = 1$). Take a word $w_1 = 10\beta_2\beta_3\cdots$; put $F(w_1) = w_1$ if $\mu_1 = 0$; put $F(w_1) = 111\cdots$ otherwise. That is, given a word $w_\ell = 1^\ell 0\gamma_{\ell+1}\gamma_{\ell+2}\cdots$, put $F(w_\ell) = w_\ell$ if $\mu_\ell = 0$; put $F(w_\ell) = 111\cdots$ otherwise. Put $F(111\cdots) = 111\cdots$.

From the construction of F, it follows that if two words u, v have common prefixes of length exactly ℓ, the words $F(u)$ and $F(w)$ either have common prefixes of length exactly ℓ or $F(u) = F(v) = 111\cdots$. Therefore, F is 1-Lipschitz.

The map F either maps each ball $\mathbf{B}(\ell)$ onto a single point $111\ldots = \mathbf{B}_0(-1)$ if $\mu_\ell = 1$ or identically onto itself if $\mu_\ell = 0$. Thus, $F(\mathbb{Z}_2) = (\cup_{\ell=0,\mu_\ell=0}^{\infty}\mathbf{B}(\ell)) \cup \mathbf{B}_0(-1)$; the measure of the full F-pre-image $F^{-1}(\mathbf{B}_0(-1))$ is exactly μ as $F^{-1}(\mathbf{B}_0(-1)) = (\cup_{\ell=0,\mu_\ell=1}^{\infty}\mathbf{B}(\ell)) \cup \mathbf{B}_0(-1)$. Denote the so-constructed map F via F_μ. Note that the map F_μ can be constructed for every $0 \leqslant \mu \leqslant 1$: e.g., for $\mu = 1$, put $\mu_1 = \mu_2 = \cdots = 1$.

In loose terms, *the causal map F_μ transforms uniform probability distribution on* $\mathbb{Z}_2$ (the one given by the probability measure $\mu_{\frac{1}{2}}$) *to another distribution on $F(\mathbb{Z}_2)$; that probability distribution is a sum of* (1) *uniform distribution on the open set* $\cup_{\ell=0,\mu_\ell=0}^{\infty}\mathbf{B}(\ell)$ *whose probability is* $1-\mu$ *and of* (2) *discrete distribution since the probability hitting the point* -1 *is exactly $\mu > 0$ rather than zero probability.*

Using the construction of the map F_μ from Example 7.9, now, given $\mu \in (0, 1)$, we construct a 1-Lipschitz (thus, causal) map $F\colon \mathbb{Z}_2 \to \mathbb{Z}_2$, which transforms uniform distribution on $\mathbb{Z}_2$ to another distribution, which is uniform on an open set C and discrete on an infinite countable set D so that $F(\mathbb{Z}_2)$ is a disjoint union of C and D.

Example 7.10 Given arbitrary $0 \leqslant \mu(\ell) \leqslant 1$, where $\ell \in \mathbb{N}_0$, for every $k \in \mathbb{N}_0$, take arbitrary word $w_{\ell,k}$, whose prefix is $1^\ell 01^k 0$ (note that 1^ℓ is an empty word if $\ell = 0$). Let $\mathbf{B}(\ell, k)$ be a ball of radius $2^{-\ell-k-2}$ centered at $w_{\ell,k}$. The ball $\mathbf{B}(\ell)$ is a disjoint union of balls $\mathbf{B}(\ell, k)$, $k = 0, 1, 2\ldots$ and of a ball of zero radius (i.e., of a one-point set) centered at the word $1^\ell 01^\infty$. For every $\ell \in \mathbb{N}_0$, define F_ℓ so that F_ℓ acts on $\mathbf{B}(\ell)$ in the same way as $F_{\mu(\ell)}$ acts on $\mathbb{Z}_2$: That is, represent $\mu(\ell) = \sum_{k=0}^{\infty}\mu_k(\ell)2^{-k-1}$, where $\mu_k(\ell) \in \{0, 1\}$, $k \in \mathbb{N}$; given a word w whose prefix is $w_{\ell,k}$, put $F_\ell(w) = w$ if $\mu_k(\ell) = 0$ and $F_\ell(w) = 1^\ell 01^\infty$ otherwise; put $F_\ell(1^\ell 01^\infty) = 1^\ell 01^\infty$.

Now, given $\mu \in (0, 1)$, represent $\mu = \sum_{\ell=0}^{\infty}2^{-\ell-1}\mu(\ell)$ for suitable $\mu(\ell) \in [0, 1]$, $\ell \in \mathbb{N}_0$; then define a map $F\colon \mathbb{Z}_2 \to \mathbb{Z}_2$, which acts on every ball $\mathbf{B}(\ell)$ by the map F_ℓ, and $F(1^\infty) = 1^\infty$. Then the map F is 1-Lipschitz by the construction. The image of the so-constructed map F is a union of two disjoints sets, namely, of a countable set $D_\mu = \{1^\ell 01^\infty\colon \ell \in \mathbb{N}_0, \mu(\ell) \neq 0\} \cup \{1^\infty\}$ and of an uncountable set C_μ, which is a disjoint union of all balls $\mathbf{B}(\ell, k)$ such that $\mu_k(\ell) = 0$. By the construction, the 2-adic measure $\mu_{\frac{1}{2}}$ of the full F-pre-image of $1^\ell 01^\infty$ is $\mu_{\frac{1}{2}}(F^{-1}(1^\ell 01^\infty)) = 2^{-\ell-1}\mu(\ell)$, whence $\mu_{\frac{1}{2}}(F^{-1}(D_\mu)) = \mu$. Since by the construction the map F acts on C_μ as identity map, we have that $\mu_{\frac{1}{2}}(C_\mu) = \mu_{\frac{1}{2}}(F^{-1}(C_\mu)) = 1 - \mu$.

That is, the *the so-constructed causal map F transforms uniform probability distribution on* $\mathbb{Z}_2$ (the one given by the probability measure $\mu_{\frac{1}{2}}$) *to another* *distribution on* $F(\mathbb{Z}_2)$ *; that probability distribution is a sum of uniform distribution on the open set*

$$C_\mu = \bigcup_{\ell,k=0,\mu_k(\ell)=0}^{\infty} \mathbf{B}(\ell,k) = \bigcup_{\ell,k=0,\mu_k(\ell)=0}^{\infty} \mathbf{B}_{2^{-\ell-1}}(-(2^\ell+1)),$$

whose probability measure is $1-\mu$ *and of discrete distribution on the set* $D_\mu = \{-(2^\ell+1)\colon \ell \in \mathbb{N}_0, \mu(\ell)=1\} \cup \{-1\} \subset \mathbb{Z}_2$ *such that the probability of hitting the point* $-(2^\ell+1) \in \mathbb{Z}_2$ (the one that corresponds to the word $1^\ell 01^\infty$) *from the set* D *is* $2^{-\ell-1}$, *so the probability of hitting* D *is* μ *since the probability of hitting the point* -1 *is zero*. Note that $F(\mathbb{Z}_2)$ is a disjoint union of the sets $C = C_\mu$ and $D = D_\mu$; the so-constructed 1-Lipschitz map $F\colon \mathbb{Z}_2 \to \mathbb{Z}_2$ acts on $C_\mu \subset \mathbb{Z}_2$ as identity map, and F compresses every ball $\mathbf{B}(\ell,k) = \mathbf{B}_{2^{-\ell-1}}(-(2^\ell+1)) \subset \mathbb{Z}_2$, which is disjoint from C_μ (i.e., such that $\mu_k(\ell)=1$) to a point $-(2^\ell+1) \in \mathbb{Z}_2$.

It is time to recall some notions from measure theory (cf., e.g., [68]). Actually, in what follows by a measure, we mostly mean a probability measure.

Definition 7.11 Given a measure space $\langle \mathbb{X}; \mu\rangle$, a measure ν on $\mathbb{X}$ is called:

- *Continuous*, if $\nu(x)=0$ for all $x \in \mathbb{X}$
- *Discrete*, if $\nu(x)=0$ for all $x \in \mathbb{X}\setminus X$, where X is a countable non-empty set of $\mathbb{X}$
- *Singular* (in symbols, $\mu \perp \nu$), if ν is continuous and $\mathbb{X} = X_1 \cup X_2$ is a disjoint union of two non-empty sets $X_1, X_2 \subset \mathbb{X}$, measurable both with respect to the measures μ and ν, and such that $\nu(X_1)=0$, $\nu(X_2)>0$, $\mu(X_1)>0$, and $\mu(X_2)=0$
- *Absolutely continuous with respect to* μ (symbolically, $\nu \ll \mu$), if $\nu(X)=0$ for every μ-measurable $X \subset \mathbb{X}$ such that $\mu(X)=0$

Definition 7.12 Let $\langle \mathbb{X}; \mu\rangle$,$\langle \mathbb{Y}; \nu\rangle$ be measure spaces; let $\Phi\colon \mathbb{X} \to \mathbb{Y}$ be a measurable map. Define a measure μ_Φ on $\mathbb{Y}$ by assuming $\mu_\Phi(S) = \mu(\Phi^{-1}(S))$ for every ν-measurable subset $S \subset \mathbb{Y}$. The measure μ_Φ on $\mathbb{Y}$ is called an *image measure* on $\Phi(\mathbb{X})$.

If both μ, ν are probability measures, the image measure is usually called an *image probability* or a *pushforward probability* (measure) or, also, image (probability) distribution, pushforward (probability) distribution on $\Phi(\mathbb{X})$. If, additionally, the image $\Phi(\mathbb{X})$ is a ν-measurable subset of $\mathbb{Y}$ such that $\nu(\Phi(\mathbb{X})) \neq 0$, then on $\Phi(\mathbb{X})$, there can be defined a yet one more probability measure ν': namely, $\nu'(S) = \nu(S)/\nu(\Phi(\mathbb{X}))$ for every ν-measurable subset $S \subset \Phi(\mathbb{X})$. The measure ν' is called the *induced probability* (measure) on $\Phi(\mathbb{X})$.

Recall that *any regular Borel probability measure* μ *on* $[0,1]$ *can be uniquely represented as*

$$\mu = \alpha_1\mu_{ac} + \alpha_2\mu_d + \alpha_3\mu_s, \text{ where } \alpha_1,\alpha_2,\alpha_3 \in \mathbb{R}_{\geqslant 0}, \alpha_1+\alpha_2+\alpha_3 = 1, \tag{7.9}$$

and with respect to the Lebesgue measure λ *on* $[0, 1]$, *the measure* μ_{ac} *is absolutely continuous, the measure* μ_d *is discrete, and the measure* μ_s *is singular* (cf., e.g., [68, p.106]). This assertion can also be stated in equivalent terms of (*cumulative*) *distribution functions* (see, e.g., [142, Chapter IV, §11, Section 11.1]). In what follows, we refer to the assertions(s) of the latter sort as to (*measure/probability/distribution*) *decomposition* (*theorem(s)*).

As the probability spaces $\langle[0, 1]; \lambda\rangle$ and $\langle\mathbb{Z}_p^n; \mu_p^n\rangle$ are isomorphic as probability spaces (cf., Sect. 7.2.1), the decomposition similar to that of (7.9) holds for the probability measure μ_p^n; but what about a decomposition of that kind on $G(\mathbb{Z}_p^n)$ where G is a 1-Lipschitz (i.e., causal) map $G: \mathbb{Z}_p^n \to \mathbb{Z}_p^m$? By Proposition 7.6, we already know that $G(\mathbb{Z}_p^n)$ is closed in $\mathbb{Z}_p^m$, thus μ_p^m-measurable. It turns out that the decomposition like (7.9) on the probability space G also holds. We will not discuss this in detail since complete examination of the probability space $G(\mathbb{Z}_p^n)$ is out of scope of the book; we only show by the way of example that the decomposition takes place.

Example 7.13 Namely, following the methods we have used in Examples 7.9 and 7.10, let us construct a 1-Lipschitz map $G: \mathbb{Z}_2 \to \mathbb{Z}_2$ such that $G(\mathbb{Z}_2)$ is a disjoint union of three μ_2^1-measurable sets $C, D, S \subset \mathbb{Z}_2$ such that:

- G acts on C as an identity map, so the G-image probability measure of C is $\mu_2^1(G^{-1}(C)) = \mu_2^1(C) > 0$
- D is infinite countable and that the image probability of every point from D is non-zero, so the total G-image probability measure of D is non-zero
- S is an uncountable closed nowhere dense subset of $\mathbb{Z}_2$, and the G-image probability measure of S is non-zero though $\mu_2^1(S) = 0$

To put it in other words, we are going to construct a 1-Lipschitz map $G: \mathbb{Z}_2 \to \mathbb{Z}_2$ such that the measure μ_G is a sum of absolutely continuous measure, discrete measure, and singular measure (cf. Definitions 7.11–7.12).

To construct this mapping $G: \mathbb{Z}_2 \to \mathbb{Z}_2$, within notation from Examples 7.9–7.10, let $k, \ell \in \mathbb{N}_0$ be such that $\mu_k(\ell) = 0$; for $j \in \mathbb{N}_0$, let $\mathbf{B}(\ell, k, j) = \mathbf{B}_{2^{-\ell-k-j-3}}(1^\ell 01^k 01^j 0)$ be a ball from T_2, which is a set of all root rays whose prefix is $1^\ell 01^k 01^j 0$. Note that for all $\ell, k \in \mathbb{N}_0$, the ball $\mathbf{B}(\ell, k)$ from Example 7.10 is a disjoint union of all balls $\mathbf{B}(\ell, k, j)$, $j = 0, 1, 2, \ldots$ and of the ball $\mathbf{B}_0(1^\ell 01^k 01^j 01^\infty) = \{1^\ell 01^k 01^j 01^\infty\}$. Recall that 1^t is empty word if $t = 0$; so, e.g., $1^0 01^0 01^0 01^\infty$ corresponds to $-8 \in \mathbb{Z}_2$.

Now for $k, \ell \in \mathbb{N}_0$ such that $\mu_k(\ell) = 0$, take arbitrarily $\nu(\ell, k) \in [0, 1]$, and let $\nu(\ell, k) = \sum_{j=0}^{\infty} \nu_j(\ell, k) 2^{-j-1}$, where $\nu_j(\ell, k) \in \{0, 1\}$, the base-2 expansion of $\nu(\ell, k)$. Let F be a 1-Lipschitz map constructed in Example 7.10. To construct the mapping G, let $G(w) = F(w)$ for every $w \in \mathbf{B}(\ell, k)$ with $\mu_k(\ell) = 1$; define the action of G only on the rest of balls $\mathbf{B}(\ell, k)$ (the ones with $\mu_k(\ell) = 0$) as follows: Given a word $w = 1^\ell 01^k 00\alpha_0\alpha_1 \cdots \in \mathbf{B}(\ell, k)$, let $G(w) = w$ if $\nu_0(\ell, k) = 0$; if $\nu_0(\ell, k) = 1$, let $G(w) = 1^\ell 01^k 00\alpha_0 0\alpha_2 0\alpha_4 0\alpha_6 0 \cdots$; that is, in this case, the automaton while reading letter by letter the word w replaces α-s with odd indices by zeros leaving α-s with even indices unchanged. Further, if $w = 1^\ell 01^k 010\alpha_0\alpha_1 \cdots$, then $G(w) = w$ whenever $\nu_1(\ell, k) = 0$; else $G(w) = 1^\ell 0 1^k 010\alpha_0 0\alpha_2 0\alpha_4 0\alpha_6 0\cdots$ whenever $\nu_1(\ell, k) = 1$. Finally, for $1^\ell 0 1^k 0 1^\infty \in \mathbf{B}(\ell, k)$ with $\mu_k(\ell) = 0$, put $G(1^\ell 0 1^k 0 1^\infty) = 1^\ell 0 1^k 0 1^\infty$. This way, the map G is defined now on all the balls $\mathbf{B}(\ell, k)$ with $\mu_k(\ell) = 0$, thus on the whole T_2. The map G is 1-Lipschitz by the construction.

Now, let D' be a (disjoint) union of all balls $\mathbf{B}(\ell, k)$ with $\mu_k(\ell) = 1$; then D' is μ_2^1-measurable, and $\mu_2^1(D') = \sum_{\ell,k=0}^{\infty} \mu_k(\ell) 2^{-\ell-k-2}$. Let S be the union of all the balls $\mathbf{B}(\ell, k, j)$ such that $\mu_k(\ell) = 0$ and $\nu_j(\ell, k) = 1$; then $\mu_2^1(S) = \sum_{\ell,k,j=0}^{\infty} \nu_j(\ell, k) 2^{-\ell-k-j-3} = \nu$. The rest set C' is the union of all the balls $\mathbf{B}(\ell, k, j)$ such that $\mu_k(\ell) = \nu_j(\ell, k) = 0$. The sets D', S', C' are pairwise disjoint, $\mu_2^1(D' \cup S' \cup C') = 1$.

Finally, provided $1 > \mu > 0$ (cf. Examples 7.9–7.10) and $1 - \mu > \nu > 0$ (i.e., not all of $\nu_j(\ell, k)$ are 0, and not all of $\nu_j(\ell, k)$ are 1), the so-constructed map G is a 1-Lipschitz map $G\colon \mathbb{Z}_2 \to \mathbb{Z}_2$ (with respect to the bijection between the set $\mathbb{Z}_2$ of 2-adic integers and the tree T_2, see Sect. 2.3.3), $\mu_2^1(G(\mathbb{Z}_2)) = \mu_2^1(G(C')) = 1 - \mu - \nu > 0$, the map G acts on C' as an identity map, and G maps D' onto a countable set $D = D_\mu$ from Example 7.10. Thus, $\mu_1^2(D) = 0$, while $\mu_2^1(G^{-1}(D)) = \mu_2^1(D') = \mu$; the set $G(S') = S$ is an uncountable closed nowhere dense subset, $\mu_2^1(S)) = 0$ and $\mu_2^1(G^{-1}(S)) = \mu_2^1(S')) = \nu > 0$; the image $G(\mathbb{Z}_2)$ is a union of pairwise disjoint sets D, S, and C. Therefore, $G(\mathbb{Z}_2)$ is a measure space such that the measure μ_G (cf., Definition 7.12) can be decomposed as $\mu_G = \mu_d + \mu_s + \mu_c$, where the measure μ_d concentrated on D is discrete, the measure μ_s concentrated on S is singular, and the measure μ_c concentrated on C is absolutely continuous (cf., Definition 7.11).

Now we can state the following theorem:

Theorem 7.14 *Given $m, n \in \mathbb{N}$, $n \leqslant m$, given $0 \leqslant p_c, p_s, p_d \leqslant 1$ such that $p_c + p_s + p_d = 1$, there exists a 1-Lipschitz map $G\colon \mathbb{Z}_p^n \to \mathbb{Z}_p^m$ such that the probability measure μ_G on $\mathbb{Z}_p^m$ is a sum of absolutely continuous probability measure μ_c, of discrete probability measure μ_d, and of singular probability measure μ_s, namely, $\mu_G = p_c\mu_c + p_d\mu_d + p_s\mu_s$.*

Proof The idea of the proof is clear from Examples 7.9–7.13 but detailed proof is involved; that is why we leave it to the interested reader. □

Chapter 8
The Main Ergodic Theorem for p-Adic 1-Lipschitz Maps

Abstract The central result of this chapter is Theorem 8.2, which completely describes measure-preserving (locally) 1-Lipschitz maps $\mathbb{Z}_p^n \twoheadrightarrow \mathbb{Z}_p^m$ for $m \leqslant n$ as well as ergodic (locally) 1-Lipschitz maps $\mathbb{Z}_p^n \twoheadrightarrow \mathbb{Z}_p^n$.

8.1 Measure-Preserving Isometries

Given a (locally) 1-Lipschitz map $F\colon \mathbb{Z}_p^n \to \mathbb{Z}_p^m$ of $\mathbb{Z}_p^n$ onto $\mathbb{Z}_p^m$, $m \leqslant n$, the map $F \bmod p^k\colon (\mathbb{Z}/p^k\mathbb{Z})^n \to (\mathbb{Z}/p^k\mathbb{Z})^m$ is well defined due to the compatibility of F with every epimorphism $\bmod p^k$, for all (sufficiently large) k (see Sect. 5.1); so we can define the notion of hereditable properties with respect to the epimorphism $\bmod p^k$ in the same way as we did in Sect. 7.1.1. As we focus on basic properties of ergodic theory, the measure-preservation and the ergodicity, the following notions are the most important further:

Definition 8.1 (Bijectivity, Transitivity, Balance Modulo p^k) A locally 1-Lipschitz function $F\colon \mathbb{Z}_p^n \to \mathbb{Z}_p^m$ is said to be *balanced modulo p^k* (respectively, *bijective modulo p^k* or *transitive modulo p^k*) if the reduced map

$$F \bmod p^k\colon (\mathbb{Z}/p^k\mathbb{Z})^n \to (\mathbb{Z}/p^k\mathbb{Z})^m$$

is well defined and balanced (respectively, bijective or transitive).

We stress once again that if F is a 1-Lipschitz function, then all reduced maps $F \bmod p^k$, $k = 1, 2, 3, \ldots$, are well defined!

Now the central result of this chapter can be stated as follows:

Theorem 8.2 *Let F be a (locally) 1-Lipschitz map of $\mathbb{Z}_p^n$ onto $\mathbb{Z}_p^m$, $m \leqslant n$. Whenever $m = n$, the map F preserves the p-adic measure μ_p (or, accordingly, is ergodic with respect to μ_p) if and only if F bijective (accordingly, transitive) modulo p^k for all (sufficiently large) $k = 1, 2, 3, \ldots$.*

V. Anashin, *Causality: The p-adic Theory*, STEAM-H: Science, Technology, Engineering, Agriculture, Mathematics & Health,
https://doi.org/10.1007/978-3-031-85818-5_8

For $n \geqslant m$, the map F preserves the measure μ_p if and only if F is balanced modulo p^k for all (sufficiently large) $k = 1, 2, 3, \ldots$.

We will prove the theorem following the lines of [12] but improving some flaws in the proof given in the said paper. Proposition 8.3 proves the "only if" part of the statement of Theorem 8.2. The rest of the statement follows directly from Propositions 8.11 and 8.12, which are stated and proved further; the propositions are of special interest themselves. For instance, as a bonus, we obtain that *a 1-Lipschitz measure-preserving transformation on $\mathbb{Z}_p^n$ is an isometry bijection.* So as not to overload the notation, we prove the first claim of the theorem only for the case $m = n = 1$ since proofs can be easily adjusted for the general case. Moreover, by the same reason, we will prove the theorem only for 1-Lipschitz maps leaving corresponding adjustments of statements and proofs to the readers.

For locally 1-Lipschitz functions, both measure-preservation and ergodicity turn out to be hereditable properties with respect to $\bmod p^k$. We recall that when configuration space is finite, measure-preservation means balance, whereas ergodicity means transitivity.

Proposition 8.3 *Given a measure-preserving (locally) 1-Lipschitz map F of $\mathbb{Z}_p^n$ onto $\mathbb{Z}_p^m$, $m \leqslant n$, the reduced map $F \bmod p^k \colon (\mathbb{Z}/p^k\mathbb{Z})^n \to (\mathbb{Z}/p^k\mathbb{Z})^m$ is balanced, for all (sufficiently large) $k = 1, 2, 3, \ldots$.*

If $m = n$, and if the map F is ergodic, the reduced map $F \bmod p^k$ is transitive for all (sufficiently large) $k = 1, 2, 3, \ldots$.

Proof During the proof, we denote $F \bmod p^k$ via $\bar{F}_k$ so as not to overload further formulae.

To prove the first claim of the proposition, let for some k there exist $\bar{x}, \bar{y} \in (\mathbb{Z}/p^k\mathbb{Z})^m = \{0, 1, \ldots, p^k - 1\}^m$ such that $\#\bar{F}_k^{-1}(\bar{x}) \neq \#\bar{F}_k^{-1}(\bar{y})$. Thus, both $\bar{F}_k^{-1}(\bar{x})$ and $F_k^{-1}(\bar{y})$ are disjoint subsets of a finite set $(\mathbb{Z}/p^k\mathbb{Z})^n = \{0, 1, \ldots, p^k - 1\}^n$. Consider two balls $\bar{x} + p^k\mathbb{Z}_p^m$ and $\bar{y} + p^k\mathbb{Z}_p^m$ in $\mathbb{Z}_p^m$. Then

$$F^{-1}(\bar{x} + p^k\mathbb{Z}_p^m) = \bigcup_{z \in \bar{F}_k^{-1}(\bar{x})} (z + p^k\mathbb{Z}_p^n),$$

$$F^{-1}(\bar{y} + p^k\mathbb{Z}_p^m) = \bigcup_{z \in \bar{F}_k^{-1}(\bar{y})} (z + p^k\mathbb{Z}_p^n),$$

as a pre-image of a ball of (sufficiently small) radius r with respect to a (locally) 1-Lipschitz map is a finite union of pairwise disjoint balls having the same radii r. Therefore, $\mu_p(F^{-1}(\bar{x} + p^k\mathbb{Z}_p^m)) \neq \mu_p(F^{-1}(\bar{y} + p^k\mathbb{Z}_p^m))$, a contradiction.

By the conditions of the second claim, $m = n$ and F is ergodic; thus, F preserves the measure μ_p. Hence, by the first claim for (a sufficiently large) $k \in \mathbb{N}$, the mapping $\bar{F}_k$ is a permutation of the elements of the residue ring $(\mathbb{Z}/p^k\mathbb{Z})^n$. If for some k the permutation $\bar{F}_k$ has more than one cycle, then there exists a proper subset $\bar{A} \subset (\mathbb{Z}/p^k\mathbb{Z})^n = \{0, 1, \ldots, p^k - 1\}^n$ such that $\bar{F}_k^{-1}(\bar{A}) = \bar{A}$.

This implies that $F^{-1}(\bar{A} + p^k\mathbb{Z}_p^n) = \bar{F}_k^{-1}(\bar{A}) + p^k\mathbb{Z}_p^n = \bar{A} + p^k\mathbb{Z}_p^n$. However, $\mu_p(\bar{A} + p^k\mathbb{Z}_p^n) = (\#\bar{A})p^{-kn}$, and $0 < (\#\bar{A})p^{-kn} < 1$ as $\bar{A}$ is a proper subset in $\{0, 1, \ldots, p^k - 1\}^n$. This contradicts the ergodicity of F. □

Now we prove that a 1-Lipschitz map $F\colon \mathbb{Z}_p^n \to \mathbb{Z}_p^n$ preserves the measure μ_p if and only if f is bijective modulo p^k, for all $k = 1, 2, \ldots$. As said, we consider the case $n = 1$ just to simplify notation; statements of Propositions 8.4 and 8.6 as well as of Notes 8.5, 8.8 and of Corollary 8.7 remain true for arbitrary $n \in \mathbb{N}$; the respective proofs are quite similar to those for the case $n = 1$. It is worth noticing here that Proposition 8.4 can be deduced also from a more general result stated in Sect. 8.2. However, we present a separate proof for that proposition to obtain some extra information about the functions the proposition deals with.

Proposition 8.4 *A 1-Lipschitz measure-preserving function $f\colon \mathbb{Z}_p \to \mathbb{Z}_p$ is a bijection of $\mathbb{Z}_p$ onto itself.*

Proof We prove that f is both injective and surjective.

Claim 1: *Under conditions of Proposition 8.4, the function f is injective.*

Indeed, if there exist $a, b \in \mathbb{Z}_p$ $(a \neq b)$ such that $f(a) = f(b) = z$, then for some k, the balls $a + p^k\mathbb{Z}_p$ and $b + p^k\mathbb{Z}_p$ are disjoint, whereas $f(a + p^k\mathbb{Z}_p), f(b + p^k\mathbb{Z}_p) \subset z + p^k\mathbb{Z}_p$. Hence, $\mu_p(f^{-1}(z + p^k\mathbb{Z}_p)) \geqslant 2p^{-k}$ since $f^{-1}(z + p^k\mathbb{Z}_p)) \supset a + p^k\mathbb{Z}_p, b + p^k\mathbb{Z}_p$; so f does not preserve μ_p.

Claim 2: *Under conditions of Proposition 8.4, the function f is bijective modulo p^k for all $k = 1, 2, \ldots$.*

Already proved; see Proposition 8.3.

Claim 3: *Under conditions of Proposition 8.4 the function f is surjective.*

Take arbitrary $z \in \mathbb{Z}_p$. Then in view of Claim 2, there exists exactly one $x_1 \in \mathbb{Z}/p\mathbb{Z}$ such that $f(x_1) \equiv z \pmod{p}$ (here and further, we identify elements of the residue ring $\mathbb{Z}/p^k\mathbb{Z}$ with non-negative rational integers $0, 1, \ldots, p^k - 1$ in an obvious way). Similarly, there exists exactly one $x_2 \in \mathbb{Z}/p^2\mathbb{Z}$ such that $f(x_2) \equiv z \pmod{p^2}$; whence necessarily $x_2 \equiv x_1 \pmod{p}$, etc.

This way we obtain a sequence $x_1, x_2, \ldots$ such that $|f(x_i) - z|_p \leqslant p^{-i}$ and $|x_{i+1} - x_i|_p \leqslant p^{-i}$ for $i = 1, 2, \ldots$. This is an exercise left to the reader to show that the sequence $x_1, x_2, \ldots$ is a Cauchy sequence (which hence converges to some $x \in \mathbb{Z}_p$) and that $f(x) = z$. □

Note 8.5 As a bonus, we have that whenever a 1-Lipschitz function $g\colon \mathbb{Z}_p \to \mathbb{Z}_p$ is bijective modulo p^k for all $k = 1, 2, \ldots$, it is a bijection of $\mathbb{Z}_p$ onto $\mathbb{Z}_p$ (see proofs of Claims 2 and 3 of the proof of Proposition 8.4).

Proposition 8.6 *Let a 1-Lipschitz function $g\colon \mathbb{Z}_p \to \mathbb{Z}_p$ be bijective modulo p^k for all $k = 1, 2, \ldots$. Then g preserves the measure μ_p.*

Proof In view of Note 8.5, the function g is a bijection of $\mathbb{Z}_p$ onto $\mathbb{Z}_p$; whence, there exists an inverse function $f = g^{-1}$, which is also a bijection of $\mathbb{Z}_p$ onto $\mathbb{Z}_p$. Moreover, f is continuous since g is continuous.

Claim 1: *f is 1-Lipschitz.*

If there are $a, b \in \mathbb{Z}_p$ such that $a \equiv b \pmod{p^k}$ and $f(a) \not\equiv f(b) \pmod{p^k}$, then assuming $a = g(u)$, $b = g(v)$ for uniquely defined $u, v \in \mathbb{Z}_p$, we have $g(u) \equiv g(v) \pmod{p^k}$, and $f(g(u)) \not\equiv f(g(v)) \pmod{p^k}$; that is, $g(u) \equiv g(v) \pmod{p^k}$, and $u \not\equiv v \pmod{p^k}$. The latter contradicts conditions of Proposition 8.6.

Claim 2: *$f(a+p^k\mathbb{Z}_p) = f(a)+p^k\mathbb{Z}_p$ for every $a \in \mathbb{Z}_p$ and every $k = 1, 2, \ldots$.*

In view of Claim 1, $f(a+p^k\mathbb{Z}_p) \subset f(a)+p^k\mathbb{Z}_p$. To prove the inverse inclusion, denote $f(a) = b$; then $g(b) = a$. Since g is 1-Lipschitz, $g(b + p^k\mathbb{Z}_p) \subset g(b) + p^k\mathbb{Z}_p$. Applying a bijection f to both sides of the inclusion, we get $b + p^k\mathbb{Z}_p \subset f(g(b) + p^k\mathbb{Z}_p)$ as f is 1-Lipschitz (see Claim 1); that is, $f(a) + p^k\mathbb{Z}_p \subset f(a + p^k\mathbb{Z}_p)$, the needed inverse inclusion.

Claim 3: *f is bijective modulo p^k for all $k = 1, 2, \ldots$.*

Assuming there exist $u, v \in \mathbb{Z}_p$ and $k \in \{1, 2, \ldots\}$ such that $u \not\equiv v \pmod{p^k}$ and $f(u) \equiv f(v) \pmod{p^k}$, one obtains that $u + p^k\mathbb{Z}_p \bigcap v + p^k\mathbb{Z}_p = \emptyset$ and $f(u) + p^k\mathbb{Z}_p = f(v) + p^k\mathbb{Z}_p$, a contradiction in view of Claim 2.

Claim 4: *f satisfies conditions of Proposition 8.6.*

See Claims 1 and 3.

Claim 5: *$g(a+p^k\mathbb{Z}_p) = g(a)+p^k\mathbb{Z}_p$ for every $a \in \mathbb{Z}_p$ and every $k = 1, 2, \ldots$.*

See Claim 4.

Claim 6: *$\mu_p(g(M)) = \mu_p(M)$, for every μ_p-measurable $M \subset \mathbb{Z}_p$.*

Since M is measurable, then

$$\mu_p(M) = \inf\{\mu_p(V) \colon V \supset M, V \text{ is open in } \mathbb{Z}_p\}$$

as the measure μ_p is regular.

Since V is open, it is a disjoint union of a countable number of balls V_j of non-zero radius each: $V = \bigcup_{j\in J} V_j$. Then $g(V) = \bigcup_{j\in J} g(V_j)$, since g is a bijection. Note that in view of Claim 5, each $g(V_j)$ is a ball of a radius that is equal to the one of the ball V_j; that is, $\mu_p(g(V_j)) = \mu_p(V_j)$, for all $j \in J$. Moreover, the balls are disjoint: $g(V_i) \bigcap g(V_j) = \emptyset$ whenever $i \neq j$ (since $f(g(V_i) \bigcap g(V_j)) = V_i \bigcap V_j$ in view of Claim 2). This implies that $\mu_p(g(V)) = \mu_p(V)$. Note that $g(V)$ is open since g is a continuous bijection. Hence,

$$\mu_p(g(M)) \leqslant \inf\{\mu_p(g(V)) \colon V \supset M, V \text{ is open in } \mathbb{Z}_p\} = \mu_p(M).$$

In view of Claim 4, one has then $\mu_p(f(R)) \leqslant \mu_p(R)$, for every measurable $R \subset \mathbb{Z}_p$. Now we take $R = g(M)$ (whence $f(R) = M$) and obtain $\mu_p(M) \leqslant \mu_p(g(M))$, thus proving the proposition. □

Corollary 8.7 *A 1-Lipschitz function $f \colon \mathbb{Z}_p\, \mathbb{Z}_p$ preserves measure if and only if it is bijective modulo p^k for all $k = 1, 2, \ldots$.*

Proof Necessity of the conditions is proved by Proposition 8.3, whereas their sufficiency is proved by Proposition 8.6. □

Note 8.8 As a bonus, we have that *every 1-Lipschitz measure-preserving function $f\colon \mathbb{Z}_p \to \mathbb{Z}_p$ is an isometry*: A distance between two points is just a radius of the smallest ball that contains them both; however, as it was shown, a measure-preserving 1-Lipschitz mapping is a bijection that merely permutes balls of pairwise equal radii.

8.2 1-Lipschitz Measure-Preserving Functions

Now we prove that a 1-Lipschitz function $F\colon \mathbb{Z}_p^n \to \mathbb{Z}_p^m$, $m \leqslant n$, preserves the measure μ_p if and only if F is balanced modulo p^k, for all $k = 1, 2, \ldots$. The "only if" part is already proved (see Proposition 8.3). To prove the "if" part, we need the following lemma.

Lemma 8.9 *Let a 1-Lipschitz function $F\colon \mathbb{Z}_p^n \to \mathbb{Z}_p^m$, $m \leqslant n$, be balanced modulo p^k, for all $k = 1, 2, \ldots$. Then for every $b \in \mathbb{Z}_p^m$, a full pre-image $F^{-1}(b + p^s\mathbb{Z}_p^m)$ is a union of $p^{s(n-m)}$ pairwise disjoint balls $a_j + p^s\mathbb{Z}_p^n$ and $j = 1, 2, \ldots, p^{s(n-m)}$.*

Proof We start with proving the lemma "modulo p^k." Recall that $\bar{F}_k$ stands for the reduced map $F \bmod p^k$.

Claim 1. *For every $\bar{b}_k \in (\mathbb{Z}/p^k\mathbb{Z})^m$, a full pre-image $\bar{F}_k^{-1}(\bar{b}_k + p^s(\mathbb{Z}/p^k\mathbb{Z})^m)$ of the coset $\bar{b}_k + p^s(\mathbb{Z}/p^k\mathbb{Z})^m \subset (\mathbb{Z}/p^k\mathbb{Z})^m$ is a disjoint union of $p^{s(n-m)}$ suitable pairwise disjoint cosets:*

$$\bar{F}_k^{-1}(\bar{b}_k + p^s(\mathbb{Z}/p^k\mathbb{Z})^m) = \bigcup_{j=1}^{p^{s(n-m)}} (\bar{a}_{k,j} + p^s(\mathbb{Z}/p^k\mathbb{Z})^n).$$

Here and further we assume that $s \leqslant k$. In this case,

$$\#(\bar{b}_k + p^s(\mathbb{Z}/p^k\mathbb{Z})^m) = p^{m(k-s)},$$

and since F is balanced modulo p^k, then

$$\#\bar{F}_k^{-1}(\bar{b}_k + p^s(\mathbb{Z}/p^k\mathbb{Z})^m) = p^{k(n-m)}p^{m(k-s)} = p^{kn-ms}. \tag{8.1}$$

Further, since F is balanced modulo p^s, then $\#\bar{F}_s^{-1}(\bar{b}_s) = p^{s(n-m)}$, for every $\bar{b}_s \in \{0, 1, \ldots, p^s - 1\}^m = (\mathbb{Z}/p^s\mathbb{Z})^m$. Take $\bar{b}_s \equiv \bar{b}_k \pmod{p^s}$, and let

$$\bar{F}_s^{-1}(\bar{b}_s) = \{\bar{a}_{s,1}, \ldots, \bar{a}_{s,p^{s(n-m)}}\} \subset (\mathbb{Z}/p^s\mathbb{Z})^n = \{0, 1, \ldots, p^s - 1\}^n.$$

For $j = 1, 2, \ldots, p^{s(n-m)}$, choose (and fix) $\bar{a}_{k,j} \in (\mathbb{Z}/p^k\mathbb{Z})^n$ so that $\bar{a}_{k,j} \equiv \bar{a}_{s,j} \pmod{p^s}$. Note that the latter congruence, in accordance with what has been agreed at the beginning of Sect. 3.2, just means that $|\bar{a}_{k,j}-\bar{a}_{s,j}|_p \leqslant p^{-s}$; that is, $\bar{a}_{k,j}^{(i)} \equiv \bar{a}_{s,j}^{(i)} \pmod{p^s}$ for each i-th component $\bar{a}_{k,j}^{(i)}$ of $\bar{a}_{k,j} \in (\mathbb{Z}/p^k\mathbb{Z})^n = \{0, 1, \dots, p^k - 1\}^n$, $i = 1, 2, \dots, n$.

Now, for $j = 1, 2, \dots, p^{s(n-m)}$, take $\hat{a}_{k,j} \in (\mathbb{Z}/p^k\mathbb{Z})^n$ so that $\hat{a}_{k,j} \equiv \bar{a}_{s,j} \pmod{p^s}$; that is, $\hat{a}_{k,j} \in \bar{a}_{k,j} + p^s(\mathbb{Z}/p^k\mathbb{Z})^n$ and vice versa. Since F is 1-Lipschitz, $\bar{F}_k(\hat{a}_{k,j}) \equiv \bar{b}_s \pmod{p^s}$; thus, $\bar{F}_k(\hat{a}_{k,j}) \in \bar{b}_k + p^s(\mathbb{Z}/p^k\mathbb{Z})^m$ (recall that $\bar{b}_s \equiv \bar{b}_k \pmod{p^s}$ by our choice). So every $\hat{a}_{k,j}$ is a $\bar{F}_k$-pre-image of a certain element of the coset $\bar{b}_k + p^s(\mathbb{Z}/p^k\mathbb{Z})^m$, and there are exactly $p^{s(n-m)}p^{n(k-s)} = p^{nk-ms}$ these $\hat{a}_{k,j}$. Comparing this number with what is given by Equation (8.1), we conclude that all these $\hat{a}_{k,j}$ constitute the full pre-image $\bar{F}_k^{-1}(\bar{b}_k + p^s(\mathbb{Z}/p^k\mathbb{Z})^m)$, which is then just the union of cosets $\bar{a}_{k,j} + p^s(\mathbb{Z}/p^k\mathbb{Z})^n$ over $j \in \{1, \dots, p^{s(n-m)}\}$. These cosets are disjoint since all $\bar{a}_{k,j}$ are different modulo p^s.

Claim 2. *For $j = 1, 2, \dots, p^{s(n-m)}$ fix $a_j \in \mathbb{Z}_p^n$ such that $a_j \equiv \bar{a}_{s,j} \pmod{p^s}$, where $\bar{a}_{s,j}$ are defined as above for $\bar{b}_k \equiv b \pmod{p^k}$. Then*

$$F^{-1}(b + p^s\mathbb{Z}_p^m) = \bigcup_{j=1}^{p^{s(n-m)}} (a_j + p^s\mathbb{Z}_p^n).$$

First note that in this setting, the definition of $\bar{a}_{s,j}$ (whence, of a_j) does not depend on k but only on b and s, since for $\bar{b}_k \equiv b \pmod{p^k}$, the set $\{\bar{a}_{s,1}, \dots, \bar{a}_{s,p^{s(n-m)}}\}$ is just a full $\bar{F}_s$-pre-image of $(b \bmod p^s)$. Here $(b \bmod p^s)$ is a unique non-negative rational integer that lays at the distance p^{-s} from the point b, an approximation of b by a non-negative rational integer with accuracy p^{-s} with respect to p-adic metric. In other words, given $b \in \mathbb{Z}_p^m$, we put $\bar{b}_s \equiv b \pmod{p^s}$, where $\bar{b}_s \in \{1, 2, \dots, p^s-1\}^m$, and then take all solutions $\bar{a}_{s,j} \in \{1, 2, \dots, p^s-1\}^n$ of the congruence $\bar{F}_s(x) \equiv \bar{b}_s \pmod{p^s}$ in indeterminate x, and after that, for each of these $p^{s(n-m)}$ solutions $\bar{a}_{s,j}$, we choose an arbitrary $a_j \in \mathbb{Z}_p^n$ so that $a_j \equiv \bar{a}_{s,j} \pmod{p^s}$.

From the definition of $\bar{a}_j$, it follows immediately that for every $h \in \mathbb{Z}_p^n$, $F(a_j + p^s h) \equiv b \pmod{p^s}$ since F is 1-Lipschitz; whence $F^{-1}(b + p^s\mathbb{Z}_p^m) \supset \bigcup_{j=1}^{p^{s(n-m)}} (a_j + p^s\mathbb{Z}_p^n)$. Thus, we must prove the inverse inclusion only.

Given $c \in b + p^s\mathbb{Z}_p^m$, for every $k \geqslant s$ from Claim 1, it follows that $F^{-1}(c) \in \bar{F}_k^{-1}(c \bmod p^k) + p^k\mathbb{Z}_p^n$, where $\bar{F}_k^{-1}(c \bmod p^k)$ is a subset of a finite set $\bigcup_{j=1}^{p^{s(n-m)}} (\bar{a}_{k,j} + p^s\{0, 1, \dots, p^{k-s} - 1\}^n)$.

Thus, applying Claim 1, we obtain:

$$F^{-1}(c) \in \bigcap_{k=s}^{\infty}(\bar{F}_k^{-1}(c \bmod p^k) + p^k\mathbb{Z}_p^n) \subset$$

$$\subset \bigcap_{k=s}^{\infty} \bigcup_{j=1}^{p^{s(n-m)}} (\bar{a}_{k,j} + p^s\{0, 1, \ldots, p^{k-s} - 1\}^n + p^k\mathbb{Z}_p^n) =$$

$$= \bigcup_{j=1}^{p^{s(n-m)}} \bigcap_{k=s}^{\infty}(\bar{a}_{k,j} + p^s\{0, 1, \ldots, p^{k-s} - 1\}^n + p^k\mathbb{Z}_p^n) =$$

$$= \bigcup_{j=1}^{p^{s(n-m)}} \bigcap_{k=s}^{\infty}(\bar{a}_{s,j} + p^s\{0, 1, \ldots, p^{k-s} - 1\}^n + p^k\mathbb{Z}_p^n) =$$

$$= \bigcup_{j=1}^{p^{s(n-m)}} (\bar{a}_{s,j} + p^s\mathbb{Z}_p^n) = \bigcup_{j=1}^{p^{s(n-m)}} (a_j + p^s\mathbb{Z}_p^n) \tag{8.2}$$

This finishes the proof of Lemma 8.9. □

Corollary 8.10 $\mu_p(F^{-1}(b+p^s\mathbb{Z}_p^m)) = \sum_{j=1}^{p^{s(n-m)}} \mu_p(a_j+p^s\mathbb{Z}_p^n) = p^{s(n-m)}p^{-sn} = p^{-sm} = \mu_p(b + p^s\mathbb{Z}_p^m))$.

Proposition 8.11 *Under conditions of Lemma 8.9, the function F preserves the measure μ_p.*

Proof Balls $b + p^s\mathbb{Z}_p^m$ constitute a base of a σ-ring of all μ_p-measurable sets of the space $\mathbb{Z}_p^m$. In view of Corollary 8.10, F is then a measurable mapping; that is, any pre-image of a measurable set is measurable. Now, let us find $\mu_p(F^{-1}(M)$ for a measurable $M \subset \mathbb{Z}_p^m$.

Any open subset $A \subset \mathbb{Z}_p^m$ is a union of balls since all balls of non-zero radii (these balls are clopen then) constitute a base of the p-adic topology: $A = \cup_{i\in I}\mathbf{B}_i$, where I is a set of indexes and all $\mathbf{B}_i$ are balls of non-zero radii. Given any two balls, they are either disjoint or one of them contains another one. As every ball from $\mathbb{Z}_p^n$ is of a form $a + p^s\mathbb{Z}_p^m$, its radius is p^{-sm}. As there are only finitely many balls of a fixed non-zero radius in $\mathbb{Z}_p^m$, and as if $a + p^s\mathbb{Z}_p^m \subset b + p^t\mathbb{Z}_p^m$ then $p^{-tm} \leqslant p^{-sm}$, then there exist maximal with respect to inclusion balls in the set $\{\mathbf{B}_i : i \in I\}$, and there are not more than countably many these maximal balls since there only countably many balls of non-zero radii in $\mathbb{Z}_p^m$. Therefore A is a disjoint union of these maximal balls; so *every open subset in $\mathbb{Z}_p^m$ is a disjoint union of a countably many balls of non-zero radii.*

As any open measurable subset $A \subset \mathbb{Z}_p^m$ is a countable disjoint union of balls of non-zero radii, then $F^{-1}(A)$ is an open measurable subset of $\mathbb{Z}_p^n$, and

$\mu_p(F^{-1}(A)) = \mu_p(A)$ in view of Corollary 8.10. As any closed subset $C \subset \mathbb{Z}_p^m$ is a complement for a suitable open subset A, i.e., $C = \mathbb{Z}_p^m \setminus A$, then $F^{-1}(C) = \mathbb{Z}_p^n \setminus F^{-1}(A)$, and therefore, $F^{-1}(C)$ is a closed measurable subset of $\mathbb{Z}_p^n$ of measure $\mu_p(F^{-1}(C)) = 1 - \mu_p(F^{-1}(A)) = 1 - \mu_p(A) = \mu_p(C)$. Further, for a measurable M, one has $\mu_p(M) = \inf\{\mu_p(V) \colon V \supset M, V \text{ is open in } \mathbb{Z}_p^m\}$ as the measure μ_p is regular; thus,

$$\mu_p(F^{-1}(M)) \leqslant \inf\{\mu_p(F^{-1}(V)) \colon V \supset M, V \text{ is open in } \mathbb{Z}_p^m\} = \mu_p(M).$$

On the other hand, $\mu_p(M) = \sup\{\mu_p(W) \colon W \subset M, W \text{ is closed in } \mathbb{Z}_p^m\}$ since μ_p is regular. But $\mu_p(F^{-1}(W))$ is a closed subset of $\mathbb{Z}_p^n$; moreover, we have already proved that $\mu_p(F^{-1}(W)) = \mu_p(W)$ for every closed $W \subset \mathbb{Z}_p^m$. Thus,

$$\mu_p(F^{-1}(M)) \geqslant \sup\{\mu_p(F^{-1}(W)) \colon W \subset M, W \text{ is closed in } \mathbb{Z}_p^m\} = \mu_p(M).$$

Finally, we conclude that $\mu_p(F^{-1}(M)) = \mu_p(M)$, thus proving the proposition.

□

8.3 1-Lipschitz Ergodic Functions

We finally characterize ergodic functions among all 1-Lipschitz functions $F \colon \mathbb{Z}_p^n \to \mathbb{Z}_p^n$.

Proposition 8.12 *A 1-Lipschitz function $F \colon \mathbb{Z}_p^n \to \mathbb{Z}_p^n$ is ergodic if and only if F is transitive modulo p^k, for all $k = 1, 2, \ldots$.*

Proof The "only if" part has already been proven (see Proposition 8.3). We need to prove the "if" part of the statement only.

By the main claim of Sect. 8.2, under conditions of Proposition 8.12, the function F is a measure-preserving isometry by Note 8.8, which holds for measure-preserving maps of $\mathbb{Z}_p^n$ to $\mathbb{Z}_p^n$ as well. Moreover, the conditions imply that every F-orbit is dense in $\mathbb{Z}_p^n$ since given arbitrary $x, y \in \mathbb{Z}_p^n$ and $k \in \mathbb{N}$, there exists $\ell \in \mathbb{N}_0$ such that $\bar{y} = \bar{F}_k^\ell(\bar{x})$ where $\bar{F}_k$ stands for the reduced map $F \bmod p^k$ and $\bar{x}, \bar{y}$ are respective residues modulo p^k in the Cartesian product $(\mathbb{Z}/p^k\mathbb{Z})^n$; so $|F^\ell(x) - y|_p \leqslant p^{-k}$. Therefore, F is a minimal isometry on a compact metric space; but all such isometries are known to be uniquely ergodic (thus ergodic) (cf., Theorem 1.19). □

For $n = 1$, summarize the results of this section in the following theorem:

Theorem 8.13 *Let $f \colon \mathbb{Z}_p \to \mathbb{Z}_p$ be an onto 1-Lipschitz map. The following propositions are equivalent:*

1. *f is minimal.*
2. *f is conjugate to the translation $t: x \mapsto x + 1$ on $\mathbb{Z}_p$.*
3. *f is uniquely ergodic.*
4. *f is ergodic.*

We haven't proved claim 2 here, but the result is well-known (see, e.g., [56, 84]). Note that the conjugating map is also a 1-Lipschitz map $\mathbb{Z}_p \to \mathbb{Z}_p$.

Concluding the current chapter, we note that ergodicity of automata maps, i.e., the 1-Lipschitz maps $\mathbb{Z}_p \to \mathbb{Z}_p$, can be studied with respect to measures other than the Haar measure on $\mathbb{Z}_p$ (see [69, 70]), but this is quite another story that we do not touch in the book. Also, problems like ergodic decompositions, repellers, attractors, and other dynamical properties of general 1-Lipschitz (thus, automata) functions on $\mathbb{Z}_p$ are out of scope of the current book; we refer the interested reader to, e.g., [63–65, 155, 166].

Chapter 9
1-Lipschitz Ergodicity on $\mathbb{Z}_p$

Abstract In this chapter, we prove a number of results on ergodicity (and measure-preservation) of 1-Lipschitz transformations on $\mathbb{Z}_p$ for various important classes of these transformations.

9.1 Ergodicity of Affine Mappings

In this section, we obtain explicit conditions for ergodicity of affine transformation $f(x) = ax + b$ on the space $\mathbb{Z}_p$, where $a, b \in \mathbb{Z}_p$. This case serves as a base for further considerations; also, it is important for applications (see Chaps. 13 and 16).

In view of Theorem 8.2, it is clear that f is measure-preserving if and only if a has a multiplicative inverse modulo p^k for all $k = 1, 2, \dots$ (i.e., a is a unit in $\mathbb{Z}_p$), in other words, if and only if $a \not\equiv 0 \pmod{p}$. The following theorem is actually a restatement of known necessary and sufficient conditions for the sequence produced by a so-called mixed congruential method to attain the longest possible period (cf., e.g., [123, Subsection 3.2.1] and references therein). We restate these conditions in dynamical terms and give another proof based on p-adic techniques.

Theorem 9.1 *The function $f(x) = ax + b$, where $a, b \in \mathbb{Z}_p$, is an ergodic transformation on $\mathbb{Z}_p$ if and only if the following conditions hold simultaneously:*

$$b \not\equiv 0 \pmod{p}; \tag{9.1}$$

$$a \equiv 1 \pmod{p}, \quad \textit{for } p \textit{ odd}; \tag{9.2}$$

$$a \equiv 1 \pmod{4}, \quad \textit{for } p = 2. \tag{9.3}$$

Proof of Theorem 9.1 In view of Theorem 8.2, we must prove that f is transitive modulo p^k if and only if conditions of Theorem 9.1 hold. We prove this by induction on k, and we state a base of induction as a lemma:

Lemma 9.2 *The function $f(x) = ax + b$ is transitive modulo p if and only if $b \not\equiv 0 \pmod{p}$ and $a \equiv 1 \pmod{p}$.*

V. Anashin, *Causality: The p-adic Theory*, STEAM-H: Science, Technology, Engineering, Agriculture, Mathematics & Health,
https://doi.org/10.1007/978-3-031-85818-5_9

Proof of Lemma 9.2 It is clear that $a \not\equiv 0 \pmod p$ (otherwise, f is a constant) and that $b \not\equiv 0 \pmod p$ (otherwise, 0 is a fixed point of f). Now, as for every $i = 1, 2 \ldots$

$$f^i(x) = a^i x + b(a^{i-1} + a^{i-2} + \cdots + a + 1) \tag{9.4}$$

we conclude that if $a \not\equiv 1 \pmod p$, then $a - 1$ is invertible in $\mathbb{Z}_p$ (cf., Sect. 2.2.2), and so $f^p(x) = a^p x + b(a^p - 1)(a - 1)^{-1}$ where $(a - 1)^{-1}$ is a multiplicative inverse of $(a - 1)$ in the ring $\mathbb{Z}_p$. Thus, as $z^p \equiv z \pmod p$ for every $z \in \mathbb{Z}$, we have $f^p(x) \equiv ax + b \pmod p$, i.e., $f^p(x) \equiv f(x) \pmod p$. However, if f is transitive modulo p, then $f^p(x) \equiv x \pmod p$. This contradiction proves that $a \equiv 1 \pmod p$.

The converse statement of Lemma 9.2 is obvious: If $a \equiv 1 \pmod p$, then (9.4) implies that $f^i(x) \equiv x + bi \pmod p$, i.e., given $x, y \in \{0, 1, \ldots, p - 1\}$ from the congruence $x + bi \equiv y \pmod p$, one finds $i \in \{0, 1, \ldots, p - 1\}$ (since $b \not\equiv 0$ mod p) such that $f^i(x) \equiv y \pmod p$. □

Now we assume that conditions of Theorem 9.1 imply transitivity of f modulo p^k; we claim then that f is transitive modulo p^{k+1}.

As f is measure-preserving, f is bijective modulo p^{k+1}; thus, as f is transitive modulo p^k, it is clear that f is transitive modulo p^{k+1} whenever $f^i(0) \equiv 0 \pmod{p^{k+1}}$ implies $i \equiv 0 \pmod{p^{k+1}}$. Note that $f^i(0) \equiv 0 \pmod{p^{k+1}}$ implies $i \equiv 0 \pmod{p^k}$ since f is transitive modulo p^k. Now we just calculate $f^i(0) \bmod p^{k+1}$ for $i = p^k\ell$.

As $a = 1 + pr$ for a suitable $r \in \mathbb{Z}_p$, from (9.4) we get

$$f^i(0) = b\frac{(1 + pr)^i - 1}{pr} = b\sum_{j=1}^{i} p^{j-1} r^{j-1} \binom{i}{j} \tag{9.5}$$

Now represent

$$\binom{i}{j} = \frac{1}{j!} i(i - 1)\cdots(i - j + 1) = \frac{i}{j}\left(\frac{i}{1} - 1\right)\left(\frac{i}{2} - 1\right)\cdots\left(\frac{i}{j - 1} - 1\right).$$

As $\frac{i}{t} \in \mathbb{Z}_p$ for $i = p^k\ell$, $t = 1, 2, \ldots, i - 1$, we conclude that $\operatorname{ord}_p \binom{p^k\ell}{j} \geqslant k - \operatorname{ord}_p j$, so from (9.5), it follows that

$$f^{p^k\ell}(0) \equiv bp^k\ell \pmod{p^{k+1}} \text{ for } p \text{ odd, and} \tag{9.6}$$

$$f^{2^k\ell}(0) \equiv b\left(2^k\ell + 2r\binom{2^k\ell}{2}\right) \pmod{2^{k+1}}, \tag{9.7}$$

since $j - \operatorname{ord}_p j < 2$ if and only if either $j = 1$ or $p = 2$ and $j = 2$.

Whenever p is odd, from (9.6), it follows that $f^{p^k\ell}(0) \equiv 0 \pmod{p^{k+1}}$ if and only if $\ell \equiv 0 \pmod p$, thus proving our claim for odd p. For $p = 2$, however, (9.7) implies that $f^{2^k\ell}(0) \equiv 0 \pmod{2^{k+1}}$ ether when ℓ is even or when both ℓ and r are odd. Yet the latter case does not hold since $a \equiv 1 \pmod 4$.

We conclude finally that conditions of Theorem 9.1 are sufficient. In view of Theorem 8.2, the above argument shows that these conditions are also necessary.

□

We stress the leading idea of the proof:

Note 9.3 Given a 1-Lipschitz measure-preserving function $f: \mathbb{Z}_p \to \mathbb{Z}_p$, which is transitive modulo p^k, the function f is transitive modulo p^{k+1} if and only if $f^{p^k\ell}(z) \equiv z \pmod{p^{k+1}}$ implies $\ell \equiv 0 \pmod p$ for some (or, equivalently, every) $z \in \mathbb{Z}_p$ (cf. Proposition 7.1).

In the sequel, we exploit this observation frequently. Note also that the assertion of Note 9.3 holds for locally 1-Lipschitz functions as well once k is sufficiently large.

9.2 Ergodicity and Measure-Preservation in Terms of Coordinate Functions

In this section, we prove criteria of measure-preservation and of ergodicity for 1-Lipschitz functions $f: \mathbb{Z}_2 \to \mathbb{Z}_2$ in terms of coordinate functions that were defined in Sect. 5.1.1. Recall that according to Proposition 5.1, every 1-Lipschitz function $f: \mathbb{Z}_2 \to \mathbb{Z}_2$ can be represented in a form

$$f\left(\sum_{i=0}^{\infty} \chi_i 2^i\right) = \sum_{i=0}^{\infty} \psi_i(\chi_0, \ldots, \chi_i) 2^i \tag{9.8}$$

where $\chi_i \in \{0, 1\}$ and each i-th coordinate function $\psi_i(\chi_0, \ldots, \chi_i) = \delta_i(f(x))$ is a Boolean function in Boolean variables $\chi_0, \ldots, \chi_i$; that is, $\psi_i: \{0, 1\}^{i+1} \to \{0, 1\}$; $i = 0, 1, 2 \ldots$.

Recall that an *algebraic normal form*, the ANF, of the Boolean function $\psi_i(\chi_0, \ldots, \chi_i)$ is a representation of this function via $\oplus$ (addition modulo 2, i.e., logical "exclusive or") and multiplication modulo 2 (i.e., logical "and," or conjunction). In other words, the ANF of the Boolean function ψ is its representation of the form

$$\psi(\chi_0, \ldots, \chi_j) = \beta \oplus \beta_0\chi_0 \oplus \beta_1\chi_1 \oplus \ldots \oplus \beta_{0,1}\chi_0\chi_1 \oplus \ldots,$$

where $\beta, \beta_0, \ldots \in \{0, 1\}$ and $\chi_0, \ldots, \chi_j$ are Boolean variables. To put it in other words, ANFs are elements of the residue ring of polynomials in variables $x_1, x_2, \ldots$ over the two-element field modulo the ideal generated by $x_i^2 - x_i$. $i = 1, 2, \ldots$.

Every Boolean function in n Boolean variables has a unique representation via ANF. In Russian mathematical literature, ANFs are known as Zhegalkin (also, Žegalkin, Gégalkine or Shegalkin) polynomials since these were introduced by the Russian mathematician Ivan Ivanovich Zhegalkin in 1927 [72].

Recall that the *weight* of the Boolean function ψ in $(j+1)$ variables is the number of $(j+1)$-bit words, which *satisfy* ψ; that is, the weight is cardinality of the truth set of ψ, and the *truth set* (also known as *satisfiability set*) of ψ is the set of all points from $\{0,1\}^{j+1}$, where ψ takes the value 1.

The following theorem belongs to mathematical folklore. That is, being widely known (at least among Russian mathematicians since not later than 1970), it is still attributed to no author. Everybody, including the author of the book, have heard the theorem from some of his colleagues, who, in turn, also learned it from somebody else. We state this theorem in dynamical terms and give a proof, which is also of dynamical flavor:

Theorem 9.4 *The function f defined by Eq.* (9.8) *is measure-preserving if and only if for every $i = 0, 1, \ldots$, the ANF of the i-th coordinate function is*

$$\psi_i(\chi_0, \ldots, \chi_i) = \chi_i \oplus \varphi_i(\chi_0, \ldots, \chi_{i-1}), \tag{9.9}$$

where φ_i is an ANF of a Boolean function in Boolean variables $\chi_0, \ldots, \chi_{i-1}$ and φ_0 is a constant from $\{0, 1\}$.

The function f is ergodic if and only if, additionally, $\varphi_0 = 1$, and every Boolean function φ_i is of odd weight, that is, takes value 1 *exactly at an odd number of points from $\{0,1\}^i$ for $i = 1, 2, \ldots$. The latter condition holds if and only if the degree of the ANF of φ_i for $i \geqslant 1$ is exactly i, that is, if and only if the ANF of φ_i contains a monomial $\chi_0 \cdots \chi_{i-1}$.*

Proof Gathering all terms of the ANF that do not contain a variable χ_j, we represent the function ψ_i in the following form:

$$\psi_i(\chi_0, \ldots, \chi_i) = \chi_i \omega_i(\chi_0, \ldots, \chi_{i-1}) \oplus \varphi_i(\chi_0, \ldots, \chi_{i-1}),$$

where both $\omega_i(\chi_0, \ldots, \chi_{i-1})$ and $\varphi_i(\chi_0, \ldots, \chi_{i-1})$ are Boolean functions in Boolean variables $\chi_0, \ldots, \chi_{i-1}$.

Obviously, whenever all $\omega_i(\chi_0, \ldots, \chi_{i-1})$ are identically 1, the function f is measure-preserving in view of Theorem 8.2 since f is bijective modulo 2^{k+1} for every $k = 0, 1, 2, \ldots$: to find a pre-image of the mapping $f \bmod 2^k$, one must solve a system of Boolean equations

$$\begin{cases} \chi_0 \oplus \varphi_0 & = \alpha_0, \\ \chi_1 \oplus \varphi_1(\chi_1) & = \alpha_1, \\ \cdots\cdots\cdots\cdots\cdots\cdots & \cdots\cdots \\ \chi_k \oplus \varphi_k(\chi_0, \ldots, \chi_{k-1}) & = \alpha_k, \end{cases}$$

which has a unique solution given any $\alpha_0, \ldots, \alpha_k \in \{0, 1\}$.

Conversely, let i be the smallest number such that $\omega_i(\chi_0, \ldots, \chi_{i-1}) = 0$ for a certain vector $(\varepsilon_0, \ldots, \varepsilon_{i-1})$ of zeros and ones. Then

$$f(\varepsilon_0+\varepsilon_1 2+\cdots\varepsilon_{i-1}2^{i-1}+0\times 2^i) \equiv f(\varepsilon_0+\varepsilon_1 2+\cdots\varepsilon_{i-1}2^{i-1}+1\times 2^i) \pmod{2^{i+1}}.$$

Whence f is not bijective modulo 2^{i+1}, thus not measure-preserving in view of Theorem 8.2.

Now, to prove the ergodicity part of the statement, we first note that f is transitive modulo 2 if and only if $\psi_0(\chi_0) = \chi_0 \oplus 1$. Further, if f is transitive modulo 2^{k+1}, then f is transitive modulo 2^j for all $j = 1, 2, \ldots, k$; so the i-th coordinate function $\delta_i(f^{2^k})(x)$ of the 2^k-th iterate of the function f is

$$\delta_i(f^{2^k}(\chi_0 + \chi_1 2 + \chi_2 4 + \cdots)) = \begin{cases} \chi_i, & \text{if } i < k; \\ \chi_k \oplus \sigma, & \text{if } i = k, \end{cases} \tag{9.10}$$

where σ is a sum modulo 2 of all values of the Boolean function φ_k at all points from $\{0, 1\}^k$; that is, σ is the weight modulo 2 of the function φ_k. From (9.10), it follows then that the transitivity of the function f modulo 2^{k+1} implies $\sigma = 1$; otherwise, $f^{2^k}(x) \equiv x \pmod{2^{k+1}}$ for every $x \in \mathbb{Z}_2$. Thus, a weight of the function φ_k must be odd.

The rest of the statement of the theorem is a well-known result from the theory of Boolean functions:*The weight of Boolean function is odd if and only if the ANF of the function has a maximum degree.*

To prove this claim, consider a Boolean function $\psi(\chi_0, \ldots, \chi_j)$ in Boolean variables $\chi_0, \ldots, \chi_j$. For $\alpha, \beta \in \{0, 1\}$, define $\alpha^\beta = 1$ whenever $\alpha = \beta$, and $\alpha^\beta = 0$ otherwise. Then we can represent the Boolean function ψ as

$$\psi(\chi_0, \ldots, \chi_j) = \bigoplus_{(\beta_0,\ldots,\beta_j)\in T(\psi)} \chi_0^{\beta_0} \cdots \chi_j^{\beta_j}, \tag{9.11}$$

where $T(\psi) \subset \{0, 1\}^{j+1}$ is a truth set of the Boolean function ψ. To obtain ANF from representation (9.11), we substitute $\chi^\beta = \chi \oplus \beta \oplus 1$ and perform all multiplications and additions modulo 2; it is obvious then that the coefficient $\mathrm{Coef}_{\chi_0\cdots\chi_j}\,\psi$ of the term $\chi_0 \cdots \chi_j$ (of degree $j + 1$, which is a maximum degree of any Boolean function in j+1 variables) in the ANF of the Boolean function ψ, is $\#T(\psi) \bmod 2$. □

9.3 Ergodicity and Measure-Preservation in Terms of Mahler Expansion

Recall that every function $f: \mathbb{Z}_p \to \mathbb{Z}_p$ can be expressed via Mahler interpolation series (4.2)

$$f(x) = \sum_{i=0}^{\infty} a_i \binom{x}{i}, \tag{9.12}$$

where $a_i \in \mathbb{Z}_p$, $i = 0, 1, 2, \ldots$. We are going now to describe how one can determine by the coefficients a_i whether f is measure-preserving or, respectively, ergodic. The central result of this section is the following theorem [9, 11]:

Theorem 9.5 *The function f defines a 1-Lipschitz measure-preserving transformation on $\mathbb{Z}_p$ whenever the following conditions hold simultaneously:*

$$a_1 \not\equiv 0 \pmod p; \tag{9.13}$$

$$a_i \equiv 0 \pmod{p^{\lfloor \log_p i \rfloor + 1}}, \quad i = 2, 3, \ldots. \tag{9.14}$$

The function f defines a 1-Lipschitz ergodic transformation on $\mathbb{Z}_p$ whenever the following conditions hold simultaneously:

$$a_0 \not\equiv 0 \pmod p; \tag{9.15}$$

$$a_1 \equiv 1 \pmod p, \quad \text{for } p \text{ odd}; \tag{9.16}$$

$$a_1 \equiv 1 \pmod 4, \quad \text{for } p = 2; \tag{9.17}$$

$$a_i \equiv 0 \pmod{p^{\lfloor \log_p (i+1) \rfloor + 1}}, \quad i = 2, 3, \ldots. \tag{9.18}$$

Moreover, in the case of $p = 2$, these conditions are necessary: Namely, if f is 1-Lipschitz and measure-preserving, then conditions (9.13) *and* (9.14) *hold simultaneously; if f is 1-Lipschitz and ergodic, then conditions* (9.15), (9.17) *and* (9.18) *hold simultaneously.*

Thus, Theorem 9.5 gives a complete description of 1-Lipschitz measure-preserving (respectively, of 1-Lipschitz ergodic) transformations on $\mathbb{Z}_p$ for $p = 2$ in terms of Mahler expansion. We also show in this section that $p = 2$ is the only case when the conditions of Theorem 9.5 are necessary.

To prove the theorem, we need some auxiliary results that are of interest by their own.

Lemma 9.6 *Given a 1-Lipschitz function $v: \mathbb{Z}_p \to \mathbb{Z}_p$ and p-adic integers c, d, $c \not\equiv 0 \pmod p$, the function $g(x) = d + cx + pv(x)$ preserves measure, and the function $h(x) = c + x + p\Delta v(x)$ is ergodic.* (Recall that Δ is a (forward) difference operator: $\Delta v(x) = v(x+1) - v(x)$ by the definition.)

Proof of Lemma 9.6 In view of Theorem 8.2, we must show that the function g (respectively, h) is bijective (respectively, transitive) modulo p^k for all $k = 1, 2, 3, \dots$.

First we prove by induction on k that g is bijective modulo p^k for all $k = 1, 2, 3, \dots$. The assertion is obviously true for $k = 1$.

Assume our claim is true for $k = 1, 2, \dots, n-1$, and prove that it holds for $k = n$. Let $g(a) \equiv g(b) \pmod{p^n}$ for some p-adic integers a, b. Then $a \equiv b \pmod{p^{n-1}}$ by induction hypothesis. Hence, $pv(a) \equiv pv(b) \pmod{p^n}$ since v is 1-Lipschitz. Further, the congruence $g(a) \equiv g(b) \pmod{p^n}$ implies the congruence $ca + pv(a) \equiv cb + pv(b) \pmod{p^n}$, and consequently, $ca \equiv cb \pmod{p^n}$. Since $c \not\equiv 0 \pmod{p}$, the latter congruence implies that $a \equiv b \pmod{p^n}$, thus proving the first assertion of Lemma 9.6.

To prove the rest of the statement, we note that the assertion we just proved implies that the function h preserves measure. To prove transitivity of h modulo p^k for all $k = 1, 2, 3, \dots$, we use induction on k.

From Lemma 9.2, it follows that h is transitive modulo p. Assume that h is transitive modulo p^{k-1}, and proceed as in Note 9.3.

We calculate successively

$$h^1(x) = c + x + pv(x+1) - pv(x),$$

$$\dots\dots\dots\dots\dots\dots\dots\dots\dots\dots\dots\dots$$

$$h^j(x) = h(h^{j-1}(x)) = c + h^{j-1}(x) + pv(h^{j-1}(x)+1) - pv(h^{j-1}(x)) =$$

$$= cj + x + p\sum_{i=0}^{j-1} v(h^i(x)+1) - p\sum_{i=0}^{j-1} v(h^i(x)),$$

and henceforth. We recall that $h^0(x) = x$ by definition. Thus,

$$h^{p^{k-1}\ell}(x) = cp^{k-1}\ell + x + p\sum_{i=0}^{p^{k-1}\ell-1} v(h^i(x)+1) - p\sum_{i=0}^{p^{k-1}\ell-1} v(h^i(x)). \tag{9.19}$$

However, as h is transitive modulo p^{k-1} and 1-Lipschitz, we see that

$$\sum_{i=0}^{p^{k-1}\ell-1} v(h^i(x)+1) \equiv \sum_{i=0}^{p^{k-1}\ell-1} v(h^i(x)) \equiv \ell \sum_{z=0}^{p^{k-1}-1} v(z) \pmod{p^{k-1}},$$

so (9.19) implies that $h^{p^{k-1}\ell}(x) \equiv cp^{k-1}\ell + x \pmod{p^k}$. Yet $c \not\equiv 0 \pmod{p}$; thus, if $h^{p^{k-1}\ell}(0) \equiv 0 \pmod{p^k}$, then necessarily, $\ell \equiv 0 \pmod{p}$. This proves Lemma 9.6 in view of Note 9.3. □

Corollary 9.7 *Under assumptions of Lemma 9.6, let* $r \equiv 1 \pmod p$ *if* p *be odd, and let* $r \equiv 1 \pmod 4$ *if* $p = 2$. *Then the function* $c + rx + p\Delta v(x)$ *is ergodic.*

Proof of the Corollary 9.7 As $r = 1 + ps$ for odd p (respectively, $r = 1 + 4s$ for $p = 2$), where $s \in \mathbb{Z}_p$, the function $u(x) = s\binom{x}{2}$ (respectively, $u(x) = 2s\binom{x}{2}$) is a polynomial over $\mathbb{Z}_p$ and, thus, 1-Lipschitz. Consequently, the function $v_1(x) = u(x) + v(x)$ is 1-Lipschitz either. Since $\Delta v_1(x) = sx + \Delta v(x)$ for odd p (respectively, $\Delta v_1(x) = 2sx + \Delta v(x)$ for $p = 2$), the proof is finished in view of Lemma 9.6. □

Proof of Theorem 9.5 Recall that according to Theorem 5.12, a function $f : \mathbb{Z}_p \to \mathbb{Z}_p$ is 1-Lipschitz if and only if it can be represented in the form

$$f(x) = b_0 + \sum_{i=1}^{\infty} b_i p^{\lfloor \log_p i \rfloor} \binom{x}{i}, \tag{9.20}$$

where $b_i \in \mathbb{Z}_p$, $i = 0, 1, 2, \ldots$. As $\lfloor \log_p i \rfloor = \lfloor \log_p(i+1) \rfloor$ for all $i = 1, 2, \ldots$ but $i = p^t - 1$, ($t = 1, 2, 3, \ldots$), and as

$$\Delta v(x) = \sum_{i=1}^{\infty} b_i p^{\lfloor \log_p i \rfloor} \binom{x}{i-1},$$

(see (1.2)) sufficiency of conditions of Theorem 9.5 follows now from Lemma 9.6 and Corollary 9.7.

To prove that for $p = 2$ conditions of Theorem 9.5 are necessary, we will express coefficients of algebraic normal forms of coordinate functions (see Sect. 9.2) via coefficients of Mahler expansion (9.20) and then apply Theorem 9.4.

During the proof, we denote $\chi_i = \delta_i(x) \in \{0, 1\}$. Then for arbitrary $n \in \mathbb{N}$ and $x \in \mathbb{Z}_2$, Lemma 5.9 implies that

$$f(x) \equiv f(\chi_0 + \chi_1 2 + \cdots + \chi_{n-1} 2^{n-1}) + 2^n \chi_n \tilde{f}_n(x) \pmod{2^{n+1}}, \tag{9.21}$$

where

$$\tilde{f}_n(x) \equiv \sum_{s=0}^{n} \frac{\Delta^{2^s} f(x)}{2^s} \pmod 2. \tag{9.22}$$

From (5.26) (see proof of Theorem 5.12), we conclude that

$$\frac{\Delta^{2^s} f(x)}{2^s} = \frac{1}{2^s} \sum_{i=2^s}^{\infty} a_i \binom{x}{i-2^s} = \sum_{i=2^s}^{\infty} b_i 2^{\lfloor \log_2 i \rfloor - s} \binom{x}{i-2^s}.$$

This, in view of Lucas' Theorem 1.2, implies that the following congruences modulo 2 hold:

$$\delta_0\left(\frac{\Delta^{2^s} f(x)}{2^s}\right) \equiv \sum_{i=2^s}^{2^{s+1}-1} \delta_0(b_i)\binom{x}{i-2^s} \equiv$$

$$\equiv \sum_{j=0}^{2^s-1} \delta_0(b_{2^s+j})\binom{x_0}{\delta_0(j)}\cdots\binom{x_{s-1}}{\delta_{s-1}(j)} \pmod 2. \tag{9.23}$$

From 9.23, it follows that $\delta_0\left(\frac{\Delta^{2^s} f(x)}{2^s}\right)$ does not depend on $\chi_s, \chi_{s+1}, \dots$ and that $\delta_0(\Delta f(x)) \equiv b_1 \equiv a_1 \pmod 2$. Now the latter congruence in view of (9.22) implies

$$\tilde{f}_n(x) \equiv \tilde{f}_n(\bar{x}_n) \equiv \sum_{s-1}^{n} \frac{\Delta^{2^s} f(\bar{x}_s)}{2^s} + a_1 \pmod 2 \tag{9.24}$$

where here and after $\bar{x}_k$ stands for $x \bmod 2^k = \chi_0 + \chi_1 2 + \cdots + \chi_{k-1}2^{k-1}$, ($k = 1, 2, \dots$).

Theorem 9.4 implies now that f preserves measure if and only if the following two conditions hold:

- f is bijective modulo 2, (9.25)
- $\tilde{f}_n(x) \equiv 1 \pmod 2$ for all $n = 1, 2, \dots$ and all $x \in \mathbb{Z}_2$. (9.26)

As $f(x) \equiv a_0 + a_1 x \pmod 2$, then condition (9.25) is equivalent to the following condition:

$$a_1 \equiv 1 \pmod 2 \tag{9.27}$$

Now, in view of (9.24) and (9.27), condition (9.26) holds if and only if the following condition

$$\frac{\Delta^{2^s} f(\bar{x}_s)}{2^s} \equiv 0 \pmod 2 \tag{9.28}$$

holds for all $s = 1, 2, 3, \dots$ and all $x \in \mathbb{Z}_2$. However, in view of (9.23), condition (9.28) holds for all $s = 1, 2, 3, \dots$ and all $x \in \mathbb{Z}_2$ if and only if the condition

$$b_i \equiv 0 \pmod 2 \tag{9.29}$$

holds for all $i = 2, 3, \dots$. As $a_i = b_i 2^{\lfloor \log_2 i \rfloor}$ for $i = 1, 2, \dots$, then (9.27) and (9.29) imply necessity of conditions (9.13) and (9.14) when $p = 2$.

Further, as an ergodic function f preserves measure, from Theorem 9.4 in view of (9.21) and condition (9.26), we conclude that the ANF of the Boolean function $\delta_i(f(x)) = \psi_i(\chi_0, \ldots, \chi_i)$ is of the following form:

$$\psi_i(\chi_0, \ldots, \chi_i) = \varphi_i(\chi_0, \ldots, \chi_{i-1}) \oplus x_i \tag{9.30}$$

where $\varphi_i(\chi_1, \ldots, \chi_{i-1}) = \delta_i(f(\chi_0 + \cdots + \chi_{i-1}2^{i-1}))$ and φ_0 is a constant. Now, from Theorem 9.4, it follows that once the function f is ergodic, $\varphi_0 = 1$, and the coefficient $\mathrm{Coef}_{\chi_0 \cdots \chi_{i-1}} \varphi_i$ of the monomial $\chi_0 \cdots \chi_{i-1}$ in the ANF φ_i must be 1 for all $i = 1, 2, \ldots$.

Since obviously $\varphi_0 \equiv a_0 \pmod 2$, we conclude now that f is a 1-Lipschitz ergodic function if and only if the following conditions (9.31)—(9.34) hold simultaneously:

$$a_0 \equiv 1 \pmod 2; \tag{9.31}$$

$$a_1 \equiv 1 \pmod 2; \tag{9.32}$$

$$a_j \equiv 0 \pmod{2^{\lfloor \log_2 j \rfloor + 1}}, \text{ for all } j = 2, 3, \ldots; \tag{9.33}$$

$$C_i = 1, \text{ for all } i = 1, 2, \ldots, \tag{9.34}$$

where $C_i = \mathrm{Coef}_{\chi_0 \cdots \chi_{i-1}} \varphi_i$. To finish the proof, we use the following recursive formula for $\mathrm{Coef}_{\chi_0 \cdots \chi_{i-1}} \varphi_i$: □

Lemma 9.8 *If a 1-Lipschitz function f preserves measure, then*

$$\mathrm{Coef}_{x_0 \cdots x_n} \varphi_{n+1} \equiv \delta_1(b_{2^{n+1}-1}) + \mathrm{Coef}_{x_0 \cdots x_{n-1}} \varphi_n \pmod 2$$

for all $n = 1, 2, \ldots$.

Proof of Lemma 9.8 We begin as in the proof of Lemma 5.9: With the use of Theorem 1.5 and taking into account that $\chi_n \in \{0, 1\}$, we conclude that

$$f(\bar{x}_{n+1}) = f(\bar{x}_n) + \chi_n \sum_{i=1}^{2^n} \Delta^i f(\bar{x}_n) \binom{2^n}{i} =$$
$$= f(\bar{x}_n) + 2^n \chi_n \sum_{k=0}^{2^n-1} \frac{\Delta^{k+1} f(\bar{x}_n)}{k+1} \binom{2^n - 1}{k}.$$

Hence,

$$\delta_{n+1}(f(\bar{x}_{n+1})) \equiv \delta_{n+1}(f(\bar{x}_n)) + \delta_1(\chi_n S_n) + \delta_n(f(\bar{x}_n))\delta_0(\chi_n S_n) \pmod 2, \tag{9.35}$$

where

$$S_n = \sum_{k=0}^{2^n-1} \frac{\Delta^{k+1} f(\bar{x}_n)}{k+1} \binom{2^n-1}{k}.$$

As by Lucas' Theorem 1.2, $\binom{2^n-1}{k} \equiv 1 \pmod 2$ for all $k \in \{0, 1, \ldots 2^n - 1\}$, we combine together Lemma 5.8 and Lemma 5.9 and conclude that

$$S_n \equiv \sum_{s=0}^{n} \frac{\Delta^{2^s} f(\bar{x}_n)}{2^s} \equiv \tilde{f}_n(\bar{x}_n) \pmod 2 \tag{9.36}$$

However, $\tilde{f}_n(\bar{x}_n) \equiv 1 \pmod 2$ since f preserves measure (see (9.26)). Then (9.36) implies that

$$\delta_0(S_n) = 1. \tag{9.37}$$

This, in view of (9.35), implies that

$$\begin{aligned} \operatorname{Coef}_{\chi_0\cdots\chi_n}\ \delta_{n+1}(f(\bar{x}_{n+1})) &\equiv \\ \operatorname{Coef}_{\chi_0\cdots\chi_n}\ \delta_1(\chi_n S_n) &+ \operatorname{Coef}_{\chi_0\cdots\chi_{n-1}}\ \delta_n(f(\bar{x}_n)) \pmod 2. \end{aligned} \tag{9.38}$$

As $\delta_1(\chi_n S_n) = \chi_n \delta_1(S_n)$, then

$$\operatorname{Coef}_{\chi_0\cdots\chi_n}\ \delta_1(\chi_n S_n) = \operatorname{Coef}_{\chi_0\cdots\chi_{n-1}}\ \delta_1(S_n). \tag{9.39}$$

Now we must calculate $\delta_1(S_n)$. By using "school-textbook" algorithms of addition and multiplication of 2-adic integers $u_k, v_k \in \mathbb{Z}_2$ represented by their 2-adic canonical forms $u_k = \delta_0(u_k) + \delta_1(u_k)2 + \delta_2(u_k)2^2 + \cdots$ and $v_k = \delta_0(v_k) + \delta_1(v_k)2 + \delta_2(v_k)2^2 + \cdots$, we get

$$\begin{aligned} \delta_1\left(\sum_{k=0}^{m} u_k v_k\right) &\equiv \sum_{k=0}^{m} \delta_0(u_k)\delta_1(v_k) + \\ &+ \sum_{k=0}^{m} \delta_1(u_k)\delta_0(v_k) + \delta_1\left(\sum_{k=0}^{m} \delta_0(u_k v_k)\right) \pmod 2. \end{aligned} \tag{9.40}$$

For $\varepsilon_k \in \{0, 1\}$, $k = 0, 1, 2, \ldots, m$, denote $\Xi(\varepsilon_0, \ldots, \varepsilon_m) = \delta_1(\sum_{k=0}^{m} \varepsilon_k)$, and then clearly

$$\Xi(\varepsilon_0, \ldots, \varepsilon_m) \equiv \left\lfloor \frac{1}{2} \operatorname{Wt}(\varepsilon_0, \ldots, \varepsilon_m) \right\rfloor \pmod 2,$$

where $\mathrm{Wt}(\varepsilon_0, \ldots, \varepsilon_m)$ is the number of nonzero coordinates of a binary vector $(\varepsilon_0, \ldots, \varepsilon_m)$. Now assuming $m = 2^n - 1$, $u_k = \frac{\Delta^{k+1} f(\bar{x}_n)}{k+1}$, $v_k = \binom{2^n-1}{k}$, we apply (9.40) to calculate $\delta_1(S_n)$.

From Lucas' Theorem 1.2, it follows that

$$\delta_0(v_k) = \delta_0\left(\binom{2^n - 1}{k}\right) = 1$$

for all $k = 0, 1, \ldots, 2^n - 1$. Hence,

$$\delta_1\left(\sum_{k=0}^{m} \delta_0(u_k v_k)\right) = \delta_1\left(\sum_{k=0}^{m} \delta_0(u_k)\delta_0(v_k)\right) = \Xi(\delta_0(u_0) \ldots, \delta_0(u_m)).$$

Further, from Lemma 5.8, it follows that for all $k \neq 2^r - 1$,

$$\delta_0(u_k) = \delta_0\left(\frac{\Delta^{k+1} f(\bar{x}_n)}{k+1}\right) = 0. \tag{9.41}$$

As f preserves measure, (9.28) holds for all $s = 1, 2, \ldots$ and all $x \in \mathbb{Z}_2$, so from (9.41), it follows that $\delta_0(u_1) = \cdots = \delta_0(u_m) = 0$, whence the function

$$\Xi(\delta_0(u_0), \ldots, \delta_0(u_m)) = \delta_1\left(\sum_{k=0}^{m} \delta_0(u_k v_k)\right)$$

in the right-hand part of (9.40) vanishes. Finally applying (9.41) and (9.28) to (9.40), we conclude that

$$\delta_1(S_n) \equiv \delta_0(\Delta f(\bar{x}_n)) + \sum_{k=0}^{2^n-1} \delta_1\left(\frac{\Delta^{k+1} f(\bar{x}_n)}{k+1}\right) \pmod{2}. \tag{9.42}$$

As f preserves measure, then (9.27) and (9.29) hold; thus, coefficients b_i of Mahler expansion (9.20) satisfy the following conditions:

$$\begin{cases} b_1 & \equiv 1 \pmod{2}; \\ b_i & = 2c_i, \text{ for appropriate } c_i \in \mathbb{Z}_2;\ i = 2, 3, \ldots. \end{cases} \tag{9.43}$$

Hence, for every $s \geqslant 2$, $s = \hat{s}2^{\mathrm{ord}_2 s}$, $\hat{s}$ odd, we have

$$\frac{\Delta^s f(x)}{s} = \frac{2}{\hat{s}} \sum_{i=s}^{\infty} c_i 2^{\lfloor \log_2 i \rfloor - \mathrm{ord}_2 s} \binom{x}{i-s}, \tag{9.44}$$

in view of (1.2) (we note that $\hat{s}$ is a unit of $\mathbb{Z}_2$; thus, $\hat{s}$ has a multiplicative inverse $\frac{1}{\hat{s}} \in \mathbb{Z}_2$). Consequently, (9.44) implies that

$$\delta_1\left(\frac{\Delta^s f(x)}{s}\right) \equiv \sum_{i=s}^{\infty} c_i 2^{\lfloor \log_2 i \rfloor - \mathrm{ord}_2 s} \binom{x}{i-s} \pmod 2. \tag{9.45}$$

As either $\lfloor \log_2 i \rfloor > \mathrm{ord}_2 s$ or $s \leqslant i$ always but the case $s = 2^r, 2^r \leqslant i \leqslant 2^{r+1} - 1$, congruence (9.45) implies that

$$\delta_1\left(\frac{\Delta^s f(x)}{s}\right) \equiv \begin{cases} \sum_{i=2^r}^{2^{r+1}-1} c_i \binom{x}{i-2^r} & \pmod 2, \text{ if } s = 2^r \text{ for } r = 1, 2, \dots; \\ 0 & \pmod 2, \text{ otherwise.} \end{cases} \tag{9.46}$$

Further, from (9.20) in view of (9.43) and (1.2), we derive that

$$\Delta f(x) \equiv b_1 \pmod 4. \tag{9.47}$$

Now from (9.42) in view of (9.43), (9.46) and (9.47), it follows that

$$\delta_1(S_n) \equiv 1 + \delta_1(b_1) + \sum_{s=1}^{n} \sum_{j=0}^{2^s-1} c_{j+2^s} \binom{\bar{x}_n}{j} \pmod 2. \tag{9.48}$$

From here with the use of Lucas' Theorem 1.2, we deduce that

$$\mathrm{Coef}_{\chi_0 \cdots \chi_{n-1}} \delta_1(S_n) \equiv c_{2^{n+1}-1} \pmod 2.$$

The latter congruence in view of (9.43), (9.38), and (9.39) finishes the proof of Lemma 9.8. □

Now we can finish our proof of Theorem 9.5. Lemma 9.8 implies that

$$\mathrm{Coef}_{\chi_0 \cdots \chi_{i-1}} \delta_1(f(\bar{x}_i)) \equiv \sum_{r=2}^{i} \delta_1(b_{2^r-1}) + \mathrm{Coef}_{\chi_0} \delta_1(f(\chi_0)) \pmod 2. \tag{9.49}$$

From (9.20), we have $f(\chi_0) = b_0 + b_1\chi_0$, so taking into account (9.43), we conclude that

$$\delta_1(f(\chi_0)) \equiv \delta_1(b_0) + \chi_0(\delta_1(b_1) + \delta_0(b_0)) \pmod 2.$$

Thus, (9.49) in view of (9.31) implies that

$$\mathrm{Coef}_{\chi_0 \cdots \chi_{i-1}} \delta_1(f(\bar{x}_i)) \equiv 1 + \sum_{r=1}^{i} \delta_1(b_{2^r-1}) \pmod 2,$$

since $a_0 = b_0$. This means that the condition (9.34) is equivalent to the following condition:

$$\sum_{r=1}^{i} \delta_1(b_{2^r-1}) \equiv 0 \pmod 2;\ i = 1, 2, 3, \ldots,$$

or, equivalently, to the condition

$$\delta_1(b_{2^r-1}) = 0;\ (r = 1, 2, 3, \ldots). \tag{9.50}$$

As $a_j = b_j 2^{\lfloor \log_2 j \rfloor}$, then, combining together (9.31), (9.32), (9.33), and (9.50), we finish the proof of Theorem 9.5. □

From Theorem 9.5 by Lemma 9.6, we immediately derive yet one more important criterion of measure-preservation/ergodicity for T-functions:

Corollary 9.9 *The following is true:*

1. *The function $f: \mathbb{Z}_2 \to \mathbb{Z}_2$ is 1-Lipschitz and measure-preserving if and only if f can be represented as*

$$f(x) = c + x + 2v(x),$$

   where $c \in \mathbb{Z}_2$ and v is a 1-Lipschitz function.
2. *The function $f: \mathbb{Z}_2 \to \mathbb{Z}_2$ is 1-Lipschitz and ergodic if and only if f can be represented as*

$$f(x) = 1 + x + 2\Delta v(x),$$

   where v is a 1-Lipschitz function.

Theorem 9.5 immediately implies the following criteria of measure-preservation/ergodicity for 2-adic $\mathcal{B}$-functions:

Corollary 9.10 *Let f be a $\mathcal{B}$-function on $\mathbb{Z}_2$; that is, let $f(x) = \sum_{i=0}^{\infty} e_i x^{\underline{i}}$, for suitable $e_i \in \mathbb{Z}_2$, $i = 0, 1, 2, \ldots$* (see Sect. 6.3). *The function f is measure-preserving if and only if*

$$e_1 \equiv 1 \pmod 2, \quad e_2 \equiv 0 \pmod 2, \quad e_3 \equiv 0 \pmod 2$$

The function f is ergodic on $\mathbb{Z}_2$ if and only if

$$e_0 \equiv 1 \pmod 2, \quad e_1 \equiv 1 \pmod 4, \quad e_2 \equiv 0 \pmod 2, \quad e_3 \equiv 0 \pmod 4$$

By using this corollary, the next corollary can also can be derived from Theorem 9.5:

Corollary 9.11 *Given a C-function $f: \mathbb{Z}_2 \to \mathbb{Z}_2$ represented via power series $f(x) = \sum_{i=0}^{\infty} c_i x^i$, where $c_i \in \mathbb{Z}_2$, $i = 0, 1, 2, \ldots$, the function f is ergodic if and only if the following conditions hold simultaneously:*

$$c_3 + c_5 + c_7 + \cdots \equiv 2c_2 \pmod 4;$$
$$c_4 + c_6 + c_8 + \cdots \equiv c_1 + c_2 - 1 \pmod 4;$$
$$c_1 \equiv 1 \pmod 2;$$
$$c_0 \equiv 1 \pmod 2.$$

Sketch Proof As $x^i = \sum_{j=0}^{i} S(i,j)x^{\underline{j}}$ and $x^{\underline{i}} = \sum_{j=0}^{i}(-1)^{i-j}s(i,j)x^j$, where $S(i,j)$ and $s(i,j)$ are Stirling numbers of the second kind and of the first kind, respectively, we can rewrite conditions for coefficients e_i from Corollary 9.10 in terms of coefficients c_i by using relations (1.1) from Sect. 1.1. This involves somewhat messy calculations that are omitted since actually the assertion of the proposition is a trivial generalization of the result due to M. V. Larin [132] on polynomials over $\mathbb{Z}$ rather than on C-functions. □

More elementary proof of the corollary is given in [1] as a consequence of the result on level transitivity of polynomial action on rooted tree that constitutes the main result of that paper. A complete analog of Theorem 9.5 for q-Mahler basis was obtained in [105]. It is important to stress here that proofs of [105] do not rely on Theorem 9.4.

For $\mathcal{B}$-functions, a general criterion of measure-preservation was proved in [107]. The latter result was extended to automata functions that are uniformly differentiable modulo p cf., [106].

For $p \neq 2$, conditions of Theorem 9.5 are sufficient, and not necessary. For $p > 2$, the conditions that are both sufficient and necessary were found in [152], which are somewhat lengthy, so we will not reproduce them in the book. Also, it is not very easy to determine if the conditions hold for a given automaton function represented via Mahler's expansion. Nevertheless, the conditions are helpful to determine ergodicity of automata functions for smaller p (see examples in [152]).

Example 9.12 For every prime p and every $a \equiv 1 \pmod p$, the function $f(x) = ax + a^x$ is a 1-Lipschitz ergodic transformation on $\mathbb{Z}_p$.

Indeed, as $a = 1 + pz$ for a suitable $z \in \mathbb{Z}_p$, we see that $f(x) = x + pzx + \sum_{i=0}^{\infty} p^i z^i \binom{x}{i} = 1 + (1 + 2pz)x + \sum_{i=2}^{\infty} p^i z^i \binom{x}{i}$, and the assertions follows immediately from Theorem 9.5.

Example 9.13 A C-function $f: \mathbb{Z}_2 \to \mathbb{Z}_2$ is ergodic on $\mathbb{Z}_2$ if and only if f is transitive modulo 8.

Follows from Corollary 9.11 and Proposition 4.7.

Example 9.14 From the preceding example, we already know that a polynomial over $\mathbb{Z}_2$ is ergodic if and only if it is transitive modulo 8. A complete list of polynomial transitive transformations on $\mathbb{Z}/8\mathbb{Z}$ is as follows:

$x+1$	$5x+1$	$2x^2+3x+1$	$2x^2+7x+1$
$x+3$	$5x+3$	$2x^2+3x+3$	$2x^2+7x+3$
$x+5$	$5x+5$	$2x^2+3x+5$	$2x^2+7x+5$
$x+7$	$5x+7$	$2x^2+3x+7$	$2x^2+7x+7$

This is proved in [132].

9.4 Ergodicity and Measure-Preservation in Terms of van der Put Expansion

In this section, we establish measure-preservation and ergodicity criteria for 1-Lipschitz transformations on $\mathbb{Z}_p$. We are mostly concerned with the case $p = 2$ for which we prove a complete description following [25, 27]. Recall that for the case $p = 2$ the 1-Lipschitz functions are known under the term T-functions. In the section, we use this term to stress that we deal with the case $p = 2$ only.

9.4.1 Measure-Preservation Criteria

In this subsection, we present two criteria of measure-preservation for 1-Lipschitz functions: The first one we prove for T-functions (i.e., for the case $p = 2$), and the second one is a general criterion for arbitrary prime p, which we state without a proof.

Theorem 9.15 *Let $f\colon \mathbb{Z}_2 \to \mathbb{Z}_2$ be a T-function represented via van der Put series* (4.10). *The T-function f is measure-preserving if and only if the following conditions hold simultaneously:*

1. $B_0 + B_1 \equiv 1 \pmod 2$.
2. $|B_m|_2 = 2^{-\lfloor \log_2 m \rfloor}$, $m = 2, 3, \ldots$.

Proof By Corollary 9.9, the T-function f is measure-preserving if and only if it can be represented in the form $f(x) = d + x + 2g(x)$, where g $\mathbb{Z}_2$ $\mathbb{Z}_2$ is a T-function and $d \in \mathbb{Z}_2$. Now, given $g(x) = \sum_{m=0}^{\infty} \tilde{B}_m \chi(m, x)$, the van der Put series for the T-function g, we find van der Put coefficients of the function $f(x) = d + x + 2g(x)$.

As the van der Puts series of the T-function $t(x) = x$ are

$$t(x) = \sum_{m=1}^{\infty} 2^{\lfloor \log_2 m \rfloor} \chi(m, x), \tag{9.51}$$

and as $d = d\chi(0, x) + d\chi(1, x)$, we get:

$$f(x) = (d + 2\tilde{B}_0)\chi(0, x) + (1 + d + 2\tilde{B}_1)\chi(1, x) + \\ + \sum_{m=2}^{\infty} \left(2^{\lfloor \log_2 m \rfloor} + 2\tilde{B}_m\right) \chi(m, x). \tag{9.52}$$

This proves necessity of conditions of the Theorem since $|\tilde{B}_m|_2 \leqslant 2^{-\lfloor \log_2 m \rfloor}$ by Theorem 5.14 and hence $\left|2^{\lfloor \log_2 m \rfloor} + 2\tilde{B}_m\right|_2 = 2^{-\lfloor \log_2 m \rfloor}$ by strong triangle inequality.

To prove sufficiency of the conditions, we note that the condition $|B_m|_2 = 2^{-\lfloor \log_2 m \rfloor}$ implies that $B_m = 2^{\lfloor \log_2 m \rfloor} + 2\tilde{B}_m$ for suitable $\tilde{B}_m \in \mathbb{Z}_2$, where

$$\left|\tilde{B}_m\right|_2 \leqslant 2^{-\lfloor \log_2 m \rfloor}, \tag{9.53}$$

and the condition $B_0 + B_1 = 1 \pmod 2$ implies that $B_0 = d + 2\tilde{B}_0$ and $B_1 = 1 + d + 2\tilde{B}_1$ for suitable $d, \tilde{B}_0, \tilde{B}_1 \in \mathbb{Z}_2$. Now from Eqs. (9.52) and (9.51), it follows that $f(x) = d + x + 2g(x)$, where $g(x) = \sum_{m=0}^{\infty} \tilde{B}_m \chi(m, x)$ is a T-function by Theorem 5.14 in view of inequality (9.53). Thus, f is measure-preserving by Corollary 9.9. □

The following criterion was established in [117]:

Theorem 9.16 *Let $f : \mathbb{Z}_p \to \mathbb{Z}_p$ be a 1-Lipschitz function; then f can be represented via van der put series as follows* (cf., (5.32))*:*

$$f(x) = \sum_{m=0}^{\infty} p^{\lfloor \log_p m \rfloor} b_m \chi(m, x),$$

where $b_m \in \mathbb{Z}_p$, $m = 0, 1, 2, \ldots$. The function f is measure-preserving if and only if the following conditions hold simultaneously:

- *The coefficients $\{b_0, b_1, \ldots, b_{p-1}\}$ constitute a complete set of residues modulo p (i.e., the function f is bijective modulo p).*
- *The coefficients $b_{m+p^k}, b_{m+2p^k}, \ldots, b_{m+(p-1)p^k}$ for any $m = 0, 1, \ldots, p^k - 1$ are all nonzero residues modulo p for $k = 2, 3, \ldots$.*

9.4.2 Ergodicity Criteria for T-functions

In this subsection, we prove the following criterion of ergodicity for T-functions:

Theorem 9.17 *A T-function $f: \mathbb{Z}_2 \to \mathbb{Z}_2$ is ergodic if and only if it can be represented as*

$$f(x) = b_0\chi(0, x) + b_1\chi(1, x) + \sum_{m=2}^{\infty} 2^{\lfloor \log_2 m \rfloor} b_m \chi(m, x)$$

for suitable $b_m \in \mathbb{Z}_2$ that satisfy the following conditions:

1. $b_0 \equiv 1 \pmod 2$.
2. $b_0 + b_1 \equiv 3 \pmod 4$.
3. $|b_m|_2 = 1$, $m \geqslant 2$.
4. $b_2 + b_3 \equiv 2 \pmod 4$.
5. $\sum_{m=2^{n-1}}^{2^n-1} b_m \equiv 0 \pmod 4$, $n \geqslant 3$.

To prove the theorem, we need the following lemma:

Lemma 9.18 *Let $f: \mathbb{Z}_2 \to \mathbb{Z}_2$ be a T-function represented by van der Put series* (4.10). *Then f is ergodic if and only there exists a sequence $a_0, a_1, \ldots$ of 2-adic integers such that*

$$B_m = \begin{cases} 1 + 2(a_1 - a_0), & \text{if } m = 0; \\ 2(1 + a_0 + 2a_2 - a_1), & \text{if } m = 1; \\ 2^{n-1} + 2^n a_{m+1} - 2^n a_m, & \text{if } 2^{n-1} \leqslant m < 2^n - 1, n \geqslant 2 \\ 2^{n-1} + 2^{n+1} a_{2^n} - 2^n a_{2^n-1} - 2^n a_{2^{n-1}}, & \text{if } m = 2^n - 1, n \geqslant 2. \end{cases} \tag{9.54}$$

Proof of Lemma 9.18 By Corollary 9.9, a T-function f is ergodic if and only if it can be represented as $f(x) = 1 + x + 2(g(x+1) - g(x))$, where $g(x)$ is a suitable T-function. That is, by Theorem 5.14,

$$g(x) = a_0\chi(0, x) + \sum_{m=1}^{\infty} 2^{n_m - 1} a_m \chi(m, x) = \sum_{m=0}^{\infty} \tilde{B}_m \chi(m, x), \tag{9.55}$$

for suitable $a_0, a_1, a_2 \ldots \in \mathbb{Z}_2$, $\tilde{B}_0, \tilde{B}_1, \tilde{B}_2, \ldots \in \mathbb{Z}_2$ (here $n_m = \lfloor \log_2 m \rfloor + 1$, $m = 1, 2, 3, \ldots$).

Now, to prove necessity of conditions of the lemma, we just need to express van der Put coefficients of the function f via the coefficients of the function g. First, we do this for van der Put coefficients $\bar{B}_m$ of the T-function

$$g(x+1) = \sum_{m=0}^{\infty} \bar{B}_m \chi(m, x). \tag{9.56}$$

If $m > 1$, then by (4.13), $\bar{B}_m = g(m+1) - g(m+1-q(m))$, where $q(m) = \delta_{n_m-1}(m)2^{n_m-1}$. If $m \neq 2^{n_m} - 1$, then $q(m) = q(m+1)$; therefore, $\bar{B}_m = g(m +$

$1) - g(m+1-q(m+1)) = \tilde{B}_{m+1} = 2^{n_m - 1}a_{m+1}$ by (9.55) as $n_m = n_{m+1}$ in this case. If $m = 2^{n_m} - 1$, then $\bar{B}_m = g(2^{n_m}) - g(2^{n_m - 1})$ as $q(2^{n_m} - 1) = 2^{n_m - 1}$. As $H_{2^n} = g(2^n) - g(0)$ and $H_{2^{n-1}} = g(2^{n-1}) - g(0)$ by (9.55), we conclude that $\bar{B}_{2^n - 1} = \tilde{B}_{2^n} - \tilde{B}_{2^{n-1}}$. Finally, the coefficients $\bar{B}_0, \bar{B}_1$ can be found directly from (9.56): $\bar{B}_0 = g(1) = \tilde{B}_1$, $\bar{B}_1 = g(2) = \tilde{B}_0 + \tilde{B}_2$. Now we can find van der Put coefficients $\hat{B}_m$ of the function $2(g(x+1) - g(x))$; they are:

$$\hat{B}_m = \begin{cases} 2(\tilde{B}_1 - \tilde{B}_0), & \text{if } m = 0; \\ 2(\tilde{B}_0 + \tilde{B}_2 - \tilde{B}_1), & \text{if } m = 1; \\ 2(\tilde{B}_{m+1} - \tilde{B}_m), & \text{if } 2^{n-1} \leqslant m < 2^n - 1, n \geqslant 2; \\ 2(\tilde{B}_{m+1} - \tilde{B}_m - \tilde{B}_{\frac{m+1}{2}}), & \text{if } m = 2^n - 1, n \geqslant 2. \end{cases} \tag{9.57}$$

As $\chi(0, x) + \chi(1, x) = 1$ for all $x \in \mathbb{Z}_2$, from (5.32), we derive the van der Put expansion for the function $x + 1$; namely,

$$x + 1 = \chi(0, x) + 2\chi(1, x) + \sum_{m=2}^{\infty} 2^{n_m - 1}\chi(m, x). \tag{9.58}$$

From (9.55), we have that $\tilde{B}_0 = a_0$, $\tilde{B}_m = 2^{n-1}a_m$ when $2^{n-1} \leqslant m \leqslant 2^n - 1$, $n = 1, 2, \ldots$. Now combining the latter expressions with (9.58) and (9.56), we conclude that van der Put coefficients B_m of the function $f(x) = 1 + x + 2(g(x+1) - g(x))$ are of the form (9.54).

To prove sufficiency of conditions of the Lemma we just remark that the above argument shows that given expressions (9.54) for the van der Put coefficients of the function f, we can represent the T-function f in the form $f(x) = 1 + x + 2(g(x + 1) - g(x))$ where the van der Put expansion for g is given by (9.55). That is, the function g is a T-function by Theorem 5.14; therefore, the T-function f is ergodic by Corollary 9.9. □

Now we are ready to prove the following proposition, which actually is a criterion of ergodicity for T-functions, in terms of van der Put coefficients:

Proposition 9.19 *Let $f : \mathbb{Z}_2 \to \mathbb{Z}_2$ be a T-function represented by the van der Put series* (4.10). *Then f is ergodic if and only if the following conditions are satisfied simultaneously:*

1. $B_0 \equiv 1 \pmod 2$.
2. $B_0 + B_1 \equiv 3 \pmod 4$.
3. $|B_m|_2 = 2^{-(n-1)}$, $n \geqslant 2$, $2^{n-1} \leqslant m < 2^n - 1$.
4. $\left|\sum_{m=2^{n-1}}^{2^n-1}(B_m - 2^{n-1})\right|_2 \leqslant 2^{-(n+1)}$, $n \geqslant 2$.

Proof of Proposition 9.19 By Lemma 9.18, if the T-function f is ergodic, then its van der Put coefficients B_m can be expressed in the form (9.54), for suitable

$a_0, a_1, a_2, \ldots \in \mathbb{Z}_2$. From (9.54), by direct calculation, we easily prove that conditions 1–4 of the proposition are true.

By Lemma 9.18, to prove sufficiency of conditions of the proposition, we must find a sequence of 2-adic integers $a_0, a_1, a_2, \ldots$ such that relations (9.54) for the van der Put coefficients B_n hold. Take arbitrarily $a_0, a_1 \in \mathbb{Z}_2$ so that

$$a_1 - a_0 = \frac{B_0 - 1}{2} \tag{9.59}$$

(cf. the first equation from (9.54) and condition 1 of the proposition); then put

$$a_2 = \frac{B_1 + 2(a_1 - a_0) - 2}{4} = \frac{B_1 + B_0 - 3}{4} \tag{9.60}$$

(cf. the second equation from (9.54)). Note that $a_2 \in \mathbb{Z}_2$ due to the condition 2 of the proposition. We construct $a_3, a_4, a_5, \ldots \in \mathbb{Z}_2$ inductively. Denote

$$\check{B}_m = \frac{B_m - 2^{n-1}}{2^n}, \tag{9.61}$$

where $n = \lfloor \log_2 m \rfloor + 1$, $m \geqslant 3$; then $\check{B}_m \in \mathbb{Z}_2$ by condition 3. Given $a_{2^{n-1}} \in \mathbb{Z}_2$, for $\alpha = 1, 2, \ldots, 2^{n-1} - 1$, put

$$a_{2^{n-1}+\alpha} = a_{2^{n-1}} + \sum_{m=2^{n-1}}^{2^{n-1}+\alpha-1} \check{B}_m, \tag{9.62}$$

$$a_{2^n} = a_{2^{n-1}} + \frac{1}{2} \sum_{m=2^{n-1}}^{2^n-1} \check{B}_m. \tag{9.63}$$

Then $a_{2^{n-1}+\alpha} \in \mathbb{Z}_2$ by condition 3 of the proposition; and $a_{2^n} \in \mathbb{Z}_2$ by condition 4. Therefore, all $a_0, a_1, a_2, \ldots$ are in $\mathbb{Z}_2$.

Now solving system of Eqs. (9.59), (9.60), (9.61), (9.62), and (9.63) with respect to unknowns B_m, $m = 0, 1, 2, 3, \ldots$, we see that the van der Put coefficients B_m satisfy conditions (9.54) of Lemma 9.18. Therefore, f is ergodic. □

Now we are able to prove Theorem 9.17:

Proof of Theorem 9.17 Consider the van der Put expansion (4.10) of the T-function f; then by Theorem 5.14, $B_m = 2^{\lfloor \log_2 m \rfloor} b_m$, for suitable $b_m \in \mathbb{Z}_2$. It is clear now that conditions 1, 2, and 3 of Proposition 9.19 are equivalent, respectively, to conditions 1, 2, and 3 of Theorem 9.17.

Take $2^{n-1} \leqslant m < 2^n$, $n \geqslant 2$; thus, $B_m = 2^{n-1} b_m$. Then condition (iv) of Proposition 9.19 is equivalent to the congruence $\sum_{m=2^{n-1}}^{2^n-1} (B_m - 2^{n-1}) \equiv 0 \pmod{2^{n+1}}$; whence is equivalent to the congruence $2^{n-1} \sum_{m=2^{n-1}}^{2^n-1} (b_m - 1) \equiv 0 \pmod{2^{n+1}}$ as $B_m = 2^{n-1} b_m$. However, the latter congruence is equivalent to the congruence $\sum_{m=2^{n-1}}^{2^n-1}(b_m - 1) \equiv 0 \pmod 4$, which in turn is equivalent either to the congruence $\sum_{m=2^{n-1}}^{2^n-1} b_m \equiv 0 \pmod 4$ (when $n \geqslant 3$) or to the congruence $\sum_{m=2^{n-1}}^{2^n-1} b_m \equiv 2 \pmod 4$ (when $n = 2$). However, the latter two congruences are, respectively, conditions 5 and 4 of Theorem 9.17. □

9.5 Special Case p=2; T-functions

The theme we address in this section is of a twofold nature: On the one hand, this section may be considered as a collection of various examples of "non-classical" 1-Lipschitz functions, where the "non-classicality" means that the functions are not 2-adic analogs of classical real functions, e.g., p-adic exponent, p-adic logarithm, p-adic analytic functions, and the likes. On the other hand, the 2-adic 1-Lipschitz functions (i.e., the T-functions) considered in this section are compositions of basic computer instructions and as a matter of fact can be judged as a straight-line program (SLP) of contemporary digital computers. From this side, this section is of practical value for computer science and may be considered as a background for applied Chap. 14. The results presented int this section were mainly published in [5, 9–11, 14]. We start with explanations of what these basic computer instructions are of contemporary digital computers and how they are related to 2-adic 1-Lipschitz maps.

9.5.1 Special T-functions: Computer Instructions

Every contemporary computer is word-oriented. That is, it works with words of zeroes and ones of a certain fixed length n (usually $n = 8, 16, 32, 64$). Each binary word $z \in \mathcal{W}_n$ of length n can be treated as the base-2 expansion of a number $z \in \{0, 1, \ldots, 2^n - 1\}$, and vice versa:

$$z = \zeta_0 + 2\zeta_1 + 2^2\zeta_2 + \cdots \longleftrightarrow \cdots \zeta_2\zeta_1\zeta_0 \in \mathcal{W}_n$$

We also can identify the set $\{0, 1, \ldots, 2^n - 1\}$ with residues modulo 2^n; that is, with the elements of the residue ring $\mathbb{Z}/2^n\mathbb{Z}$ modulo 2^n. Actually, *arithmetical* (numerical) instructions of a processor are just *operations of the residue ring* $\mathbb{Z}/2^n\mathbb{Z}$: An n-bit word processor performing a single instruction of addition (or multiplication) of two n-bit numbers just deletes more significant digits of a sum (or of a product) of these numbers, thus merely reducing the result modulo 2^n. Note that to calculate a sum of two integers (i.e., without reducing the result modulo 2^n), any "standard" processor uses not a single instruction but a program that consists of basic instructions.

Another type of basic instructions of a processor are *bitwise logical* instructions: XOR, OR, and AND, NOT. These instructions are clear from their definitions, but nonetheless below, we define them in a formal way to show that actually *all these instructions can be uniquely expanded to 2-adic 1-Lipschitz maps.*

The third type of basic instructions may be called *machine* ones, since they depend on a processor. But usually they include such standard instructions as shifts (left and right) and circular rotations of an n-bit word.

Now, let

$$z = \delta_0(z) + \delta_1(z)2 + \delta_2(z)2^2 + \delta_3(z)2^3 + \cdots$$

be the base-2 expansion of $z \in \mathbb{N}_0 = \{0, 1, 2, \ldots\}$. Then formal definitions of bitwise-logical and machine instructions are as follows:

- y XOR z is a bitwise addition modulo 2, the bitwise "exclusive or" instruction: $\delta_j(y \text{ XOR } z) = \delta_j(y) + \delta_j(z) - 2\delta_j(y)\delta_j(z) = \delta_j(y) \oplus \delta_j(z) \equiv \delta_j(y) + \delta_j(z) \pmod 2$.
- y AND z is a bitwise multiplication modulo 2, the bitwise conjunction: $\delta_j(y \text{ AND } z) = \delta_j(y) \wedge \delta_j(z) = \delta_j(y)\delta_j(z)$.
- y OR z is bitwise "or", the bitwise disjunction: $\delta_j(y \text{ OR } z) = \delta_j(y) \vee \delta_j(z) = \delta_j(y) + \delta_j(z) - \delta_j(y)\delta_j(z)$.
- NOT (z) is a bitwise logical negation: $\delta_j(\text{NOT } (z)) = \delta_j(z) \oplus 1$.
- SHR$(z) = \left[\frac{z}{2}\right]$, the integral part of $\frac{z}{2}$, is a shift toward low-order bits: $\delta_j(2z) = \delta_{j+1}(z)$, $j = 0, 1, 2, \ldots$.
- SHL$(z) = 2z$ is a shift toward high-order bits: $\delta_0(2z) = 0$, $\delta_j(2z) = \delta_{j-1}(z)$, $j = 1, 2, \ldots$.
- MASK$(z, y) = y$ AND z is *masking* of z with the mask y (here y is a constant and z variable).
- z mod $2^k = z$ AND $(2^k - 1)$ is a reduction of z modulo 2^k.

The above considerations serve as a reasoning for the following conclusion:

Contemporary Digital Computers Think 2-Adically The computers live in the non-Archimedean 2-adic world; they work with approximations of 2-adic integers with respect to 2-adic metric, and most of basic instructions are continuous 2-adic functions: All the mentioned bitwise logical and machine instructions, not speaking of arithmetical ones, are well-defined on 2-adic integers and are continuous 2-adic functions; moreover, all of them but a shift toward low-order bits (SHR) are 2-adic 1-Lipschitz functions.

Therefore, all compositions of computer instructions that involve arithmetical instructions (addition, multiplication), bitwise logical instructions (XOR, AND, OR, NOT), and machine instructions (except for SHR) are, on the one hand, straight-line computer programs (SLP) and, on the other hand, 2-adic 1-Lipschitz functions. This observation makes it possible to analyze behavior of these straight-line programs with the use of 2-adic dynamics. In this chapter, we justify the above maxim on the ultrametricity of the world where the computers reside, by rigorous mathematical statements.

First, we observe that from the definitions of the computer instructions, the following identities can immediately be derived: for all $u, v \in \mathbb{Z}_2$, it holds that

$$\begin{aligned} &\mathsf{NOT}\,(u) = u\ \mathsf{XOR}\ (-1); \\ &\mathsf{NOT}\,(u) + u = -1; \\ &u\ \mathsf{XOR}\ v = u + v - 2(u\ \mathsf{AND}\ v); \\ &u\ \mathsf{OR}\ v = u + v - (u\ \mathsf{AND}\ v); \\ &u\ \mathsf{OR}\ v = (u\ \mathsf{XOR}\ v) + (u\ \mathsf{AND}\ v). \end{aligned} \tag{9.64}$$

Next, from Sect. 6.3.3, it immediately follows that the following arithmetical instructions are also T-functions (and moreover, $\mathcal{B}$-functions) of $u, v \in \mathbb{Z}_2$:

$$\begin{aligned} &\text{Exponentiation, } \uparrow\!: \ (u, v) \mapsto u \uparrow v = (1 + 2u)^v; \\ &\text{Raising to negative powers, } \ u \uparrow (-r) = (1 + 2u)^{-r}, r \in \mathbb{N}; \\ &\text{Division, } / : u/v = u(v \uparrow (-1)) = \frac{u}{1 + 2v}. \end{aligned} \tag{9.65}$$

These T-functions may also be used in compositions of SLPs; therefore, the following wild-looking T-function may serve as an example of how complex the compositions could be:

$$(1 + z)\ \mathsf{XOR}\ 4\left(1 - 2\frac{z\ \mathsf{AND}\ z^2 + z^3\ \mathsf{OR}\ z^4}{3 - 4(5 + 6z^5)^{z^6 \mathsf{XOR} z^7}}\right)^{7 - \frac{8z^8}{9 + 10z^9}}$$

Moreover, this wild-looking T-function is uniformly differentiable modulo 4 on $\mathbb{Z}_2$.

Further, a number of computer instructions are not only continuous with respect to 2-adic metric, but they are differentiable with respect to this metric. By using definitions of uniform differentiability modulo p^k, it is not difficult to obtain the following "table of derivatives" of bitwise-logical instructions:

Example 9.20 Let $c \in \mathbb{Z} = \{0, \pm 1, \pm 2, \ldots\}$. The following is true:

1. The function $z\,\mathsf{AND}\,c$ is uniformly differentiable on $\mathbb{Z}_2$; $(z\,\mathsf{AND}\,c)' = 0$ if $c \geqslant 0$, and $(z\ \mathsf{AND}\ c)' = 1$ if $c < 0$.
2. The function $f(z) = z\ \mathsf{XOR}\ c$ is uniformly differentiable on $\mathbb{Z}_2$; $f'(z) = 1$ if $c \geqslant 0$, and $f'(z) = -1$ if $c < 0$.
3. The function $x\ \mathsf{OR}\ c$ is uniformly differentiable on $\mathbb{Z}_2$; $(z\ \mathsf{OR}\ c)' = 1$ if $c \geqslant 0$, $(z\ \mathsf{OR}\ c)' = 0$ if $c < 0$.

4. Given $n \in \mathbb{N}$, the function $x \bmod 2^n$ is uniformly differentiable on $\mathbb{Z}_2$; $(z \bmod 2^n)' = 0$.
5. The function $\mathsf{NOT}\, z$ is uniformly differentiable on $\mathbb{Z}_2$; $(\mathsf{NOT}\, z)' = -1$.

Indeed, from the definition of differentiability and its $\mathsf{mod}\ p^k$-counterpart (see Sects. 3.1.2 and 3.1.3), it immediately follows that if $f(z) = z\ \mathsf{AND}\ c$, then, $f(z + 2^n s) = f(z)$ whenever $c \geqslant 0$, and $f(z + 2^n s) = f(z) + 2^n s$ whenever $c < 0$, for all $n \in \mathbb{N}$ large enough: Recall that for all j large enough, the digits $\delta_j(c)$ in the canonical 2-adic representation of $c \in \mathbb{Z}$ are all 0 if c is non-negative and are all 1 otherwise. This proves assertion 1.

The rest of the assertions of Example 9.20 can be proved in a similar way: For instance, from the identity $u\ \mathsf{XOR}\ v = u + v - 2(u\ \mathsf{AND}\ v)$ (cf., (9.64)) in view of the already proven assertion 1, assertion 2 follows immediately by the rules of derivations, etc. We leave proofs of other assertions as an exercise to the reader.

We note, however, that the functions XOR, OR, AND are *not* differentiable as functions of two 2-adic variables; nonetheless, the function XOR is uniformly differentiable modulo 2 (but not modulo 4): See Example 3.18, Note 3.19, and Example 3.23.

9.5.2 Measure-Preservation/Ergodicity Criteria for T-functions

In this subsection, we prove some criteria of measure-preservation/ergodicity for T-functions with special focus on T-functions, which are constructed from computer instructions described in preceding subsection. These considerations demonstrate how the techniques developed earlier in the book.

We first show how the techniques of coordinate functions, the ANF-techniques, work. With the use of the ANF technique, it is possible to give a short proof of the main result of paper [121] (see further Example 9.21). Note that that recurrence laws of several cryptographic generators are based on the corresponding mapping introduced in [121]. In the following example, we restate the main result of [121, Theorem 3] in dynamical terms and prove it by using Theorem 9.4:

Example 9.21 The mapping $f\colon \mathbb{Z}_2 \to \mathbb{Z}_2$ defined as $f(x) = x + (x^2\ \mathsf{OR}\ C)$, where $C \in \mathbb{Z}_2$, is measure-preserving if and only if $C \equiv 1 \pmod 2$. The mapping f is ergodic if and only if $C \equiv 5 \pmod 8$ or $C \equiv 7 \pmod 8$.

Proof of Example 9.21 Denote $c_i = \delta_i(C)$; for $x \in \mathbb{Z}_2$ and $i = 0, 1, 2, \ldots$, denote $\chi_i = \delta_i(x) \in \{0, 1\}$. To calculate ANF of the Boolean function $\delta_i(x + (x^2\ \mathsf{OR}\ C))$ in variables $\chi_0, \chi_1, \ldots$, we start with the following easy claims (all the equations below are in 2-element field $\mathbb{F}_2$):

- $\delta_0(x^2) = \chi_0$, $\delta_1(x^2) = 0$, $\delta_2(x^2) = \chi_0\chi_1 + \chi_1$.
- $\delta_n(x^2) = \chi_{n-1}\chi_0 + \psi_n(\chi_0, \ldots, \chi_{n-2})$ for all $n \geqslant 3$, where ψ_n is a Boolean function in $n - 1$ Boolean variables $\chi_0, \ldots, \chi_{n-2}$.

The first of these claims can be easily verified by direct calculations. To prove the second, represent $x = \bar{x}_{n-1} + 2^{n-1}s_{n-1}$ for $\bar{x}_{n-1} = x \bmod 2^{n-1}$, calculate $x^2 \equiv (\bar{x}_{n-1} + 2^{n-1}s_{n-1})^2 \equiv \bar{x}_{n-1}^2 + 2^n s_{n-1}\bar{x}_{n-1} + 2^{2n-2}s_{n-1}^2 \equiv \bar{x}_{n-1}^2 + 2^n\chi_{n-1}\chi_0 \pmod{2^{n+1}}$ for $n \geqslant 3$, and note that $\bar{x}_{n-1}^2$ depends only on $\chi_0, \dots, \chi_{n-2}$. This implies that the following equalities hold in the 2-element field $\mathbb{F}_2$:

1. $\delta_0(x^2 \text{ OR } C) = \chi_0 + c_0 + \chi_0 c_0$.
2. $\delta_1(x^2 \text{ OR } C) = c_1$.
3. $\delta_2(x^2 \text{ OR } C) = \chi_0\chi_1 + \chi_1 + c_2 + c_2\chi_1 + c_2\chi_0\chi_1$.
4. $\delta_n(x^2 \text{ OR } C) = \chi_{n-1}\chi_0 + \psi_n + c_n + c_n\chi_{n-1}\chi_0 + c_n\psi_n$ for $n \geqslant 3$.

From here, it follows that if $n \geqslant 3$, then $\delta_n(x^2 \text{ OR } C) = \lambda_n(\chi_0, \dots, \chi_{n-1})$, and $\deg \lambda_n \leqslant n - 1$ since ψ_n depends only on $\chi_0, \dots, \chi_{n-2}$.

Now we successively calculate $\gamma_n = \delta_n(x + (x^2 \text{ OR } C))$ for $n = 0, 1, 2, \dots$. We have $\delta_0(x + (x^2 \text{ OR } C)) = c_0 + \chi_0 c_0$, so necessarily $c_0 = 1$ since otherwise f is not bijective modulo 2. Proceeding further with $c_0 = 1$, we obtain $\delta_1(x + (x^2 \text{ OR } C)) = c_1 + \chi_0 + \chi_1$ since χ_1 is a carry. Then $\delta_2(x + (x^2 \text{ OR } C)) = (c_1\chi_0 + c_1\chi_1 + \chi_0\chi_1) + (\chi_0\chi_1 + \chi_1 + c_2 + c_2\chi_1 + c_2\chi_0\chi_1) + \chi_2 = c_1\chi_0 + c_1\chi_1 + \chi_1 + c_2 + c_2\chi_1 + c_2\chi_0\chi_1 + \chi_2$; here $c_1\chi_0 + c_1\chi_1 + \chi_0\chi_1$ is a carry. From here in view of Theorem 9.4, we immediately deduce that $c_2 = 1$ since otherwise f is not transitive modulo 8. Now for $n \geqslant 3$, one has $\gamma_n = \alpha_n + \lambda_n + \chi_n$, where α_n is a carry, and $\alpha_{n+1} = \alpha_n\lambda_n + \alpha_n\chi_n + \lambda_n\chi_n$. But if $c_2 = 1$, then $\deg \alpha_3 = \deg(\mu\nu + \chi_2\mu + \chi_2\nu) = 3$, where $\mu = c_1\chi_0 + c_1\chi_1 + \chi_0\chi_1$, $\nu = (\chi_0\chi_1 + \chi_1 + c_2 + c_2\chi_1 + c_2\chi_0\chi_1) = 0$. This implies inductively in view of Claim 4 above that $\deg \alpha_{n+1} = n + 1$ and that $\gamma_{n+1} = \chi_{n+1} + \xi_{n+1}(\chi_0, \dots, \chi_n)$, $\deg \xi_{n+1} = n + 1$. So conditions of Theorem 9.4 are satisfied, and we are done. □

The ANF techniques turned out to be useful to determine measure-preservation/ergodicity of T-functions, which involve generalized multiplicative inverses $\text{inv}_2(x)$ defined by (3.3). It is clear that the mapping $x \mapsto \text{inv}_p x$ is a 1-Lipschitz bijection of $\mathbb{Z}_p$ onto $\mathbb{Z}_p$; therefore, the mapping is measure-preserving cf. Sect. 8.1. The following Proposition 9.22 is the ergodicity criterion of inv_2-based T-functions.

Proposition 9.22 *Let f be any 1-Lipschitz transformation on $\mathbb{Z}_2$. If f is ergodic, then both compositions $f(\text{inv}_2(x))$ and $\text{inv}_2(f(x))$ are ergodic. Vice versa, if either of transformations $f(\text{inv}_2(x))$ or $\text{inv}_2(f(x))$ is ergodic, then f is ergodic.*

To prove the proposition, we need the following lemma:

Lemma 9.23 *The ANF of the i-th coordinate function $\delta_i(\text{inv}_2(x))$ is of the form*

$$\delta_i(\text{inv}_2(x)) = \chi_i \oplus \varphi_i(\chi_0, \dots, \chi_{i-1}),$$

where $\chi_i = \delta_i(x)$, $\varphi_0 = 0$, and the weight of every Boolean function $\varphi_i(\chi_0, \dots, \chi_{i-1})$ in Boolean variables $\chi_0, \dots, \chi_{i-1}$ is even, $i = 0, 1, 2, \dots$.

Recall that the weight of Boolean function $\varphi_i(\chi_0, \ldots, \chi_{i-1})$ in Boolean variables $\chi_0, \ldots, \chi_{i-1}$ is even if and only if the ANF of this Boolean function does not contain the monomial $\chi_0 \cdots \chi_{i-1}$ (see Theorem 9.4).

Proof of Lemma 9.23 As $\mathrm{inv}_2 \colon \mathbb{Z}_2 \to \mathbb{Z}_2$ is a 1-Lipschitz measure-preserving transformation on $\mathbb{Z}_2$, then in view of Eq. (9.8) of Sect. 9.2 and of Theorem 9.4, the Boolean function $\delta_i(\mathrm{inv}_2(x))$ depends only on Boolean variables $\chi_0, \ldots, \chi_i$, and $\delta_i(\mathrm{inv}_2(x))$ is linear with respect to variable χ_i: $\delta_i(\mathrm{inv}_2(x)) = \chi_i \oplus \varphi_i(\chi_0, \ldots, \chi_{i-1})$ for a suitable Boolean function $\varphi_i(\chi_0, \ldots, \chi_{i-1})$ in Boolean variables $\chi_0, \ldots, \chi_{i-1}$, for all $i = 0, 1, 2, \ldots$ (recall that a Boolean function on empty set of variables is a constant).

Now by induction on i, we prove that the weight of the Boolean function $\varphi_i(\chi_0, \ldots, \chi_{i-1})$ is even, for all $i = 0, 1, 2, \ldots$; that is, the number of Boolean i-dimensional vectors on which the Boolean function $\varphi_i(\chi_0, \ldots, \chi_{i-1})$ takes value 1 is even. Direct calculations show that $\mathrm{inv}_2(x) \equiv x \pmod{2^n}$ for $n = 1, 2, 3$; so $\varphi_0 = \varphi_1 = \varphi_2 = 0$; for $n = 4$, we have $\mathrm{inv}_2(x) \not\equiv x \pmod{2^n}$ if and only if x is congruent 3,5,11, or 14 modulo 16, so the weight of the Boolean function $\varphi_3(\chi_0, \chi_1, \chi_2)$ is 2. Let our claim be true for Boolean functions $\varphi_0, \ldots, \varphi_{i-1}$; let us prove it for the Boolean function $\varphi_i(\chi_0, \ldots, \chi_{i-1})$.

For a Boolean function ψ, denote via $\bar{\psi}$ its negation; that is, $\bar{\psi} = \psi \oplus 1$. Now take arbitrary $x \equiv 1 \pmod 2$ (in other words, put $\chi_0 = 1$), and consider $\delta_i(\mathrm{inv}_2(1 + \mathsf{NOT}\,(x))$. Since $x = 1 + 2z$, where $z = \chi_1 + 2\chi_2 + 4\chi_3 + \cdots$, then $\mathrm{inv}_2(1 + \mathsf{NOT}\,(x)) = (1 + 2\mathsf{NOT}\,(z))^{-1} = (1 - 2(1+z))^{-1} = -(1+2z)^{-1} = 1 + \mathsf{NOT}\,((1 + 2z)^{-1})$ (we used the second formula from (9.64) during these conversions). It is obvious that if we denote $(1 + 2\mathsf{NOT}\,(z))^{-1} = 1 + \sum_{j=1}^{\infty} 2^j \zeta_j$, then $1 + \mathsf{NOT}\,((1 + 2z)^{-1}) = 1 + \sum_{j=1}^{\infty} 2^j \bar{\zeta}_j$, where $\zeta_j \in \{0, 1\}$ $(j = 1, 2, \ldots)$. By this reason, the just proven equality $(1 + 2\mathsf{NOT}\,(z))^{-1} = 1 + \mathsf{NOT}\,((1 + 2z)^{-1})$ implies that

$$\varphi_i(1, \chi_1, \ldots, \chi_{i-1}) = \varphi_i(1, \bar{\chi}_1, \ldots, \bar{\chi}_{i-1}), \tag{9.66}$$

for all $\chi_1, \ldots, \chi_{i-1} \in \{0, 1\}$, since $\zeta_i = \delta_i(\mathrm{inv}_2(x)) = \chi_i \oplus \varphi_i(\chi_0, \ldots, \chi_{i-1})$, $i = 1, 2, \ldots$. Further, since $\mathrm{inv}_2(ab) = \mathrm{inv}_2(a)\,\mathrm{inv}_2(b)$ for all $a, b \in \mathbb{Z}_2$, then $\mathrm{inv}_2(2z) = 2\,\mathrm{inv}_2(z)$, so $\varphi_i(0, \chi_1, \ldots, \chi_{i-1}) = \varphi_{i-1}(\chi_1, \ldots, \chi_{i-1})$; however, by induction hypothesis, the weight of the Boolean function $\varphi_{i-1}(\chi_1, \ldots, \chi_{i-1})$ in Boolean variables $\chi_1, \ldots, \chi_{i-1}$ is even. This, together with Eq. (9.66), completes induction and proves the Lemma. □

Now we are able to prove Proposition 9.22.

Proof of Proposition 9.22 For $i = 0, 1, 2, \ldots$, denote $\delta_i(x) = \chi_i$. If f is ergodic, then by Theorem 9.4,

$$\delta_i(f(x)) = \chi_i \oplus \chi_0 \cdots \chi_{i-1} \oplus \psi_i(\chi_0, \ldots, \chi_{i-1}), \tag{9.67}$$

where the ANF of the Boolean function $\psi_i(\chi_0, \cdots, \chi_{i-1})$ does not contain the monomial $\chi_0 \cdots \chi_{i-1}$, $\psi_0 = 0$, $i = 0, 1, 2 \ldots$ (we recall that product over empty set is 1). By Lemma 9.23,

$$\delta_i(\mathrm{inv}_2(x)) = \chi_i \oplus \varphi_i(\chi_0, \dots, \chi_{i-1}), \tag{9.68}$$

where $\varphi_0 = 0$ and ANF of the Boolean function $\varphi_i(\chi_0, \dots, \chi_{i-1})$ does not contain the monomial $\chi_0 \cdots \chi_{i-1}$, $i = 0, 1, 2, \dots$. Whence ANF of the Boolean function $\delta_i(u(x))$, where $u(x)$ is either of functions $f(\mathrm{inv}_2(x))$ or $\mathrm{inv}_2(f(x))$, is of the form

$$\delta_i(u(x)) = \chi_i \oplus \chi_0 \cdots \chi_{i-1} \oplus \vartheta_i(\chi_0, \dots, \chi_{i-1}), \tag{9.69}$$

where the ANF of the Boolean function $\vartheta_i(\chi_0, \cdots, \chi_{i-1})$ does not contain the monomial $\chi_0 \cdots \chi_{i-1}$, $\vartheta_0 = 1$, $i = 0, 1, 2 \dots$. Thus, by Theorem 9.4, both $f(\mathrm{inv}_2(x))$ and $\mathrm{inv}_2(f(x))$ are ergodic.

To prove the converse statement, note that if f is not ergodic, then by Theorem 9.4, the ANF of some Boolean function $\delta_i(f(x))$ in representation (9.67) does not contain the monomial $\chi_0 \cdots \chi_{i-1}$. Thus, in view of (9.68), representation (9.69) of $\delta_i(u(x))$ does not contain the monomial $\chi_0 \cdots \chi_{i-1}$ either. Therefore, $u(x)$ is not ergodic by virtue of Theorem 9.4. □

The following theorem serves as an example of how the techniques of p-adic series works in the design of "computer-oriented" ergodic T-functions cf. [14].

Theorem 9.24 *The following is true:*

1. *The function $f\colon \mathbb{Z}_2 \to \mathbb{Z}_2$ of the form*

$$f(x) = a + \sum_{i=1}^{n} a_i(x \,\mathsf{XOR}\, b_i),$$

where $a, a_i, b_i \in \mathbb{Z}_2$, $i = 1, 2, 3, \dots$, is measure-preserving (respectively, ergodic) if and only if it is bijective (respectively, transitive) modulo 2 (respectively, modulo 4).

2. *The function $f\colon \mathbb{Z}_2 \to \mathbb{Z}_2$ of the form*

$$f(x) = a + \sum_{i=0}^{\infty} a_i \delta_i(x),$$

where $a, a_i \in \mathbb{Z}_2$, $i = 0, 1, 2, \dots$, is 1-Lipschitz and ergodic if and only if the following conditions hold simultaneously:

$$a \equiv 1 \pmod 2;$$

$$a_0 \equiv 1 \pmod 4;$$

$$|a_i|_2 = 2^{-i},$$

for $i = 1, 2, 3, \ldots$. The function f is 1-Lipschitz and measure-preserving if and only if

$$|a_i|_2 = 2^{-i} \quad (i = 0, 1, 2, 3, \ldots).$$

Proof of Theorem 9.24 Consider Mahler expansion for the function $\delta_i(x)$, $i = 0, 1, 2, \ldots$:

$$\delta_i(x) = \sum_{j=0}^{\infty} \sigma_i(j)\binom{x}{j}. \tag{9.70}$$

To apply Theorem 9.5, we must estimate 2-adic norms of coefficients $\sigma_i(j)$ first. To do this, we need several lemmas. □

Lemma 9.25 *For all $i, j = 1, 2, 3, \ldots$, the following equations hold:*

$$\sigma_i(0) = 0;$$

$$\sigma_0(j) = (-1)^{j+1}2^{j-1};$$

$$\sigma_i(j) = (-1)^{j+1}\sum_{k=1}^{\infty}(-1)^k\binom{j-1}{k2^i-1}.$$

Proof of Lemma 9.25 As $\delta_i(0) = 0$ for all $i = 0, 1, 2, \ldots$, then $\sigma_i(0) = 0$. From Mahler expansion for $\delta_i(x)$, see (9.70), by inversion formula (see Theorem 1.6), we obtain that

$$\sigma_i(j) = (-1)^j\sum_{k=0}^{\infty}(-1)^k\delta_i(k)\binom{j}{k}.$$

Hence, in view of the definition of the function $\delta_i(j)$,

$$\sigma_i(j) = (-1)^j\sum_{s=1}^{\infty}\sum_{k=(2s-1)2^i}^{s2^{i+1}-1}(-1)^k\binom{j}{k}.$$

From here, using the following well-known identity (see, e.g., [78, Chapter 5])

$$\sum_{k=m}^{n}(-1)^k\binom{a}{k} = (-1)^m\binom{a-1}{m-1} + (-1)^n\binom{a-1}{n}, \tag{9.71}$$

we conclude that

$$\sigma_i(j) = (-1)^j \sum_{s=1}^{\infty} \left(\binom{j-1}{(2s-1)2^i - 1} - \binom{j-1}{2s2^i - 1} \right).$$

This proves the Lemma since the latter identity implies that

$$\sigma_i(j) = \begin{cases} (-1)^{j+1} 2^{j-1}, & \text{if } i = 0; \\ (-1)^{j+1} \sum_{k=1}^{\infty} (-1)^k \binom{j-1}{k2^i - 1} & \text{otherwise.} \end{cases}$$

□

Lemma 9.26 *For all $m, t, r = 0, 1, 2, \ldots$ that satisfy simultaneously two conditions $0 \leqslant t \leqslant 2^m - 1$ and $m \geqslant r$, the following congruence holds:*

$$\binom{2^m - 1}{t} \equiv (-1)^{t - \lfloor t2^{-r} \rfloor} \binom{2^{m-r} - 1}{\lfloor t2^{-r} \rfloor} \pmod{2^{m-r+1}}.$$

In particular, for all $m, s, j \in \mathbb{N}$ that satisfy simultaneously two conditions $m > s \geqslant 1$ and $j \leqslant 2^{m-s} - 1$, the following congruence holds:

$$\binom{2^m - 2}{2^s j - 1} \equiv (-1)^j 2^s j \binom{2^{m-s} - 1}{j - 1} \pmod{2^{m-s+1}}.$$

Proof of Lemma 9.26 We recall that every $s \in \mathbb{Z}_2$ has a unique representation of the form $s = 2^{\operatorname{ord}_2 s} \hat{s}$, where $\hat{s}$ is the unit of $\mathbb{Z}_2$; that is, $\hat{s}$ is odd, meaning $\delta_0(\hat{s}) = 1$, and henceforth s has a multiplicative inverse $\hat{s}^{-1}$ in $\mathbb{Z}_2$ (see Note 2.9). Put $M = \{i : i = 1, 2, \ldots, t;\ \operatorname{ord}_2 i \geqslant r\}$, and let M' be a complement of M in $\{1, 2, \ldots, t\}$; then

$$\binom{2^m - 1}{t} = \prod_{i=1}^{t} \frac{2^m - i}{i} = \prod_{i=1}^{t} \left(\frac{2^{m - \operatorname{ord}_2 i}}{\hat{i}} - 1 \right) \equiv$$

$$\equiv (-1)^{\#M'} \prod_{i \in M} \left(\hat{s}^{-1} 2^{m - \operatorname{ord}_2 i} - 1 \right) \pmod{2^{m-r+1}}. \tag{9.72}$$

The condition $\operatorname{ord}_2 i \geqslant r$ for $i = 1, 2, \ldots, t$ holds if and only if $i = j2^r$ for $j = 1, 2, \ldots, \lfloor 2^{-r} t \rfloor$. This means that $\#M' = t - \lfloor 2^{-r} t \rfloor$. So, the product in the right-hand side of congruence (9.72) is equal to

$$(-1)^{\#M'} \prod_{j=1}^{\lfloor 2^{-r} t \rfloor} \left(\hat{j}^{-1} 2^{m - r - \operatorname{ord}_2 j} - 1 \right) = (-1)^{t - \lfloor t2^{-r} \rfloor} \binom{2^{m-r} - 1}{\lfloor t2^{-r} \rfloor}.$$

This proves the first part of assertion of the Lemma. The second part now becomes obvious since

$$\binom{2^m-2}{2^s j-1} = \frac{2^m-2^s j}{2^m-1}\binom{2^m-1}{2^s j-1} \equiv 2^s j\binom{2^m-1}{2^s j-1} \pmod{2^{m-s+1}}.$$

□

Lemma 9.27 *For $s, k = 1, 2, 3, \ldots$, the following is true:*

1. $|\sigma_s(k)|_2 \leqslant 2^{-\lfloor \log_2 k\rfloor+s-1}$, *whenever* $k \neq 2^s, 2^{s+1}$.
2. $|\sigma_s(2^s)|_2 = 1$, $|\sigma_s(2^{s+1})|_2 = \frac{1}{2}$.
3. $|\sigma_s(2^m-1)|_2 \leqslant 2^{-m+s-1}$, *whenever* $m > s \geqslant 1$.

Proof of Lemma 9.27 Represent k as $k = 2^m + t$, where $m = \lfloor \log_2 k\rfloor$, $0 \leqslant t < 2^m$. We may assume that $m \geqslant s$ since otherwise $\sigma_s(k) = 0$ in view of Lemma 9.25. Further, Lemma 9.25 implies that

$$\sigma_s(2^m+t) = (-1)^{t+1}\sum_{j=1}^{\infty}(-1)^j\binom{2^m+t-1}{2^s j-1}. \tag{9.73}$$

Now by the following well-known identity (see, e.g., [78, Chapter 5])

$$\sum_{k=0}^{n}\binom{a}{k}\binom{b}{n-k} = \binom{a+b}{n},$$

we conclude that

$$\begin{aligned}\binom{2^m-1+t}{2^s j-1} &= \sum_{k=0}^{\infty}\binom{t}{k}\binom{2^m-1}{2^s j-k-1} = \\ &= \sum_{n=0}^{\infty}\sum_{r=0}^{2^s-1}\binom{t}{2^s n+r}\binom{2^m-1}{2^s(j-n-1)+(2^s-r-1)}.\end{aligned} \tag{9.74}$$

Here, as usual, we assume that $\binom{a}{b} = 0$ for $b < 0$. In view of Lemma 9.26, Eq. (9.74) implies that

$$\begin{aligned}\sum_{n=0}^{\infty}\sum_{r=0}^{2^s-1}(-1)^{n+r+j}\binom{t}{2^s n+r}\binom{2^{m-s}-1}{j-n-1} &\equiv \\ &\equiv \binom{2^m-1+t}{2^s j-1} \pmod{2^{m-s+1}}.\end{aligned} \tag{9.75}$$

Now (9.73) in view of (9.75) implies that

$$\sigma_s(2^m+t) \equiv (-1)^{t+1}\sum_{n=0}^{\infty}\sum_{r=0}^{2^s-1}(-1)^{n+r}\binom{t}{2^s n+r}\sum_{j=1}^{\infty}\binom{2^{m-s}-1}{j-n-1} \equiv$$

$$\equiv 2^{2^{m-s}-1}(-1)^{t+1}\sum_{n=0}^{\infty}\sum_{r=0}^{2^s-1}(-1)^{n+r}\binom{t}{2^s n+r} \pmod{2^{m-s+1}}. \qquad (9.76)$$

Now applying identity (9.71) and assuming that $t \neq 0$, in view of Lemma 9.25, we conclude that

$$(-1)^{t+1}\sum_{n=0}^{\infty}\sum_{r=0}^{2^s-1}(-1)^{n+r}\binom{t}{2^s n+r} =$$

$$= (-1)^{t+1}\sum_{n=0}^{\infty}(-1)^n\binom{t}{2^s n+r}\left(\binom{t-1}{2^s n-1} - \binom{t-1}{2^s(n+1)-1}\right) =$$

$$= 2(-1)^{t+1}\sum_{n=1}^{\infty}(-1)^n\binom{t-1}{2^s n-1} = 2\sigma_s(t).$$

The left-hand side of this equation is equal to -1 when $t = 0$. So, taking all these arguments into account, from (9.76), we conclude that

$$\sigma_s(2^m+t) \equiv \begin{cases} 2^{2^{m-s}}\sigma_s(t) \pmod{2^{m-s+1}}, & \text{if } t \neq 0; \\ -2^{2^{m-s-1}} \pmod{2^{m-s+1}}, & \text{if } t = 0. \end{cases}$$

The latter congruence proves claims 1 and 2 of the Lemma since it easily implies that

$$\sigma_s(2^m+t) \equiv \begin{cases} 1 \pmod{2}, & \text{if } m = s, t = 0; \\ 2 \pmod{4}, & \text{if } m = s+1, t = 0; \\ 0 \pmod{2^{m-s+1}}, & \text{in all other cases.} \end{cases}$$

Finally, if $m > s \geqslant 1$, then combining together Lemmas 9.25 and 9.26, we conclude that

$$\sigma_s(2^m-1) \equiv 2^s\sum_{j=1}^{\infty}\binom{2^{m-s}-1}{j-1} \pmod{2^{m-s+1}}.$$

From here by a well-known identity $\sum_{k=1}^{n} k\binom{n}{k} = 2^{n-1}n$ (see, e.g., [78, Chapter 5]), we deduce that

$$\sigma_s(2^m - 1) \equiv 2^{2^{m-s}-1+s}(2^{m-s} - 1) \pmod{2^{m-s+1}}.$$

This proves claim 3 and the Lemma. □

Now we are ready to prove Theorem 9.24. We start with claim 1. The operation XOR and, consequently, the function f are 1-Lipschitz. Further, for all $u, v \in \mathbb{Z}_2$, the following identity holds (cf., (9.64)):

$$u \,\mathsf{XOR}\, v = \sum_{k=0}^{\infty} 2^k(\delta_k(u) + \delta_k(v) - 2\delta_k(u)\delta_k(v)) = u + v - \sum_{k=0}^{\infty} 2^{k+1}\delta_k(u)\delta_k(v).$$

Consequently,

$$f(x) = a + \sum_{i=1}^{n} a_i b_i + \sum_{i=1}^{n} a_i x - 2\sum_{i=1}^{n}\sum_{k=0}^{\infty} 2^k \delta_k(x)\delta_k(b_i).$$

Now, considering interpolation series for $\delta_k(x)$ and taking into account that (in view of Lemma 9.25) $\sigma_0(1) = 1$ and $\sigma_i(1) = 0$ for $i = 1, 2, 3, \ldots$, we conclude that

$$f(x) = a + \sum_{i=1}^{n} a_i b_i + x\left(\sum_{i=1}^{n} a_i - 2\sum_{i=1}^{n} \delta_0(b_i)\right) - \\ - \sum_{j=2}^{\infty}\binom{x}{j}\sum_{i=1}^{n}\sum_{k=0}^{\infty} 2^{k+1}\sigma_k(j)\delta_k(b_i).$$

Lemma 9.27 immediately implies that for $k \geqslant 2$,

$$2^{k+1}\sigma_k(j) \equiv \begin{cases} 0 \pmod{2^{\lfloor \log_2 j \rfloor + 1}}, & \text{if } j = 2^k, 2^{k+1}; \\ 0 \pmod{2^{\lfloor \log_2 j \rfloor + 2}}, & \text{otherwise.} \end{cases}$$

Now, Theorem 9.5 implies that f is measure-preserving (respectively, ergodic) if and only if $\sum_{i=1}^{n} a_i \equiv 1 \pmod 2$ (respectively, if and only if $a + \sum_{i=1}^{n} a_i b_i \equiv 1 \pmod 2$ and $\sum_{i=1}^{n} a_i + 2\sum_{i=1}^{n} b_i \equiv 1 \pmod 4$). This is obviously equivalent to claim 1 of Theorem 9.24.

To prove claim 2 of the theorem, we first note that functions δ_i for $i > 0$ are not 1-Lipschitz. As $\sigma_i(0) = 0$ for $i > 0$ (see Lemma 9.25), we have

$$f(x) = a + \sum_{j=1}^{\infty}\binom{x}{j}\sum_{i=0}^{\infty} a_i\sigma_i(j).$$

Theorem 9.5 implies now that the function f is measure-preserving if and only if the following congruences hold simultaneously:

$$\begin{cases} \sum_{i=0}^{\infty} a_i \sigma_i(1) \equiv 1 \pmod 2; \\ \sum_{i=0}^{\infty} a_i \sigma_i(j) \equiv 0 \pmod{2^{\lfloor \log_2 j \rfloor + 1}}, \quad j = 2, 3, \ldots \end{cases} \tag{9.77}$$

In view of Lemma 9.25, the first of conditions (9.77) is equivalent to the congruence

$$a_0 \equiv 1 \pmod 2. \tag{9.78}$$

Moreover, Lemma 9.25 implies that $\sigma_i(j) = 0$ for $i \geqslant \lfloor \log_2 j \rfloor$. Hence, the second of conditions (9.77) is equivalent to the following system of congruences:

$$\sum_{i=0}^{\lfloor \log_2 j \rfloor} a_i \sigma_i(j) \equiv 0 \pmod{2^{\lfloor \log_2 j \rfloor + 1}}, \quad j = 2, 3, \ldots. \tag{9.79}$$

Consider the following subsystem of system (9.79) for $j = 2^k$, $k = 1, 2, 3, \ldots$:

$$\sum_{i=0}^{k} a_i \sigma_i(2^k) \equiv 0 \pmod{2^{k+1}}, \quad k = 1, 2, 3, \ldots \tag{9.80}$$

We claim that 2-adic integers a_i satisfy system of congruences (9.80) if and only if $a_i = 2^i \pmod{2^{i+1}}$, $i = 0, 1, 2, \ldots$. We proceed with induction on i. If $i = 1$, we, by Lemma 9.25 (for $k = 1$), conclude that

$$2a_0 + a_1 \sigma_1(2) \equiv 0 \pmod 4. \tag{9.81}$$

In view of claim 2 of Lemma 9.27, the 2-adic integer $\sigma_1(2)$ has a multiplicative inverse in $\mathbb{Z}_2$, so in view of (9.78), congruence (9.81) is equivalent to the congruence

$$a_1 \equiv 2 \pmod 4.$$

Now let our claim be true for $k < n$; consider the congruence

$$\sum_{i=0}^{n} a_i \sigma_i(2^n) \equiv 0 \pmod{2^{n+1}}. \tag{9.82}$$

By induction hypothesis, $a_i = 2^i + s_i 2^{i+1}$ $(i = 0, 1, \ldots, n-1)$ for suitable $s_i \in \mathbb{Z}_2$. Then, taking into account claim 2 of Lemma 9.27, we conclude that $a_i \sigma_i(2^n) \equiv 2^{n+1} \pmod{2^{n+2}}$ for $i = 0, 1, \ldots, n-2$ and $a_{n-1}\sigma_{n-1}(2^n) \equiv 2^n \pmod{2^{n+1}}$. Hence, congruence (9.82) is equivalent to the congruence $2^n + a_n \sigma_n(2^n) \equiv 0$

(mod 2^{n+1}). As $\sigma_n(2^n)$ is a unit of $\mathbb{Z}_2$ (in force of claim 2 of Lemma 9.27), the latter congruence implies that $a_n \equiv 2^n \pmod{2^{n+1}}$.

From claim 1 of Lemma 9.27, it easily follows that if $a_i \equiv 2^i \pmod{2^{i+1}}$, then a_i also satisfy each congruence of the system (9.79) for those j that are not powers of 2. This means that conditions (9.77) are equivalent to the following set of congruences:

$$a_i \equiv 2^i \pmod{2^{i+1}}, \quad i = 0, 1, 2, 3, \ldots.$$

So we have proved the second part of claim 2 of Theorem 9.24. To prove the first part of this claim, we note that since $\lfloor \log_2(i+1) \rfloor + 1 = \lfloor \log_2 i \rfloor + 1$ for $i \neq 2^k - 1$, the sufficient and necessary conditions for ergodicity of function f from Theorem 9.5 in the case under consideration can be rewritten in the following form:

$$a \equiv 1 \pmod 2; \tag{9.83}$$

$$\sum_{i=0}^{\infty} a_i \sigma_i(1) \equiv 0 \pmod 4; \tag{9.84}$$

$$\sum_{i=0}^{\infty} a_i \sigma_i(j) \equiv 0 \pmod{2^{\lfloor \log_2 j \rfloor + 1}}, \quad j = 2, 3, 4, \ldots; \tag{9.85}$$

$$\sum_{i=0}^{\infty} a_i \sigma_i(2^k - 1) \equiv 0 \pmod{2^{k+1}}, \quad k = 2, 3, 4, \ldots. \tag{9.86}$$

As $\sigma_i(1) = 0$ for $i \neq 0$ (see Lemma 9.25), then (9.84) is equivalent to the following condition:

$$a_0 \equiv 1 \pmod 2. \tag{9.87}$$

During the proof of the second part of claim 2, we have established that if $a_0 \equiv 1 \pmod 2$ (and, in particular, if (9.87) is satisfied), then conditions (9.85) are equivalent to the following conditions:

$$a_i \equiv 2^i \pmod{2^{i+1}}, \quad (i = 1, 2, 3, \ldots). \tag{9.88}$$

Finally, combining together claim 1 of Lemmas 9.27 and 9.25, we conclude that if 2-adic integers a_i $(i = 0, 1, 2, \ldots)$ satisfy conditions (9.88) and (9.87) simultaneously, then a_i also satisfy conditions (9.86). Thus, the union of conditions (9.83)–(9.86) is equivalent to the union of conditions (9.83), (9.87), and (9.88). This proves the first part of claim 2 and Theorem 9.24. □

Chapter 10
Measure-Preservation and Ergodicity of Uniformly Differentiable Functions

Abstract In this chapter, we find ergodicity and/or measure-preservation conditions for functions $F\colon \mathbb{Z}_p^n \to \mathbb{Z}_p^m$ that are uniformly differentiable modulo p and that have integer-valued derivatives modulo p. Recall that in view of Theorem 5.5, all these functions are locally 1-Lipschitz (or, in other terms, locally compatible), that is, these are the functions that are 1-Lipschitz on all sufficiently small balls. Therefore, for these functions F, the reduced maps $F \bmod p^k\colon (\mathbb{Z}/p^k\mathbb{Z})^n \to (\mathbb{Z}/p^k\mathbb{Z})^m$ are well defined, provided k is sufficiently large, say, $k \geqslant N_1(F)$ (cf. Definition 3.22). Thus, we can apply Theorem 8.2 to study measure-preservation and ergodicity of F. As 1-Lipschitz uniformly differentiable functions are a special case of functions under consideration, the theory that follows can be applied to various important classes of functions, e.g., for analytic functions on $\mathbb{Z}_p$, C-functions, $\mathcal{B}$-functions, $\mathcal{A}$-functions (in particular, for twice integer-valued polynomials over $\mathbb{Q}_p$), etc.

10.1 Conditions of Measure-Preservation

In this section, we answer the question when is a uniformly differentiable modulo p function is measure-preserving, provided all its derivatives modulo p are integer-valued? We modify the proofs from [11].

Theorem 10.1 *Let the function $F\colon \mathbb{Z}_p^n \to \mathbb{Z}_p^m$, $m \leqslant n$, be uniformly differentiable modulo p, and let all partial derivatives modulo p of the function F be integer-valued. Then F is measure-preserving whenever the following two conditions hold simultaneously:*

1. *F is balanced modulo p^k for some $k \geqslant N_1(F)$.*
2. *The rank $\operatorname{rk} F_1'(\mathbf{y})$ of Jacobi matrix $F_1'(\mathbf{y})$ modulo p is m at all points $\mathbf{y} \in \mathbb{Z}_p^n$.*

Moreover, in the case when $m = n$, the mentioned conditions are also necessary: If $F\colon \mathbb{Z}_p^n \to \mathbb{Z}_p^n$ is measure-preserving, then F is bijective modulo p^k for all $k \geqslant N_1(F)$, and $\det F_1'(\mathbf{y}) \not\equiv 0 \pmod p$ for all $\mathbf{y} \in \mathbb{Z}_p^n$.

V. Anashin, *Causality: The p-adic Theory*, STEAM-H: Science, Technology, Engineering, Agriculture, Mathematics & Health,
https://doi.org/10.1007/978-3-031-85818-5_10

Finally, the function $F: \mathbb{Z}_p^n \to \mathbb{Z}_p^n$ *is measure-preserving if and only if* F *is bijective modulo* p^k *for some* $k \geqslant N_1(f) + 1$.

Proof During the proof, we consider elements of the ring $(\mathbb{Z}/p^r\mathbb{Z})^\ell$ (the ℓ-th Cartesian power of the residue ring $\mathbb{Z}/p^r\mathbb{Z}$) as ordered strings of ℓ numbers from $\{0, 1, \ldots, p^r - 1\}$. With this in mind, for $\mathbf{y} \in (\mathbb{Z}/p^s\mathbb{Z})^m$, denote $F_s^{-1}(\mathbf{y}) = \{\mathbf{v} \in (\mathbb{Z}/p^s\mathbb{Z})^n : F(\mathbf{v}) \equiv \mathbf{y} \pmod{p^s}\}$, a pre-image of $\mathbf{y}$ with respect to the reduced map $F \bmod p^s : (\mathbb{Z}/p^s\mathbb{Z})^n \to (\mathbb{Z}/p^s\mathbb{Z})^m$. Let $s \geqslant k \geqslant N_1(F)$. As F is locally 1-Lipschitz, F is a sum of a 1-Lipschitz function with a step function of order not greater than $N_1(F)$ (see Theorem 5.5)); so we conclude that if $\mathbf{u} \in F_{s+1}^{-1}(\mathbf{w})$, then $\bar{\mathbf{u}} \in F_s^{-1}(\bar{\mathbf{w}})$. Here and further in the proof, $\bar{\mathbf{a}} = (\bar{a}_1, \ldots, \bar{a}_m) \in (\mathbb{Z}/p^s\mathbb{Z})^m$ stands for $\mathbf{a} \bmod p^s = (a_1 \bmod p^s, \ldots, a_m \bmod p^s)$, where $\mathbf{a} = (a_1, \ldots, a_m) \in (\mathbb{Z}/p^{s+1}\mathbb{Z})^m$. Put $\mathbf{z} = \bar{\mathbf{u}} + p^s\mathbf{h} \in (\mathbb{Z}/p^{s+1}\mathbb{Z})^n$, where $\mathbf{h} \in (\mathbb{Z}/p\mathbb{Z})^n$. As F is uniformly differentiable modulo p, then

$$F(\mathbf{z}) \equiv F(\bar{\mathbf{u}}) + p^s\mathbf{h}F_1'(\bar{\mathbf{u}}) \pmod{p^{s+1}} \tag{10.1}$$

(cf. Definition 3.22). As $F(\bar{\mathbf{u}}) \equiv \bar{\mathbf{w}} + p^s\mathbf{b} \pmod{p^{s+1}}$ and $\mathbf{w} = \bar{\mathbf{w}} + p^s\mathbf{c}$ for suitable $\mathbf{b}, \mathbf{c} \in (\mathbb{Z}/p\mathbb{Z})^m$, from (10.1) we deduce that $\mathbf{z} \in F_{s+1}^{-1}(\mathbf{w})$ if and only if $\bar{\mathbf{z}} \in F_s^{-1}(\bar{\mathbf{w}})$ (i.e., $\bar{\mathbf{u}} \in F_s^{-1}(\bar{\mathbf{w}})$), and $\mathbf{h}$ satisfies the following system of linear equations over a field $\mathbb{Z}/p\mathbb{Z}$:

$$\mathbf{b} + \mathbf{h}F_1'(\bar{\mathbf{u}}) = \mathbf{c}. \tag{10.2}$$

Consequently, if columns of the matrix $F_1'(\bar{\mathbf{u}})$ are linearly independent over $\mathbb{Z}/p\mathbb{Z}$, then linear system (10.2) has exactly p^{n-m} pairwise distinct solutions $\mathbf{h} \in (\mathbb{Z}/p\mathbb{Z})^n$ given arbitrary $\mathbf{b}, \mathbf{c} \in (\mathbb{Z}/p\mathbb{Z})^m$. From here it follows that

$$\#F_{s+1}^{-1}(\mathbf{w}) = (\#F_s^{-1}(\bar{\mathbf{w}}))p^{n-m}. \tag{10.3}$$

Hence, if F is balanced modulo p^s (i.e., if $\#F_s^{-1}(\bar{\mathbf{w}})$ does not depend on $\bar{\mathbf{w}}$) and if the rank of the matrix $F_1'(\bar{\mathbf{w}})$ is m for all $\bar{\mathbf{w}} \in (\mathbb{Z}/p^s\mathbb{Z})^n$, then (10.3) implies that F is balanced modulo p^{s+1}. However, in view of Proposition 3.26, the matrix $F_1'(\bar{\mathbf{w}})$ depends only on $\bar{\mathbf{w}} \bmod p^{N_1(F)}$. This in view of Theorem 8.2 proves the first claim of Theorem 10.1.

To prove the second claim, take $m = n$, and suppose that $F: \mathbb{Z}_p^n \to \mathbb{Z}_p^n$ is a measure-preserving function. In view of Theorem 8.2, this implies that F is bijective modulo p^k for all $k \geqslant N_1(F)$. Definition 3.22 of uniform differentiability modulo p implies that

$$F(\mathbf{u} + p^k\mathbf{h}) \equiv F(\mathbf{u}) + p^k\mathbf{h}F_1'(\mathbf{u}) \pmod{p^{k+1}} \tag{10.4}$$

for all $\mathbf{u}, \mathbf{h} \in \mathbb{Z}_p$. Here $F_1'(\mathbf{u})$ is an $n \times n$ matrix over a field $\mathbb{Z}/p\mathbb{Z}$. If $\det F_1'(\mathbf{u}) \equiv 0 \pmod{p}$ for some $\mathbf{u} \in \mathbb{Z}_p^n$ (or, equivalently, for some $\mathbf{u} \in \{0, 1, \ldots, p^{N_1(F)} - 1\}^n$

as partial derivatives modulo p are step functions; see Proposition 3.26), then there exists $\mathbf{h} \in \{0, 1, \ldots, p-1\}^n, \mathbf{h} \not\equiv (0, \ldots, 0) \pmod p$ such that $\mathbf{h}F_1'(\mathbf{u}) \equiv (0, \ldots, 0) \pmod p$. However, then (10.4) implies that $F(\mathbf{u} + p^k\mathbf{h}) \equiv F(\mathbf{u}) \pmod{p^{k+1}}$, in contradiction to bijectivity modulo p^{k+1} of the function F as $\mathbf{u} + p^k\mathbf{h} \not\equiv \mathbf{u} \pmod{p^{k+1}}$.

Finally, if F is bijective modulo p^k for some $k \geqslant N_1(F) + 1$, then F is bijective modulo p^{k-1} due to the compatibility of F (see Proposition 7.1), and $\det F_1'(\mathbf{u}) \equiv 0 \pmod p$ nowhere on $\mathbb{Z}_p$ since otherwise the above argument shows that F is not bijective modulo p^k. Therefore, F is measure-preserving by the first assertion of Theorem 10.1. □

Note 10.2 The bound given by Theorem 10.1 is sharp: That is, there exists a function $f : \mathbb{Z}_p \to \mathbb{Z}_p$ such that:

- f is uniformly differentiable modulo p
- A derivative f_1' is integer-valued
- f is bijective modulo $p^{N_1(f)}$
- f is not bijective modulo $p^{N_1(f)+1}$
- f is not measure-preserving

For instance, a polynomial $f(x) = 1 + x^p$ is bijective modulo p, $N_1(f) = 1$; however, the polynomial f is not measure-preserving since $f'(z) \equiv 0 \pmod p$ for all $z \in \mathbb{Z}_p$.

Note 10.3 Due to periodicity of partial derivatives modulo p (see Proposition 3.26), *in order to verify whether the condition* $\operatorname{rk} F_1'(\mathbf{y}) = m$ *from the statement of Theorem 10.1 (or, respectively, the condition* $\det F_1'(\mathbf{y}) \not\equiv (0, \ldots, 0) \pmod p$) *holds for all* $\mathbf{y} \in \mathbb{Z}_p^n$, *it is sufficient to verify these conditions only for* $\mathbf{y} \in \{0, 1, \ldots, p^{N_1(F)} - 1\}^n$.

The following obvious corollary of Theorem 10.1 holds:

Corollary 10.4 *Under assumptions of Theorem 10.1, let* $m = 1$. *Then* F *if measure-preserving whenever* F *is balanced modulo* p^k *for some* $k \geqslant N_1(F)$, *and all partial derivatives modulo* p *of the function* F *vanish simultaneously at no point of* $(\mathbb{Z}/p^k\mathbb{Z})^n$. *If additionally* $n = 1$, *then* F *is measure-preserving if and only if it is bijective modulo* $p^{N_1(F)}$ *and its derivative modulo* p *vanishes at no point of* $\{0, 1, \ldots, p^{N_1(F)} - 1\}$. *Equivalently, if* $m = n = 1$, *then* F *is measure-preserving if and only if* F *is bijective modulo* $p^{N_1(F)+1}$.

From Corollary 10.4, it immediately follows that:

- A polynomial from $\mathbb{Z}_p[x_1, \ldots, x_n]$ is measure-preserving whenever it is balanced modulo p and all its partial derivatives vanish modulo p simultaneously at no point from $(\mathbb{Z}/p\mathbb{Z})^n = \{0, 1, \ldots, p-1\}^n$
- In particular, a polynomial from $\mathbb{Z}_p[x]$ is measure-preserving if and only it is bijective modulo p and its derivative vanishes modulo p nowhere
- Moreover, a polynomial from $\mathbb{Z}_p[x]$ is measure-preserving if and only if it is bijective modulo p^2

It is worth noticing here that these results about polynomials over $\mathbb{Z}_p$ (as well as analogs of these results for polynomials over commutative rings) are well-known in the theory of polynomials over universal algebras (see, e.g., [133]); however, it turns out that these results remain true for a class of functions, which is much wider than polynomials, namely, the results hold for $\mathcal{B}$-functions also. We postpone a proof of these results (see Corollary 10.23).

A natural question arises whether the sufficient conditions of measure-preservation for polynomials, which are proved above, are also necessary. The answer is negative: A counter-example that follows is based on ideas from [108].

Example 10.5 Consider the polynomial $f(x, y) = 2x + y^3$ over $\mathbb{Z}_2$, in variables x, y. As $\frac{\partial f(x,y)}{\partial x} = 2$, $\frac{\partial f(x,y)}{\partial y} = 3y^2$, both partial derivatives are 0 modulo 2 whenever $y \equiv 0 \pmod 2$. Nevertheless, f is a measure-preserving mapping from $\mathbb{Z}_2^2$ onto $\mathbb{Z}_2$. Here is a proof.

By induction on ℓ, we prove that f is balanced modulo p^ℓ for all $k = 1, 2, \ldots$. The claim follows then from Theorem 8.2. For $\ell = 1$, we have that $f(x, y) \equiv y \pmod 2$, that is, f is balanced modulo 2.

Let $\ell > 1$. We will show that for every $z \in \mathbb{Z}/2^\ell\mathbb{Z}$, there exist exactly 2^ℓ pairs (x, y) such that $f(x, y) \equiv z \pmod{2^\ell}$ and $(x, y) \in \{0, 1, \ldots, 2^\ell - 1\}^2$.

Indeed, if $z = 1 + 2r$ for some $r \in \{0, 1, \ldots, 2^{\ell-1} - 1\}$, then it follows that $y = 1 + 2k$ for some $k \in \{0, 1, \ldots, 2^{\ell-1} - 1\}$. So $2x + (1 + 2k)^3 \equiv 1 + 2r \pmod{2^\ell}$ implies $x+3k+6k^2+4k^3 \equiv r \pmod{2^{\ell-1}}$. The left-hand part of the latter congruence is a polynomial $g(x, k)$ in x, k. The polynomial $g(x, k)$ is measure-preserving in view of Theorem 10.1, a balanced modulo 2. This implies that the congruence $g(x, k) \equiv r \pmod{2^{\ell-1}}$ in unknowns x, k has exactly $2^{\ell-1}$ solutions in $\{0, 1, \ldots, 2^{\ell-1} - 1\}^2$.

If $z = 2r$ for some $r \in \{0, 1, \ldots, 2^{\ell-1} - 1\}$, then it follows that $y = 2k$ for some $k \in \{0, 1, \ldots, 2^{\ell-1} - 1\}$; consequently, the congruence $f(x, y) \equiv z \pmod{2^\ell}$ implies the congruence $x+4k^3 \equiv r \pmod{2^{\ell-1}}$. The polynomial $d(x, k) = x+4k^3$ is measure-preserving in view of Theorem 10.1.

Now using an argument similar to that of the case $z = 1 + 2r$, we conclude that the congruence $f(x, y) \equiv 2r \pmod{2^\ell}$ in unknowns x, y has exactly 2^ℓ solutions in $\{0, 1, \ldots, 2^\ell - 1\}^2$. This proves that f is measure-preserving.

We remark that given a function $F : \mathbb{Z}_p^n \to \mathbb{Z}_p^m$, $m < n$, which is uniformly differentiable modulo p and such that all its partial derivatives modulo p are integer-valued, still no necessary and sufficient conditions of measure-preservation for the function $F : \mathbb{Z}_p^n \to \mathbb{Z}_p^m$ are known in general case. These conditions are not known even if F is a multivariate polynomial over $\mathbb{Z}_p$ (or over $\mathbb{Z}$).

10.2 No Uniformly Differentiable 1-Lipschitz Ergodic Transformations on $\mathbb{Z}_p^n$ for $n \geqslant 2$

Now we are going to find conditions for ergodicity of functions that are uniformly differentiable modulo p and have integer-valued derivatives modulo p. This class of functions contains all locally 1-Lipschitz functions that are uniformly differentiable modulo p (cf. Theorem 5.5). It turns out that ergodic functions of this kind can be univariate only; namely, the following theorem holds [11]:

Theorem 10.6 *Let a function $F: \mathbb{Z}_p^n \to \mathbb{Z}_p^n$ be uniformly differentiable modulo p, let all its partial derivatives modulo p be integer-valued, and let F be ergodic. Then* $n = 1$.

To prove Theorem 10.6, we need two lemmas.

Lemma 10.7 *Let the function $f: \mathbb{Z}_p^n \to \mathbb{Z}_p$ be uniformly differentiable modulo p, let it have integer-valued derivatives modulo p, and let f be an identity modulo p^k for some $k > N_1(f)$. Then every partial derivative modulo p of the function f is an identity modulo p* (cf., Definition 4.6).

Proof of Lemma 10.7 Fix arbitrary $x_0, x_1, \ldots, x_{i-1}, x_{i+1}, \ldots, x_n \in \mathbb{Z}_p$, and consider a function $g_i(x_0, x_1, \ldots, x_n) = x_i + x_0 f(x_1, \ldots, x_n)$ of variable x_i. It is clear that g_i is uniformly differentiable modulo p^k, its derivative modulo p^k is integer-valued, and g_i is bijective modulo p^k. As $k > N_1(f)$, the function g_i is measure-preserving by Theorem 10.1; so everywhere on $\mathbb{Z}_p$ the derivative modulo p of the function g_i is not zero modulo p :

$$\frac{\partial_1}{\partial_1 x_i} g_i(u_0, \ldots, u_n) = 1 + u_0 \frac{\partial_1}{\partial_1 x_i} f(u_1, \ldots, u_n) \not\equiv 0 \pmod{p} \tag{10.5}$$

for all $u_0, \ldots, u_n \in \mathbb{Z}_p$. If

$$\frac{\partial_1}{\partial_1 x_i} f(u_1, \ldots, u_n) \equiv d \not\equiv 0 \pmod{p}$$

for some $u_1, \ldots, u_n \in \mathbb{Z}_p$, then taking u_0 such that $u_0 d \equiv -1 \pmod{p}$, we get a contradiction to (10.5). This proves Lemma 10.5. □

Lemma 10.8 *Let the function $H: \mathbb{Z}_p^n \to \mathbb{Z}_p^n$ be uniformly differentiable modulo p, and let H have integer-valued derivatives modulo p. If H is bijective modulo p^k and if H induces a trivial permutation modulo p^{k-1}* (i.e., an identity transformation on $(\mathbb{Z}/p^{k-1}\mathbb{Z})^n$) *for some $k > N_1(H) + 1$, then H induces on $(\mathbb{Z}/p^k\mathbb{Z})^n$ either a trivial permutation or a permutation whose multiplicative order is p. That is, either this permutation is a unit element of* $\mathrm{Sym}(p^{kn})$*, the finite symmetric group of all permutations on a p^{kn}-element set, or the order of this permutation (as an element from* $\mathrm{Sym}(p^{kn})$*) is p.*

Proof of Lemma 10.8 Let G be an arbitrary function that satisfies conditions of Lemma 10.8; let $N_1(G) = N_1(H)$. Represent both H and G in the following form:

$$H(x_1, \ldots, x_n) = (x_1, \ldots, x_n) + U(x_1, \ldots, x_n);$$
$$G(x_1, \ldots, x_n) = (x_1, \ldots, x_n) + V(x_1, \ldots, x_n).$$

Then both U and V are uniformly differentiable modulo p and have integer-valued derivatives modulo p, and $N_1(U) = N_1(V) = N_1(H)$. Moreover, both U and V are identities modulo p^{k-1} whenever $k-1 > N_1(H)$. Then Lemma 10.7 implies that $U_1' = V_1' = 0$ everywhere on $\mathbb{Z}_p^n$. As $|U|_p \leqslant p^{-k+1}$ and $|V|_p \leqslant p^{-k+1}$ everywhere on $\mathbb{Z}_p^n$, and as both U and V are uniformly differentiable modulo p, from (3.12), we conclude that

$$\begin{aligned}
H(G(h_1, \ldots, h_n)) = H((h_1, \ldots, h_n) + V(h_1, \ldots, h_n)) \equiv \\
H(h_1, \ldots, h_n) + V(h_1, \ldots, h_n) H_1'(h_1, \ldots, h_n) \equiv \\
H(h_1, \ldots, h_n) + V(h_1, \ldots, h_n) + V(h_1, \ldots, h_n) U_1'(h_1, \ldots, h_n) \equiv \\
(h_1, \ldots, h_n) + U(h_1, \ldots, h_n) + V(h_1, \ldots, h_n) \pmod{p^k}
\end{aligned}$$

for all $h_1, \ldots, h_n \in \mathbb{Z}_p$. This implies, in particular, that for all $s \in \mathbb{N}$, the following congruence for iterates of H holds:

$$H^s(h_1, \ldots, h_n) \equiv (h_1, \ldots, h_n) + sU(h_1, \ldots, h_n) \pmod{p^k}.$$

As U is an identity modulo p^{k-1}, the latter congruence implies that $H^p(h_1, \ldots, h_n) \equiv (h_1, \ldots, h_n) \pmod{p^k}$ for all $h_1, \ldots, h_n \in \mathbb{Z}_p$. This proves Lemma 10.8 since in view of Theorem 10.1, the function H is measure-preserving and, thus in view of Theorem 8.2, induces a permutation of elements of $(\mathbb{Z}/p^k\mathbb{Z})^n$. □

Now we are able to prove Theorem 10.6:

Proof of Theorem 10.6 As F is an ergodic locally 1-Lipschitz function, by Theorem 8.2, there exists $k > N_1(F) + 1$ such that F is transitive modulo p^n for all $n \geqslant k-1$. The function F then permutes elements of $(\mathbb{Z}/p^k\mathbb{Z})^n$; we denote the corresponding permutation via $\sigma_k(F)$. Consider the permutation $\sigma = \sigma_k(F)^{p^{(k-1)n}}$. As F is transitive modulo p^k, the multiplicative order of the permutation σ is p^n; hence, σ is not a trivial permutation; that is, σ is not a unit element of the group $\mathrm{Sym}(p^{kn})$.

On the other hand, $\sigma = \sigma_k(F^{p^{(k-1)n}})$. However, $F^{p^{(k-1)n}}$ is bijective modulo p^k and induces a trivial permutation modulo p^{k-1} (the latter claim follows from the transitivity of F modulo p^{k-1}). As σ is not trivial, in view of Lemma 10.8, the multiplicative order of σ must be p. However, according to the preceding argument, the multiplicative order of σ is p^n, so necessarily $n = 1$. □

Of course, there exist non-differentiable 1-Lipschitz ergodic transformations on $\mathbb{Z}_p^n$ for every $n > 1$. For instance, given a 1-Lipschitz ergodic transformation f on $\mathbb{Z}_p$, one can construct a 1-Lipschitz ergodic transformation on $\mathbb{Z}_p^n$ for every $n > 1$ as it is explained below.

Let $B\colon \mathbb{Z}_p^n \rightleftarrows \mathbb{Z}_p$ be a bijection defined by the rule $\delta_k(B(x_0, \dots, x_{n-1})) = \delta_\ell(x_r)$, where $r \in \{0, 1, \dots, n-1\}$ is the least non-negative residue of $k \in \{0, 1, 2, \dots\}$ modulo n, $k = \ell n + r$, $(x_0, \dots, x_{n-1}) \in \mathbb{Z}_p^n$, and $\delta_j(u)$ is the j-th digit of canonical 2-adic representation of $u \in \mathbb{Z}_2$. In simple words, we consider every element of $\mathbb{Z}_p^n$ as of a table of n one-side infinite rows (say, stretching from left to right) of symbols from $\{0, 1, \dots, p-1\}$, and we put into a correspondence to this table an infinite string of symbols from $\{0, 1, \dots, p-1\}$ (i.e., an element from $\mathbb{Z}_p$) obtained by reading successively entries of each column of the table, from top to bottom and from left to right.

Now take a 1-Lipschitz transformation $H\colon \mathbb{Z}_p \to \mathbb{Z}_p$ and a conjugate transformation $H^B(x_0, \dots, x_{n-1}) = B^{-1}(H(B(x_0, \dots, x_{n-1})))$

$$H^B(x_0, \dots, x_{n-1}) = (f_0(x_0, \dots, x_{n-1}), \dots, f_{n-1}(x_0, \dots, x_{n-1}))\colon \mathbb{Z}_p^n \to \mathbb{Z}_p^n.$$

Obviously, in force of Theorem 8.2, the conjugate mapping H^B is 1-Lipschitz and ergodic whenever the mapping H is ergodic: Given a univariate triangular mapping H (see Sect. 5.1.1 about these)

$$x = \sum_{i=0}^{\infty} \chi_i p^i = (\chi_0, \chi_1, \chi_2, \dots) \overset{H}{\mapsto} (\psi_0(\chi_0); \psi_1(\chi_0, \chi_1); \psi_2(\chi_0, \chi_1, \chi_2); \dots),$$

we just construct a n-variate triangular mapping

$$\begin{array}{cccccccc}
\chi_0 & \chi_n & \chi_{2n} \dots & \overset{f_0}{\mapsto} & \psi_0(x) & \psi_n(x) & \psi_{2n}(x) \dots \\
\chi_1 & \chi_{n+1} & \chi_{2n+1} \dots & \overset{f_1}{\mapsto} & \psi_1(x) & \psi_{n+1}(x) & \psi_{2n+1}(x) \dots \\
\dots & \dots & \dots & \dots & \dots & \dots & \dots \\
\chi_{n-1} & \chi_{2n-1} & \chi_{3n-1} \dots & \overset{f_{n-1}}{\mapsto} & \psi_{n-1}(x) & \psi_{2n-1}(x) & \psi_{3n-1}(x) \dots
\end{array}$$

where $\chi_0, \chi_1, \dots \in \{0, 1, \dots, p-1\}$, $\psi_m(x) = \psi_m(\chi_0, \dots, \chi_m) \in \{0, 1, \dots, p-1\}$, $m = 0, 1, 2, \dots$. Now assuming that rows in the left-hand side are new variables, $x_j = (\chi_j, \chi_{n+j}, \chi_{2n+j}, \dots) = \sum_{i=0}^{\infty} \chi_{in+j} p^i$ $(j = 0, 1, \dots, n-1)$ we see that the n-variate mapping $H^B = (f_0, f_1, \dots, f_{n-1})$, where $f_j(x_0, \dots, x_{n-1}) = \sum_{i=0}^{\infty} \psi_{in+j}(x) p^i$ for $j = 0, 1, \dots, n-1$, is transitive modulo p^k for all $k = 1, 2, \dots$ whenever H is transitive modulo p^k for all $k = 1, 2, \dots$.

This easy construction of multivariate ergodic transformation is of importance in computer science (see Chap. 14).

10.3 Differentiable mod p^k Ergodic Transformations on $\mathbb{Z}_p$

In this section, we establish conditions for ergodicity of some uniformly differentiable modulo p^k transformations on $\mathbb{Z}_p$ since according to Theorem 10.6, the univariate case is the only case when these transformations exist. The central result of this subsection is Theorem 10.9, which gives sufficient and necessary conditions of ergodicity for functions that are uniformly differentiable modulo p^2, [11].

We note that to prove Theorem 10.9, we have to significantly generalize the method that was originally used by M. V. Larin in the proof of his criterion of transitivity modulo p^n of a polynomial with rational integer coefficients [132]. Although Larin's criterion of transitivity modulo p^n of polynomials over $\mathbb{Z}$ was cited since the beginning of the 1990s (see, e.g., [5, 9, 10]), it was published only in 2002 (see [132]); for odd p, the criterion was also obtained by D. L. Desjardins and M. E. Zieve (see [47]).

Theorem 10.9 *Let the function $f\colon \mathbb{Z}_p \to \mathbb{Z}_p$ be uniformly differentiable modulo p^2, and let the derivative modulo p^2 of the function f be integer-valued. Then f is ergodic if and only if it is transitive modulo p^n for some (equivalently, for every) $n \geqslant N_2(f) + 1$ whenever p is odd or, respectively, for some (equivalently, for every) $n \geqslant N_2(f) + 2$ whenever $p = 2$.*

To prove the theorem, we need a lemma.

Lemma 10.10 *Let the function $f\colon \mathbb{Z}_p \to \mathbb{Z}_p$ be uniformly differentiable modulo p, let its derivative modulo p be integer-valued, and let the function f be transitive modulo p^k for some $k \geqslant N_1(f) + 1$. Then f induces on $\mathbb{Z}/p^{k+1}\mathbb{Z}$, a permutation that is either a single cycle of length p^{k+1} or a product of p pairwise disjoint cycles of length p^k each.*

Proof of Lemma 10.10 The idea of the proof is as follows: As f is transitive (whence, bijective) modulo p^k for some $k \geqslant N_1(f) + 1$, then in view of Theorem 10.1, f is bijective modulo p^{k+1}. The corresponding permutation of elements of the residue ring $\mathbb{Z}/p^{k+1}\mathbb{Z}$ is a product of disjoint cycles, and a reduction modulo p^k maps every cycle on the whole residue ring $\mathbb{Z}/p^k\mathbb{Z}$ since f is transitive modulo p^k. Thus, a length of a cycle must be a multiple of p^k. Further, as f is locally 1-Lipschitz (see the very beginning of Sect. 10.3), f maps balls (of radii less than $p^{-N_1(f)}$) into balls; thus, as p-adic ball are cosets in the ring $\mathbb{Z}_p$ with respect to ideals generated by powers of p, the iterate $f^{p^k} \bmod p^{k+1}$ permutes cosets of the ring $\mathbb{Z}/p^{k+1}\mathbb{Z}$ with respect to ideal generated by p^k. Moreover, as $f^{p^k} \bmod p^k$ is an identity transformation on $\mathbb{Z}/p^k\mathbb{Z}$, every coset must be invariant with respect to the action of $f^{p^k} \bmod p^{k+1}$. Now it is clear that whenever this action is transitive on the coset, then f is transitive on $\mathbb{Z}/p^{k+1}\mathbb{Z}$. However, it turns out that $f^{p^k} \bmod p^{k+1}$ acts on the coset by an affine transformation; that is, the action is conjugate to an affine transformation on the finite field of p elements. Here, Lemma 9.2 comes into play.

We proceed with all this in mind. Remind that for $x \in \mathbb{Z}_p$, we denote via $\chi_i = \delta_i(x) \in \{0, 1, \ldots, p-1\}$ the coefficient of the i-th term in p-adic canonical representation of x; $i = 0, 1, 2, \ldots$ (see Sect. 2.3.1). Now Definition 3.22 of uniform differentiability modulo p^k implies that for arbitrary $x \in \mathbb{Z}_p$ and $s \geqslant N_1(f) = N$, the following congruence holds:

$$f(\chi_0 + \chi_1 p + \cdots + \chi_{s-1} p^{s-1} + \chi_s p^s) \equiv f(\chi_0 + \chi_1 p + \cdots + \chi_{s-1} p^{s-1}) + \\ + \chi_s p^s f_1'(\chi_0 + \chi_1 p + \cdots + \chi_{s-1} p^{s-1}) \pmod{p^{s+1}}. \quad (10.6)$$

The latter congruence implies that the s-th coordinate function $\delta_s(f(x))$ of the function f is of the following form

$$\delta_s(f(x)) \equiv \Phi_s(\chi_0, \ldots, \chi_{s-1}) + \chi_s f_1'(x) \pmod{p}, \quad (10.7)$$

where $\Phi_s(\chi_0, \ldots, \chi_{s-1}) = \delta_s(f(\chi_0 + \chi_1 p + \cdots + \chi_{s-1} p^{s-1}))$. As a derivative $f_1'(x)$ modulo p is a step function of order not greater than N by Proposition 3.26, the derivative $f_1'(x)$ depends only on $\chi_0, \ldots, \chi_{N-1}$; so we can rewrite (10.7) in the form of

$$\delta_s(f(x)) \equiv \Phi_s(\chi_0, \ldots, \chi_{s-1}) + \chi_s \Psi(\chi_0, \ldots, \chi_{N-1}) \pmod{p}, \quad (10.8)$$

where $\Psi(\chi_0, \ldots, \chi_{N-1}) = f_1'(x)$.

Further, as a chain rule holds for derivatives modulo p as well (cf. Proposition 3.24), we conclude that for a derivative modulo p of the r-th iterate of f ($r = 1, 2, \ldots$), the following congruence holds:

$$(f^r(x))_1' \equiv \prod_{j=0}^{r-1} f_1'(f^j(x)) \pmod{p}. \quad (10.9)$$

As f is uniformly differentiable modulo p and its derivative modulo p is integer-valued, f is locally 1-Lipschitz (cf. Theorem 5.5); so transitivity of f modulo p^k for some $k \geqslant N$ implies transitivity of f modulo p^n for all $k \geqslant n \geqslant N$ (see Proposition 7.1). However, as f_1' depends only on $\chi_0, \ldots, \chi_{N-1}$, and as f is transitive modulo p^N, from (10.9), we deduce that

$$(f^{p^n}(x))_1' \equiv \left(\prod_{\gamma_0, \ldots, \gamma_{N-1}=0}^{p-1} \Psi(\gamma_0, \ldots, \gamma_{N-1}) \right)^{p^{n-N}} \pmod{p}. \quad (10.10)$$

Denote the product in parentheses in the right-hand side of (10.10) via Π. Then, as the function f^{p^n} is uniformly differentiable modulo p and its derivative modulo p is integer-valued, from (10.8) and (10.10), we get that

$$\delta_n(f^{p^n}(x)) \equiv \Xi_n(\chi_0, \ldots, \chi_{n-1}) + \chi_n \Pi^{p^{n-N}} \pmod{p}, \quad (10.11)$$

where $\Xi_n(\chi_0, \ldots, \chi_{n-1}) = \delta_n(f^{p^n}(\chi_0 + x_1 p + \cdots + \chi_{n-1}p^{n-1}))$. As a locally 1-Lipschitz function f is transitive modulo p^{n+1} for $k > n \geqslant N$, the function f^{p^n}, on the one hand, induces a trivial permutation modulo p^n and, on the other hand, induces on each coset $a + p^n(\mathbb{Z}/p^{n+1}\mathbb{Z})$ of the residue ring $\mathbb{Z}/p^{n+1}\mathbb{Z}$ a permutation that is a cycle of length p. This in particular implies that the right-hand side of (10.11), considered as a function in a variable χ_n, must be a permutation, moreover—a cycle of length p on $\{0, 1, \ldots, p-1\}$. However, as this function is an affine transformation on a finite field $\mathbb{Z}/p\mathbb{Z}$, from Lemma 9.2, it follows that $\Pi^{p^{n-N}} \equiv 1 \pmod p$; whence $\Pi \equiv 1 \pmod p$ (since $z^p \equiv z \mod p$; see Sect. 1.2.3). Finally, we conclude that

$$f^{p^k}(x) \equiv f^{p^k}(\chi_0 + \chi_1 p + \cdots + \chi_k p^k) \equiv$$
$$\equiv \chi_0 + \chi_1 p + \cdots + \chi_{k-1}p^{k-1} + p^k\left(\Xi_k(\chi_0, \ldots, \chi_{k-1}) + \chi_k\right) \pmod{p^{k+1}}. \tag{10.12}$$

The latter congruence implies that f induces a permutation modulo p^{k+1}, which we denote as σ. We claim that if

$$\Xi_k(\chi_0, \ldots, \chi_{k-1}) \not\equiv 0 \pmod p$$

for some (equivalently, for all) $\chi_0, \ldots, \chi_{k-1} \in \{0, 1, \ldots, p-1\}$, then f is transitive modulo p^{k+1}; otherwise, the permutation σ is a product of p disjoint cycles of length p^k each.

To prove the latter claim, take arbitrary $\gamma_0, \ldots, \gamma_k \in \{0, 1, \ldots, p-1\}$, and denote C a cycle of the permutation σ that contains a point $\gamma_0 + \gamma_1 p + \cdots + \gamma_{k-1}p^{k-1} + \chi_k p^k \in \mathbb{Z}/p^{k+1}\mathbb{Z}$. As f is transitive modulo p^k, then $C \bmod p^k = \mathbb{Z}/p^k\mathbb{Z}$; thus, p^k is a factor of $\#C$, the length of the cycle C.

Now, if $\Xi_k(\gamma_0, \ldots, \gamma_{k-1}) \not\equiv 0 \pmod p$, then (10.12) implies that

$$f^{p^k}(\gamma_0 + \gamma_1 p + \cdots + \gamma_{k-1}p^{k-1} + \chi_k p^k) \not\equiv$$
$$\not\equiv \gamma_0 + \gamma_1 p + \cdots + \gamma_{k-1}p^{k-1} + \chi_k p^k \pmod{p^{k+1}}, \tag{10.13}$$

i.e., that $\#C > p^k$. On the other hand, (10.12) implies that $\#C$ is a factor of p^{k+1}. Finally, we conclude that in this case, $\#C = p^{k+1}$; that is, f is transitive modulo p^{k+1}.

Now let the congruence $\Xi_k(\gamma_0, \ldots, \gamma_{k-1}) \equiv 0 \pmod p$ hold for some $\gamma_0, \ldots, \gamma_k \in \{0, 1, \ldots, p-1\}$. Then this congruence must hold for all $\gamma_0, \ldots, \gamma_k \in \{0, 1, \ldots, p-1\}$, since otherwise in view of the preceding argument, the function f is transitive modulo p^{k+1}, so (10.13) holds for all $\gamma_0, \ldots, \gamma_k \in \{0, 1, \ldots, p-1\}$; this, in view of (10.12), implies that $\Xi_k(\gamma_0, \ldots, \gamma_{k-1}) \not\equiv 0 \pmod p$ for all $\gamma_0, \ldots, \gamma_k \in \{0, 1, \ldots, p-1\}$, a contradiction. Thus, within the case under consideration, (10.12) implies that σ^{p^k} is an identity permutation; hence, $\#C = p^k$ as p^k is a factor of $\#C$. This finally proves Lemma 10.10. □

Note 10.11 During the proof of Lemma 10.10, we have shown that whenever the function f is transitive modulo $p^{N_1(f)+1}$ (in particular, whenever f is ergodic), then necessarily, $\prod_{i=0}^{p^{N_1(f)}-1} f_1'(f^i(x)) \equiv 1 \pmod p$ for every $x \in \mathbb{Z}_p$.

Proof of Theorem 10.9 In the proof of Lemma 10.10, it is established that if f is transitive modulo p^k for some $k \geqslant N_1(f)$, then f is transitive modulo p^n for all $k \geqslant n \geqslant N_1(f)$. This, in view of Theorem 8.2, proves the "only if" part of the assertion of Theorem 10.9 as f is locally 1-Lipschitz and $N_2(f)+1 > N_1(f)$.

To prove the "if" part of the assertion, in view of Theorem 8.2, it is sufficient to show that if $n \geqslant N_2(f)+1$ (resp., if $n \geqslant N_2(f)+2$ for $p=2$) and if f is transitive modulo p^n, then f is transitive modulo p^{n+1}. In turn, to prove the latter claim, in view of Lemma 10.10, it is sufficient to prove that not every element of the residue ring $\mathbb{Z}/p^{n+1}\mathbb{Z}$ is a fixed point of the transformation $f^{p^n} \bmod p^{n+1}$:

$$f^{p^n}(x) \not\equiv x \pmod{p^{n+1}}. \tag{10.14}$$

for some $x \in \mathbb{Z}_p$.

As transitivity of f modulo p^n implies transitivity of f modulo p^{n-1} (see Proposition 7.1), we get that

$$f^{p^{n-1}}(x) = x + p^{n-1}\xi(x), \tag{10.15}$$

where $\xi\colon \mathbb{Z}_p \to \mathbb{Z}_p$; note that $\xi(x) \not\equiv 0 \pmod p$ for all $x \in \mathbb{Z}_p$ since otherwise Lemma 10.10 implies that f is not transitive modulo p^n. Further, as f is uniformly differentiable modulo p^2 and its derivative modulo p^2 is integer-valued, the r-th iterate f^r is uniformly differentiable modulo p^2, and its derivative modulo p^2 is integer-valued, for all $r = 1, 2, \ldots$; moreover, $(f^r(x))_2' \equiv \prod_{j=0}^{r-1} f_2'(f^j(x)) \pmod{p^2}$ cf. (10.9). Now, as $n-1 \geqslant N_2(f)$, then using chain rule for derivatives modulo p^2 and an obvious equality $f^{sp^{n-1}}(x) = f^{(s-1)p^{n-1}}(x + p^{n-1}\xi(x))$ ($s = 1, 2, \ldots$), which follows from (10.15), we successively calculate

$$f^{p^n}(x) \equiv f^{(p-1)p^{n-1}}(x) + p^{n-1}\xi(x) \prod_{j=0}^{(p-1)p^{n-1}-1} f_2'(f^j(x)) \equiv$$

$$\equiv f^{(p-2)p^{n-1}}(x) + p^{n-1}\xi(x) \left(\prod_{j=0}^{(p-2)p^{n-1}-1} f_2'(f^j(x)) \right.$$

$$+ \prod_{j=0}^{(p-1)p^{n-1}-1} f_2'(f^j(x))\Bigg) \equiv$$

$$\equiv \ldots \equiv x + p^{n-1}\xi(x)\left(1 + \sum_{i=1}^{p-1} \prod_{j=0}^{(p-i)p^{n-1}-1} f_2'(f^j(x))\right) \pmod{p^{n+1}}. \tag{10.16}$$

As f_2' is a step function of order not greater than $N_2(f)$ (see Proposition 3.26) and f is transitive modulo p^{n-1}, we conclude that for arbitrary $i, j \in \mathbb{N}$, the following congruence holds:

$$f_2'(f^j(x)) \equiv f_2'(f^{j+ip^{n-1}}(x)) \pmod{p^2}.$$

In view of transitivity of f modulo p^{n-1}, the latter congruence implies that

$$\prod_{j=0}^{(p-i)p^{n-1}-1} f_2'(f^j(x)) \equiv \alpha(x)^{p-i} \pmod{p^2},$$

where

$$\alpha(x) = \prod_{j=0}^{p^{n-1}-1} f_2'(f^j(x)).$$

In view of (10.16), we now conclude that

$$f^{p^n}(x) \equiv x + p^{n-1}\xi(x)\left(1 + \sum_{i=1}^{p-1} \alpha(x)^i\right) \pmod{p^{n+1}}. \tag{10.17}$$

Again, as f_2' is a periodic function with a period of length $p^{N_2(f)}$, and as f is transitive modulo p^{n-1} for $n-1 \geqslant N_2(f)$, then $\alpha(x) \bmod p^2$ does not depend on x; namely,

$$\alpha(x) = \prod_{j=0}^{p^{n-1}-1} f_2'(f^j(x)) \equiv \left(\prod_{z=0}^{p^{N_2(f)}-1} f_2'(z)\right)^{p^{n-1-N_2(f)}} \pmod{p^2}. \tag{10.18}$$

We claim that $\alpha(x) \equiv 1 \pmod p$. Indeed, during the proof of Lemma 10.10, we have already established that if $k \geqslant N_1(f)$ and if f is transitive modulo p^k, then

$$\prod_{j=0}^{p^{N_1(f)}-1} f_1'(f^j(x)) \equiv 1 \pmod p \tag{10.19}$$

for all $x \in \mathbb{Z}_p$ (see the proof of (10.12)). From Definition 3.22 of a derivative modulo p^ℓ, it follows that $f_2'(x) = f_1'(x) \pmod p$; consequently,

$$\alpha(x) \equiv 1 + p\beta \pmod{p^2} \tag{10.20}$$

for some $\beta \in \mathbb{N}_0$. In view of (10.19) and (10.20), from (10.17), we deduce now that

$$f^{p^n}(x) \equiv x + p^{n-1}\xi(x)\left(p + p\beta\sum_{i=1}^{p-1} i\right) \pmod{p^{n+1}}; \tag{10.21}$$

so for $p \neq 2$, we conclude that

$$f^{p^n}(x) \equiv x + p^n\xi(x) \pmod{p^{n+1}}.$$

This, in view of Lemma 10.10, proves Theorem 10.9 in the case $p \neq 2$ since $\xi(x) \not\equiv 0 \pmod p$ (see (10.15) and the text thereafter).

For the case $p = 2$, congruence (10.21) implies that

$$f^{2^n}(x) \equiv x + 2^n(1+\beta) \pmod{2^{n+1}}; \tag{10.22}$$

so to finish the proof, it suffices to show that β is even.

For $n \geqslant N_2(f) + 2$, the transitivity of f modulo 2^n implies that f is transitive modulo $2^{N_2(f)+2}$, so in view of the definition of a derivative modulo p^2, we have that

$$f^{2^N}(x + 2^N\xi) \equiv f^{2^N}(x) + 2^N\xi \prod_{j=0}^{2^N-1} f_2'(f^j(x)) \pmod{2^{N+2}} \tag{10.23}$$

where $N = N_2(f)$, $\xi \in \mathbb{Z}_2$. As f is transitive modulo 2^{N+2}, we conclude that for every $x \in \{0, 1, \ldots, 2^N - 1\}$, the mapping

$$\varphi_x : \xi \mapsto \delta_N(f^{2^N}(x + 2^N\xi)) + 2\delta_{N+1}(f^{2^N}(x + 2^N\xi)) \quad (\xi \in \{0, 1, 2, 3\})$$

is a cycle of length 4 on the residue ring $\mathbb{Z}/4\mathbb{Z}$. From (10.20) and (10.18), we now conclude that

$$\prod_{j=0}^{2^N-1} f_2'(f^j(x)) \equiv 1 + 2\beta \pmod 4;$$

thus, (10.23) implies now that

$$\varphi_x(\xi) \equiv c(x) + \xi(1 + 2\beta) \pmod 4, \tag{10.24}$$

where $c(x) = \delta_N(f^{2^N}(x)) + 2\delta_{N+1}(f^{2^N}(x))$. However, for every x, the mapping φ_x is transitive modulo 4, so (10.24), in view of Theorem 9.1, implies that $\beta \equiv 0 \pmod 2$. This ends the proof of Theorem 10.9. □

Note 10.12 (The Bound Given by Theorem 10.9 Is Sharp) For instance, if p is odd, then there exists a function $f : \mathbb{Z}_p \to \mathbb{Z}_p$ such that:

- f is uniformly differentiable modulo p^2
- A derivative f_2' is integer-valued
- f is transitive modulo $p^{N_2(f)}$
- f is not transitive modulo $p^{N_2(f)+1}$
- f is not ergodic

The 1-Lipschitz function $f(x) = \delta_0(x + 1)$ serves as a respective example: The function f is uniformly differentiable, its derivative is 0 everywhere on $\mathbb{Z}_p$, and $N_2(f) = 1$, f is transitive modulo p. However, f is not even bijective (not speaking of transitivity) modulo p^2; thus, f is not ergodic in view of Theorem 8.2.

Note 10.13 (A Straightforward Analog of Theorem 10.9 for Functions that Are Uniformly Differentiable Modulo p Is Not True) Namely, for every $n \in \mathbb{N}$, there exists a 1-Lipschitz function $f : \mathbb{Z}_2 \to \mathbb{Z}_2$ such that f is uniformly differentiable modulo 2, $f_1' = 1$ everywhere on $\mathbb{Z}_2$, $N_1(f) = 1$, f is transitive modulo 2^k for $k = 1, 2, \ldots, n$, and f is not transitive modulo 2^k for all $k > n$. By the argument similar to that which follows, one can construct a counterexample for $p \neq 2$ as well.

Indeed, for $x \in \mathbb{Z}_2$, consider its canonical 2-adic expansion $x = \chi_0 + \chi_1 2 + \chi_2 2^2 + \ldots$, where $\chi_0, \chi_1, \chi_2 \ldots \in \{0, 1\}$. Consider a function

$$f(x) = \sum_{i=0}^{\infty} \psi_i(\chi_0, \ldots, \chi_i) 2^i,$$

where every $\psi_i(x_0, \ldots, x_i)$ is a Boolean function that is linear with respect to the Boolean variable x_i; that is, the algebraic normal form (ANF) of the function $\psi_i(x_0, \ldots, x_i)$ is

$$\psi_i(\chi_0, \ldots, \chi_i) = \varphi_i(\chi_0, \ldots, \chi_{i-1}) \oplus \chi_i,$$

(see Sect. 9.2). The function f is 1-Lipschitz. Moreover, direct calculations show that for arbitrary $s \in \mathbb{N}$ and $h \in \mathbb{Z}_2$, there holds a congruence $f(x + 2^s h) \equiv f(x)+2^s h \pmod{2^{s+1}}$; whence, the function f is uniformly differentiable modulo 2, $f_1'(x) = 1$ for all $x \in \mathbb{Z}_2$, and $N_1(f) = 1$.

Now, given $n \in \mathbb{N}$, take a function f such that $\varphi_0 = 1$, all Boolean functions $\varphi_i(\chi_0, \dots, \chi_{i-1})$ are of odd weight for all $i = 1, 2, \dots$ but $i = n$, and $\varphi_n(\chi_0, \dots, \chi_{n-1})$ is of even weight. Then, according to Theorem 9.4, f is transitive modulo 2^k for $k = 1, 2, \dots, n$, but f is not transitive modulo 2^{n+1}; thus, f is not ergodic.

10.4 Measure-Preservation and Ergodicity of $\mathcal{A}$-, $\mathcal{B}$-, and C-functions

Theorems 10.1 and 10.9 exhibit a "Hensel's-Lemma-like" phenomenon that often occurs in p-adic dynamics: Behavior of a dynamical system on the whole continual space is determined by its "behavior modulo p^k," i.e., on a finite space (cf. Hensel's Lemma, 3.14).

The phenomenon is of practical value, e.g., in computer science cf. Chap. 14: for instance, to determine whether a dynamical system is ergodic (i.e., transitive) on a large finite space, it is sufficient to determine whether it is ergodic on a relatively small finite space, for a smaller space, one may use computers but cannot for a larger space. Thus, it is important to estimate $N_1(f)$ and $N_2(f)$ with the highest possible accuracy to reduce computational costs. Moreover, although both Theorems 10.1 and Theorem 10.9 give sharp bounds for cardinality of these smaller spaces where one must verify measure preservation (respectively, ergodicity) of a dynamical system (see Notes 10.2 and 10.12), these bounds are sharp only in the class of all functions that are uniformly differentiable modulo p (respectively, modulo p^2). However, for narrower classes of functions, these bounds can obviously be sharpened, e.g., for affine functions: Theorem 9.1 together with Lemma 9.2 implies that an affine function $f(x) = ax + b$ is ergodic if and only if it is transitive modulo p whenever p is odd or modulo 4 whenever $p = 2$, and *not* modulo p^2 and modulo 8, respectively, as follows from Theorem 10.9.

In this section, we show that for some important classes of functions, the said bounds can be significantly reduced. Moreover, we calculate these bound explicitly in a contrast to those given by Theorems 10.1 and 10.9: In general case, it might be not an easy problem to find $N_1(f)$ and $N_2(f)$ given arbitrary f, but for $\mathcal{A}$-, $\mathcal{B}$-, and C-functions, this problem is tractable.

We start with $\mathcal{A}$-functions. Let $f \in \mathcal{A}$; then, according to Definition 6.15 of $\mathcal{A}$-functions, $p^n f \in \mathcal{B}$ for a suitable $n \in \mathbb{N}_0$. Given $f \in \mathcal{A}$, denote $\rho(f) = \min\{n \in \mathbb{N}_0 \colon p^n f \in \mathcal{B}\}$; put

$$\lambda(f) = \min\left\{k \in \mathbb{N} \colon 2\frac{p^k - 1}{p - 1} - k > \rho(f)\right\}.$$

The following theorem is true [11].

Theorem 10.14 *An $\mathcal{A}$-function f is measure-preserving if and only if it is bijective modulo $p^{\lambda(f)+1}$. The function f is ergodic if and only if it is transitive modulo $p^{\lambda(f)+1}$ whenever $p \notin \{2, 3\}$ or modulo $p^{\lambda(f)+2}$ whenever $p \in \{2, 3\}$.*

Basically, our proof of Theorem 10.14 will follow lines of the proof of Theorem 10.9; however, we will need more than 2 terms in decomposition of the function $f(x + p^k h)$ modulo a power of p cf. (10.6). According to Theorem 6.16, we can develop any $\mathcal{A}$-function f into Taylor series; unfortunately, coefficients $\frac{f^{(j)}(x)}{j!}$ in the series are not necessarily p-adic integers if $j > 1$. So we need to use more delicate techniques to calculate $f(x + p^k h)$ modulo the power of p; thus we start with some technical results; the first of which is actually obvious:

Lemma 10.15 *The numbers $\varkappa(i) = \operatorname{ord}_p i! - \lfloor \log_p i \rfloor$ $(i = 1, 2, 3, \ldots)$ constitute a monotone nondecreasing sequence.*

Proof of Lemma 10.15 If $\lfloor \log_p i \rfloor = \lfloor \log_p (i-1) \rfloor$, then $\varkappa(i-1) \leqslant \varkappa(i)$ as $\operatorname{ord}_p i! \geqslant \operatorname{ord}_p (i-1)!$. Assume now that $\lfloor \log_p j \rfloor > \lfloor \log_p (j-1) \rfloor$ for some $j \in \mathbb{N}$. Evidently, $\lfloor \log_p j \rfloor + 1$ is a number of digits in the base-p expansion of j. Hence, our assumption holds if and only if $j - 1 = (p-1) + (p-1)p + \cdots + (p-1)p^n = p^{n+1} - 1$ for some $n \in \mathbb{N}_0$. But then $\operatorname{ord}_p j! = \operatorname{ord}_p (j-1)! + n$, $\lfloor \log_p (j-1) \rfloor = n$, $\lfloor \log_p j \rfloor = n + 1$, and thus $\varkappa(j) > \varkappa(j-1)$. □

As f is 1-Lipschitz, in view of Theorem 5.12, it can be represented in the following form:

$$f(x) = b_0 + \sum_{i=1}^{\infty} b_i p^{\lfloor \log_p i \rfloor} \binom{x}{i}, \tag{10.25}$$

where $b_j \in \mathbb{Z}_p$, $j = 0, 1, 2, \ldots$. Everywhere during the proof of Theorem 10.14, we assume that f is represented in this form. In the sequel, we denote $\lambda(f)$ via λ, $\rho(f)$ via ρ.

Lemma 10.16 *Under assumptions of Theorem 10.14, let p be an odd prime; then the following is true:*

$$b_i \equiv \begin{cases} 0 \pmod{p}, & \text{for } i \geqslant 2p^{\lambda}; \\ 0 \pmod{p^2}, & \text{for } i \geqslant 3p^{\lambda}. \end{cases}$$

Proof of Lemma 10.16 Represent f as

$$f(x) = b_0 + \sum_{i=1}^{\infty} \frac{1}{i!} b_i p^{\lfloor \log_p i \rfloor} x^{\underline{i}},$$

where, we recall, $x^{\underline{i}} = x(x-1)\cdots(x-i+1)$ is the i-th falling factorial power of x, $x^{\underline{0}} = 1$. As $f \in \mathcal{A}$, i.e., as $\left| b_i p^{\lfloor \log_p i \rfloor} \right|_p \leqslant p^\rho |i!|_p$, we conclude that

$$\operatorname{ord}_p b_i \geqslant \operatorname{ord}_p i! - \lfloor \log_p i \rfloor - \rho, \tag{10.26}$$

for all $i = 1, 2, \ldots$.

In view of Lemma 10.15, to finish the proof of Lemma 10.16, it is sufficient to show only that $\varkappa(2p^\lambda) - \rho \geqslant 1$ and $\varkappa(3p^\lambda) - \rho \geqslant 2$. We recall that $\operatorname{ord}_p i! = \frac{1}{p-1}(i - \operatorname{wt}_p i)$ (see Lemma 2.1).

As $p \neq 2$, we conclude that $\varkappa(2p^\lambda) - \rho = \frac{1}{p-1}(2p^\lambda - 2) - \lambda - \rho \geqslant 1$ in view of the definition of $\lambda = \lambda(f)$. Hence, if $p \neq 3$, then

$$\varkappa(3p^\lambda) - \rho = \frac{1}{p-1}(3p^\lambda - 3) - \lambda - \rho = \varkappa(2p^\lambda) + \frac{1}{p-1}(p^\lambda - 1) - \rho \geqslant 2.$$

This proves Lemma 10.16 for $p \neq 3$.

Finally, let $p = 3$. Then

$$\varkappa(3p^\lambda) - \rho = \varkappa(3^{\lambda+1}) - \rho = \frac{1}{2}(3^{\lambda+1} - 1) - \lambda - 1 - \rho \geqslant 2;$$

otherwise in view of the inequality $3^\lambda - 1 - \lambda > \rho$ (which follows directly from the definition of λ), we conclude that

$$\frac{1}{2}(3^{\lambda+1} - 1) - \lambda - 1 - 3^\lambda + 1 + \lambda < 1,$$

i.e., that $3^\lambda - 1 < 2$; so $\lambda < 1$, a contradiction. This concludes the proof of Lemma 10.16. □

Corollary 10.17 *Under assumptions of Theorem 10.14, let p be an odd prime; then for every $i \in \mathbb{N}$, the following is true:*

$$\frac{\Delta^i f(x)}{i} \equiv \begin{cases} 0 \pmod{p^2}, & \text{if } i \geqslant 2p^\lambda + 1; \\ 0 \pmod{p}, & \text{if } i \geqslant p^\lambda + 1. \end{cases}$$

Proof of Corollary 10.17 As $\Delta^j\binom{x}{i} = \binom{x}{i-j}$ if $i \geqslant j$ and $\Delta^j\binom{x}{i} = 0$ if $i < j$ (see (1.2)), the following equality holds:

$$\frac{\Delta^i f(x)}{i} = \frac{1}{\hat{\imath}}\sum_{j=i}^{\infty} b_j p^{\lfloor \log_p j\rfloor - \operatorname{ord}_p j}\binom{x}{j-i},$$

where $\hat{\imath} = ip^{-\operatorname{ord}_p i} \in \mathbb{Z}_p$, $\operatorname{ord}_p \hat{\imath} = 0$. Now the assertion is obvious in view of Lemma 10.16. □

Recall that every $\mathcal{A}$-function is infinitely many times differentiable on $\mathbb{Z}_p$, and its derivative f' is integer-valued (see Sect. 6.4).

Proposition 10.18 *Under assumptions of Theorem 10.14, let p be an odd prime; then $N_1(f) \leqslant \lambda$, $N_2(f) \leqslant \lambda + 1$, and*

$$f'(x) \equiv \sum_{i=1}^{p^\lambda}(-1)^{i-1}\frac{\Delta^i f(x)}{i} \pmod{p},$$

$$f'(x) \equiv \sum_{i=1}^{2p^\lambda}(-1)^{i-1}\frac{\Delta^i f(x)}{i} \pmod{p^2},$$

where $\lambda = \lambda(f)$.

Proof of Proposition 10.18 To prove Proposition 10.18, we show that for all $x, h \in \mathbb{Z}_p$,

$$f(x + p^m h) \equiv f(x) + p^m h f'(x) \pmod{p^{m+2}} \tag{10.27}$$

whenever $m \geqslant \lambda + 1$ and that

$$f(x + p^m h) \equiv f(x) + p^m h f'(x) \pmod{p^{m+1}} \tag{10.28}$$

whenever $m \geqslant \lambda$.

Since f is 1-Lipschitz, it is sufficient to prove congruences (10.27) and (10.28) only for $h \in \{1, 2, \ldots, p^2 - 1\}$. By Theorem 1.5,

$$f(x + n) = \sum_{i=0}^{n}\binom{n}{i}\Delta^i f(x);$$

thus, for $n = p^m h$, we obtain that

$$f(x + p^m h) = f(x) + p^m h \varphi_m(x, h), \tag{10.29}$$

where

$$\varphi_m(x,h) = \sum_{i=1}^{p^m h} \binom{p^m h - 1}{i-1} \frac{\Delta^i f(x)}{i}. \tag{10.30}$$

Now from Corollary 10.17, we deduce that

$$\varphi_m(x,h) \equiv \sum_{i=1}^{p^\lambda} \binom{p^m h - 1}{i-1} \frac{\Delta^i f(x)}{i} \pmod p \tag{10.31}$$

whenever $m \geqslant \lambda$ and that

$$\varphi_m(x,h) \equiv \sum_{i=1}^{2p^\lambda} \binom{p^m h - 1}{i-1} \frac{\Delta^i f(x)}{i} \pmod{p^2} \tag{10.32}$$

whenever $m \geqslant \lambda + 1$.

In view of Corollary 1.3, from (10.31), it follows that

$$\varphi_m(x,h) \equiv \sum_{i=1}^{p^\lambda} (-1)^{i-1} \frac{\Delta^i f(x)}{i} \pmod p,$$

for $m \geqslant \lambda$, thus proving assertions of Proposition 10.18 on estimates of $\mathbb{N}_1(f)$ and on the residue $f'(x) \bmod p$.

To prove the rest of the statement of Proposition 10.18, we first note that for $i = 1, 2, \ldots, 2p^\lambda$, the following obvious equality holds:

$$\binom{p^m h - 1}{i-1} = \prod_{k=0}^{i-2} \frac{p^m h - (k+1)}{k+1} = \prod_{j=1}^{i-1} \left(\frac{h}{\hat{j}} p^{m - \operatorname{ord}_p j} - 1 \right), \tag{10.33}$$

where $\hat{j} = j p^{-\operatorname{ord}_p j}$ is a unit of $\mathbb{Z}_p$, (i.e., $\hat{j}$ has a multiplicative inverse $\frac{1}{\hat{j}}$ in $\mathbb{Z}_p$); hence, every term of the product in the right-hand side of (10.33) is a p-adic integer.

If $i \leqslant p^\lambda$, then $m - \operatorname{ord}_p j \geqslant 2$ for all $j = 1, 2, \ldots, i-1$; so (10.33) implies that

$$\binom{p^m h - 1}{i-1} \equiv (-1)^{i-1} \pmod{p^2}. \tag{10.34}$$

If $p^\lambda+1 \leqslant i \leqslant 2p^\lambda$ and $j \in \{1, 2, \dots, i-1\}$, then $m - \mathrm{ord}_p\, j = 1$ only in the case when simultaneously $j = p^\lambda$ and $m = \lambda+1$; otherwise, $m - \mathrm{ord}_p\, j \geqslant 2$. However, if $m - \mathrm{ord}_p\, j = 1$, then

$$\frac{\Delta^i f(x)}{i} \equiv 0 \pmod{p}$$

(see Corollary 10.17); hence, in both cases, we have that

$$\left(\frac{h}{\hat{j}} p^{m-\mathrm{ord}_p j} - 1\right) \frac{\Delta^i f(x)}{i} \equiv -\frac{\Delta^i f(x)}{i} \pmod{p^2}.$$

From here, in view of (10.33), we deduce that

$$\binom{p^m h - 1}{i-1} \frac{\Delta^i f(x)}{i} \equiv (-1)^{i-1} \frac{\Delta^i f(x)}{i} \pmod{p^2}. \tag{10.35}$$

for all $i = 1, 2, \dots, 2p^\lambda$. Now combining together (10.32), (10.34), and (10.35), we conclude that

$$\varphi_m(x, h) \equiv \sum_{i=1}^{2p^\lambda} (-1)^{i-1} \frac{\Delta^i f(x)}{i} \pmod{p^2}.$$

This, in view of (10.29), (10.30), and (10.32), completes the proof of Proposition 10.18. □

Lemma 10.19 *Under assumptions of Theorem 10.14, let p be an odd prime; then the function*

$$\theta(x) = \sum_{j=2}^{p-1} (-1)^j \sum_{i=1}^{j-1} \frac{\Delta^{jp^{\lambda-1}} f(x)}{ijp^{\lambda-1}} + \sum_{k=1}^{p-1} (-1)^{k-1} \frac{\Delta^{kp^{\lambda-1}+p^\lambda} f(x)}{kp^\lambda} + \frac{\Delta^{2p^\lambda} f(x)}{2p^{\lambda+1}}.$$

is integer-valued, $\theta(a) \equiv \theta(b) \pmod{p}$ *whenever* $a \equiv b \pmod{p^\lambda}$, *and*

$$f(x + p^\lambda h) \equiv f(x) + p^\lambda h f'(x) + p^{\lambda+1} h^2 \theta(x) \pmod{p^{\lambda+2}}$$

for all $x, h \in \mathbb{Z}_p$.

Proof of Lemma 10.19 First we prove that θ is integer-valued, i.e., that θ maps $\mathbb{Z}_p$ into $\mathbb{Z}_p$. As f is 1-Lipschitz, every fraction $\frac{\Delta^s f(x)}{s}$ ($s = 1, 2, 3, \dots$) is a p-adic integer (see Proposition 5.4); so it suffices to show only that the following functions $\alpha(x)$ and $\beta_k(x)$ are integer-valued

$$\alpha(x) = \frac{\Delta^{2p^\lambda} f(x)}{2p^{\lambda+1}};$$

$$\beta_k(x) = \frac{\Delta^{kp^{\lambda-1}+p^\lambda} f(x)}{kp^\lambda},$$

for all $k \in \{1, 2, \dots, p-1\}$. As

$$\Delta^i f(x) = \sum_{j=i}^{\infty} b_j p^{\lfloor \log_p j \rfloor} \binom{x}{j-i} \tag{10.36}$$

for $i = 1, 2, 3, \dots$, and as

$$b_j p^{\lfloor \log_p j \rfloor} \equiv 0 \pmod{p^{\lambda+1}} \tag{10.37}$$

for all rational integers $j \geqslant 2p^\lambda$, in force of (10.25) and Lemma 10.16, then $\alpha(x) \in \mathbb{Z}_p$ for all $x \in \mathbb{Z}_p$. If $j \geqslant kp^{\lambda-1} + p^\lambda$, then $\lfloor \log_p j \rfloor \geqslant \lambda$; so (10.36) implies that $\beta_k(x) \in \mathbb{Z}_p$ for all $x \in \mathbb{Z}_p$.

Now we prove that for all $a, b \in \mathbb{Z}_p$, the congruence $a \equiv b \pmod{p^\lambda}$ implies a congruence $\theta(a) \equiv \theta(b) \pmod{p}$. From (10.36), it follows that

$$\alpha(x) \equiv \frac{1}{2} \sum_{j=2p^\lambda}^{3p^\lambda - 1} \frac{b_j}{p} \binom{x}{j-2p^\lambda} \pmod{p}. \tag{10.38}$$

Note that in (10.38), every fraction $\frac{b_j}{p}$ is a p-adic integer in force of Lemma 10.19. Now, as $a \equiv b \pmod{p^\lambda}$, then Lucas' Theorem 1.2 implies that for all $j = 2p^\lambda, 2p^\lambda + 1, \dots, 3p^\lambda - 1$, the following congruence holds:

$$\binom{a}{j-2p^\lambda} \equiv \binom{b}{j-2p^\lambda} \pmod{p}.$$

Thus, (10.38) implies that

$$\alpha(a) \equiv \alpha(b) \pmod{p}. \tag{10.39}$$

Further, combining (10.36) with Lemma 10.16, we conclude that the following congruence holds for all $k = 1, 2, \dots, p-1$:

$$\beta_k(x) \equiv \frac{1}{k} \sum_{j=kp^{\lambda-1}+p^\lambda}^{2p^\lambda - 1} b_j \binom{x}{j-kp^{\lambda-1}-p^\lambda} \pmod{p}.$$

From this congruence in force of Lucas' Theorem 1.2, it follows that

$$\beta_k(a) \equiv \beta_k(b) \pmod{p} \tag{10.40}$$

whenever $a \equiv b \pmod{p^\lambda}$.

Further, denote

$$\gamma_k(x) = \frac{\Delta^{kp^{\lambda-1}} f(x)}{kp^{\lambda-1}};$$

then in view of (10.36), we conclude that

$$\gamma_k(x) \equiv \frac{1}{k} \sum_{j=kp^{\lambda-1}}^{p^\lambda - 1} b_j \binom{x}{j - kp^{\lambda-1}} \pmod{p}$$

for all $k = 1, 2, \ldots, p-1$. Now applying Lucas' Theorem 1.2 once again, we conclude that

$$\gamma_k(a) \equiv \gamma_k(b) \pmod{p} \tag{10.41}$$

whenever $a \equiv b \pmod{p^\lambda}$. Now from (10.39)–(10.41), it follows that the congruence $a \equiv b \pmod{p^\lambda}$ implies the congruence $\theta(a) \equiv \theta(b) \pmod{p}$.

Now we prove final assertion of Lemma 10.19. Our proof will follow the lines of the proof of Proposition 10.18; however, now, we are considering the case $m = \lambda$ rather than $m \geqslant \lambda + 1$. Actually, we will derive a congruence for $f(x + p^\lambda h)$ modulo $p^{\lambda+2}$ from equality (10.29) with $m = \lambda$. In order to do this, we must find a residue of $\varphi_\lambda(x, h)$ (see (10.30)) modulo p^2. Again, as f is 1-Lipschitz, during the proof, we may assume that $h \in \mathbb{N}$.

In view of Lemma 5.8, from (10.33), it follows that if $i \in \{1, 2, \ldots, 2p^\lambda\}$ and either $i \leqslant p^{\lambda-1}$ or $p^{\lambda-1} < i < p^\lambda$, $p^{\lambda-1}$ is not a factor of i, then

$$\binom{p^\lambda h - 1}{i - 1} \frac{\Delta^i f(x)}{i} \equiv (-1)^{i-1} \frac{\Delta^i f(x)}{i} \pmod{p^2}. \tag{10.42}$$

Let now $i = kp^{\lambda-1}$ for $k \in \{2, 3, \ldots, p-1\}$; then (10.33) implies that

$$\binom{p^\lambda h - 1}{i - 1} \equiv (-1)^{kp^{\lambda-1} - 1} + (-1)^k ph \sum_{j=1}^{k-1} \frac{1}{j} \pmod{p^2}. \tag{10.43}$$

Further, if $p^\lambda \leqslant i \leqslant 2p^\lambda$ and $\operatorname{ord}_p i \neq \lambda, \lambda-1$, then combining (10.36) together with (10.37), we see that

$$\frac{\Delta^i f(x)}{i} \equiv 0 \pmod{p^2}. \tag{10.44}$$

Now we find residues modulo p^2 of terms $\binom{p^\lambda h-1}{i-1}\frac{\Delta^i f(x)}{i}$ of the function $\varphi_\lambda(u,h)$ (see (10.30)) in two remaining cases, when $i = \nu p^\lambda$ and $\nu \in \{1,2\}$, respectively, and when $i = kp^{\lambda-1} + p^\lambda$, $k \in \{1, 2, \ldots, p-1\}$. In the latter case in view of Corollary 10.17 and (10.33), the following congruence holds:

$$\binom{p^\lambda h-1}{i-1}\frac{\Delta^i f(x)}{i} \equiv (-1)^{i-1}\frac{\Delta^i f(x)}{i} + (-1)^{k-1}h\frac{\Delta^i f(x)}{i} \pmod{p^2}. \tag{10.45}$$

It is obvious that for all $k = 1, 2, \ldots, p-1$, the following trivial equality holds:

$$\left(1+\frac{p}{k}\right)\frac{\Delta^{kp^{\lambda-1}+p^\lambda} f(x)}{kp^{\lambda-1}+p^\lambda} = \frac{\Delta^{kp^{\lambda-1}+p^\lambda} f(x)}{kp^{\lambda-1}}. \tag{10.46}$$

As, in view of Corollary 10.17,

$$\frac{\Delta^{kp^{\lambda-1}+p^\lambda} f(x)}{kp^{\lambda-1}+p^\lambda} \equiv 0 \pmod{p},$$

then, since $\frac{p}{k} \in \mathbb{Z}_p$ and $\operatorname{ord}_p \frac{p}{k} = 1$, the equality (10.46) implies that

$$\frac{\Delta^{kp^{\lambda-1}+p^\lambda} f(x)}{kp^{\lambda-1}+p^\lambda} \equiv \frac{\Delta^{kp^{\lambda-1}+p^\lambda} f(x)}{kp^{\lambda-1}} \pmod{p^2}.$$

From here, substituting $i = kp^{\lambda-1} + p^\lambda$ to (10.45), we deduce that

$$\begin{aligned}\binom{p^\lambda h-1}{kp^{\lambda-1}+p^\lambda-1}\frac{\Delta^{kp^{\lambda-1}+p^\lambda} f(x)}{kp^{\lambda-1}+p^\lambda} &\equiv \\ \equiv (-1)^{kp^{\lambda-1}+p^\lambda-1}\frac{\Delta^{kp^{\lambda-1}+p^\lambda} f(x)}{kp^{\lambda-1}+p^\lambda} &+ (-1)^{k-1}ph\beta_k(x) \pmod{p^2}.\end{aligned} \tag{10.47}$$

In the case $i = p^\lambda$, the equality (10.33) implies that

$$\binom{p^\lambda h-1}{p^\lambda-1} \equiv (-1)^{p^\lambda-1} - ph\sum_{j=1}^{p-1}\frac{1}{j} \equiv (-1)^{p^\lambda-1} \pmod{p^2} \tag{10.48}$$

since $\sum_{j=1}^{p-1}\frac{1}{j} \equiv \sum_{j=1}^{p-1} j \equiv 0 \pmod p$ for $p \neq 2$. Finally, for $i = 2p^\lambda$ from (10.33) in view of Corollary 10.17, we conclude that

$$\binom{p^\lambda h - 1}{2p^\lambda - 1}\frac{\Delta^{2p^\lambda} f(x)}{2p^\lambda} \equiv (-1)^{2p^\lambda - 1}\frac{\Delta^{2p^\lambda} f(x)}{2p^\lambda} + h\frac{\Delta^{2p^\lambda} f(x)}{2p^\lambda} \equiv$$

$$\equiv (-1)^{2p^\lambda - 1}\frac{\Delta^{2p^\lambda} f(x)}{2p^\lambda} + hp\alpha(x) \pmod{p^2}. \tag{10.49}$$

Now collecting together (10.42), (10.44), and (10.47)–(10.49), we finish the proof of Lemma 10.19 in the same way with the proof of Proposition 10.18. □

Lemma 10.20 *Under assumptions of Theorem 10.14, let p be an odd prime; then for all $x, h \in \mathbb{Z}_p$, the following congruence holds:*

$$f'(x + p^\lambda h) \equiv f'(x) + 2ph\theta(x) \pmod{p^2}.$$

Here θ is the function defined in the statement of Lemma 10.19.

Proof of Lemma 10.20 From Proposition 10.18, it follows that

$$f'(x + p^\lambda h) \equiv \sum_{i=1}^{2p^\lambda}(-1)^{i-1}\frac{\Delta^i f(x + p^\lambda h)}{i} \pmod{p^2}. \tag{10.50}$$

For $i = 1, 2, \ldots, 2p^\lambda$ Lemma 10.19 implies that

$$\frac{\Delta^i f(x + p^\lambda h)}{i} \equiv \frac{\Delta^i f(x)}{i} + hp^{\lambda - \mathrm{ord}_p}\frac{\Delta^i f'(x)}{\hat{\imath}} +$$

$$+ h^2 p^{\lambda + 1 - \mathrm{ord}_p i}\frac{\Delta^i \theta(x)}{\hat{\imath}} \pmod{p^2}, \tag{10.51}$$

where $\hat{\imath} = ip^{-\mathrm{ord}_p i}$ is a unit in $\mathbb{Z}_p$; that is, $\hat{\imath}$ has a multiplicative inverse $\frac{1}{\hat{\imath}} \in \mathbb{Z}_p$.

We show now that the term of order 2 (with respect to h) in (10.51) is 0 modulo p^2. If this term is not 0 modulo p^2, then necessarily $i \in \{p^\lambda, 2p^\lambda\}$. However, from (10.36), it follows that in this case,

$$\frac{\Delta^{i+kp^{\lambda-1}} f(x)}{kp^{\lambda-1}} \equiv 0 \pmod p,$$

$$\frac{\Delta^{i+kp^{\lambda-1}+p^\lambda} f(x)}{kp^\lambda} \equiv 0 \pmod p, \tag{10.52}$$

$$\frac{\Delta^{i+2p^\lambda} f(x)}{2p^\lambda} \equiv 0 \pmod p,$$

for all $k \in \{1, 2, \dots, p-1\}$. Now, by definition of θ, from (10.52), we see that $\frac{\Delta^i \theta(x)}{\hat{i}} \equiv 0 \pmod{p}$ for $i \in \{p^\lambda, 2p^\lambda\}$, and thus

$$h^2 p^{\lambda+1-\mathrm{ord}_p i} \frac{\Delta^i \theta(x)}{\hat{i}} \equiv 0 \pmod{p^2}; \qquad i = 1, 2, \dots, 2p^\lambda. \tag{10.53}$$

Now we consider a term of order 1 in (10.51). If this term is not 0 modulo p^2, then necessarily, $i \in \{1, 2, \dots, 2p^\lambda\}$, and $\mathrm{ord}_p i \geqslant \lambda - 1$; that is, $i \in \{p^\lambda, 2p^\lambda, kp^{\lambda-1}, kp^{\lambda-1} + p^\lambda : k = 1, 2, \dots, p-1\}$. Combining together Corollary 10.17, Proposition 10.18, and Lemma 5.8 we conclude that

$$f'(x) \equiv \frac{\Delta^{p^\lambda} f(x)}{p^\lambda} + \sum_{t=0}^{\lambda-1} \sum_{\tau=1}^{p-1} (-1)^{\tau-1} \frac{\Delta^{\tau p^t} f(x)}{\tau p^t} \pmod{p}; \tag{10.54}$$

whence,

$$\Delta^i f'(x) \equiv \frac{\Delta^{i+p^\lambda} f(x)}{p^\lambda} + \sum_{t=0}^{\lambda-1} \sum_{\tau=1}^{p-1} (-1)^{\tau-1} \frac{\Delta^{i+\tau p^t} f(x)}{\tau p^t} \pmod{p}. \tag{10.55}$$

The latter congruence in force of (10.36) and Lemma 10.16 implies that $\Delta^i f'(x) \equiv 0 \pmod{p}$ when $i \in \{kp^{\lambda-1} + p^\lambda : k - 1, 2, \dots, p-1\}$; consequently,

$$hp \frac{\Delta^{kp^{\lambda-1}+p^\lambda} f'(x)}{k+p} = 0 \pmod{p^2}; \qquad k - 1, 2, \dots, p-1, \tag{10.56}$$

since a multiplicative inverse $\frac{1}{k+p}$ of $k+p$ is in $\mathbb{Z}_p$ for $k = 1, 2, \dots, p-1$.

If $i \in \{kp^{\lambda-1} : k = 1, 2, \dots, p-1\}$, then in view of Lemma 10.16 from (10.36) and (10.55), we deduce that

$$\Delta^{kp^{\lambda-1}} f'(x) \equiv \frac{\Delta^{kp^{\lambda-1}+p^\lambda} f(x)}{p^\lambda} + \sum_{\tau=1}^{p-k-1} (-1)^{\tau-1} \frac{\Delta^{(\tau+k)p^{\lambda-1}} f(x)}{\tau p^{\lambda-1}} \pmod{p}. \tag{10.57}$$

If $i = 2p^\lambda$, then Proposition 10.18 implies that

$$\Delta^{2p^\lambda} f'(x) \equiv \sum_{j=1}^{2p^\lambda} (-1)^{j-1} \frac{\Delta^{j+2p^\lambda} f(x)}{j} \pmod{p^2}.$$

This, in view of (10.36) and Lemma 10.16, implies that

$$\Delta^{2p^\lambda} f'(x) \equiv 0 \pmod{p^2}. \tag{10.58}$$

Now we consider the case $i = p^\lambda$. Proposition 10.18 implies that

$$\Delta^{p^\lambda} f'(x) \equiv \sum_{j=1}^{1+p^\lambda} (-1)^{j-1} \frac{\Delta^{j+p^\lambda} f(x)}{j} \pmod{p^2}, \tag{10.59}$$

since for $j = p^\lambda + 1, \ldots, 2p^\lambda$ from (10.36) in view of Lemma 10.16, it follows that

$$\frac{\Delta^{j+p^\lambda} f(x)}{j} \equiv 0 \pmod{p^2}.$$

Moreover, (10.36) implies that the latter congruence holds also for all $j \leqslant p^\lambda - 1$ such that $j \neq kp^{\lambda-1}$, where $k = 1, 2, \ldots, p-1$. Thus, from (10.59), we deduce that

$$\Delta^{p^\lambda} f'(x) \equiv \frac{\Delta^{2p^\lambda} f(x)}{p^\lambda} + \sum_{k=1}^{p-1} (-1)^{k-1} \frac{\Delta^{kp^{\lambda-1}+p^\lambda} f(x)}{kp^{\lambda-1}} \pmod{p^2}. \tag{10.60}$$

Now, substituting (10.53), (10.56), (10.57), (10.58), and (10.60) to (10.51) and summing up all obtained congruences for i ranging from 1 to $2p^\lambda$, in view of (10.50) and Proposition 10.18, we conclude that

$$\begin{aligned} f'(x + p^\lambda h) \equiv f'(x) + hp \left(\sum_{k=1}^{p-1} \frac{(-1)^{k-1}}{k} \sum_{\tau=1}^{p-k-1} (-1)^{\tau-1} \frac{\Delta^{(\tau+k)p^{\lambda-1}} f(x)}{\tau p^{\lambda-1}} + \right. \\ \left. + \sum_{k=1}^{p-1} (-1)^{k-1} \frac{\Delta^{kp^{\lambda-1}+p^\lambda} f(x)}{kp^\lambda} \right) + \\ + h \sum_{k=1}^{p-1} (-1)^{k-1} \frac{\Delta^{kp^{\lambda-1}+p^\lambda} f(x)}{kp^{\lambda-1}} + h \frac{\Delta^{2p^\lambda} f(x)}{p^\lambda} \pmod{p^2}. \end{aligned} \tag{10.61}$$

Easy calculations in $\mathbb{Q}_p$ prove that the following equality for $k, \tau \in \{1, 2, \ldots, p-1\}$ is true:

$$\sum_{k+\tau=m} \frac{1}{k\tau} = \sum_{k+\tau=m} \frac{1}{(m-\tau)\tau} = \frac{1}{m} \sum_{k+\tau=m} \left(\frac{1}{\tau} + \frac{1}{m-\tau} \right) = \frac{2}{m} \sum_{\tau=1}^{m-1} \frac{1}{\tau}$$

From here, it follows that

$$\sum_{k=1}^{p-1}\frac{(-1)^{k-1}}{k}\sum_{\tau=1}^{p-k-1}(-1)^{\tau-1}\frac{\Delta^{(\tau+k)p^{\lambda-1}}f(x)}{\tau p^{\lambda-1}}=$$

$$=\sum_{m=1}^{p-1}(-1)^m\sum_{k+\tau=m}\frac{\Delta^{mp^{\lambda-1}}f(x)}{k\tau p^{\lambda-1}}=2\sum_{m=1}^{p-1}(-1)^m\sum_{\tau=1}^{m-1}\frac{\Delta^{mp^{\lambda-1}}f(x)}{\tau mp^{\lambda-1}} \tag{10.62}$$

As it was shown in the proof of Lemma 10.19, both $\alpha(x)$ and $\beta_k(x)$ are p-adic integers for $k = 1, 2, \dots, p-1$ and $x \in \mathbb{Z}_p$; thus,

$$\begin{aligned} 2hp\alpha(x) &= h\frac{\Delta^{2p^{\lambda}}f(x)}{p^{\lambda}}, \\ hp\beta_k(x) &= h\frac{\Delta^{kp^{\lambda-1}+p^{\lambda}}f(x)}{kp^{\lambda-1}}, \end{aligned} \tag{10.63}$$

and fractions in the right-hand side are p-adic integers. Finally, the assertion of Lemma 10.20 follows from (10.61), (10.62), and (10.63) and from the definition of the function θ. □

Now we have all the tools ready to prove Theorem 10.14.

Proof of Theorem 10.14 For $p = 2$, Theorem 10.14 follows from Theorem 9.5 in view of Lemma 10.15. Indeed, under conditions of Theorem 10.14, coefficients a_j of Mahler expansion $f(x) = \sum_{j=0}^{\infty} a_j\binom{x}{j}$ of the function f satisfy the following congruence: $2^{\rho}a_i \equiv 0 \pmod{2^{\mathrm{ord}_2(i!)}}$. However, from the definition of λ in view of Lemma 10.15, it follows that $\mathrm{ord}_2(i!) - \lfloor\log_2 i\rfloor \geqslant \rho + 1$ for all $i \geqslant 2^{\lambda+1}$, as $\mathrm{ord}_2(2^{\lambda+1}!) = 2^{\lambda+1} - 1$ (see Lemma 2.1). That is, $a_i \equiv 0 \pmod{2^{\lfloor\log_2 i\rfloor+1}}$ for all $i \geqslant 2^{\lambda+1}$. A similar argument proves that $a_i \equiv 0 \pmod{2^{\lfloor\log_2(i+1)\rfloor+1}}$ for all $i \geqslant 2^{\lambda+2}$. In view of Theorem 9.5, this proves Theorem 10.14 in the case $p = 2$.

Now let $p \neq 2$. The first assertion of Theorem 10.14 in this case immediately follows from Theorem 10.1 and Proposition 10.18. Further, if $p = 3$, then, as $N_2(f) \leqslant \lambda + 1$ according to Proposition 10.18, the second assertion of Theorem 10.14 follows from Theorem 10.9.

Thus, we only must prove the second assertion of Theorem 10.14 for $p \notin \{2, 3\}$. As $N_2(f) \leqslant \lambda + 1$ according to Proposition 10.18, in force of Theorem 10.9, it is sufficient to show that f is transitive modulo $p^{\lambda+2}$ whenever f is transitive modulo $p^{\lambda+1}$. For this purpose, in view of Lemma 10.10, it is sufficient only to prove that $f^{p^{\lambda+1}}(x) \not\equiv x \pmod{p^{\lambda+2}}$ for at least one $x \in \mathbb{Z}_p$. Now we merely calculate $f^{p^{\lambda+1}}(x) \bmod p^{\lambda+2}$.

Under our assumptions, f is transitive modulo p^λ since f is 1-Lipschitz. Then by Lemma 10.10, we conclude that

$$f^{p^\lambda}(x) = x + p^\lambda \xi(x), \quad \xi(x) \not\equiv 0 \pmod p, \tag{10.64}$$

for all $x \in \mathbb{Z}_p$; here, $\xi : \mathbb{Z}_p \to \mathbb{Z}_p$ is a function defined everywhere on $\mathbb{Z}_p$.

We claim that for all $i = 0, 1, 2, \ldots$, the following congruence holds:

$$\begin{aligned} f^{p^\lambda+i}(x) &\equiv f^i(x) + p^\lambda \xi(x) \prod_{j=0}^{i-1} f'(f^j(x)) + \\ &+ p^{\lambda+1}\xi(x)^2 \prod_{j=0}^{i-1} f'(f^j(x)) \sum_{k=0}^{i-1} \frac{\theta(f^k(x))}{f'(f^k(x))} \prod_{\tau=0}^{k-1} f'(f^\tau(x)) \pmod{p^{\lambda+2}}. \end{aligned} \tag{10.65}$$

Recall that a sum (respectively, a product) over an empty set of indices is assumed to be 0 (respectively, 1). Note also that as f is transitive modulo $p^{\lambda+1}$, f is bijective modulo $p^{\lambda+1}$. Then (as $\lambda + 1 \geqslant N_1(f) + 1$ by Proposition 10.18) f is measure-preserving and that $f'(z) \not\equiv 0 \pmod p$ for all $z \in \mathbb{Z}_p$. Thus, denominators of all fractions in (10.65) have multiplicative inverses in $\mathbb{Z}_p$; so during the proof of (10.65) and further on, we assume that all calculations are performed in $\mathbb{Z}_p$.

To prove (10.65), we note that according to the chain rule, $\prod_{j=0}^{i-1} f'(f^j(x)) = (f^i(x))'$, so (10.65) can be rewritten in the form

$$\begin{aligned} f^{p^\lambda+i}(x) &\equiv f^i(x) + p^\lambda \xi(x)(f^i(x))' + \\ &+ p^{\lambda+1}\xi(x)^2 (f^i(x))' \sum_{k=0}^{i-1} \frac{(f^k(x))'}{f'(f^k(x))} \theta(f^k(x)) \pmod{p^{\lambda+2}} \end{aligned}$$

and then proved by induction on i. Indeed, for $i = 0$, our claim trivially follows from (10.64). Now we substitute the above expression for $f^{p^\lambda+i}(x) \bmod p^{\lambda+2}$ into the equation $f^{p^\lambda+i+1}(x) = f(f^{p^\lambda+i}(x))$, and with the use of Lemma 10.19 and obvious direct calculations, we prove the demanded congruence for $f^{p^\lambda+i+1}(x)$. We omit details.

Now we apply (10.65) to calculate $f^{p^{\lambda+1}}(x) \bmod p^{\lambda+2}$. We have

$$f^{p^\lambda+i}(x) \equiv f^i(x) + p^\lambda \xi(x) A_i(x) + p^{\lambda+1}\xi(x)^2 B_i(x) \pmod{p^{\lambda+2}}, \tag{10.66}$$

where

$$A_i(x) = (f^i(x))' = \prod_{j=0}^{i-1} f'(f^j(x));$$

$$B_i(x) = (f^i(x))' \sum_{k=0}^{i-1} \frac{(f^k(x))'}{f'(f^k(x))} \theta(f^k(x)) =$$

$$= \left(\prod_{j=0}^{i-1} f'(f^j(x))\right) \left(\sum_{k=0}^{i-1} \frac{\theta(f^k(x))}{f'(f^k(x))^2} \prod_{\tau=0}^{k} f'(f^\tau(x))\right).$$

Lemma 10.20 implies that $f'(a + p^\lambda h) \equiv f'(a) \pmod{p}$. From here we deduce that $f'(f^k(x)) \equiv f'(f^r(x)) \pmod{p}$ whenever $k \equiv r \pmod{p^\lambda}$, as f is transitive modulo p^λ. By the latter reason, $\theta(f^k(x)) \equiv \theta(f^r(x)) \pmod{p}$ whenever $k \equiv r \pmod{p^\lambda}$, in view of Lemma 10.19. Further, $N_1(f) \leqslant \lambda$ by Proposition 10.18, and f is transitive modulo $p^{\lambda+1}$ by our assumption, so necessarily,

$$\prod_{\tau=0}^{p^\lambda - 1} f'(f^\tau(x)) \equiv 1 \pmod{p} \tag{10.67}$$

(see the proof of Lemma 10.10); consequently, $\prod_{\iota=0}^{k} f'(f^\tau(x)) = \prod_{\tau=0}^{r} f'(f^\tau(x)) \pmod{p}$ whenever $k \equiv r \pmod{p^\lambda}$. Finally, we conclude that

$$B_{tp^\lambda}(x) \equiv t \sum_{\tau=0}^{p^\lambda - 1} \frac{\theta(f^\tau(x))}{f'(f^\tau(x))^2} \prod_{\nu=0}^{\tau} f'(f^\nu(x)) \equiv t B_{p^\lambda}(x) \pmod{p}, \tag{10.68}$$

for every $t \in \mathbb{N}$.

Now we calculate $A_{tp^\lambda}(x) \bmod p^2$ for $t \in \mathbb{N}$. Congruence (10.66) in view of (10.67) implies that

$$f^{kp^\lambda + \tau}(x) \equiv f^\tau(x) + kp^\lambda \xi(x) \prod_{j=0}^{\tau-1} f'(f^j(x)) \pmod{p^{\lambda+1}}, \tag{10.69}$$

for all $k \in \mathbb{N}$ and all $\tau \in \{0, 1, \ldots, p^\lambda - 1\}$. As Lemma 10.20 implies that $f'(u) \equiv f'(v) \pmod{p^2}$ whenever $u \equiv v \pmod{p^{\lambda+1}}$, and as

$$A_{tp^\lambda}(x) = \prod_{k=0}^{t-1} \prod_{\tau=0}^{p^\lambda - 1} f'(f^{kp^\lambda + \tau}(x)),$$

we conclude in view of congruence (10.69) that

$$A_{tp^\lambda}(x) = \prod_{k=0}^{t-1}\prod_{\tau=0}^{p^\lambda-1} f'\left(f^\tau(x) + kp^\lambda\xi(x)\prod_{j=0}^{\tau-1} f'(f^j(x))\right) \pmod{p^2}.$$

This implies in view of Lemma 10.20, that

$$A_{tp^\lambda}(x) = \prod_{k=0}^{t-1}\prod_{\tau=0}^{p^\lambda-1}\left(f'(f^\tau(x)) + 2kp\xi(x)\theta(f^\tau(x))\prod_{j=0}^{\tau-1} f_2'(f^j(x))\right) \equiv$$

$$\equiv \prod_{k=0}^{t-1}\left(\prod_{\tau=0}^{p^\lambda-1} f'(f^\tau(x)) + 2kp\xi(x)B_{p^\lambda}(x)\right) \pmod{p^2}. \qquad (10.70)$$

According to (10.67),

$$\prod_{j=0}^{p^\lambda-1} f'(f^j(x)) = 1 + p\varepsilon$$

for a suitable $\varepsilon \in \mathbb{Z}_p$; consequently, (10.70) implies that

$$A_{tp^\lambda}(x) \equiv \prod_{k=0}^{t-1}\left(1 + p\varepsilon + 2kp\xi(x)B_{p^\lambda}(x)\right) \equiv$$

$$\equiv 1 + tp\varepsilon + pt(t-1)\xi(x)B_{p^\lambda}(x) \pmod{p^2}. \qquad (10.71)$$

Now combining together (10.66), (10.68), and (10.71), we conclude that

$$f^{(t+1)p^\lambda}(x) = f^{p^\lambda + tp^\lambda}(x) \equiv$$

$$\equiv f^{tp^\lambda}(x) + p^\lambda\xi(x) + \varepsilon tp^{\lambda+1}\xi(x) + p^{\lambda+1}t^2\xi(x)^2 B_{p^\lambda}(x) \pmod{p^{\lambda+2}}. \qquad (10.72)$$

Finally, by obvious induction on n, from (10.72) and (10.64), we deduce that

$$f^{np^\lambda}(x) \equiv x + np^\lambda\xi(x) + \varepsilon p^{\lambda+1}\xi(x)\frac{n(n-1)}{2} +$$

$$+ p^{\lambda+1}\xi(x)^2 B_{p^\lambda}(x)\frac{n(n-1)(2n-1)}{6} \pmod{p^{\lambda+2}}.$$

From here, it follows in particular that $f^{p^{\lambda+1}}(x) \equiv x + p^{\lambda+1}\xi(x) \pmod{p^{\lambda+2}}$ since $p \neq 2, 3$. However, the latter congruence in view of (10.64) implies that $f^{p^{\lambda+1}}(x) \not\equiv x \pmod{p^{\lambda+2}}$. This finally proves Theorem 10.14. □

Note 10.21 With the use of Theorem 10.14, we can determine *whether a given integer-valued and compatible polynomial $f(x) \in \mathbb{Q}_p[x]$ is ergodic.*

Namely, represent $f(x)$ as $f(x) = \frac{g(x)}{r}$, where $r \in \mathbb{Z}_p$, $g(x) \in \mathbb{Z}_p[x]$ and at least one of the coefficients of $g(x)$ is co-prime to p. Actually, r is the least common denominator of all coefficients of $f(x)$ represented by irreducible fractions: We assume that $f(x)$ is represented in a falling factorial basis $x^{\underline{0}} = 1, x^{\underline{1}} = x, x^{\underline{2}} = x(x-1), \ldots$ or in a standard basis $1, x, x^2, \ldots$. Then $\rho(f) = \operatorname{ord}_p r$; note that r does not depend on a choice of the basis. Now we easily find $\lambda(f)$ and determine (e.g., by direct calculations) whether f is transitive modulo $p^{\lambda(f)+1}$ in the case of $p \neq 2, 3$ or, respectively, modulo $p^{\lambda(f)+2}$ whenever $p = 2$ or $p = 3$.

To determine whether a polynomial $f(x) \in \mathbb{Q}_p[x]$ induces a 1-Lipschitz measure-preserving (respectively, ergodic) transformation on $\mathbb{Z}_p$, it is sufficient to evaluate f at $\approx p^3 \deg f$ points:

Proposition 10.22 *A polynomial $f(x) \in \mathbb{Q}_p[x]$ induces a 1-Lipschitz measure-preserving (respectively, ergodic) transformation on $\mathbb{Z}_p$ if and only if the mapping $z \mapsto f(z) \bmod p^{\lfloor \log_p(\deg f) \rfloor + 3}$ is a compatible and bijective (respectively, transitive) transformation on the residue ring $\mathbb{Z}/p^{\lfloor \log_p(\deg f) \rfloor + 3}\mathbb{Z}$.*

Proof We prove only the ergodicity claim; the proof of the measure-preservation claim goes along similar lines and thus is omitted.

The coefficients $a_i \in \mathbb{Q}_p$ $(i = 0, 1, \ldots, d)$ in Mahler expansion of the polynomial $f(x)$ of degree d are completely determined by values of $f(x)$ at the points $0, 1, \ldots, d$ (cf., (4.2)). In particular, all values $f(0), f(1), \ldots, f(d)$ are p-adic integers if and only if all coefficients $a_i \in \mathbb{Q}_p$ $(i = 0, 1, \ldots, d)$ are p-adic integers, i.e., if and only if the polynomial $f(x)$ is integer-valued.

As $\Delta^i f(x) = 0$ for $i > \deg f = d$, in view of Theorem 5.12 from the proof of Proposition 5.4, it follows that f is a 1-Lipschitz transformation on $\mathbb{Z}_p$ if and only if f induces a compatible transformation on the residue ring $\mathbb{Z}/p^k\mathbb{Z}$ for some (arbitrarily fixed) $k \geqslant \lfloor \log_p d \rfloor + 1$.

In force of Theorem 10.14, an integer-valued polynomial $f(x) \in \mathbb{Q}_p[x]$, which induces a 1-Lipschitz transformation on $\mathbb{Z}_p$ is ergodic (on $\mathbb{Z}_p$) if and only if f is transitive modulo p^k for any (arbitrarily fixed) $k \geqslant \lambda(f) + 2$. Considering now Mahler expansion (10.25) for $f(x)$; that is, $f(x) = b_0 + \sum_{i=1}^{d} b_i p^{\lfloor \log_p i \rfloor} \binom{x}{i}$, where $b_j \in \mathbb{Z}_p$ for $j = 0, 1, 2, \ldots$, we conclude that $\rho(f)$ is the least nonnegative rational integer that is not smaller than any of $\operatorname{ord}_p(i!) - \lfloor \log_p i \rfloor - \operatorname{ord}_p b_i$ $(i = 1, 2, \ldots, d)$. Thus, since the function $\operatorname{ord}_p(i!) - \lfloor \log_p i \rfloor$ is monotone nondecreasing by Lemma 10.15, every $k \in \mathbb{N}$ that satisfies the inequality $2\frac{p^k-1}{p-1} -$

$k > \mathrm{ord}_p(d!) - \lfloor \log_p d \rfloor$ cannot be smaller than $\lambda(f)$. However, $\mathrm{ord}_p(d!) = \frac{1}{p-1}(d - \mathrm{wt}_p\, d)$ by Lemma 2.1; so taking arbitrary $k \in \mathbb{N}$ that satisfies the inequality

$$2\frac{p^k - 1}{p - 1} - k > \frac{d}{p - 1}, \tag{10.73}$$

we conclude that $k \geqslant \lambda(f)$. Elementary considerations show that $k = \lfloor \log_p d \rfloor + 1$ satisfies inequality (10.73), thus ending the proof. □

It is obvious that in some cases, conditions of Theorem 10.14 and of Proposition 10.22 can be relaxed; e.g., it is obvious that whenever $p > 3$, the proposition remains true after replacing $p^{\lfloor \log_p(\deg f) \rfloor + 3}$ by $p^{\lfloor \log_p(\deg f) \rfloor + 2}$. However, the point is that for some important classes of functions, these bounds can be tightened significantly so that the conditions depend only on the whole class rather than on a concrete function from the class:

Corollary 10.23 *A $\mathcal{B}$-function* (and thus a C-function) f *is measure-preserving if and only if f is bijective modulo p^2. The function f is ergodic if and only if f is transitive modulo p^2 whenever $p \notin \{2, 3\}$ or modulo p^3 whenever $p \in \{2, 3\}$.*

Proof By definition of the class $\mathcal{B}$, $\rho(f) = 0$ for every $f \in \mathcal{B}$; whence, $\lambda(f) = 1$, and the conclusion follows from Theorem 10.14. □

From here, we immediately deduce the following known result:

Corollary 10.24 (Cf. [47, 132]) *A polynomial $f \in \mathbb{Z}_p[x]$ is ergodic if and only if f is transitive modulo p^2 whenever $p \notin \{2, 3\}$ or modulo p^3 whenever $p \in \{2, 3\}$.*

Note 10.25 The bounds given by Corollary 10.24 (and therefore by Corollary 10.23) are sharp: the polynomial $2x^3 + 3x + 5$ is transitive modulo 4, and is not transitive modulo 8 (whence is not ergodic on $\mathbb{Z}_2$); the polynomial

$$1 + x - x(x - 1)(x - 2)(x - 3)(x - 4)(x - 6)(x - 7)$$

is transitive modulo 9, and is not transitive modulo 27 (whence is not ergodic on $\mathbb{Z}_3$); the polynomial $1 + x^p$ is transitive modulo p, and is not transitive (it is not even bijective in view of Note 10.3) modulo p^2, whence is not measure-preserving on $\mathbb{Z}_p$. The first two examples are due to M. V. Larin, [132].

Example 10.26 For every prime p, the function $f(x) = 1 + x + \frac{p}{1+p^2x}$ is a 1-Lipschitz ergodic transformation on $\mathbb{Z}_p$.

This is obvious in view of Corollary 10.23 and Theorem 9.1 since $\frac{p}{1+p^2x} = p - p^3x + p^5x^2 - p^7x^3 + \cdots$.

Chapter 11
1-Lipschitz Ergodicity on Subspaces

Abstract In this chapter, we address the ergodicity of 1-Lipschitz transformations on certain subspaces S of the space $\mathbb{Z}_p$, namely, on balls and on spheres of non-zero radii. The subspaces are endowed with a natural probability measure $\hat{\mu}_p$ induced on S by the probability measure μ_p on the whole space $\mathbb{Z}_p$; i.e., $\hat{\mu}_p(A) = \mu_p(A)/\mu_p(S)$ for each μ_p-measurable subset $A \subset S$, so $\hat{\mu}_p(S) = 1$. Now, if $f : S \to S$ is a 1-Lipschitz map, we can speak of ergodicity of this map with respect to the measure $\hat{\mu}_p$. In the sequel, speaking of ergodicity (and of measure preservation) of a map f on a subspace S, we mean that S is invariant under action of f and the measure is $\hat{\mu}_p$.

11.1 1-Lipschitz Ergodic Transformations on Balls

In many cases, the problem to determine ergodicity of a 1-Lipschitz transformation on a ball $B_{p^{-k}}(a) = a + p^k\mathbb{Z}_p$ from $\mathbb{Z}_p$ can be reduced to the one on the whole space $\mathbb{Z}_p$. Indeed, if f is a 1-Lipschitz transformation such that $f(a + p^k\mathbb{Z}_p) \subset a + p^k\mathbb{Z}_p$, then necessarily $f(a) = a + p^k y$ for a suitable $y \in \mathbb{Z}_p$. Thus, $f(a + p^k z) = a + p^k g(z)$ for any $z \in \mathbb{Z}_p$; so we can relate to f the following 1-Lipschitz transformation on $\mathbb{Z}_p$

$$g \colon z \mapsto g(z) = \frac{1}{p^k}(f(a + p^k z) - a);\quad z \in \mathbb{Z}_p.$$

It is clear that the transformation f is ergodic on the ball $B_{p^{-k}}(a)$ if and only if the transformation g is ergodic on $\mathbb{Z}_p$.

For instance, 1-Lipschitz transformations that involve taking multiplicative inverses usually cannot be well-defined on the whole space $\mathbb{Z}_p$, but they can be well-defined on a ball centered at a unit of $\mathbb{Z}_p$. Respective transformations are used, e.g., in the design of generators of pseudorandom numbers, which we discuss in detail in Chap. 14, so here we just present a couple of examples of this kind to demonstrate how the approach works. Namely, we give a short proofs of two known

V. Anashin, *Causality: The p-adic Theory*, STEAM-H: Science, Technology, Engineering, Agriculture, Mathematics & Health,
https://doi.org/10.1007/978-3-031-85818-5_11

results on the so-called inversive generators of the longest possible period. It can be easily shown that either of the generators attains the longest possible period if and only if the law of recursion is an ergodic transformation of $\mathbb{Z}_2$; we prove the main results of papers [57, 110] in this form, as criteria of ergodicity for respective maps.

Example 11.1 ([57]) The function $f(x) = ax^{-1}+b$, where $a, b \in \mathbb{Z}$ and $a+b \equiv 1 \pmod 2$, is ergodic on the ball $1 + 2\mathbb{Z}_2$ if and only if $a \equiv 1 \pmod 4$ and $b \equiv 2 \pmod 4$.

Indeed, the condition $a + b \equiv 1 \pmod 2$ implies that the 2-adic ball $1 + 2\mathbb{Z}_2$ is invariant under action of f. We have then that $1 + 2g(z) = a(1 + 2z)^{-1} + b = a + b - 2az + 4az^2 - 8az^3 + \cdots$, so $g(z) = \frac{a+b-1}{2} - az + 2az^2 - 4az^3 + \cdots$ is a C-function of variable z. However, in view of Corollary 9.11, the function g is ergodic on $\mathbb{Z}_2$ if and only if $\frac{a+b-1}{2} \equiv 1 \pmod 2$ (Condition 4), $-a \equiv 1 \pmod 2$ (Condition 3), and $0 \equiv -a - 2a - 1 \pmod 4$ (Condition 2). This concludes the proof.

Example 11.2 ([110]) The function $f(x) = ax^{-1} + b + cx$, where $a + b + c \equiv 1 \pmod 2$, is ergodic on the ball $1+2\mathbb{Z}_2$ if and only if $a+c \equiv 1 \pmod 4$ and $b \equiv 2 \pmod 4$.

Only minor modifications of the above proof of the Example 11.1 are needed: Actually, in this case, $1 + 2g(z) = a(1 + 2z)^{-1} + b + c(1 + 2z) = a + b + c - 2(a-c)z+4az^2-8az^3+\cdots$; so $g(z) = \frac{a+b+c-1}{2} - (a-c)z+2az^2-4az^3+\cdots$, and the result follows.

If $p = 2$, then the question about ergodicity of 1-Lipschitz transformations on p-adic spheres can be reduced to the case of balls since a 2-adic sphere is a ball. Now we prove ergodicity criterion for 1-Lipschitz maps on 2-adic spheres following [26]; so to the rest of the section $p = 2$ and $f : \mathbb{Z}_2 \to \mathbb{Z}_2$ is a 1-Lipschitz function.

Let $\mathbf{S}_{2^{-r}}(a)$ be a sphere of radius 2^{-r} with a center at the point $a \in \{0, \ldots, 2^r - 1\}$, and let the sphere $\mathbf{S}_{2^{-r}}(a)$ be invariant under action of f; that is, let $f(\mathbf{S}_{2^{-r}}(a)) \subset \mathbf{S}_{2^{-r}}(a)$. As $p = 2$, the sphere $\mathbf{S}_{2^{-r}}(a)$ coincides with the ball $\mathbf{B}_{2^{-r-1}}(a + 2^r)$ of radius 2^{-r-1} centered at the point $a + 2^r$: $\mathbf{S}_{2^{-r}}(a) = \{a + 2^r + 2^{r+1}x : x \in \mathbb{Z}_2\} = \mathbf{B}_{2^{-r-1}}(a + 2^r)$. Therefore, the sphere $\mathbf{S}_{2^{-r}}(a)$ is f-invariant if and only if $f(a + 2^r + 2^{r+1}\mathbb{Z}_p) \subset a + 2^r + 2^{r+1}\mathbb{Z}_p$; that is, if and only if

$$f(a + 2^r) \equiv a + 2^r \pmod{2^{r+1}} \tag{11.1}$$

as a 1-Lipschitz function maps a ball of radius $2^{-\ell}$ into a ball of radius $2^{-\ell}$.

Further, as f is 1-Lipschitz, we can represent $f : \mathbb{Z}_2 \to \mathbb{Z}_2$ as

$$f(a + 2^r + 2^{r+1}x) = f(a + 2^r) + 2^{r+1}g(x); \tag{11.2}$$

then $g : \mathbb{Z}_2 \to \mathbb{Z}_2$ is a 1-Lipschitz function. The following proposition holds:

Proposition 11.3 *The function f is ergodic on the sphere $\mathbf{S}_{2^{-r}}(a)$ if and only if $f(a+2^r) \equiv a+2^r \pmod{2^{r+1}}$, and the function g defined by*

$$g(x) = \frac{f(a+2^r+2^{r+1}x)-(a+2^r)}{2^{r+1}} \tag{11.3}$$

is a 1-Lipschitz ergodic transformation on $\mathbb{Z}_2$.

Proof The first of conditions of the proposition just means that the sphere $\mathbf{S}_{2^{-r}}(a)$ is f-invariant (see (11.1)). Due to this condition, we see then that Condition 2 of Lemma 11.8 holds if and only if the function g is transitive modulo 2^t for all $t = 1, 2, 3, \ldots$ (we state and prove this lemma in the next section, for arbitrary prime p rather than for $p = 2$). However, the latter condition is equivalent to the ergodicity of the function g on $\mathbb{Z}_2$ by Theorem 8.2. □

Now given a 1-Lipschitz function $f\colon \mathbb{Z}_2 \to \mathbb{Z}_2$, in view of Theorem 5.14 and (5.32), f has a unique representation via van der Put series:

$$f(x) = \sum_{m=0}^{\infty} B_f(m)\chi(m,x) = \sum_{m=0}^{\infty} 2^{\lfloor \log_2 m \rfloor} b_f(m)\chi(m,x), \tag{11.4}$$

where $b_f(m) \in \mathbb{Z}_2$; so $B_f(m) = 2^{\lfloor \log_2 m \rfloor} b_f(m)$ for all $m = 0, 1, 2, \ldots$.

Theorem 11.4 *The function f represented by van der Put series* (11.4) *is ergodic on the sphere $\mathbf{S}_{2^{-r}}(a)$ if and only if the following conditions hold simultaneously:*

1. $f(a+2^r) \equiv a+2^r+2^{r+1} \pmod{2^{r+2}}$.
2. $\left|b_f(a+2^r+m2^{r+1})\right|_2 = 1$, *for* $m \geqslant 1$.
3. $b_f(a+2^r+2^{r+1}) \equiv 1 \pmod 4$.
4. $b_f(a+2^r+2^{r+2}) + b_f(a+2^r+3\times 2^{r+1}) \equiv 2 \pmod 4$.
5. $\sum_{m=2^{n-1}}^{2^n-1} b_f(a+2^r+m2^{r+1}) \equiv 0 \pmod 4$, *for* $n \geqslant 3$.

Proof Represent f by (11.2) and g by (11.3), and then calculate van der Put coefficients B_m of the function $g(x) = \sum_{m=0}^{\infty} B_m\chi(m,x)$ as follows. Given $m \in \{2, 3, \ldots\}$, denote $\acute{m} = m - 2^{\lfloor \log_2 m \rfloor}$; then using formula (4.13) for van der Put coefficients B_m and $B_f(m)$, we find that

$$B_0 = g(0) = \frac{f(a+2^r)-(a+2^r)}{2^{r+1}}; \tag{11.5}$$

$$B_1 = g(1) = \frac{f(a+2^r+2^{r+1})-(a+2^r)}{2^{r+1}}; \tag{11.6}$$

$$B_m = B_{\acute{m}+2^n} = g(\acute{m}+2^n) - g(\acute{m}) =$$
$$= \frac{f(a+2^r+2^{r+1}\acute{m}+2^{n+r+1})-(a+2^r)}{2^{r+1}} -$$

$$
\begin{aligned}
&= \frac{f(a+2^r+2^{r+1}\acute{m})-(a+2^r)}{2^{r+1}} = \\
&= \frac{f(a+2^r+2^{r+1}\acute{m}+2^{n+r+1})-f(a+2^r+2^{r+1}\acute{m})}{2^{r+1}} = \\
&= \frac{B_f(a+2^r+2^{r+1}\acute{m}+2^{n+r+1})}{2^{r+1}}, \quad (m \geqslant 2, n=\lfloor \log_2 m \rfloor).
\end{aligned} \tag{11.7}
$$

As g is 1-Lipschitz, Theorem 5.14 yields that for every $k = 0, 1, 2, \ldots$, there exists $b_k \in \mathbb{Z}_p$ such that $B_k = b_k 2^{\lfloor \log_2 k \rfloor}$. So from (11.7), it follows that if $m \geqslant 2$ and $n = \lfloor \log_2 m \rfloor$, then

$$
b_m = \frac{B_m}{2^n} = \frac{B_f(a+2^r+2^{r+1}\acute{m}+2^{n+r+1})}{2^{n+r+1}} = b_f(a+2^r+m2^{r+1}). \tag{11.8}
$$

Now apply Theorem 9.17. In view of (11.8), condition 3 (for $m \geqslant 2$) of Theorem 9.17 is equivalent to condition 2 of the theorem under proof. By the same reason, conditions 4 and 5 of Theorem 9.17 accordingly are equivalent to conditions 4 and 5 of the theorem under proof. Combining (11.3) with (4.13), we see that condition 1 of Theorem 9.17 is equivalent to condition 1 of the theorem under proof.

Finally, combining (11.5) and (11.6) with (4.13), we see that

$$
\begin{aligned}
b_1 - b_0 = g(1) - g(0) &= \frac{f(a+2^r+2^{r+1})-f(a+2^r)}{2^{r+1}} = \\
&= \frac{B_f(a+2^r+2^{r+1})}{2^{r+1}} = b_f(a+2^r+2^{r+1})
\end{aligned} \tag{11.9}
$$

It is an elementary exercise to show that condition 2 in the statement of Theorem 9.17 can be replaced by the condition $b_1 - b_0 \equiv 1 \pmod 4$; therefore, (11.9) proves condition 3 (as well as condition 1 for $m = 1$) of the theorem under proof. □

In [151], explicit conditions of ergodicity in terms of coefficients of polynomials on the sphere $\mathbf{S}_{1/2}(0) = \mathbf{B}_{1/2}(1) = \mathbb{Z}_2 \setminus 2\mathbb{Z}_2$ were found, whereas in [150], explicit conditions of ergodicity on the sphere in terms of linear relations on coefficients were found for rational functions. Note that the 2-adic sphere $\mathbf{S}_{1/2}(0)$ is the group of all units (invertible elements) of the ring $\mathbb{Z}_2$; so the results of [150, 151] are of special interest. We do not present the results here since corresponding conditions of ergodicity are somewhat lengthy.

The situation with ergodicity on "genuine" spheres, the ones which are not balls, is more complicated. We address this issue in the next section.

11.2 1-Lipschitz Ergodic Transformations on Spheres

Dynamics on spheres and circles is a classical area of dynamical system theory (cf., e.g., [111]). In this section, we address ergodic properties (with respect to the measure μ_p) of p-adic 1-Lipschitz maps, which act on p-adic spheres. The results obtained below not only constitute a contribution to general theory but also are of interest for applications (see Chaps. 14 and 16).

Let $\mathbf{S}_{p^{-r}}(y)$ be a sphere of radius $\frac{1}{p^r}$, $r \geqslant 1$, with a center at $y \in \mathbb{Z}_p$; that is,

$$\mathbf{S}_{p^{-r}}(y) = \left\{z \in \mathbb{Z}_p : |z - y|_p = \frac{1}{p^r}\right\}$$

We remind that the sphere is a disjoint union of balls of radius $\frac{1}{p^{r+1}}$ each,

$$\mathbf{S}_{p^{-r}}(y) = \bigcup_{s=1}^{p-1} (y + p^r s + p^{r+1}\mathbb{Z}_p), \tag{11.10}$$

since $\mathbf{S}_{p^{-r}}(y)$ is a set-theoretic complement of the ball $y + p^{r+1}\mathbb{Z}_p$ in the ball $y + p^r\mathbb{Z}_p$ (cf. Sect. 2.2.1). So $\mathbf{S}_{p^{-r}}(y)$ is a clopen (whence, a μ_p-measurable) subset of $\mathbb{Z}_p$. The following easy proposition is true:

Proposition 11.5 *If $\mathbf{S}_{p^{-r}}(y)$ is invariant under action of a 1-Lipschitz map f, then $f(y) \equiv y \pmod{p^r}$.*

Proof Since $\mathbf{S}_{p^{-r}}(y)$ is invariant, and since f maps balls into balls, $f(y + p^r s + p^{r+1}\mathbb{Z}_p) \subset y + p^r \hat{s} + p^{r+1}\mathbb{Z}_p$ for a suitable $\hat{s} \in \{1, 2, \ldots, p-1\}$ (see (11.10)). However, $f(y + p^r s) \equiv f(y) \pmod{p^r}$ since f is 1-Lipschitz, and the conclusion follows. □

From this proposition, the following corollary can immediately be derived:

Corollary 11.6 *Let all spheres around $y \in \mathbb{Z}_p$ of radii less than $\varepsilon > 0$ be invariant under action of a 1-Lipschitz map f. Then y is a fixed point of f.*

The analogue of Theorem 8.2 for a sphere (rather than for the whole space $\mathbb{Z}_p$) remains true. Moreover, as it can be seen from the respective proofs, all results of Chap. 8 hold not only for the whole space $\mathbb{Z}_p$ but (up to a proper re-statement) for any finite disjoint union of balls of pairwise equal radii as well. We summarize this (without giving a proof) as the following proposition:

Proposition 11.7 *A 1-Lipschitz map $f : \mathbb{Z}_p \to \mathbb{Z}_p$ is ergodic on the sphere $\mathbf{S}_{p^{-r}}(y)$ if and only if it induces on the residue ring $\mathbb{Z}/p^{k+1}\mathbb{Z}$, a mapping which is transitive* *on the subsets*

$$\mathbf{S}_{p^{-r}}(y) \bmod p^{k+1} = \{y + p^r s + p^{r+1}\mathbb{Z}\colon s = 1, 2, \ldots, p-1\} \subset \mathbb{Z}/p^{k+1}\mathbb{Z}$$

for all $k = r, r+1, \ldots$.

To put it in other words, the map f is ergodic on the sphere $\mathbf{S}_{p^{-r}}(y)$ if and only if the reduced map $f \bmod p^{k+1}$ permutes cyclically elements of the subsets $\mathbf{S}_{p^{-r}}(y) \bmod p^{k+1}$ (cf. Sect. 7.1.1), for all $k = r, r+1, \ldots$.

It is worth noticing also that whenever a 1-Lipschitz map f is ergodic on the sphere $S_{p^{-r}}(y)$, the map f is a bijection of this sphere onto itself; moreover, it is an isometry on this sphere (see Notes 8.5 and 8.8). The same holds for balls. From these notices, we deduce the following Lemma:

Lemma 11.8 *A 1-Lipschitz map* $f\colon \mathbb{Z}_p \to \mathbb{Z}_p$ *is ergodic on the sphere* $\mathbf{S}_{p^{-r}}(y)$ *if and only if the following two conditions hold simultaneously:*

1. *The mapping* $z \mapsto f(z) \bmod p^{r+1}$ *is transitive on the set*

$$\mathbf{S}_{p^{-r}}(y) \bmod p^{r+1} = \{y + p^r s\colon s = 1, 2, \ldots, p-1\} \subset \mathbb{Z}/p^{r+1}\mathbb{Z}.$$

2. *The mapping* $z \mapsto f^{p-1}(z) \bmod p^{r+t+1}$ *is transitive on the set*

$$\mathbf{B}_{p^{-(r+1)}}(y + p^r s) \bmod p^{r+t+1} = \{y + p^r s + p^{r+1}S\colon S = 0, 1, 2, \ldots, p^t - 1\},$$

for all $t = 1, 2, \ldots$ *and some (equivalently, all)* $s \in \{1, 2, \ldots, p-1\}$.

Condition 2 holds if and only if f^{p-1} *is an ergodic transformation on the ball* $\mathbf{B}_{p^{-(r+1)}}(y + p^r s) = y + p^r s + p^{r+1}\mathbb{Z}_p$ *of radius* $\frac{1}{p^{r+1}}$ *centered at* $y + p^r s$*, for some (equivalently, all)* $s \in \{1, 2, \ldots, p-1\}$.

Proof As every 1-Lipschitz ergodic transformation f of the sphere is bijective on this sphere, and f is an isometry on this sphere as well (see the notions above), $f(a + p^k\mathbb{Z}_p) = f(a) + p^k\mathbb{Z}_p$, for all $a \in \mathbb{Z}_p$ and all $k = 1, 2, \ldots$. Thus, the mapping $z \mapsto f(z) \bmod p^{k+1}$ $(k > r)$ permutes cyclically elements of the set

$$\begin{aligned}&\mathbf{S}_{p^{-r}}(y) \bmod p^{k+1}\\&\quad = \{y + p^r s + p^{r+1}S\colon s = 1, 2, \ldots, p-1;\ S = 0, 1, 2, \ldots, p^{k-r} - 1\}\end{aligned}$$

if and only if conditions 1 and 2 hold simultaneously for $t = k - r$. This proves the first part of the statement of the lemma, in view of Proposition 11.7. The second part of the statement is just an analogue of Proposition 11.7 for balls rather than for spheres. □

Note 11.9 Obviously, Lemma 11.8 holds for spheres of radii 1 as well, in the following form: *A 1-Lipschitz transformation* $f : \mathbf{S}_1(y) \to \mathbf{S}_1(y)$ *on the sphere*

$$\mathbf{S}_1(y) = \bigcup_{s\in\{0,\dots,p-1\}\setminus\{y\}} s + p\mathbb{Z}_p$$

is ergodic if and only if $f \bmod p$ *is transitive on the set* $\{0, \dots, p-1\} \setminus \{y\}$ *and* f^{p-1} *is an ergodic transformation on every (equivalently, some) ball* $\mathbf{B}_{p^{-1}}(s)$, $s \in \{0, \dots, p-1\} \setminus \{y\}$.

11.2.1 Ergodicity of $\mathcal{B}$-functions and of Analytic Functions

Recall that p-adic number $z \in \mathbb{Z}_p$ is called *primitive modulo* p^k if $z \bmod p^k$ generates the whole group $(\mathbb{Z}/p^k\mathbb{Z})^*$ of invertible elements of the residue ring $\mathbb{Z}/p^k\mathbb{Z}$ (cf. Sect. 1.2.2). Note that when $k > 2$, we speak on primitivity modulo p^k only for odd p (see Proposition 1.14).

Theorem 11.10 *Let the function* f *lie in* $\mathcal{B}$. *The function* f *is ergodic on the sphere* $S_{p^{-r}}(y)$ *of sufficiently small radius* p^{-r} *if and only if one of the following alternatives holds:*

1. *Whenever* p *is odd, then simultaneously:*
   - $f(y) \equiv y \pmod{p^{r+1}}$.
   - *The derivative* $f'(y)$ *is primitive modulo* p^2.
2. *Whenever* $p = 2$*, then simultaneously:*
   - $f(y) \equiv y \pmod{2^{r+1}}$.
   - $f(y) \not\equiv y \pmod{2^{r+2}}$.
   - $f'(y) \equiv 1 \pmod 4$.

Note 11.11 Within context of the theorem, the "sufficiently small" means that $r \geqslant 2$ if $p > 3$ or $r \geqslant 3$ if $p \leqslant 3$.

Proof of Theorem 11.10 As it immediately follows from Theorem 6.11, for every $g \in \mathcal{B}$ and all $k \in \mathbb{Z}_p$, $k = 1, 2, 3, \dots$ the equality

$$g(a + p^k h) = g(a) + g'(a)p^k h + p^{2k}h^2\hat{g}(h), \tag{11.11}$$

holds for a suitable C-function $\hat{g}$ of variable h. Of course, coefficients of series (6.5), which represents the function $p^{2k}g \in \mathcal{B}$, depend also on a and k, but this is of no importance at the moment.

Since $f(y) = y + p^r z$ for a suitable $z \in \mathbb{Z}_p$ in view of Proposition 11.5, from (11.11), we deduce the following equality

$$f(y+p^r s+p^{r+1}S) = f(y)+(p^r s+p^{r+1}S)f'(y)+p^{2r}(s+pS)^2\hat{w}(s+pS) = \\ = y + p^r z + p^r s f'(y) + p^{r+1}Sf'(y) + p^{2r}v(s) + p^{2r+1}w(S), \qquad (11.12)$$

where v, $\hat{w}$, and w are C-functions in respective variables and $r \geqslant 1$. Note that we have used (11.11) twice; with $g = f, a = y, p^k h = p^r s + p^{r+1}S$, for the first time, and with $g = w, a = s, p^k h = pS$, for the second time. Note also that w depends on s, yet this is of no importance in the argument that follows.

Iterating (11.12), we conclude that

$$f^{p-1}(y + p^r s + p^{r+1}S) = y + p^r z \sum_{i=0}^{p-2}(f'(y))^i \\ + p^r s(f'(y))^{p-1} + p^{r+1}S(f'(y))^{p-1} + \\ + p^{2r}\breve{v}(s) + p^{2r+1}\breve{w}(S) \qquad (11.13)$$

for suitable $\breve{v}$ and $\breve{w}$ from the class $\mathcal{B}$ since the both are compositions of C-functions.

Now, to satisfy condition 2 of Lemma 11.8, the ball $y + p^r s + p^{r+1}\mathbb{Z}_p$ must be invariant under action of f^{p-1}, and f^{p-1} must act ergodically on this ball. However, (11.13) implies that the ball is invariant if and only if

$$\sigma(z, s) = z\sum_{i=0}^{p-2}(f'(y))^i + s(f'(y))^{p-1} \equiv s \pmod{p}. \qquad (11.14)$$

Assuming the ball is invariant, we conclude that $\sigma(z, s) = s + p\gamma(z, s)$ for a suitable p-adic integer $\gamma(z, s)$. So, having s fixed, from (11.13), we conclude that under this assumption, the following equality holds:

$$f^{p-1}(y + p^r s + p^{r+1}S) = y + p^r s + p^{r+1}(\gamma(z, s) + (f'(y))^{p-1}S + \\ + p^{r-1}\breve{v}(s) + p^r\breve{w}(S));$$

Thus, to satisfy condition 2 of Lemma 11.8, the following $\mathcal{B}$-function

$$G_{z,s}(S) = \gamma(z, s) + S(f'(y))^{p-1} + p^{r-1}\breve{v}(s) + p^r\breve{w}(S) \qquad (11.15)$$

in variable S must be ergodic on $\mathbb{Z}_p$. Now, whenever $r > 1$ and $p > 3$ or whenever $r > 2$ and $p \leqslant 3$, from Corollary 10.23, we deduce that the $\mathcal{B}$-function $G_{z,s}(S)$

from (11.15) is ergodic on $\mathbb{Z}_p$ if and only if the polynomial

$$L_{z,s}(S) = \gamma(z, s) + p^{r-1}\breve{v}(s) + (f'(y))^{p-1}S \tag{11.16}$$

of degree 1 in variable S is transitive modulo p^2 for $p > 3$ or modulo p^3 for $p \leqslant 3$. But this, in view of Theorem 9.1 and (11.16), implies that $f'(y) \not\equiv 0 \pmod p$. Now, as $f(y) = y + p^r z$, from (11.12), it follows that to satisfy condition 1 of Lemma 11.8, the mapping $s \mapsto z + sf'(y) \pmod p$ must be transitive on the multiplicative group (i.e., on the whole group of units) $(\mathbb{Z}/p\mathbb{Z})^*$ of the field $\mathbb{Z}/p\mathbb{Z}$. Hence, $z \equiv 0 \pmod p$ (i.e., $f(y) \equiv y \pmod{p^{r+1}}$) since otherwise $s \mapsto 0 \pmod p$ for $s \equiv -\frac{z}{f'(y)} \pmod p$. From this moment, we consider the cases $p = 2$ and $p > 2$ separately.

Case 1: $p > 2$. In this case, the mapping $s \mapsto sf'(y) \pmod p$ is transitive on $(\mathbb{Z}/p\mathbb{Z})^*$ if and only if $f'(y)$ is a primitive element of the field $\mathbb{Z}_p$ (i.e., $f'(y)$ generates the cyclic group $(\mathbb{Z}/p\mathbb{Z})^*$).

Whenever this holds, every ball $y + p^r s + p^{r+1}\mathbb{Z}_p$, $s \in \{1, 2, \ldots, p-1\}$ is invariant under action of f^{p-1} in view of (11.14). Moreover, since $z \equiv 0 \pmod p$, in the case when $f'(y)$ is primitive modulo p, we have that $\sigma(z, s) \equiv s(f'(y))^{p-1} \pmod{p^2}$ and whence $\gamma(z, s) \equiv bs \pmod p$, where $(f'(y))^{p-1} = 1 + pb$, $b \in \mathbb{Z}_p$ (see (11.14) and the text thereafter for the definition of $\sigma(z, s)$ and $\gamma(z, s)$).

Now, the polynomial $L_{z,s}(S)$ (see (11.16)) in variable S is ergodic on $\mathbb{Z}_p$ (and thus condition 2 of Lemma 11.8 is satisfied) if and only if $b \not\equiv 0 \pmod p$ (see Theorem 9.1). Yet this means that $f'(y)$ must be a generator of the multiplicative group $(\mathbb{Z}/p^2\mathbb{Z})^*$.

Case 2: $p = 2$. In this case, the sphere $S_{2^{-r}}(y) = y + 2^r + 2^{r+1}\mathbb{Z}_2$ is a ball (see (11.10)). Moreover, the above condition $f'(y) \not\equiv 0 \pmod p$ means that $f'(y) \equiv 1 \pmod 2$, and so the condition that the mapping $s \mapsto sf'(y) \pmod p$ is transitive on the multiplicative group $(\mathbb{Z}/p\mathbb{Z})^*$, which just means that $z + f'(y) \equiv 1 \pmod 2$ in this case, is automatically satisfied since we have already proved that $z \equiv 0 \pmod p$ (i.e., that $z = pc$ for suitable $c \in \mathbb{Z}_p$), for any p.

Further, if the polynomial $L_{z,s}(S)$ in variable S is transitive modulo p^3, then $f'(y) \equiv 1 \pmod 4$ (see (11.16) and Theorem 9.1). That is, $f'(y) = 1 + 4b$ for some $b \in \mathbb{Z}_2$. Hence, $\gamma(z, s) = c + 2b$ (see (11.14) and the text thereafter), so in view of (11.16) and Theorem 9.1, if $L_{z,s}(S)$ is transitive modulo 8, then $c \equiv 1 \pmod 2$; that is, $f(y) = y + 2^r z = y + 2^{r+1}c \not\equiv y \pmod{2^{r+2}}$. This proves Theorem 11.10. □

Corollary 11.12 *Let $y \in \mathbb{Z}_p$ be a fixed point of the function $f \in \mathcal{B}$, and let p be odd. Then, f is ergodic on all spheres around y of sufficiently small radii if and only if f is ergodic on some sphere around y of a sufficiently small radius.*

From Theorem 11.10, we immediately derive complete characterization of C-functions, which are ergodic on p-adic spheres.

Theorem 11.13 *Let f be a C-function. Whenever p is odd, the mapping $z \mapsto f(z)$ is an ergodic transformation on every sufficiently small sphere centered at $y \in \mathbb{Z}_p$ if and only if the following two conditions hold simultaneously:*

- $f(y) = y$.
- *The derivative $f'(y)$ of the function f at the point $y \in \mathbb{Z}_p$ is primitive modulo p^2.*

In the case $p = 2$, no C-function exists such that the mapping $z \mapsto f(z)$ is ergodic on all spheres around $y \in \mathbb{Z}_2$ of radii less than ε, whatever $\varepsilon > 0$ is taken.

Proof This is obvious in view of Theorem 11.10 and Corollary 11.6. □

Corollary 11.14 (Cf., [33]) *Let p be an odd prime.*

- *Let map $M_{a,\ell}\colon \mathbb{Z}_p \to \mathbb{Z}_p$, be a map such that $M_{a,\ell}(z) = az^\ell$, where $\ell \in \mathbb{N}$, $\ell \not\equiv 1 \pmod p$, and $a \in \mathbf{B}_{p^{-1}}(1)$. The map $M_{a,\ell}$ has a unique fixed point $x_0 \in \mathbf{B}_{p^{-1}}(1)$; the map $M_{a,\ell}$ is ergodic on $\mathbf{S}_{p^{-r}}(x_0)$ if and only if ℓ is primitive modulo p^2.*
- *The affine map $T_{a,b}\colon z \mapsto az + b$, where $a, b \in \mathbb{Z}_p$, $a \neq 1$, has a fixed point $y = \frac{b}{1-a} \in \mathbb{Q}_p$. In the case when $y \in \mathbb{Z}_p$, the map $T_{a,b}$ is ergodic on $\mathbf{S}_{p^{-r}}(y)$ if and only if a is primitive modulo p^2.*

Here are some more examples of ergodic transformations on spheres.

Example 11.15 Let p be an odd prime. The function $f(x) = ax + (x^p - x)u(x)$, where u is a $\mathcal{B}$-function, is ergodic on all sufficiently small spheres around 0 if and only if $a - u(0)$ is primitive modulo p^2 (cf., [159]).

Example 11.16 Let p be an odd prime; let $\ell \in \mathbb{N}$ be primitive modulo p^2. Then the following assertions are true:

- The functions $f(x) = 1 + \ell(-1 + x + p^2 v(x))$ and $g(x) = \ell(ax + a^x - 2a) + 1$ are ergodic on all (sufficiently small) spheres around 1, for every $a \in 1 + p^2\mathbb{Z}_p$ and every $\mathcal{B}$-function v (cf., Example 9.12).
- (cf., Examples 4.3 and 10.26) The functions $f(x) = \ell x + \ln_p(1 + p^2 x)$ and $g(x) = \frac{\ell x}{1+p^2 x}$ are ergodic on all (sufficiently small) spheres around 0. Recall that $\ln_p$ is the p-adic logarithm.

The *perturbed monomial mapping* is the mapping of the form $x \mapsto x^\ell + u(x)$ where $u\colon \mathbb{Z}_p \to \mathbb{Z}_p$ is a "p-adically small" polynomial over $\mathbb{Z}_p$; that is, $|u(x)|_p \leqslant p^{-r}$ for some "sufficiently large" $r \in \mathbb{Z}_p$. In [86–88, 118, 140], the problem of ergodicity of perturbed monomial dynamics on p-adic spheres around 1 was posed. The next proposition from [12] gives a complete solution of the problem.

Proposition 11.17 *The perturbed monomial mapping $f\colon x \mapsto x^\ell + q(x)$, where $q(x) = p^{r+1}u(x)$ for some function $u \in \mathcal{B}$* (e.g., for a polynomial $u(x) \in \mathbb{Z}_p[x]$), *is ergodic on the sphere $\mathbf{S}_{p^{-r}}(1)$* (where $r > 1$) *if and only if ℓ is primitive modulo p^2.*

Proof The proof immediately follows from Theorem 11.10 in all cases but the case when $p = 3$ and $r = 2$. To handle this latter case, some extra efforts must be undertaken. Namely, for $p = 3$, in view of Theorem 6.11, we conclude that

$$f^2(1 + 3^r s + 3^{r+1} S) = f^2(1) + (3^r s + 3^{r+1} S) f'(f(1)) f'(1) + \\ + \frac{1}{2}(3^r s + 3^{r+1} S)^2 (f''(f(1))(f'(1))^2 + f'(f(1)) f''(1)) + 3^{3r+1} \hat{w}(S), \tag{11.17}$$

where $\hat{w}(S)$ is a $\mathcal{B}$-function in variable S. Now taking $f(x) = x^\ell + 3^{r+1} q(x)$, from (11.17), we derive that

$$f^2(1 + 3^r s + 3^{r+1} S) = 1 + (\ell + 1) 3^{r+1} u(1) + (3^r s + 3^{r+1} S) \ell^2 + \\ + \frac{1}{2}(3^r s + 3^{r+1} S)^2 \ell^2 (\ell - 1)(\ell + 1) + 3^{2r+1} v(s) + 3^{2r+2} w(S), \tag{11.18}$$

where v and w are $\mathcal{B}$-functions in variables s and S, respectively. However, ℓ must be primitive modulo 3 (see Case 2 of the proof of Theorem 11.10); so $\ell \equiv 2 \pmod 3$. Hence, $\ell^2 = 1 + 3b$ for a suitable $b \in \mathbb{Z}$. Also, $\ell(\ell - 1)(\ell + 1)$ is a multiple of 3; combining all this stuff with (11.18), we conclude that

$$f^2(1 + 3^r s + 3^{r+1} S) = 1 + 3^r s + 3^{r+1}(b + (\ell + 1) u(1) + S\ell^2 + \\ + 3^r \breve{v}(s) + 3^{r+1} \breve{w}(S)), \tag{11.19}$$

for suitable $\mathcal{B}$-functions $\breve{v}$ and $\breve{w}$. Now we must check whether the $\mathcal{B}$-function

$$L(S) = b + (\ell + 1) u(1) + S\ell^2 + 3^r \breve{v}(s) + 3^{r+1} \breve{w}(S)$$

is ergodic on $\mathbb{Z}_3$ (cf. (11.15)), where the residue term is $p^r \breve{w}(S)$ rather than $3^{r+1} \breve{w}(S)$ as in the case under consideration. The reason for this is that now extra factor 3 in the fourth term of (11.18) arises because of the multiplier $\ell(\ell - 1)(\ell + 1)$.

Applying Corollary 10.23 and Theorem 9.1 to the $\mathcal{B}$-function L in variable S, we see that L is ergodic on $\mathbb{Z}_p$ if and only if $b \not\equiv 0 \pmod 3$ since $(\ell + 1) q(1) \equiv 0 \pmod 3$; we remind that $\ell \equiv 2 \pmod 3$. Thus we finally conclude that ℓ must be primitive modulo p^2. □

Note that Proposition 11.17 remains true for $r = 1$ (see the proof for this case in [161]).

11.2.2 Ergodicity of $\mathcal{A}$-functions on Spheres

Some important functions (for instance, some compatible integer-valued polynomials over $\mathbb{Q}_p$, i.e., those polynomials that do not have necessarily integer p-adic coefficients, that map $\mathbb{Z}_p$ into itself, and that are 1-Lipschitz everywhere on $\mathbb{Z}_p$) and do not lie in $\mathcal{B}$, but they lie in a wider class $\mathcal{A}$ (see Sects. 6.4 and 6.3). Therefore, Theorem 11.10 is not applicable to these functions; nonetheless, we can adjust the proof of Theorem 11.10 in order to determine whether a $\mathcal{A}$-function is ergodic on a p-adic sphere.

Theorem 11.18 *The assertion of Theorem 11.10 remains true for $f \in \mathcal{A}$.*

Proof The definition of an $\mathcal{A}$-function implies that $f = \frac{1}{p^n}\bar{f}$ for a suitable $\mathcal{B}$-function $\bar{f}$ and suitable $n \in \mathbb{N}_0$ (see Sect. 6.4). Then with the use of Theorem 6.16, we can rewrite the key equation (11.11) of Theorem 11.10 in the following form:

$$g(a + p^k h) = g(a) + g'(a)p^k h + p^{2k-n}h^2\hat{g}(h), \tag{11.20}$$

where $g \in \mathcal{A}$, $p^n g \in \mathcal{B}$, $\hat{g} \in C$, and k is sufficiently large (so that $2k - n$ is positive). Then from (11.12), we obtain (for a sufficiently large r) that

$$f(y+p^r s+p^{r+1}S) = f(y)+(p^r s+p^{r+1}S)f'(y)+p^{2r-n}(s+pS)^2\hat{w}(s+pS) =$$
$$y + p^r z + p^r sf'(y) + p^{r+1}Sf'(y) + p^{2r-n}v(s) + p^{2r+1-n}w(S), \tag{11.21}$$

where v, $\hat{w}$, and w are C-functions in the respective variables. Now we assume that r is so large that the inequality $2r - n \geqslant r + 3$ holds and finish the proof in a manner similar to that of Theorem 11.10. □

It is worth noticing here that by virtue of Theorem 11.10, the perturbed monomial mappings on spheres are ergodic whenever the perturbations are "p-adically small" $\mathcal{B}$-functions (and even $\mathcal{A}$-functions), and not only "p-adically small" polynomials over $\mathbb{Z}_p$ as it is assumed in the original statement of the problem: For instance, whenever ℓ is primitive modulo p^2, perturbed monomial $x^\ell + \frac{1}{p}(x^p - x)^2$ is ergodic on all sufficiently small spheres though the perturbed monomial is an integer-valued polynomial over $\mathbb{Q}_p$ and *not* a polynomial over $\mathbb{Z}_p$. Here are some more examples of $\mathcal{A}$-functions that are *not* $\mathcal{B}$-functions and are ergodic on all sufficiently small spheres around 0, provided ℓ is primitive modulo p^2:

$$\ell x + \ln_p(1 + p^2 x) + \frac{1}{p}(x^p - x)^2; \quad \frac{\ell x}{1 + p^2 x} + \frac{1}{p}(x^p - x)^2$$

(cf., Example 11.16).

Chapter 12
Plots of 1-Lipschitz Functions in Euclidean Space

Abstract In this chapter, we introduce yet one more kind of representations of the p-adic 1-Lipschitz functions, namely, as plots in Euclidean space: in the unit n-dimensional hypercube $[0, 1]^n$ or, when convenient, on the surface of n-dimensional unit torus $\mathbb{T}^n$. Moreover, the representations may at the same time be considered in dual space, as functions on complex characters of $\mathbb{Q}_p$. The representations carry important information about measure properties of 1-Lipschitz functions; the information will be needed in Chaps. 13 and 16.

12.1 The Pontryagin-Type Duality for 1-Lipschitz Maps

We start with recalling (following [186]) some facts about *complex characters* of $\mathbb{Q}_p$: The character is a continuous group homomorphism of the additive group of $\mathbb{Q}_p$ to the circle group on a unit circle $\mathbb{S}$, which is a multiplicative group of all complex numbers whose moduli (i.e., absolute values) are 1. The latter group can be considered as a group of all complex numbers of the form $\mathrm{e}^{2\pi ir}$ where i is an imaginary unit (i.e., a complex number whose square is -1: $\mathrm{i}^2 = -1$) and $r \in [0, 1] \subset \mathbb{R}$. Actually, as $\mathrm{e}^{2\pi \mathrm{i}0} = \mathrm{e}^{2\pi \mathrm{i}1} = 1$, we assume that $r \in [0, 1) \subset \mathbb{R}$ (or $r \in (0, 1] \subset \mathbb{R}$, if convenient); however, we may also "identify" $\mathbb{S}$ and a unit interval $\mathbb{I} = [0, 1]$ by using the map $r \mapsto \mathrm{e}^{2\pi ir}$, which is bijective at all points r from $[0, 1]$ but the terminal ones, 0 and 1.

We will need a particular character, the continuous homomorphism (which actually is "onto," i.e., an epimorphism), which is given by the map $s \mapsto \mathrm{e}^{2\pi \mathrm{i}]s[_p}$, where $s \in \mathbb{Q}_p$ and $]s[_p$ is fractional part of s: Recall that the p-adic canonical representation of s is $s = \sum_{j=k}^{\infty} \xi_j p^j$ where $k \in \mathbb{Z}$ and $\xi_j \in \{0, 1, \dots, p-1\}$; so $]s[_p = \sum_{j=k}^{-1} \xi_j p^j$ (cf., Sect. 2.3.1).

Given a 1-Lipschitz map $f : \mathbb{Z}_p \to \mathbb{Z}_p$, we are going to define a "dual map" $\check{f}$ on $\mathbb{S}$. To do this, we first expand f to the whole domain $\mathbb{Q}_p$ as follows. Recall that any 1-Lipschitz map $f : \mathbb{Z}_p \to \mathbb{Z}_p$ admits a unique representation via coordinate functions (cf., Sect. 5.1.1), i.e., in the form $f(z) = \sum_{j=0}^{\infty} \delta_j(f(z))p^j$, where

V. Anashin, *Causality: The* p*-adic Theory*, STEAM-H: Science, Technology, Engineering, Agriculture, Mathematics & Health,
https://doi.org/10.1007/978-3-031-85818-5_12

$\delta_j(f(z)) = \psi_j(\delta_0(z), \delta_1(z), \ldots, \delta_j(z))$ is a map $\psi_j\colon \{0, 1, \ldots, p-1\}^{j+1} \to \{0, 1, \ldots, p-1\}$, $j = 0, 1, 2, \ldots$, provided the p-adic canonical expansion of z is $z = \sum_{j=0}^{\infty} \delta_j(z) p^j$ (cf., Definition 5.3 and Proposition 5.1).

Now, given $s \in \mathbb{Q}_p \setminus \mathbb{Z}_p$, let in representations $s = \sum_{j=k}^{\infty} \xi_j p^j$ the number k is taken the smallest such that $\xi_k \neq 0$; then this representation is unique for each $s \in \mathbb{Q}_p \setminus \mathbb{Z}_p$. Denote this $k < 0$ via $k(s)$; for $z \in \mathbb{Z}_p$, put $k(z) = 0$; define a map $\tilde{f}\colon \mathbb{Q}_p \to \mathbb{Q}_p$ by assuming

$$\tilde{f}(s) = \sum_{j=k(s)}^{\infty} \psi_{j-k(s)}(\xi_{-k(s)}, \xi_{-k(s)+1}, \ldots, \xi_j) p^j \tag{12.1}$$

Speaking simply, we just consider p-adic numbers from $\mathbb{Q}_p \setminus \mathbb{Z}_p$ as the ones having a finite fractional part and infinite integral part:

$$\cdots \xi_3 \xi_2 \xi_1 \xi_0 . \xi_{-1} \xi_{-2} \cdots \xi_k, \text{ where } \xi_k \neq 0 \tag{12.2}$$

and define the action of the automaton $\mathfrak{A}_f$ whose automaton function is f as the action on left-infinite words (12.2) without a dot, i.e., as if they were the words representing p-adic integers from $\mathbb{Z}_p$. Of course, the dot should be placed back in the former position in the output word produced by the automaton. If $s \in \mathbb{Z}_p$, the automaton processes the word according to the 1-Lipschitz map $f\colon \mathbb{Z}_p \to \mathbb{Z}_p$. So all the words that represent p-adic numbers from $\mathbb{Q}_p$ are processed by the causal map $\tilde{f}$ as described in Sect. 1.6.2.

Now we can define the *dual map* $\check{f}\colon \mathbb{S} \to \mathbb{S}$; namely

$$\check{f}(\mathrm{e}^{2\pi \mathrm{i} \{s\}_p}) = \mathrm{e}^{2\pi \mathrm{i} \{\tilde{f}(s)\}_p}, \text{ where } s \in \mathbb{Q}_p \tag{12.3}$$

This map can also be judged as an "order-reversal" Monna map (cf., the Monna map defined by (7.4)). Namely, under the bijection between $\mathbb{S}$ and $[0, 1) \subset \mathbb{R}$ mentioned above, the map $\check{f}$ acts on $[0, 1)$ as follows:

$$\check{f}\colon \sum_{j=k}^{-1} \xi_j p^j \mapsto \sum_{j=k}^{-1} \psi_{j-k}(\xi_k, \xi_{k+1}, \ldots, \xi_j) p^j, \tag{12.4}$$

where $k = -1, -2, -3, \ldots$, $\psi_\ell\colon \{0, 1, \ldots, p-1\}^{\ell+1} \to \{0, 1, \ldots, p-1\}$ are coordinate functions of f, $\xi_\ell \in \{0, 1, \ldots, p-1\}$, $\ell \in \mathbb{N}_0$. In the rest of the book, we use the same symbol $\check{f}$ for the dual map of a 1-Lipschitz map $f\colon \mathbb{Z}_p \to \mathbb{Z}_p$ in both cases, as a map $\mathbb{S} \to \mathbb{S}$ or as a map $[0, 1) \to [0, 1)$, without risk of ambiguity.

Finally, we give one more definition of the dual map $\check{f}$, defined as follows:

$$\frac{z \bmod p^n}{p^n} \mapsto \frac{f(z) \bmod p^n}{p^n}, \quad \text{where } n \in \mathbb{N}, z \in \mathbb{Z}_p \tag{12.5}$$

Note that the both maps (12.4) and (12.5) are defined only on rational numbers from $[0, 1)$, which are expressed by simple fractions from c/p^n, where $c \in \{0, 1, \ldots, p^n - 1\}$ and $n \in \mathbb{N}$; these points are in one-to-one correspondence with p^n-the roots of unity on which the map (12.3) is defined. With respect to this correspondence, all the three maps coincide.

The important fact about the duals of 1-Lipschitz maps is that they are in a one-to-one correspondence with the 1-Lipschitz maps; namely, $\check{f} = \check{g}$ for 1-Lipschitz maps $f, g\colon \mathbb{Z}_p \to \mathbb{Z}_p$ if and only if $f = g$. Thus, we may judge the dual $\check{f}$ of a 1-Lipschitz map $f\colon \mathbb{Z}_p \to \mathbb{Z}_p$ as a Fourier-image of f.

This is evident since the 1-Lipschitz maps $f, g\colon \mathbb{Z}_p \to \mathbb{Z}_p$ coincide on $\mathbb{Z}_p$ if and only if they coincide as maps on $\mathbb{N}_0$ or, in other terms, as maps on finite words $\mathcal{W}$ over the alphabet $\mathcal{A} = \{0, 1, \ldots, p-1\}$, but from the definitions given above, it immediately follows that actually, the duals are uniquely defined, namely, by their actions on these finite words.

The duality presented in this section will serve as an important tool in the rest of the book, mostly in Chaps. 13 and 16. In the next section, we will address only one application of this tool related to distributions produced by 1-Lipschitz maps, which we discussed in Sect. 7.2.2. To do this, we need yet one more important notion that we will use further in the rest of the book, the notion of a *plot* of a 1-Lipschitz map.

Given a 1-Lipschitz function $f\colon \mathbb{Z}_p \to \mathbb{Z}_p$, the maps (12.4) (or, which is the same, (12.5)) and (12.3) define the following sets of points in the Euclidean unit square $\mathbb{I}^2 = [0, 1] \times [0, 1] \subset \mathbb{R}^2$

$$\mathcal{E}(f) = \left\{ e_k^f(z) = \left(\frac{z \bmod p^k}{p^k}; \frac{f(z) \bmod p^k}{p^k} \right) \in \mathbb{I}^2 \colon z \in \mathbb{Z}_p, k = 1, 2, \ldots \right\}$$

and, respectively, on the unit torus $\mathbb{T}^2 = \mathbb{S} \times \mathbb{S}$

$$\check{\mathcal{E}}(f) = \left\{ \check{e}_k^f(z) = \left(e^{2\pi i]p^{-k} z[_p}; e^{2\pi i]p^{-k} f(z)[_p} \right) \in \mathbb{T}^2 \colon z \in \mathbb{Z}_p, k = 1, 2, \ldots \right\}$$

The unit square $\mathbb{I}^2$ is a universal cover of the torus $\mathbb{T}^2$; this way, the points $e_k^f(z) \in \mathbb{I}^2$ are identified with the points $\check{e}_k^f(z) \in \mathbb{T}^2$; so in what follows, we do not differ the both point sets $\check{\mathcal{E}}(f)$, $\mathcal{E}(f)$ and speak either of the points on the surface of the torus $\mathbb{T}^2$ or in the square $\mathbb{I}^2$ when and what is more convenient.

Definition 12.1 Given a 1-Lipschitz map $f\colon \mathbb{Z}_p \to \mathbb{Z}_p$, the closure $\mathcal{P}(f)$ of the point set $\mathcal{E}(f)$ in the square $\mathbb{I}^2$ (or of the point set $\check{\mathcal{E}}(f)$ in the torus $\mathbb{T}^2$) is called a (one-dimensional) *plot* of the 1-Lipschitz function f. The set $\mathcal{P}'(f)$ of all the limit points of the plot, the *derived set* of the set $\mathcal{P}(f)$, is called the *limit plot* of the 1-Lipschitz function f.

Along with the set $\mathcal{E}(f)$, we will consider the sets $\mathcal{E}_k(f)$, where $k \in \mathbb{N}$ is fixed:

$$\mathcal{E}_k(f) = \left\{\left(\frac{z \bmod p^k}{p^k}; \frac{f(z) \bmod p^k}{p^k}\right) \in \mathbb{I}^2 \colon z \in \mathbb{Z}_p\right\} \tag{12.6}$$

Therefore, $\mathcal{E}(f) = \cup_{k=1}^{\infty}\mathcal{E}_k(f)$.

As any 1-Lipschitz map $f\colon \mathbb{Z}_p \to \mathbb{Z}_p$ is an automaton function of some automaton $\mathfrak{A}$ whose input/output alphabet is $\{0, 1, \dots, p-1\}$, and vice versa, any automaton function of such automata is 1-Lipschitz maps $\mathbb{Z}_p \to \mathbb{Z}_p$ (cf., Theorem 1.25), we will also speak of plots $\mathcal{P}(\mathfrak{A}) = \mathcal{P}(f)$ (respectively, of limit plots $\mathcal{P}'(\mathfrak{A}) = \mathcal{P}'(f)$) of automata $\mathfrak{A}$, whose automaton function is the 1-Lipschitz map $f\colon \mathbb{Z}_p \to \mathbb{Z}_p$.

Recall that *limit point*, *accumulation point*, or *cluster point* is synonymic notions of the point such that every neighborhood of which contains points other than that point. Recall also that derived set of a closed set is also closed, so $\mathcal{P}'(f) = \mathcal{P}'(\mathfrak{A})$ is closed. *Being closed, the set $\mathcal{P}(\mathfrak{A})$ is measurable with respect to the Lebesgue measure on $\mathbb{R}^2$.* Respective notions in the general case, for 1-Lipschitz map $F = (f_1, \dots, f_m)\colon \mathbb{Z}_p^n \to \mathbb{Z}_p^m$, are defined as follows: For $\mathbf{z} = (z_1, \dots, z_n) \in \mathbb{Z}_p^n$, $k \in \mathbb{N}$, denote

$$e_k^F(\mathbf{z}) = \left(\frac{z_1 \bmod p^k}{p^k}, \dots, \frac{z_n \bmod p^k}{p^k}, \frac{f_1(\mathbf{z}) \bmod p^k}{p^k}, \dots, \frac{f_m(\mathbf{z}) \bmod p^k}{p^k}\right) \tag{12.7}$$

Then the notion of the (limit) plot can be defined via obvious analogy with Definition 12.1 by taking $\mathcal{E}(F) = \{e_k^F(\mathbf{z})\colon \mathbf{z} \in \mathbb{Z}_p^n, k \in \mathbb{N}\}$ instead of $\mathcal{E}(f)$. Respective notation in that case is $\mathcal{P}(F) = \mathcal{P}(\mathfrak{A})$, $\mathcal{P}'(F) = \mathcal{P}(\mathfrak{A})$, etc. Note that we use semicolon in the notation of points in the case $m = n = 1$ to avoid possible ambiguity with notation of open real interval and we use commas instead of semicolons in the rest of the cases since there is no risk of ambiguity in these cases.

The plots are much more informative compared to the graphs of Monna maps since in the latter case, more significant digits of the base-p representation of $f(z)$ play the leading role: For instance, Figs. 12.1, 12.2, 12.3, and 12.4, which are graphs of Monna maps of respective functions, are looking somewhat alike, whereas the plots of respective functions are quite different visually (cf. Figs. 12.5, 12.6, 12.7, and 12.8): we can see various geometrical structures there, such as straight lines, parabolas, stripes, etc. Note that actually the figures are graphs of $\mathcal{E}_k(f)$ for $k \in \mathbb{N}$ and therefore are pictures of how $\mathcal{E}_k(f)$ approaches $\mathcal{P}'(f)$ as k increases. This also

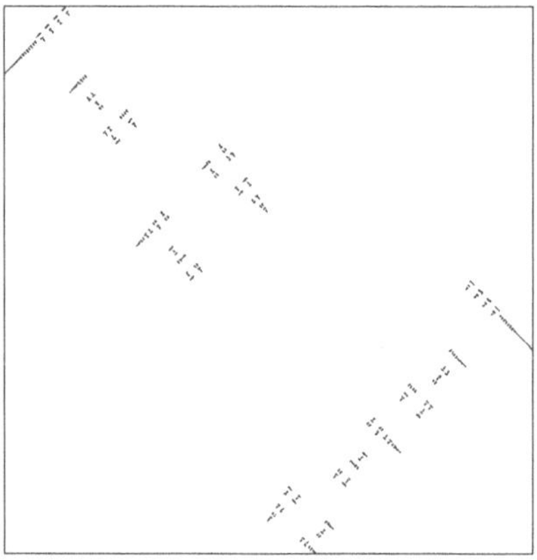

Fig. 12.1 The function $f\colon \mathbb{Z}_2 \to \mathbb{Z}_2$, $f(z) = z + z^2 \text{ OR } C$, $C = -131065$

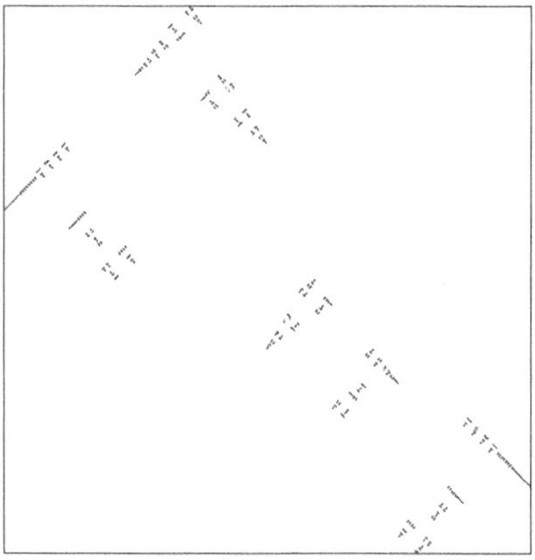

Fig. 12.2 The function $f(x) = x + x^2 \text{ OR } C$, $C = 5 = (101)_2$

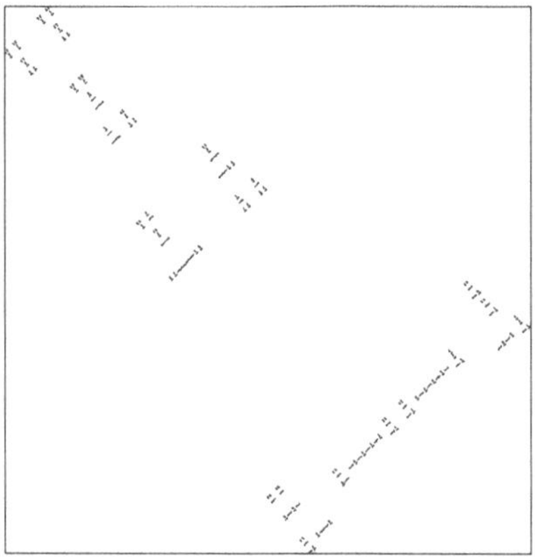

Fig. 12.3 Same function $f(z) = z + z^2 \text{ OR } C$, $C = (11101010100001001)_2$

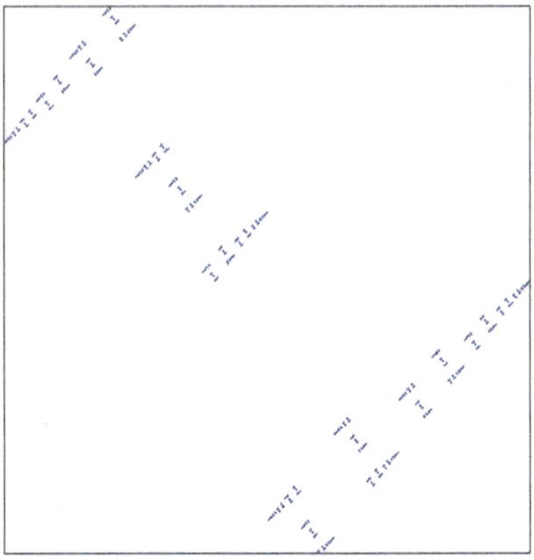

Fig. 12.4 Affine function $f: \mathbb{Z}_2 \to \mathbb{Z}_2$, $f(z) = 3 + 5z$

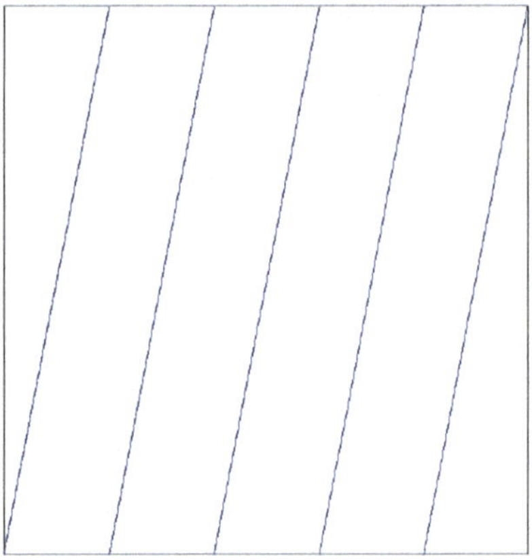

Fig. 12.5 Affine map $f: \mathbb{Z}_2 \to \mathbb{Z}_2$, $f(z) = 3 + 5z$

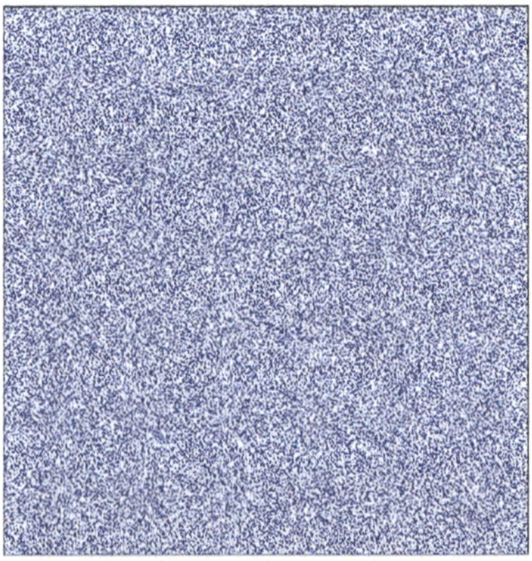

Fig. 12.6 Polynomial map $\mathbb{Z}_2 \to \mathbb{Z}_2$ of degree 8

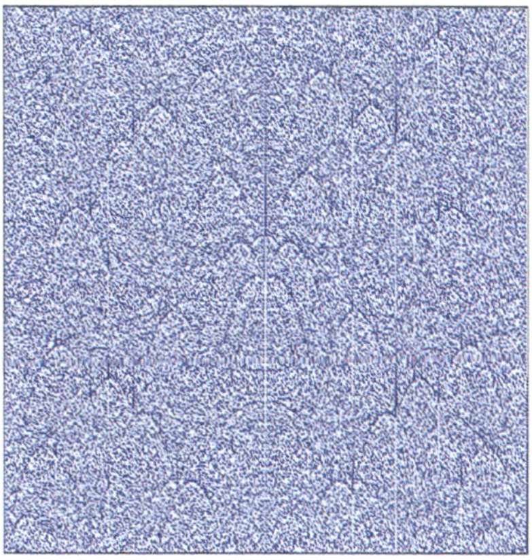

Fig. 12.7 The map $f : \mathbb{Z}_2 \to \mathbb{Z}_2$, $f(z) = z + z^2 \operatorname{OR} C$, $C = (101)_2$

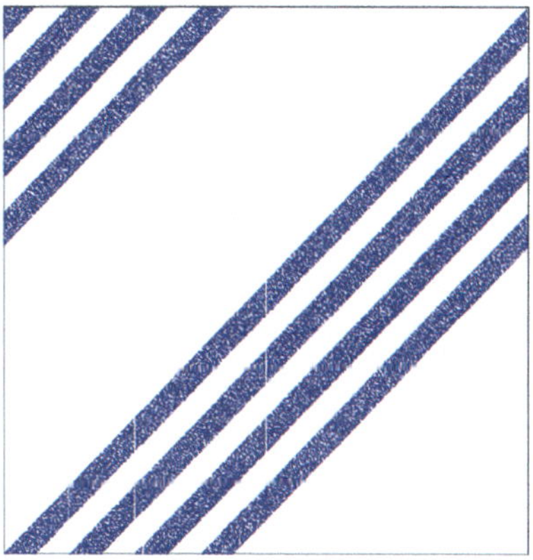

Fig. 12.8 Same map, $C = (10101010100001001)_2$

gives us valuable information about behavior of f. In what follows, we use notation $C = (\xi_0\xi_1\xi_2 \cdots \chi_\ell)_p$ if the base-p of $C \in \mathbb{N}_0$ is $\sum_{j=0}^{\ell} \chi_j p^j$; for instance, $C = (101)_2$ stands for the base-2 expansion of 5.

Moreover, some of these graphs exhibit strong dependence on k (see, e.g., Figs. 12.9, 12.10, 12.11, 12.12). From these figures, one may guess that the limit plot of the function $f(z) = z + z^2 \operatorname{OR} (-131065))$, whose domain and range are 2-adic integers, consists only of a segment of straight line $y = x$ in the unit square, and that guess is true: This is not difficult to deduce from the definition of the limit plot, taking into account that all the digits in the 2-adic canonical expansion of the number -131065 starting with 17-th are all 1.

In the next section, we address a concrete problem in order to demonstrate that the plots are capable to reveal details of behavior of 1-Lipschitz maps, which cannot be revealed by the graphs of the respective Monna maps. For many other examples, how the plots are working in application to various problems in automata theory, computer science, and quantum theory foundations, see Chaps. 13 and 16.

Fig. 12.9 The function $f(z) = z + z^2$ OR $(-131065))$, $k = 16$

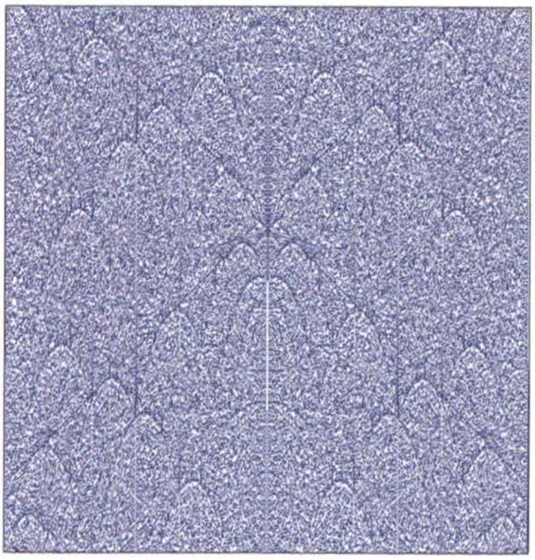

Fig. 12.10 The function $f(z) = z + z^2$ OR $(-131065))$, $k = 17$

Fig. 12.11 Same function, $k = 18$

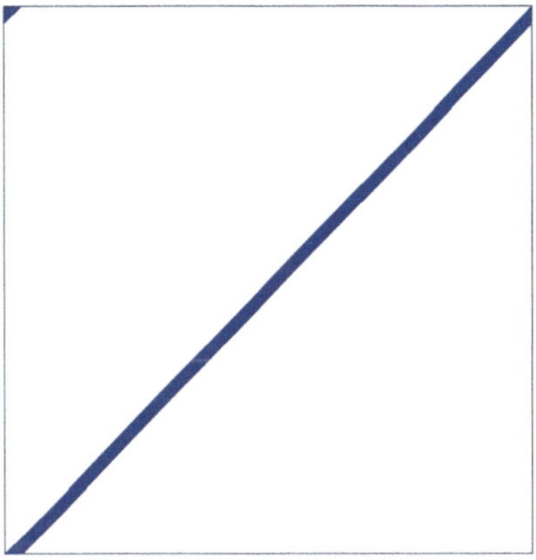

Fig. 12.12 Same function, $k = 22$

12.2 Points Falling on Hyperplanes

In this section, we explain, loosely speaking, what the straight lines in the graphs mentioned above mean. In more rigorous terms, given a 1-Lipschitz map $f: \mathbb{Z}_p \to \mathbb{Z}_p$, we would like to know what the *linear complexity* of the sequence of iterations $z, f(z), f^2(z), \ldots$ is. Here is the definition:

Definition 12.2 Let $\mathcal{Z} = (z_i)_{i=0}^{\infty}$ be a sequence over a commutative ring R. The *linear complexity* $\mathrm{L}_R(\mathcal{Z})$ of the sequence $\mathcal{Z}$ over R is the smallest $r \in \mathbb{N}_0$ having the property that there exist $c, c_0, c_1, \ldots, c_{r-1} \in R$ (not all equal to 0) such that the following equation holds for all $i = 0, 1, 2, \ldots$:

$$c + \sum_{j=0}^{r-1} c_j z_{i+j} = 0. \tag{12.8}$$

We say that $\mathrm{L}_R(\mathcal{Z}) = \infty$ if no such r exists.

From Chap. 8, we already know that a 1-Lipschitz map $f: \mathbb{Z}_p \to \mathbb{Z}_p$ is ergodic if and only if the orbit $(f^j(z))_{j=0}^{\infty}$ is dense, but it turns out that the linear complexity of the dense orbit strongly depends on f; namely, the linear complexity may be very small (e.g., equal to 2 if $f(z) = z + 1$ or high or even equal to infinity. Speaking loosely, the linear complexity of dense orbits of ergodic dynamical system $\langle \mathbb{Z}_p; f \rangle$ shows to what extent one ergodic dynamical system is more (or less) "chaotic" than another ergodic system. We postpone for Chaps. 13 and 16 a more general investigation of various aspects of "chaoticity" of dynamical systems $\langle \mathbb{Z}_p; f \rangle$, where f is ergodic; in the current section, we only show that ergodic polynomials of degree greater than 1 are "quite chaotic" since the respective orbits have infinite linear complexity over $\mathbb{Z}_p$ and over $\mathbb{Q}_p$.

We note that in this section, we use the notion of linear complexity of a sequence over a ring in a broader sense than it is commonly used (see, e.g., [62]). More often, the linear complexity of the sequence (x_n) of elements of a commutative ring R is understood as the smallest $r > 0$ such that there exist $c_0, \ldots, c_{r-1} \in R$, which satisfy simultaneously all equations $x_{n+r} = \sum_{j=0}^{r-1} c_j x_{n+j}$ for $n = 0, 1, 2, \ldots$. We, in distinction to the latter, consider non-homogeneous relations (i.e., with a nonzero constant term), as well as relations where all coefficients may be zero divisors (however, not all 0 simultaneously; in the assertion of Theorem 12.6 below, the latter restriction is not important). If R is a field, then both notions basically do not differ one from another: If a sequence satisfies the relation $c + \sum_{i=0}^{r} c_i x_{n+i} = 0$, where $c_r \neq 0$, then it satisfies the relation $x_{n+r+1} = c_r^{-1} c_0 x_n - \sum_{j=0}^{r-1} c_r^{-1}(c_j - c_{j+1})x_{n+j+1}$. Our definition is more convenient for geometric interpretations.

For instance, if $R = \mathbb{Z}/p^k\mathbb{Z}$; then geometrically, Eq. (12.8) means that all points $(\frac{z_i}{p^k}, \frac{z_{i+1}}{p^k}, \ldots, \frac{z_{i+r-1}}{p^k})$, $i = 0, 1, 2, \ldots$, of the unit r-dimensional Euclidean hypercube fall into parallel hyperplanes. Given a 1-Lipschitz ergodic transformation f on $\mathbb{Z}_p$, with the use of linear complexity over the residue ring $\mathbb{Z}/p^k\mathbb{Z}$, we can study distribution of r-tuples of the sequence $(f^i(x))_{i=0}^{\infty}$ modulo p^k. From Theorem 8.2, we know that *independently on what concrete transformation f is taken*, this sequence is strictly uniformly distributed as the sequence of elements from $\mathbb{Z}/p^k\mathbb{Z}$: The length of the shortest period is p^k, and every element from $\mathbb{Z}/p^k\mathbb{Z}$ occurs at the period exactly once. However, distribution of consecutive pairs of elements in this sequence (triples, etc.) varies depending on f. For example, although every orbit of affine ergodic transformation $f(z) = az + b$ on $\mathbb{Z}_p$ is strictly uniformly distributed modulo p^n for all $n \in \mathbb{N}$, the linear complexity over $\mathbb{Z}/p^k\mathbb{Z}$ of the orbit is only 2, as it immediately follows from (12.8). Hence, the points that correspond to pairs of consecutive residues fall into a small number of parallel straight lines in a unit square, and this picture does not depend on k (cf. Fig. 12.5).

Yet one more example: The transformation $f(z) = z + z^2 \ \mathsf{OR}\ C$ on $\mathbb{Z}_2$ (first considered in [121]) is ergodic if and only if $C \equiv 5 \pmod 8$ or $C \equiv 7 \pmod 8$ (k is fixed). However, the distribution of pairs in orbit degrade when k is increasing as the number of 1-s in higher-order bit positions of C increases (cf. Figs. 12.7 and 12.8).

Moreover, in some cases (e.g., when C is a negative rational integer), the distribution degenerates, whereas k unboundedly increases (see Figs. 12.9, 12.10, 12.11, 12.12); note that the limit plot (as $k \to \infty$) in this case will be the same as for the linear transformation $f(x) = x - 1$.

The central result of the section is that for ergodic polynomials of degree greater than 1, the respective linear complexities tend to infinity as $k \to \infty$ (cf. Fig. 12.6). In other words, the orbits have infinite linear complexities over $\mathbb{Z}_p$ (and over $\mathbb{Q}_p$). The following theorem is true, [11]:

Theorem 12.3 *Let $f(x) \in \mathbb{Q}_p[x]$ be an integer-valued 1-Lipschitz ergodic polynomial* (cf., Proposition 10.22) *of degree $\geqslant 2$. Then the linear complexity $\mathrm{L}_{\mathbb{Z}/p^k\mathbb{Z}}(\mathcal{X}_k)$ of the sequence $\mathcal{X}_k = (f^i(x_0) \bmod p^k)_{i=0}^{\infty}$ over $\mathbb{Z}/p^k\mathbb{Z}$ increases unboundedly as $k \to \infty$; i.e., $\lim_{k\to\infty} \mathrm{L}_{\mathbb{Z}/p^k\mathbb{Z}}(\mathcal{X}_k) = \infty$.*

We split the proof of this theorem into several assertions, which are of interest by themselves.

Proposition 12.4 *Let $f \in \mathbb{Q}_p[x]$ be an integer-valued 1-Lipschitz ergodic polynomial of degree d over a field $\mathbb{Q}_p$ of p-adic numbers; let $r \in \mathbb{N}$ be such that for each $k = 1, 2, \ldots$, there exist $c, c_0, \ldots, c_r \in \mathbb{Z}_p$ (not all congruent to 0 modulo p), which satisfy the following congruences:*

$$c + \sum_{i=0}^{r} c_i x_{n+i} \equiv 0 \pmod{p^k}, \qquad (n = 0, 1, 2, \ldots) \tag{12.9}$$

where $x_j = f^j(x_0)$, $x_0 \in \mathbb{Z}_p$, and $j = 0, 1, 2, \ldots$. Then $d = 1$.

To prove the proposition, we need the following lemma:

Lemma 12.5 *Under assumptions of Proposition 12.4, let $c, c_0, \ldots, c_r \in \mathbb{Z}_p$ not depend on k; that is, let there exist $c, c_0, \ldots, c_r \in \mathbb{Z}_p$, which satisfy (12.9) for all $k \in \mathbb{N}$ simultaneously. Then $d = 1$.*

Proof of Lemma 12.5 As f is ergodic, $d \neq 0$. Assume that $d > 1$. Consider $w(x) = c + \sum_{i=0}^{r} c_i f^i(x)$. As $w(x)$ is a composition of integer-valued 1-Lipschitz polynomials over $\mathbb{Q}_p$, $w(x) \in \mathbb{Q}_p[x]$ is an integer-valued 1-Lipschitz polynomial over $\mathbb{Q}_p$. However, $\deg f^i(x) = d^i$; whence, as $d > 1$, we conclude that $w(x)$, being a sum of polynomials of pairwise distinct degrees, must be a polynomial of a nonzero degree.

On the other hand, since $x_{n+i} \equiv f^i(f^n(x_0)) \pmod{p^k}$, assumptions of the lemma imply that $w(x_n) \equiv 0 \pmod{p^k}$ for all $n = 0, 1, 2, \ldots$. In other words, $w(z) \equiv 0 \pmod{p^k}$ for all $z \in \mathbb{Z}_p$ since x_n takes all values in $\{0, 1, \ldots, p^k - 1\}$ in view of the ergodicity of f and the 1-Lipschitzness of w. Assumptions of the lemma now imply that $w(z) \equiv 0 \pmod{p^k}$ for all $z \in \mathbb{Z}_p$ and all $k = 1, 2, \ldots$. Consequently, $w(z) = 0$ for all $z \in \mathbb{Z}_p$, and hence, the polynomial $w(x)$ must be 0 in the ring $\mathbb{Q}_p[x]$. This contradiction proves the lemma. □

Proof of Proposition 12.4 By assumption, for each $k \in \mathbb{N}$, the set $\mathcal{L}_k$ of all $\mathbf{c} = (c, c_0, \ldots, c_r) \in \mathbb{Z}_p^{r+2}$, such that $|\mathbf{c}|_p = 1$ and $c, c_0, \ldots, c_r$ satisfy (12.9), is not empty. Obviously, $\mathcal{L}_1 \supset \mathcal{L}_2 \supset \ldots$ since f is 1-Lipschitz.

Further, we assert that each set $\mathcal{L}_k$ is closed in the topology of the metric space $\mathbb{Z}_p^{r+2}$. Indeed, if $\mathbf{c} \in \mathcal{L}_k$, $\mathbf{c}' \in \mathbb{Z}_p^{r+2}$, $|\mathbf{c} - \mathbf{c}'|_p \leqslant p^{-s}$, $s \geqslant k$, then $\mathbf{c}' = \mathbf{c} + p^s\mathbf{z}$ for a suitable $\mathbf{z} \in \mathbb{Z}_p^{r+2}$. Hence, $|\mathbf{c}'|_p = 1$, and $\mathbf{c}'$ satisfies (12.9); consequently, $\mathbf{c}' \in \mathcal{L}_k$.

So we have a nested sequence $\mathcal{L}_1 \supset \mathcal{L}_2 \supset \ldots$ of non-empty closed subsets in $\mathbb{Z}_p^{r+2}$. It is well-known that in compact metric spaces, every nested sequence of non-empty closed subsets has a non-empty intersection (actually this property, the spherical completeness, is equivalent to compactness); thus, as $\mathbb{Z}_p^{r+2}$ is compact, the intersection of the sets $\mathcal{L}_1 \supset \mathcal{L}_2 \supset \ldots$ is not empty. That is, there exists $\mathbf{c}'' \in \mathbb{Z}_p^{r+2}$, which satisfies assumptions of Lemma 12.5. Yet then $d = 1$. □

Now we are able to prove the following theorem:

Theorem 12.6 *Let $f \in \mathbb{Q}_p[x]$ be an integer-valued 1-Lipschitz ergodic polynomial, let $\deg f > 1$, and let there exists $r \in \mathbb{N}$ such that for each $k \in \mathbb{N}$, the linear complexity over the ring $\mathbb{Z}/p^k\mathbb{Z}$ of the recurrence sequence $(x_n)_{n=0}^{\infty}$ defined by the recursion $x_{n+1} \equiv f(x_n) \pmod{p^k}$ does not exceed r. In other words, let there exist $c^{(k)}, c_0^{(k)}, \ldots, c_r^{(k)} \in \mathbb{Z}_p$ such that the following congruences hold:*

$$c^{(k)} + \sum_{i=0}^{r} c_i^{(k)} x_{n+i} \equiv 0 \pmod{p^k} \qquad (n = 0, 1, 2, \ldots). \tag{12.10}$$

Then $\lim\limits_{k\to\infty}^{p} c^{(k)} = \lim\limits_{k\to\infty}^{p} c_1^{(k)} = \ldots = \lim\limits_{k\to\infty}^{p} c_r^{(k)} = 0.$

Proof of Theorem 12.6 To start with, we note that from the proofs of both Lemma 12.5 and Proposition 12.4, it follows that they remain true if we let k under their assumptions range over arbitrary infinite subset of $\mathbb{N}$ rather than over the whole set $\mathbb{N}$.

Now for each $k \in \mathbb{N}$, take (and fix) $c^{(k)}, c_0^{(k)}, c_1^{(k)}, \ldots, c_r^{(k)} \in \mathbb{Z}_p^{r+2}$ that satisfy (12.10). Put $\mathbf{c}_k = (c^{(k)}, c_0^{(k)}, c_1^{(k)}, \ldots, c_r^{(k)}) \in \mathbb{Z}_p^{r+2}$. In view of Proposition 12.4, we have then $|\mathbf{c}_k|_p < 1$ for all sufficiently large $k \in \mathbb{N}$. Denote $\mathcal{N} = \{k \in \mathbb{N} \colon 1 > |\mathbf{c}_k|_p > p^{-k}\}$. In other words, $k \notin \mathcal{N}$ if and only if (12.10) is equivalent to the congruence $0 \equiv 0 \pmod{p^k}$. It is obvious that if $\mathcal{N}$ is finite, then the conclusion of the theorem is true. Let $\mathcal{N}$ be infinite.

For $k \in \mathcal{N}$, put $\hat{\mathbf{c}}_k = |\mathbf{c}_k|_p \mathbf{c}_k$ and denote by $\hat{\mathcal{N}}$ the set of all $m \in \mathbb{N}$ such that $p^k|\mathbf{c}_k|_p = p^m$ for a suitable $k \in \mathcal{N}$. In other words, we replace every set of congruences (12.10) with the equivalent system of congruences

$$\hat{c}^{(k)} + \sum_{i=0}^{r} \hat{c}_i^{(k)} x_{n+i} \equiv 0 \pmod{p^m} \qquad (n = 0, 1, 2, \ldots),$$

where $(\hat{c}^{(k)}, \hat{c}_0^{(k)}, \hat{c}_1^{(k)}, \ldots, \hat{c}_r^{(k)}) = \hat{\mathbf{c}}_k$, $p^m = p^k|\mathbf{c}_k|_p$.

If the set $\hat{\mathcal{N}}$ is finite, the conclusion of the theorem is obviously true. If $\hat{\mathcal{N}}$ is infinite, then, since $|\hat{\mathbf{c}}_k|_p = 1$, in view of Proposition 12.4 and the note at the beginning of the proof, we conclude that $\deg f = 1$, a contradiction. □

Now the proof of the theorem becomes easy:

Proof of Theorem 12.3 We just note that by Lemma 12.5, every orbit of the dynamical system $\langle \mathbb{Z}_p; f\rangle$ has infinite linear complexity over the ring $\mathbb{Z}_p$, provided $f \in \mathbb{Q}_p[x]$ is integer-valued 1-Lipschitz ergodic polynomial of degree $d > 1$, thus proving Theorem 12.3. □

This assertion can be slightly strengthened.

Corollary 12.7 *Under conditions of Theorem 12.3, every orbit of the dynamical system $\langle \mathbb{Z}_p; f\rangle$ has infinite linear complexity over the field $\mathbb{Q}_p$.*

Proof If for suitable $c, c_0, \ldots, c_r \in \mathbb{Q}_p$, which are not 0 simultaneously, the equality $c + \sum_{j=0}^{r} c_j x_{n+j} = 0$ holds for all $n = 0, 1, 2, \ldots$, then the equality $hc + \sum_{j=0}^{r} hc_j x_{n+j} = 0$, where $h = 1$ if $c, c_0, \ldots, c_r \in \mathbb{Z}_p$, and $h = |(c, c_0, \ldots, c_r)|_p$ otherwise holds either. As f is 1-Lipschitz, the conclusion now follows from Lemma 12.5. □

Note 12.8 In the statement of Theorem 12.3, the assumption that f is a polynomial over the field $\mathbb{Q}_p$ is essential.

Indeed, for instance, let $p = 2$, and let $f(x) = 1 + x + 4(-1)^{1+x}$. By Proposition 6.12, $(-1)^x$ (and thus f) are $\mathcal{B}$-functions, so by Corollary 9.9, the function f is ergodic since $\Delta(-1)^x = 2(-1)^{x+1}$. However, it is easy to see that $f^2(x) = x + 2$, so points on every orbit $(x_n = f^n(x_0))_{n=0}^{\infty}$ satisfy the relation $x_{n+2} = x_n + 2$; whence, the points satisfy the homogeneous linear relation $x_{n+4} - 2x_{n+2} + x_n = 0$. Therefore, the "non-homogeneous" linear complexity from Definition 12.2, both over the ring $\mathbb{Z}_2$ and over the field $\mathbb{Q}_2$ of the orbit, is 2; the standard "homogeneous" linear complexity of the orbit is 4.

Part III
Applications

In Part III, we present applications of the techniques developed in previous parts of the book to automata theory, computer science, combinatorics, and foundations of quantum theory.

Chapter 13
Applications to Automata Theory

Abstract In this chapter we show how the theory developed in preceding chapters can be effectively applied to examine behavior of automata as well as their "computational capabilities." Also, results of this chapter are used significantly in Chap. 16.

13.1 The p-Adic 1-Lipschitzness of Automata Maps

We start with a formal proof that the class of all automata functions $f_{\mathfrak{A}(s_0)}$ of initial automata $\mathfrak{A}(s_0) = \langle \mathbb{F}_p, \mathcal{S}, \mathbb{F}_p, S, O, s_0 \rangle$ whose initial state is s_0 (cf. Definition 1.22) coincides with the class of all 1-Lipschitz functions $f: \mathbb{Z}_p \to \mathbb{Z}_p$; that is, we prove Theorem 1.25 for the case when both input and output alphabets of the automaton consists of p symbols $\{0, 1, \ldots, p-1\}$. The result is not new: Seemingly the earliest proof of Theorem 1.25 was given in [167, Chapter IV, Theorem 8.2]. The theorem can also be derived from general result on asynchronous automata [84, Theorem 2.4, Proposition 3.7]; in special case $p = 2$, the result was proved in [183]. To make the book as self-consistent as possible, we give (following [16]) a direct "p-adic" proof here as in the book we consider only synchronous automata and for arbitrary p.

Theorem 13.1 *The automaton function $f_{\mathfrak{A}(s_0)}: \mathbb{Z}_p \to \mathbb{Z}_p$ of the automaton $\mathfrak{A}(s_0) = \langle \mathbb{F}_p, \mathcal{S}, \mathbb{F}_p, S, O, s_0 \rangle$ (cf. Sect. 1.6.1) is 1-Lipschitz. Conversely, for every 1-Lipschitz function $f: \mathbb{Z}_p \to \mathbb{Z}_p$ there exists an automaton $\mathfrak{A}(s_0) = \langle \mathbb{F}_p, \mathcal{S}, \mathbb{F}_p, S, O, s_0 \rangle$ such that $f = f_{\mathfrak{A}(s_0)}$.*

Proof Given a p-adic integer $z \in \mathbb{Z}_p$, denote via $\delta_i(z) \in \mathbb{F}_p$ the i-th digit of the canonical p-adic expansion of z; that is, the coefficient of the i-th term in the canonical representation $z = \sum_{i=0}^{\infty} \delta_i(z) p^i$ (cf. Sect. 2.3.1). As $s_i = S(\delta_{i-1}(z), s_{i-1})$ for every $i = 1, 2 \ldots$, the i-th output symbol $\xi_i = \delta_i(f_{\mathfrak{A}}(z))$ depends only on input symbols $\chi_0, \chi_1, \ldots, \chi_i$; that is,

$$\delta_i(f_{\mathfrak{A}}(z)) = \psi_i(\delta_0(z), \delta_1(z), \ldots, \delta_i(z))$$

V. Anashin, *Causality: The p-adic Theory*, STEAM-H: Science, Technology, Engineering, Agriculture, Mathematics & Health,
https://doi.org/10.1007/978-3-031-85818-5_13

for all $i = 0, 1, 2, \ldots$ and for suitable mappings $\psi_i : \mathbb{F}_p^{i+1} \to \mathbb{F}_p$. That is, $f = f_{\mathfrak{A}(s_0)} : \mathbb{Z}_p \to \mathbb{Z}_p$ is of the form

$$f : x = \sum_{i=0}^{\infty} \chi_i p^i \mapsto f(x) = \sum_{i=0}^{\infty} \psi_i(\chi_0, \ldots, \chi_i) p^i. \tag{13.1}$$

This means that the function $f_{\mathfrak{A}(s_0)}$ is 1-Lipschitz by Proposition 5.1 as the mentioned proposition in application to the mappings we consider here can be restated as follows: A mapping $f : \mathbb{Z}_p \to \mathbb{Z}_p$ is 1-Lipschitz if and only if f can be represented in the form (13.1) for suitable mappings $\psi_i : \mathbb{F}_p^{i+1} \to \mathbb{F}_p$, $i = 0, 1, 2, \ldots$.

Conversely, let $f : \mathbb{Z}_p \to \mathbb{Z}_p$ be a 1-Lipschitz mapping; then by Proposition 5.1, f can be represented in the form (13.1) for suitable mappings $\psi_i : \mathbb{F}_p^{i+1} \to \mathbb{F}_p$, $i = 0, 1, 2, \ldots$. We now construct an automaton $\mathfrak{A}(s_0) = \langle \mathbb{F}_p, \mathcal{S}, \mathbb{F}_p, S, O, s_0 \rangle$ such that $f_{\mathfrak{A}(s_0)} = f$.

Let $\mathbb{F}_p^\star$ be a set of all non-empty finite words over the alphabet $\mathbb{F}_p$. We consider these words as the base-p expansions of numbers from $\mathbb{N} = \{1, 2, 3, \ldots\}$ and enumerate all these words by integers $1, 2, 3, \ldots$ in radix order in accordance with the natural order on $\mathbb{F}_p$, $0 < 1 < 2 < \cdots < p - 1$:

$$\begin{aligned} &0 < 1 < 2 < \ldots < p - 1 < 00 < 01 < 02 \\ &< \ldots < 0(p-1) < 10 < 11 < 12 < \ldots; \end{aligned}$$

so that $\nu(0) = 1$, $\nu(1) = 2$, $\nu(2) = 3, \ldots, \nu(p-1) = p$, $\nu(00) = p + 1$, $\nu(01) = p + 2, \ldots$. This way we establish a one-to-one correspondence between the words $w \in \mathbb{F}_p^\star$ and integers $i \in \mathbb{N}$: $w \leftrightarrow \nu(w)$, $i \leftrightarrow \omega(i)$ ($\nu(w) \in \mathbb{N}$, $\omega(i) \in \mathbb{F}_p^\star$). Note that $\nu(\omega(i)) = i$, $\omega(\nu(w)) = w$ for all $i \in \mathbb{N}$ and all non-empty words from $w \in \mathbb{F}_p^\star$. Define $\omega(0)$ to be the empty word.

Now put $\mathcal{S} = \mathbb{N}_0 = \{0, 1, 2, 3, \ldots\}$, the set of all states of the automaton $\mathfrak{A}(s_0)$ under construction, and take the initial state $s_0 = 0$. The state transition function S is defined as follows:

$$S(r, i) = \nu(r \circ \omega(i)), \tag{13.2}$$

where $i = 0, 1, 2, \ldots$ and $r \in \mathbb{F}_p$. That is, $S(r, i)$ is the number of the word $r \circ \omega(i)$ which is a concatenation of the word $\omega(i)$ (the word whose number is i); the prefix, with the single-letter word r; and the suffix (cf. Sect. 1.4.1 for definitions and notation).

Now consider a one-to-one mapping $\theta_n(\chi_{n-1} \cdots \chi_1 \chi_0) = (\chi_0, \chi_1, \ldots, \chi_{n-1})$ from the n-letter words onto $\mathbb{F}_p^n$, and define the output function of the automaton $\mathfrak{A}(0)$ as follows:

$$O(r, i) = \psi_{\Lambda(\omega(i))}(\theta_{\Lambda(\omega(i))+1}(r \circ \omega(i))), \tag{13.3}$$

Fig. 13.1 Moore diagram of the automaton $\mathfrak{A}(0)$, $p = 2$; so $\omega(0)$ is the empty word, $\omega(1) = 0$, $\omega(2) = 1$, $\omega(3) = 00$, $\omega(4) = 01$, $\omega(5) = 10$, $\omega(6) = 11, \ldots$

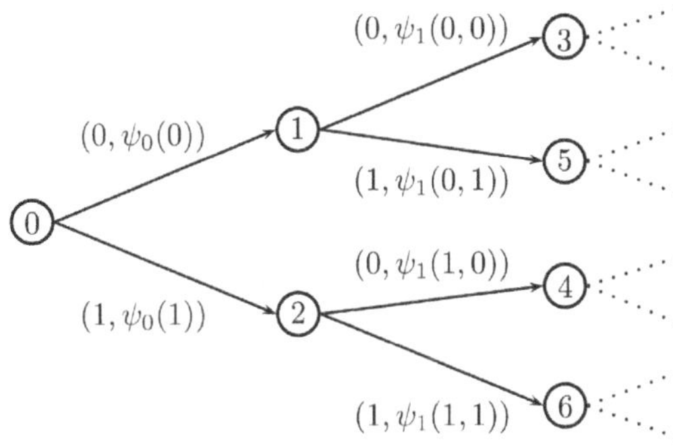

where $i = 0, 1, 2, \ldots$ and $r \in \mathbb{F}_p$. Recall that we denote via $\Lambda(w)$ the length of the word w.

The idea of the construction is illustrated by Fig. 13.1 which depicts Moore diagram of the automaton $\mathfrak{A}(0)$ for the case $p = 2$: In the figure, the edges are labeled by pairs (input letter, output letter).

Now, as both f and $f_{\mathfrak{A}(s_0)}$ are 1-Lipschitz, thus continuous with respect to the p-adic metric, and as $\mathbb{N}_0$ is dense in $\mathbb{Z}_p$, to prove that $f = f_{\mathfrak{A}(s_0)}$ suffices to show that

$$f_{\mathfrak{A}(s_0)}(\tilde{w}) \equiv f(\tilde{w}) \pmod{p^{\Lambda(w)}}) \tag{13.4}$$

for all finite non-empty words $w \in \mathbb{F}_p^\star$, where $\tilde{w} \in \mathbb{N}_0$ stands for the integer whose base-p expansion is w. We prove that (13.4) holds for all $w \in \mathbb{F}_p^\star$ once $\Lambda(w) = n > 0$ by induction on n.

If $n = 1$ then $\tilde{w} \in \mathbb{F}_p$; so once w is fed to the automaton $\mathfrak{A}$, the automaton reaches the state $S(w, 0) = \nu(w)$ (cf. (13.2)) and produces the output $O(w, 0) = \psi_0(\theta_1(w)) \equiv f(\tilde{w}) \pmod p$ (cf. (13.3)), see (13.1). Thus, (13.4) holds in this case.

Assume that (13.4) holds for all $w \in \mathbb{F}_p^\star$ such that $\Lambda(w) = n < k$ and prove that (13.4) holds also when $\Lambda(w) = n = k$. Represent $w = r \circ v$, where $r \in \mathbb{F}_p$ and $\Lambda(v) = n - 1$. By the induction hypothesis, after the word v has been fed to $\mathfrak{A}$, the automaton reaches the state $\nu(v)$ and produces the output word v_1 of length $n - 1$ such that $\tilde{v}_1 \equiv f(\tilde{v}) \bmod p^{n-1}$. Next, being feeded by the letter r, the automaton (which is in the state $\nu(v)$ now) outputs the letter $O(r, \nu(v)) = \psi_{\Lambda(\omega(\nu(v)))}(\theta_{\Lambda(\omega(\nu(v)))+1}(r \circ \omega(\nu(v)))) = \psi_{\Lambda(v)}(\theta_{\Lambda(v)+1}(r \circ v))$. This means that if being fed by w, the automaton $\mathfrak{A}(s_0)$ produces the output word $v_2 = (\psi_{\Lambda(v)}(\theta_{\Lambda(v)+1}(r \circ v))) \circ v_1$. Now note that $\tilde{v}_2 \equiv f(\tilde{w}) \pmod{p^n}$ by (13.1). □

Further in the chapter, given a 1-Lipschitz function $f : \mathbb{Z}_p \to \mathbb{Z}_p$, via $\mathfrak{A}_f(s_0)$ we denote an initial automaton $\langle \mathbb{F}_p, \mathcal{S}, \mathbb{F}_p, S, O, s_0 \rangle$ (cf. Definition 1.22) whose initial state is s_0 and whose automaton function is f; that is, $f_{\mathfrak{A}_f(s_0)} = f : \mathbb{Z}_p \to \mathbb{Z}_p$. Recall that given f, the automaton $\mathfrak{A}_f(s_0)$ is *not* unique (cf. Sect. 1.6); nonetheless, this non-uniqueness will not cause misunderstanding since in the chapter we are mostly interested with automata functions rather than with "internal structure" of respective automata such as state sets, state transition, output functions, etc. That is, further in this chapter we address properties of word transformations performed by automata, that is, by 1-Lipschitz maps $\mathbb{Z}_p^n \to \mathbb{Z}_p^n$.

We note that the p-adic approach (and wider, the non-Archimedean one) has already been successfully applied to automata theory. Seemingly the paper [144] is the first one where the p-adic techniques is applied to study automata functions; the paper deals with linearity conditions of automata maps. For application of the non-Archimedean methods to automata and formal languages, see expository paper [160] and references therein; for applications to automata and group theory (as well as to dynamics, spectral theory, ergodic theory, C^*-algebras, graph theory, random walks, actions on rooted trees), see [82, 84]. In [183–185], the 2-adic methods are used to study binary automata (the ones whose input/output alphabet is $\mathbb{F}_2$), in particular, to obtain the finiteness criterion for these automata in completely different terms than Theorems 13.19 and 13.22; see Sect. 13.5 further.

We stress once again that in literature, the term "automaton" is used in a very broad meaning, much more broad that the one we discuss in the book. Speaking of classical automata, we mean , the least powerful algorithms, from computational view. There exists a hierarchy of classes of automata over discrete time, with respect to their computational capabilities, the Turing machines (and, which is equivalent, cellular automata) being the most powerful; they can perform any algorithm. In between, there are automata with delays, push-down automata, asynchronous automata, etc. In the paper, we do not consider automata over discrete time other than the classical automata, the letter-to-letter transducers. We note, however, that some results on classical automata which are considered in the paper can be expanded to the classes of automata which are wider than the letter-to-letter transducers; see, e.g., [37, 176–178] an asynchronous automata and automata with delays. To emphasize the difference between synchronous and asynchronous automata, we mention that the mapping $\mathbb{Z}_p \to \mathbb{Z}_p$ is continuous (and not necessarily 1-Lipschitz) with respect to p-adic metric if and only if it is defined by some non-degenerate asynchronous automaton, [84, Theorem 2.4]; note that the class of all mappings defined in a similar way by Turing machines over a p-letter input alphabet constitute functions $\mathbb{Z}_p \to \mathbb{Z}_p$ which need not be defined everywhere on $\mathbb{Z}_p$, not speaking of other properties. Also, we do not touch numerous results on so-called p-adic scaling mappings and the likes which are continuous but not 1-Lipschitz mappings $\mathbb{Z}_p \to \mathbb{Z}_p$ and therefore are not automata functions defined by letter-to-letter transducers; we only mention here some of important papers in that area: [71, 119, 120, 192].

Concluding the subsection, we recall a yet one more class of automata, the so-called timed automata which are used in modeling of the hybrid systems, of smart contracts, verification, and other applications where computer programs interact with other devices or other programs in physical time which is usually assumed to be represented by real numbers. For instance, the smart contracts are computer programs representing legal commercial contracts, and the latter can be represented by classical automata, cf., [67]. The main problem of modeling behavior of smart contracts in digital economy as a whole is that smart contracts interact with each other in physical time; so that is why timed automata are currently considered as relevant mathematical models of the latter interaction. However, it is impossible to use real numbers in computer simulations, only approximations of the reals by rational numbers. Therefore a natural question arises what is precision of such models since actually in the models the automata over continuous time are replaced by classical automata over discrete time. This question was addressed in [20] where it was shown that timed automata can be approximated with any desirable accuracy by classical automata with multiple inputs/outputs. We refer the interested reader to paper [21] and references therein for further reading on this theme.

13.1.1 Automata Word Transformations

. We first note that from the proof of Theorem 13.1, it is clear that the automaton $\mathfrak{A}(s_0)$ maps n-letter words to n-letter words (which are treated as the base-p expansions of numbers from $\{0, 1, \dots, p^n - 1\} = \mathbb{Z}/p^n\mathbb{Z}$) according to the map $f_{n,\mathfrak{A}(s_0)}\colon \mathbb{Z}/p^n\mathbb{Z} \to \mathbb{Z}/p^n\mathbb{Z}$ which is just a reduction modulo p^n of the automaton function $f_{\mathfrak{A}(s_0)}\colon \mathbb{Z}_p \to \mathbb{Z}_p$. Namely, $f_{n,\mathfrak{A}(s_0)} = f_{\mathfrak{A}(s_0)} \bmod p^n$, where $n = 1, 2, 3, \dots$.

We consider only *accessible* automata; that is, given a state u, there exists a (finite) input word w such that after w has been fed to the automaton, the automaton reaches the state u. To every automaton of this kind, we associate a *family* $\mathfrak{A}(s)$, $s \in \mathcal{S}$, of *sub-automata* of the automaton which are the automata $\mathfrak{A}(s_0)$ which are the automata obtained by taking states $s \in \mathcal{S}$ of the automaton $\mathfrak{A}(s_0)$ as initial.

To put it in other words, along with initial automata $\mathfrak{A}(s_0)$, we will also consider *non-initial automata* $\mathfrak{A} = \langle \mathbb{F}_p, \mathcal{S}, \mathbb{F}_p, S, O \rangle$, cf. [82] where automata of this kind are also called *Mealy-type*. In the book, to avoid possible confusion, the non-initial automata will be referred also as to *discrete systems* or, for brevity, just as to *systems* (cf. Sect. 1.6). The justification of the latter term is as follows: According to the most general definition of a system in general mathematical systems theory (see, e.g., [109]), by a discrete system, people usually understand a stationary dynamical system with discrete time, that is, a five-tuple $\mathfrak{A} = \langle \mathcal{I}, \mathcal{S}, \mathcal{O}, S, O \rangle$ where $\mathcal{I}$ is a non-empty finite set, the input alphabet; $\mathcal{O}$ is a non-empty finite set, the output alphabet; $\mathcal{S}$ is a non-empty (possibly, infinite) set of states; $S\colon \mathcal{I} \times \mathcal{S} \to \mathcal{S}$ is a state transition function; and $O\colon \mathcal{I} \times \mathcal{S} \to \mathcal{O}$ is an output function.

That is, the stationary dynamical systems with discrete time are just non-initial automata in our terminology. Obviously, the system $\mathfrak{A}$ corresponds to a family $\mathcal{F}(\mathfrak{A})$ of all automata $\mathfrak{A}(s) = \langle \mathcal{I}, \mathcal{S}, \mathcal{O}, S, O, s \rangle$, $s \in \mathcal{S}$. To the latter family, we relate a family of automata functions $f_{\mathfrak{A}(s)}$, $s \in \mathcal{S}$.

Recall that a family $\mathcal{F}$ of transformations $M \to M$ of a finite non-empty set M is called *transitive* whenever given a pair $(a, b) \in M \times M$, there exists $f \in \mathcal{F}$ such that $f(a) = b$. Recall also that a single transformation $f\colon M \to M$ is said to be transitive whenever f is bijective and the family $\{\mathrm{id}, f, f^2, \ldots\}$ is transitive, where id is an identity map: $\mathrm{id}(r) = r$ for every $r \in M$; cf., Sect. 7.1.1.

The properties we are focused at in this chapter are mainly related to such properties of maps $f_{n,\mathfrak{A}(s_0)} = f_{\mathfrak{A}(s_0)} \bmod p^n$ on n-letter words as bijectivity and transitivity, as well as the respective properties of the families of maps defined by $\mathcal{F}_{\mathfrak{A}}$.

Also, we will pay special attention to the properties of finite automata maps, i.e., to word transformations which can be performed by finite automata, i.e., the automata having a finite set of states. This finiteness condition of the set of states of an automaton results in a number of important properties of the automaton function which we consider further; now we mention only a well-known one which immediately follows from the finiteness of the number of states of the automaton: If an input sequence of automaton $\mathfrak{A}$ having finite number of states is eventually periodic, then respective output sequence of $\mathfrak{A}$ is eventually periodic. In terms of properties of the automaton function $f_{\mathfrak{A}}$, this means that the automaton function maps rational p-adic integers to rational p-adic integers: $f_{\mathfrak{A}}(\mathbb{Z}_p \cap \mathbb{Q}) \subset \mathbb{Z}_p \cap \mathbb{Q}$ (cf. Proposition 2.10).

13.1.2 Reversibility of Automata

An initial automaton $\mathfrak{A} = \mathfrak{A}(s_0) = \langle \mathbb{F}_p, \mathcal{S}, \mathbb{F}_p, S, O, s_0 \rangle$ is called *n-reversible* (or, reversible on words of length n) whenever the reduction $f_{n,\mathfrak{A}} = f_{\mathfrak{A}} \bmod p^n$ modulo p^n of the automaton function $f_{\mathfrak{A}} \colon \mathbb{Z}_p \to \mathbb{Z}_p$ to the residue ring $\mathbb{Z}/p^n\mathbb{Z}$ is bijective. In other words, the automaton is n-reversible if and only if it performs an invertible (i.e., bijective) mapping of n-letter words to n-letter words. The automaton $\mathfrak{A}$ is called *reversible* (or, *invertible*, [82]) whenever its automaton function $f = f_{\mathfrak{A}}$ is bijective. The reversibility is also called *automata weak invertibility* [175]; we will not use this term in the book.

Speaking loosely, the reversibility of the automaton $\mathfrak{A}$ means that there exists another automaton $\mathfrak{B}$ which can work "in the opposite direction", that is, converts output words of the automaton $\mathfrak{A}$ back to input words of $\mathfrak{A}$; or, $f_{\mathfrak{A}} = f_{\mathfrak{B}}^{-1}$.

From Theorem 8.2, it follows that the automaton $\mathfrak{A}$ is reversible if and only if it is n-reversible for all $n = 1, 2, \ldots$, that is, if and only if the automaton function $f = f_{\mathfrak{A}} \colon \mathbb{Z}_p \to \mathbb{Z}_p$ is measure-preserving. Now, given an automaton function, to determine whether the automaton is reversible one may use various techniques we considered in Part II of the book.

It is worth mentioning here that the techniques can be also applied to determine reversibility of automata whose input and output alphabets are of prime power (rather than of just prime) order. In simple words, an automaton $\mathfrak{A}$ with input/output alphabet p^k can be considered as an automaton $\mathfrak{A}'$ with k input "channels" and k output "channels" over alphabet $\mathbb{F}_p$ up to a natural correspondence between letters from the p^k-letter alphabet and k-dimensional vectors $\alpha^{\downarrow}$ over $\mathbb{F}_p$. Theorem 13.1 in

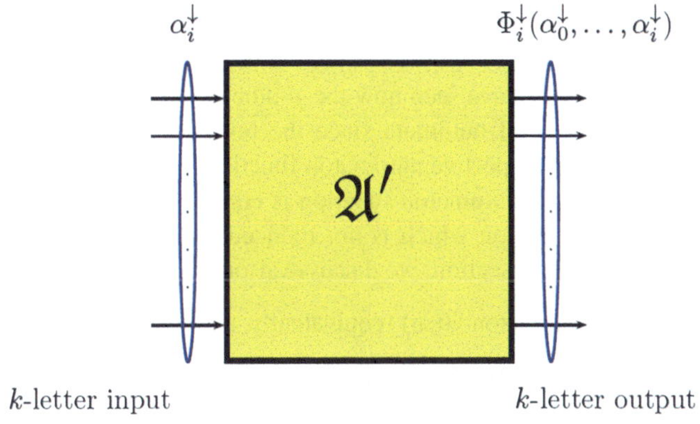

Fig. 13.2 Automaton with k input channels and k output channels over a p-letter alphabet

this case yields that the automata function is triangular (cf. Definition 5.3); that is, F can be considered as a k-variate 1-Lipschitz map $F: \mathbb{Z}_p^k \to \mathbb{Z}_p^k$. Thus, Theorem 8.2 holds for these automata maps; so the reversibility of the automaton is equivalent to measure-preservation of its automaton function (Fig. 13.2). Therefore to determine reversibility of the automaton, one may apply various techniques from Part II.

For instance, the following theorem is just a restatement of claim 3 of Theorem 10.1:

Theorem 13.2 *Let the automaton function $F = F_{\mathfrak{A}'}: \mathbb{Z}_p^k \to \mathbb{Z}_p^k$ be uniformly differentiable modulo p. Then the automaton $\mathfrak{A}'$ is reversible if and only if it is n-reversible for a sufficiently large n* (actually, for $n \geqslant N_1(F) + 1$).

Example 13.3 Let $\mathfrak{A}$ be an automaton over the alphabet $F = \{0, 1\}$ whose automaton function is $f = f_{\mathfrak{A}} = x + x^2 \ \mathsf{OR}\ C$, where $C \in \mathbb{Z}$. The automaton $\mathfrak{A}$ is reversible if and only if $C \equiv 1 \pmod 2$.

Actually the claim of this example is already proved; see Example 9.21. But there is another proof which does use differentiability of $f = f_{\mathfrak{A}}$ rather than Theorem 9.4. Namely, combining together Example 9.20 and Definition 3.5, it is easy to see that the function $f(x) = x + x^2 \ \mathsf{OR}\ C$ is uniformly differentiable on $\mathbb{Z}_2$, and its derivative is 1 and $N_1(f) = 1$. It is trivial to verify directly that the function f is bijective modulo 2 if and only is $C \equiv 1 \pmod 2$. Now the claim follows immediately from Corollary 10.4.

Yet one more example: The *Thue-Morse automaton* is an initial transducer $\mathfrak{A} = \langle \mathbb{F}_2, \mathbb{F}_2, \mathbb{F}_2, S, O, 0\rangle$, where $S = O$, and $S(0,0) = 0$, $S(1,0) = 1$, $S(0,1) = 1$, $S(1,1) = 0$; that is, S is just an addition in the two-element field $\mathbb{F}_2$: $S(x, y) = x + y$ for $x, y \in \mathbb{F}_2$. The Thue-Morse automaton is reversible; this is not difficult to prove by using either Theorem 9.4 or Theorem 13.2.

13.2 Transitivity of Automata

In Sect. 13.1.2 we already have seen how the p-adic ergodic theory can be applied to determine reversibility of automata since the reversibility is equivalent to the measure-preservation of respective automaton function. The same approach can be applied to determine if an automaton function is ergodic; however, the transitivity of automata is a wider notion which is *not* reduced only to the ergodicity of the automaton function. In this section, we discuss various aspects of transitivity.

Definition 13.4 An automaton $\mathfrak{A}(s_0)$ (equivalently, a system $\mathfrak{A}$) is called:

- n-*word transitive*, if the mapping $f_{n,\mathfrak{A}(s_0)}$ is transitive on the set $\mathcal{W}_n$ of all words of length n;
- *word transitive*, if $\mathfrak{A}(s_0)$ is n-word transitive for all $n \in \mathbb{N}$;
- *completely transitive*, if for every $n \in \mathbb{N}$, the family $f_{n,\mathfrak{A}(s)}$, $s \in \mathcal{S}$, is transitive on $\mathcal{W}_n$;
- *absolutely transitive*, if for every $s \in \mathcal{S}$ the automaton $\mathfrak{A}(s)$ is completely transitive; that is, if for every $n \in \mathbb{N}$, the family $f_{n,\mathfrak{A}(t)}$, $t \in \mathcal{S}_{\mathfrak{A}(s)}$, is transitive on $\mathcal{W}_n$, where $\mathcal{S}_{\mathfrak{A}(s)}$ is the set of all reachable states of the automaton $\mathfrak{A}(s)$.

The transitivity properties may be defined in equivalent way.

Definition 13.5 (Automata Transitivity, Equivalent)

1. The *word transitivity* means that given two finite words w, w' whose lengths are equal one to another, $\Lambda(w) = \Lambda(w') = n$, the word w can be transformed into w' by a sequential composition of a sufficient number of copies of $\mathfrak{A}(s_0)$ (cf. Fig. 13.3).
2. The *complete transitivity* means that given finite words w, w', $\Lambda(w) = \Lambda(w')$, there exists a finite word y (may be of length other than that of w and w') such that the automaton $\mathfrak{A}(s_0)$ transforms the input word $w \circ z$ (with the prefix y) to the output word $w' \circ y'$ that has a suffix w' (cf. Fig. 13.4).

Fig. 13.3 Word transitivity of automata, illustration

Fig. 13.4 Complete transitivity of automata, illustration

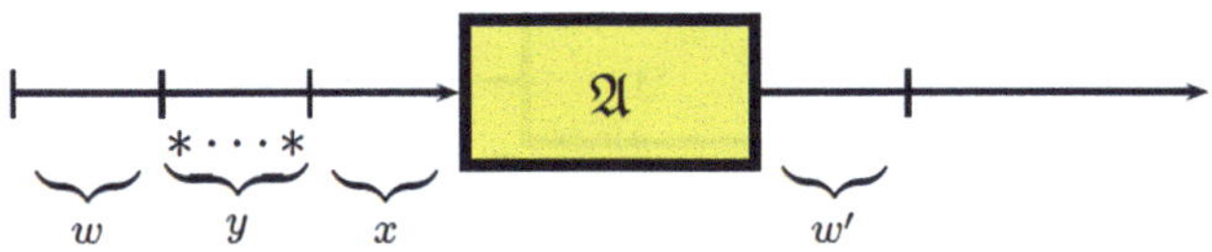

Fig. 13.5 Absolute transitivity of automata, illustration

3. The *absolute transitivity* means that given finite words x, w, w' such that $\Lambda(w) = \Lambda(w')$ (maybe, $\Lambda(x) \neq \Lambda(w)$), there exists a finite word y such that the automaton $\mathfrak{A}(s_0)$ transforms the input word $w \circ y \circ x$ to the output word $w' \circ y' \circ x'$ (cf. Fig. 13.5).

By Theorem 8.2, an automaton $\mathfrak{A} = \mathfrak{A}(s_0)$ is word transitive if and only if its automation function $f_{\mathfrak{A}}$ is ergodic; so to determine if the automaton is word transitive, one may apply various techniques developed in Part II. For instance, the following theorem is just a restatement of Theorem 10.9.

Theorem 13.6 *Let the automaton function $f = f_{\mathfrak{A}}: \mathbb{Z}_p \to \mathbb{Z}_p$ be uniformly differentiable modulo p^2. Then the automaton $\mathfrak{A}$ is word transitive if and only if it is n-word transitive for a sufficiently large n.*

For instance, by using the theorem, we immediately obtain a proof of main result of [121]:

Example 13.7 Let $p = 2$. Then an automaton whose automaton function is $f(x) = x + x^2 \mathsf{OR}\, 5$ is word transitive.

Indeed, the automaton function f is measure-preserving since the automaton is reversible, and $N_1(f) = 1$ (cf. Example 13.3); therefore by Theorem 10.9 it is sufficient to verify that the automaton is five-word transitive, i.e., the function $f \bmod 32$ is a permutation on the residue ring modulo 32 which is a single cycle (i.e., of length 32). This can be done by direct calculations of values of f modulo 32.

To determine complete/absolute transitivity of automata, we develop extra techniques in the sections which follow. These techniques exploit the dual map $\check{f}$ of the automaton function f (cf. Sect. 12.1).

13.3 Automata 0-1 Law

First, given the automaton function $f = f_{\mathfrak{A}}$ of an initial automaton $\mathfrak{A}$, we illustrate the notion of the dual map $\check{f}$ introduced in Sect. 12.1. Given an input word $w = \chi_{k-1} \cdots \chi_1 \chi_0$ and corresponding output word $w' = \xi_{k-1} \cdots \xi_1 \xi_0$, we consider in

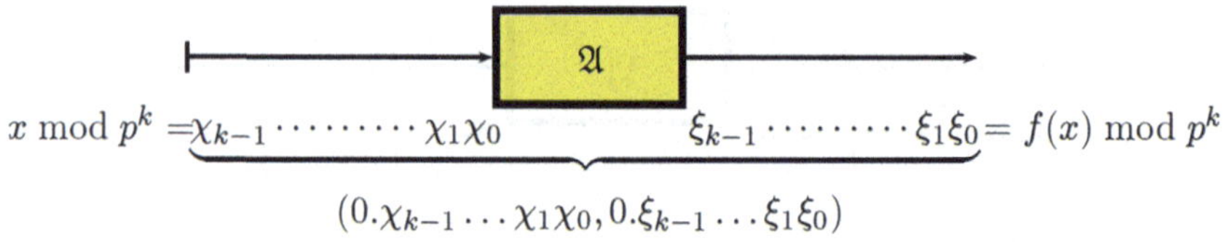

Fig. 13.6 The dual map of automaton $\mathfrak{A}$, illustration

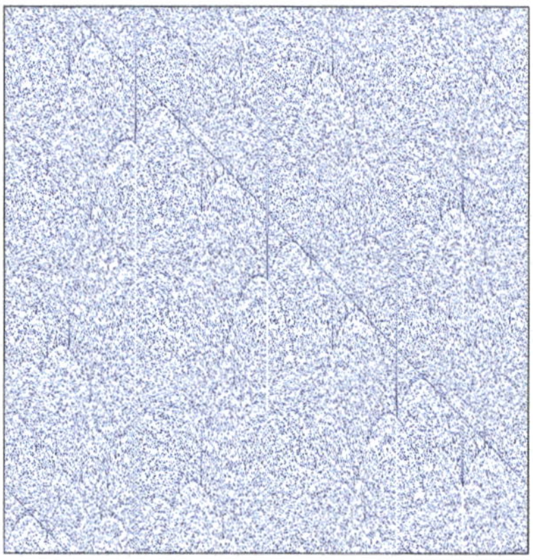

Fig. 13.7
$f(x) = 2x^2 + 3x + 1, k = 16$

the unit square $\mathbb{I}^2$ (where $\mathbb{I} = [0, 1]$ is the unit real segment) the set of all points

$$(\chi_{k-1}p^{-1} + \cdots + \chi_1 p^{-k+1} + \chi_0 p^{-k}; \xi_{k-1}p^{-1} + \cdots + \xi_1 p^{-k+1} + \xi_0 p^{-k}) = \\ = (0.\chi_{k-1} \ldots \chi_1\chi_0, 0.\xi_{k-1} \ldots \xi_1\xi_0)$$

for all pairs $(w; w')$ of input/output words of length k (cf. Fig. 13.6):

Taking $k = 1, 2, 3, \ldots$ we obtain the point sets $\mathcal{E}_k(f)$, cf. Sect. 12.1. It can be observed that the behavior of these sets while $k \to \infty$, i.e., when $\mathcal{E}_k(f)$ are approaching the plot $\mathcal{P}(f)$ (cf. Definition 12.1), can basically be of two types only:

1. The sets $\mathcal{E}_k(f)$ are getting more and more dense (cf. Figs. 13.7, 13.8, 13.9, 13.10, $p = 2$)
2. The sets $\mathcal{E}_k(f)$ are getting less and less dense while $k \to \infty$ (cf. Fig. 12.9–12.12 $(p = 2)$).

Now we explain this experimental phenomenon. By Definition 12.1, the plot $\mathcal{P}(f)$ is the closure of the set $\mathcal{E}(f) = \bigcup_{k=1}^{\infty} E_k(f)$ in the topology of real plane $\mathbb{R}^2$; therefore the plot is measurable with respect to the Lebesgue measure on the real

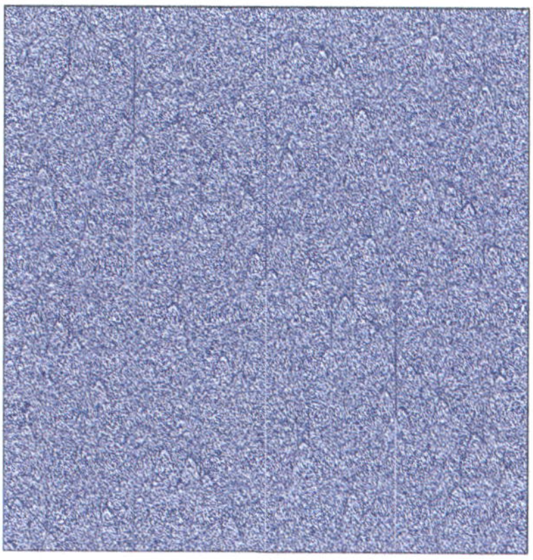

Fig. 13.8 Same function f, $k = 18$

Fig. 13.9 Same function f, $k = 20$

Fig. 13.10 Same function f, $k = 23$

plane $\mathbb{R}^2$. Let $\alpha(f)$ be the Lebesgue measure of the plot $\mathcal{P}(f)$, then obviously $0 \leqslant \alpha(f) \leqslant 1$. It turns out that the only two extreme cases occur:

Either the Lebesgue measure of the plot $\mathcal{P}(f)$ is zero, or the measure is 1. This is the *automata 0-1 law.*

Now we prove the automata 0-1 law following [16]:

Theorem 13.8 *For a 1-Lipschitz* (i.e., automaton) *map $f : \mathbb{Z}_p \to \mathbb{Z}_p$, the following alternative holds: either $\alpha(f) = 0$ (equivalently, $\mathcal{P}(f)$ is nowhere dense in $\mathbb{I}^2$) or $\alpha(f) = 1$ (equivalently, $\mathcal{P}(f) = \mathbb{I}^2$). Moreover, $\alpha(f) = 1$ if and only if the automaton $\mathfrak{A}_f$* (whose automaton function is f) *is completely transitive.*

Recall that nowhere dense sets can nevertheless have positive Lebesgue measures, (cf. fat Cantor sets (e.g., the Smith-Volterra-Cantor set), also known as ε-Cantor sets); see, for instance, [2].

Proof of Theorem 13.8 Let $\alpha(f) > 0$; we prove that then $\alpha(f) = 1$. Consider the following two cases separately:

- Case 1: There is a point in the plot $\mathcal{P}(f)$ such that some open neighborhood (in the unit square $\mathbb{I}^2 = [0, 1]^2$) of this point lies completely in $\mathcal{P}(f)$.
- Case 2: Any open neighborhood U of any point from $\mathcal{P}(f)$ contains points from $[0, 1]^2 \setminus \mathcal{P}(f)$ (i.e., $\mathcal{P}(f)$ is nowhere dense in $\mathbb{I}^2 = [0, 1]^2$).

Case 1: In this case, there exist u, v, u', v', $0 \leqslant u < v \leqslant 1$, $0 \leqslant u' < v' \leqslant 1$ such that the square $[u, v] \times [u', v'] \subset \mathbb{I}^2$ lies completely in $\mathcal{E}(f)$, and every point from the open real interval (u', v') is a limit (with respect to standard Archimedean metric in $\mathbb{R}$) of some sequence of fractions $u' < \frac{f(a_m) \bmod p^m}{p^m} < v'$, where $u < \frac{a_m}{p^m} < v$, $m = 1, 2, \ldots$. (recall that $z \bmod p^m$ stands for the least non-negative residue of z modulo p^m, i.e., $z \bmod p^m \in \{0, 1, \ldots, p^m - 1\}$). Thus, we can take $n \in \mathbb{N}$ and $w = \omega_0 + \omega_1 p + \cdots + \omega_{n-1} p^{n-1}$, where $\omega_i \in \{0, 1, \ldots, p-1\}$, $i = 0, 1, \ldots, n-1$, so that the square

$$S = \left[\frac{w}{p^n}, \frac{w}{p^n} + \frac{1}{p^n}\right] \times \left[\frac{f(w) \bmod p^n}{p^n}, \frac{f(w) \bmod p^n}{p^n} + \frac{1}{p^n}\right] \subset \mathbb{I}^2$$

lies completely in $\mathcal{P}(f)$, and every inner point $(x; y)$ of the square S (i.e., the one which has an open neighborhood that lies completely in S) is a limit as $j \to \infty$

(with respect to the standard Archimedean metric in $\mathbb{R}^2$) of a sequence of inner points

$$(r_j; t_j) = \left(\frac{z_j + p^{N_j} w}{p^{N_j+n}}; \frac{f(z_j + p^{N_j} w) \operatorname{mod} p^{N_j+n}}{p^{N_j+n}} \right) \in S,$$

where $N_j \in \mathbb{N}$, $z_j \in \{0, 1, \dots, p^{N_j} - 1\}$.

However, as f is a 1-Lipschitz map from $\mathbb{Z}_p$ to $\mathbb{Z}_p$, for every $z \in \{0, 1, \dots, p^N - 1\}$ we have that $f(z + p^N w) \equiv (f(z) \operatorname{mod} p^N) + p^N \xi_N(z) \pmod{p^{N+n}}$ for suitable $\xi_N(z) \in \{0, 1, \dots, p^n - 1\}$; thus,

$$\frac{f(z + p^N w) \operatorname{mod} p^{N+n}}{p^{N+n}} = \frac{f(z) \operatorname{mod} p^N}{p^{N+n}} + \frac{\xi_N(z)}{p^n}.$$

Hence, $\xi_{N_j}(z_j) = f(w) \operatorname{mod} p^n$ for all $j = 1, 2, \dots$ as all $(r_j; t_j)$ are inner points of S. Therefore, every inner point $(x; y) \in S$, which is then can be represented as

$$(x; y) = \left(\frac{w}{p^n} + \frac{\chi}{p^n}; \frac{f(w) \operatorname{mod} p^n}{p^n} + \frac{\gamma}{p^n} \right),$$

where χ and γ are real numbers, $0 < \chi < 1, 0 < \gamma < 1$, is a limit (as $j \to \infty$) of the point sequence

$$(r_j; t_j) = \left(\frac{w}{p^n} + \frac{z_j}{p^{N_j}} \times \frac{1}{p^n}; \frac{f(w) \operatorname{mod} p^n}{p^n} + \frac{f(z_j) \operatorname{mod} p^{N_j}}{p^{N_j}} \times \frac{1}{p^n} \right) \in S.$$

From here it follows that every inner point $(\chi; \gamma) \in \mathbb{I}^2$ is a limit point of the corresponding sequence of points $\left(\frac{z_j}{p^{N_j}}; \frac{f(z_j) \operatorname{mod} p^{N_j}}{p^{N_j}} \right)$ as $j \to \infty$. This means that $\mathcal{P}(f) = \mathbb{I}^2$ and thus $\alpha(f) = 1$.

Case 2: No point from $\mathcal{P}(f)$ has an open neighborhood that lies completely in $\mathcal{P}(f)$; i.e., any open neighborhood U of any point from $\mathcal{P}(f)$ contains points from the subset $\mathbb{I}^2 \setminus \mathcal{P}(f)$ which is open in $\mathbb{I}^2$.

Hence, U contains an open subset that lies completely in $\mathbb{I}^2 \setminus \mathcal{P}(f)$ (we assume that $\mathbb{I}^2 \setminus \mathcal{P}(f) \neq \emptyset$ since otherwise $\alpha(f) = 1$ and there is nothing to prove). Then there exists an open square

$$T_m(a, b) = \left(\frac{a}{p^m}, \frac{a}{p^m} + \frac{1}{p^m} \right) \times \left(\frac{b}{p^m}, \frac{b}{p^m} + \frac{1}{p^m} \right),$$

where $a, b \in \{0, 1, \ldots, p^m - 1\}$, that lies completely in $\mathbb{I}^2 \setminus \mathcal{P}(f)$. That is, $T_m(a, b)$ contains no points of the form

$$\left(\frac{x \bmod p^k}{p^k}; \frac{f(x) \bmod p^k}{p^k}\right),$$

where $x \in \mathbb{Z}_p$ and $k \in \mathbb{N}$.

In other words this means that there exist words $\tilde{a}, \tilde{b}$ of length m in the alphabet $\mathbb{F}_p$ (which are just the base-p representations of a and b, respectively) such that, whenever the automaton $\mathfrak{A} = \mathfrak{A}_f$ is fed by any input word $\tilde{w}$ having suffix $\tilde{a}$, i.e., $w = p^\ell a + u$ where $u \in \{0, 1, \ldots, p^\ell - 1\}$, the respective output word $f(w) \bmod p^{\ell+m} = p^\ell t + v$, $v \in \{0, 1, \ldots, p^\ell - 1\}$, newer has the suffix $\tilde{b}$, i.e., $t \neq b$ for all $\ell \in \mathbb{N}_0$ and all $u \in \{0, 1, \ldots, p^\ell - 1\}$ (u is the empty word if $\ell = 0$).

It is clear now that given any numbers $a', b' \in \{0, 1, \ldots, p^{m'} - 1\}$, $m' \geqslant m$, such that $a' \equiv a \pmod{p^m}$, $b' \equiv b \pmod{p^m}$, the corresponding open square $T_{m'}(a', b')$ lies completely outside of $\mathcal{P}(f)$, i.e., contains no points of the form $\left(\frac{x \bmod p^k}{p^k}; \frac{f(x) \bmod p^k}{p^k}\right)$, where $x \in \mathbb{Z}_p$ and $k \in \mathbb{N}$. Indeed, otherwise some input word w' with the suffix a' results in the output word with the suffix b'; however, this means that the corresponding initial subword (whose suffix is a) of the word w' results in output word whose suffix is b. The latter case contradicts our choice of a, b.

Now take $m' = im$ for $i = 1, 2, \ldots$ and construct inductively a disjoint family $\mathcal{T}_i$ of $(p^{2m} - 1)^{i-1}$ open squares $T_{m'}(a', b')$. The family $\mathcal{T}_1$ consists of the only member $T_m(a, b)$.

Given the family $\mathcal{T}_{i-1}$, the family $\mathcal{T}_i$ consists of all open squares $T_{im}(a', b')$, where $a', b' \in \{0, 1, \ldots, p^{im} - 1\}$, $a' \equiv a \pmod{p^m}$, $b' \equiv b \pmod{p^m}$, that are disjoint from all members of the families $\mathcal{T}_1, \ldots \mathcal{T}_{i-1}$.

That is, at the first step we obtain a family $\mathcal{T}_1$ which consists of the only $p^{-m} \times p^{-m}$ square $T_1(a, b)$; at the second step we obtain a family $\mathcal{T}_2$ that consists of $p^{2m} - 1$ disjoint $p^{-2m} \times p^{-2m}$ squares; on the third step we obtain a family $\mathcal{T}_3$ that consists of $(p^{2m} - 1)p^{2m} - (p^{2m} - 1) = (p^{2m} - 1)^2$ disjoint $p^{3m} \times p^{3m}$ squares, etc.

The union T of all these open squares from $\mathcal{T}_1, \mathcal{T}_2, \ldots$ is open, whence, measurable, and the Lebesgue measure of T is

$$\frac{1}{p^{2m}} + (p^{2m} - 1)\frac{1}{p^{4m}} + (p^{2m} - 1)^2\frac{1}{p^{6m}} + \cdots = 1$$

However, by the construction, T contains no points of the form

$$\left(\frac{x \bmod p^k}{p^k}; \frac{f(x) \bmod p^k}{p^k}\right),$$

where $x \in \mathbb{Z}_p$ and $k \in \mathbb{N}$. Consequently, $T \cap \mathcal{P}(f) = \emptyset$; in turn, this implies that the Lebesgue measure of $\mathcal{P}(f)$ must be 0, i.e., that $\alpha(f) = 0$ which contradicts our assumption; see the very beginning of the proof. It remains to note that the claim about absolute transitivity of the theorem under the proof immediately follows from the proof of Case 1, in view of equivalent definition of complete transitivity (cf. Definition 13.5). The proof of Theorem 13.8 is now completed. □

Note 13.9 Nowhere in the proof of Theorem 13.8 we used that p is a prime; so the theorem is true without this limitation.

In what follows, we will say for brevity that an *automaton/automaton function is of measure 0* (*or 1*) if Lebesgue measure of the plot is, respectively, 0 (or 1).

13.4 Conditions of Complete Transitivity

In this section we address the question when an automaton is completely transitive. According to Theorem 13.8, to determine if an automaton is completely transitive, one needs to determine whether the automaton is of measure 1 or not; so below we prove some sufficient conditions for an automaton to be of measure 0 or 1.

13.4.1 Finite Automata Are All of Measure 0

During the proofs we say that a 1-Lipschitz function $f : \mathbb{Z}_p \to \mathbb{Z}_p$ *has lacunas* whenever there exists a non-empty subset U of the unit square $\mathbb{I}^2$ which is open (in standard topology of $\mathbb{R}^2$) and contains no points of the form

$$\left(\frac{x \bmod p^n}{p^n}; \frac{f(x) \bmod p^n}{p^n}\right), \quad \text{where } x \in \mathbb{Z}_p, n = 1, 2, 3, \ldots$$

We call this open subset U an *f-lacuna*. We omit f- when this does not lead to ambiguity. According to Theorem 13.8, f has lacunas if and only if f is of measure 0. Thus, to prove that a 1-Lipschitz function is of measure 0, it suffices to demonstrate that it has a lacuna (cf. [24, Theorem 11.10]).

Theorem 13.10 *If a 1-Lipschitz function $f : \mathbb{Z}_p \to \mathbb{Z}_p$ is an automaton function of a finite automaton, then f is of measure 0.*

Proof To prove the theorem, it suffices to construct an f-lacuna. As f is 1-Lipschitz, it is clear that given $k \in \mathbb{N}$, for all $x \in \mathbb{Z}_p$, we can represent $f(x)$ as

$$f(x) = (f(x \bmod p^k)) \bmod p^k + p^k g_z(y), \tag{13.5}$$

where $y = \frac{1}{p^k}(x - (x \operatorname{mod} p^k)) \in \mathbb{Z}_p$, $z = x \operatorname{mod} p^k$, and $g_z \colon \mathbb{Z}_p \to \mathbb{Z}_p$ is a 1-Lipschitz function. Now, as f is the automaton function of a finite automaton $\mathfrak{A} = \langle \mathbb{F}_p, \mathcal{S}, \mathbb{F}_p, S, O, s_0 \rangle$, the number of pairwise distinct functions g_z is finite as actually g_z is an automaton function of the automaton $\mathfrak{A}_z = \langle \mathbb{Z}_p, \mathcal{S}, \mathbb{Z}_p, S, O, s(z) \rangle$, where $s(z) \in \mathcal{S}$ is the state the automaton $\mathfrak{A}$ reaches after being fed by the input word $z = x \operatorname{mod} p^k$. Therefore there exists $N \in \mathbb{N}$ such that for all $k > N$ the function $z = z_k(x)$ in the equality (13.5) takes values in a finite set, that is, the finite number of values the function $z_k(x)$ takes does not depend on k. Clearly, this number does not exceed p^N, where $N = \lceil \# \mathcal{S} \rceil$; we recall that all states of the automaton $\mathfrak{A}$ are assumed to be reachable from the initial state s_0.

Now we take $n > N$, fix arbitrarily $\alpha_0, \ldots, \alpha_{n-1} \in \{0, 1, \ldots, p-1\}$, and denote $a = \alpha_0 + \alpha_1 p + \cdots \alpha_{n-1} p^{n-1}$. There exist not more than p^N pairwise distinct numbers $g_z(a) \operatorname{mod} p^n \in \{0, 1, \ldots, p^n - 1\}$ as there exist not more than p^N pairwise distinct functions g_z. As $n > N$, there exists $b \in \{0, 1, \ldots, p^n - 1\}$ different from all these $g_z(a) \operatorname{mod} p^n$:

$$b = \beta_0 + \beta_1 p + \cdots + \beta_{n-1} p^{n-1},$$

for suitable $\beta_i \in \mathbb{F}_p$, $i = 0, 1, 2, \ldots, n-1$. That is, as $\mathfrak{A}$ is a finite automaton, given a sufficiently long word $\alpha_{n-1} \cdots \alpha_0$ over the alphabet $\mathbb{F}_p$, there exists a word $\beta_{n-1} \cdots \beta_0$ such that no output word (of length $n + K$, $K \geqslant N$) of the automaton $\mathfrak{A}$ has a suffix $\beta_{n-1} \cdots \beta_0$ whenever the input word (of length $n + K$) of the automaton has a suffix $\alpha_{n-1} \cdots \alpha_0$. Therefore given a number $a = \alpha_0 + \alpha_1 p + \cdots + \alpha_{n-1} p^{n-1}$ we have that if for some $x \in \mathbb{Z}_p$ and $L \geqslant N + n$

$$\frac{x \operatorname{mod} p^L}{p^L} \in I(a) = \left[\frac{p^N a}{p^{N+n}}, \frac{p^N - 1 + p^N a}{p^{N+n}} \right] = \left[\frac{a}{p^n}, \frac{a}{p^n} + \frac{p^N - 1}{p^{N+n}} \right],$$

then

$$\frac{f(x) \operatorname{mod} p^L}{p^L} \notin I(b) = \left[\frac{b}{p^n}, \frac{b}{p^n} + \frac{p^N - 1}{p^{N+n}} \right],$$

where $I(a)$, $I(b)$ are shortcuts for corresponding closed real segments. As only a finite number of rational numbers of the form $\frac{x \operatorname{mod} p^L}{p^L}$ (where $x \in \mathbb{Z}_p$) from the real closed segment $I(a)$ are such that $\frac{f(x) \operatorname{mod} p^L}{p^L} \in I(b)$ (may be, only those with $L < N + n$), the open real interval $I'(a) = \left(\frac{a}{p^n}, \frac{a}{p^n} + \frac{1}{p^{N+n}} \right) \subset I(a)$ contains no points of this kind. So $I'(a) \times I'(b) \subset \mathbb{I}^2$, where $I'(b)$ stands for the open real interval $\left(\frac{b}{p^n}, \frac{b}{p^n} + \frac{1}{p^{N+n}} \right)$, is an f-lacuna. □

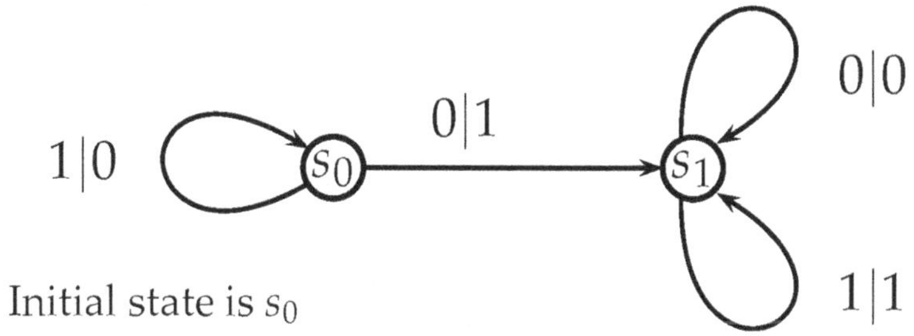

Fig. 13.11 The state transition diagram of the 2-adic odometer

We finally conclude from Theorems 13.10 and 13.8:

> There are no completely transitive automata among finite ones. The plots of finite automata are Lebesgue measure zero nowhere dense uncountable subsets of the unit square.

Nonetheless a finite automaton can be word transitive; the 2-adic odometer $x \mapsto x+1$ on $\mathbb{Z}_2$ serves as an example since this function is ergodic by Theorem 9.1 and is an automaton function of an automaton having only two states (cf. Fig. 13.11 and Sect. 1.6).

13.4.2 Complete Transitivity: Absolute Transitivity

In this subsection we, following [16], find sufficient conditions for a 1-Lipschitz function to be of measure 1 (thus, sufficient conditions for complete transitivity of respective automata), then we prove that automata whose automata functions are polynomials of degree greater than 1 over $\mathbb{Z}$ are all absolutely transitive (and thus all of measure 1). We keep same conventions on terminology as before.

Theorem 13.11 *Let $f: \mathbb{Z}_p \to \mathbb{Z}_p$ be a 1-Lipschitz function, and let f be differentiable everywhere in a ball $\mathbf{B} \subset \mathbb{Z}_p$ of a non-zero radius. The function f is of a measure 1 whenever the following two conditions hold simultaneously:*

1. $f(\mathbf{B} \cap \mathbb{N}_0) \subset \mathbb{N}_0$*;*
2. *f is two times differentiable at some point $v \in \mathbf{B} \cap \mathbb{N}_0$, and $f''(v) \neq 0$.*

Proof We will show that for every sufficiently large k and every $z, u \in \{0, 1, \ldots, p^k - 1\}$ there exists $M = M(k)$ and $a \in \{0, 1, \ldots, p^M - 1\}$ such that

$$\left|\frac{a}{p^M} - \frac{u}{p^k}\right| < \frac{1}{p^k} \text{ and } \left|\frac{f(a) \bmod p^M}{p^M} - \frac{z}{p^k}\right| < \frac{1}{p^k}. \tag{13.6}$$

This will prove the theorem as every point from the unit square $\mathbb{I}^2$ can be approximated by points of the form $\left(\frac{u}{p^k}; \frac{z}{p^k}\right)$.

Briefly, the idea of the proof is as follows: As $v \in \mathbb{N}_0$, there exists $k \in \mathbb{N}_0$ such that all coefficients $v_i \in \{0, 1, \ldots, p-1\}$ in the p-adic expansion $v = \sum_{i=0}^{\infty} v_i p^i$ are zero, for all $i \geqslant k$. We then replace zeros in this expansion at positions starting with ℓ-th, $\ell > k$, by certain other coefficients from $\{0, 1, \ldots, p-1\}$ so that the resulting natural number $a = v + p^\ell t$ satisfies inequalities (13.6) for some M.

As f is differentiable everywhere in $\mathbf{B}$, for $x \in \mathbf{B}$, we have that given arbitrary $K \in \mathbb{N}$, the following congruence holds for all $h \in \mathbb{Z}_p$ and all sufficiently large $L \in \mathbb{N}$:

$$f(x + p^L h) \equiv f(x) + p^L h f'(x) \pmod{p^{K+L}}. \tag{13.7}$$

Let $|f''(v)|_p = p^{-s}$; that is, $f''(v) = p^s \xi$, where $s \in \mathbb{Z}$ and ξ is a unity of $\mathbb{Z}_p$ (in other words, ξ has a multiplicative inverse in $\mathbb{Z}_p$). Note that s is not necessarily non-negative since $f''(v)$ is in $\mathbb{Q}_p$, and not necessarily in $\mathbb{Z}_p$; however, further in the proof we assume that $k + s > 0$ as we may take k large enough.

Now let $r \in \mathbb{N}$ be an arbitrary number such that $r > s$, $p^r > v$, and p^{-r} is less than the radius of the ball $\mathbf{B}$ (it is clear that there are infinitely many choices of r). Given r, consider $n \in \mathbb{N}$ such that $n > \max\{\log_p f(v + p^{k+r} t) \colon t = 0, 1, 2, \ldots, p^k - 1\}$ and $n > 2k + 2r + 2s$ (we remind that in view of condition 2 of the theorem, all $f(v + p^{k+r} t)$ are in $\mathbb{N}_0$ due to our choice of n). Put

$$\tilde{u} = 1 + p^{k+r+s} u \tag{13.8}$$

$$\tilde{z} = f'(v) + p^{k+r+s} \hat{z}, \tag{13.9}$$

where $\hat{z} \in \{0, 1, \ldots, p^k - 1\}$ is such that $\lfloor \frac{\tilde{z}}{p^{k+r+s}} \rfloor \bmod p^k = z$. In other words, we choose $\hat{z}$ in such a way that the number whose base-p expansion stands in positions from $(k + r + s)$-th to $(2k + r + s - 1)$-th in the canonical p-adic expansion of $\tilde{z}$, is equal to z. Obviously, given $f'(v)$ and z, there exists a unique $\hat{z}$ that satisfy this condition: $\hat{z} \equiv z - \lfloor \frac{f'(v)}{p^{k+r+s}} \rfloor \pmod{p^k}$; so

$$\tilde{z} \bmod p^{2k+r+s} = (f'(v) \bmod p^{k+r+s}) + p^{k+r+s} z. \tag{13.10}$$

As f is two times differentiable at v, for every $\zeta \in \{0, 1, \ldots p^k - 1\}$ we conclude that

$$f'(v + p^{r+k}\zeta) \equiv f'(v) + p^{r+k}\zeta f''(v) \pmod{p^{2k+r+s}} \tag{13.11}$$

for all sufficiently large r (formally, we just substitute f' for f, v for x, ζ for h, $k+s$ for K, and $r+k$ for L in (13.7)). From here we deduce that as f is differentiable in $\mathbf{B}$, the following congruence holds for all sufficiently large n:

$$\begin{aligned} f(v + p^{r+k}\zeta + p^n\tilde{u}) \equiv f(v + p^{r+k}\zeta) + \\ + p^n\tilde{u}(f'(v) + p^{r+k}\zeta f''(v)) \pmod{p^{n+2k+r+s}}. \end{aligned} \tag{13.12}$$

Note that the latter congruence is obtained by combination of congruence (13.7) where $K = 2k+r+s$, $x = v+p^{r+k}\zeta$, $h = \tilde{u}$ and $L = n$, with congruence (13.11).

We claim that there exists $\zeta \in \{0, 1, \ldots p^k - 1\}$ such that

$$\tilde{u}(f'(v) + p^{r+k}\zeta f''(v)) \equiv \tilde{z} \pmod{p^{2k+r+s}}. \tag{13.13}$$

Indeed, in view of (13.8)–(13.9) this congruence is equivalent to the congruence $(1 + p^{k+r+s}u)(f'(v) + p^{r+k}\zeta f''(v)) \equiv f'(v) + p^{k+r+s}\hat{z} \pmod{p^{2k+r+s}}$, and the latter congruence is equivalent to the congruence $f'(v) + p^{r+k}\zeta f''(v) \equiv (1 - p^{k+r+s}u)(f'(v) + p^{k+r+s}\hat{z}) \pmod{p^{2k+r+s}}$ as $(1 + p^{k+r+s}u)^{-1} \equiv 1 - p^{k+r+s}u \pmod{p^{2k+r+s}}$. That is, congruence (13.13) is equivalent to the congruence $p^{k+r}\zeta f''(v) \equiv p^{k+r+s}\hat{z} - p^{k+r+s}uf'(v) \pmod{p^{2k+r+s}}$. However, as $f''(v) = p^s\xi$, the latter congruence is equivalent to the congruence $\zeta\xi \equiv \hat{z} - uf'(v) \pmod{p^k}$. From here we find $\zeta \equiv \xi^{-1}(\hat{z} - uf'(v)) \pmod{p^k}$, thus proving our claim (we remind that ξ is a unity of $\mathbb{Z}_p$; hence, ξ has a multiplicative inverse ξ^{-1} modulo p^k).

Now we put $M = n + 2k + r + s$ and $a = v + p^{r+k}\zeta + p^n(1 + p^{k+r+s}u)$; then

$$\frac{a}{p^M} = \frac{u}{p^k} + \frac{v + p^{r+k}\zeta + p^n}{p^{n+2k+r+s}},$$

so $\left|\frac{a}{p^M} - \frac{u}{p^k}\right| < \frac{1}{p^k}$, since $v < p^r$, $\zeta < p^k$, and $n > 2r + 2s + 2k$. However, at the same time, combining (13.13), (13.9), (13.10), and (13.12), we see that

$$\frac{f(a) \operatorname{mod} p^M}{p^M} = \frac{z}{p^k} + \frac{f(v + p^{r+k}\zeta)}{p^n} \times \frac{1}{p^{2k+r+s}} + \frac{f'(v) \operatorname{mod} p^{k+r+s}}{p^{k+r+s}} \times \frac{1}{p^k}, \tag{13.14}$$

since $f(a) \operatorname{mod} p^M = f(v + p^{r+k}\zeta) + p^n(f'(v) \operatorname{mod} p^{k+r+s}) + p^{n+k+r+s}z$ (the number in the right-hand side is less than p^M due to our choice of n). Now from

(13.14) it follows that $\left|\frac{f(a) \bmod p^M}{p^M} - \frac{z}{p^k}\right| < \frac{1}{p^k}$ since $0 \leqslant f(v + p^{r+k}\zeta) \leqslant p^n - 1$ due to our choice of n. □

Note 13.12 We note that $\alpha(f(x)) = \alpha(-f(x)) = \alpha(f(-x))$ for every 1-Lipschitz function $f\colon \mathbb{Z}_p \to \mathbb{Z}_p$ of variable x; so we may replace condition 1 of Theorem 13.11 by either of conditions $f(\mathbf{B} \cap -\mathbb{N}_0) \subset \mathbb{N}_0$, $f(\mathbf{B} \cap \mathbb{N}_0) \subset -\mathbb{N}_0$, or $f(\mathbf{B} \cap -\mathbb{N}_0) \subset -\mathbb{N}_0$, where $\mathbb{N}_0 = \{0, -1, -2, \ldots\}$.

Indeed, for every $c \in \mathbb{N}$ and every $n \in \mathbb{N}$, we have that $\frac{-c \bmod p^n}{p^n} = \frac{p^n - (c \bmod p^n)}{p^n} = 1 - \frac{c \bmod p^n}{p^n}$. Thus, a symmetry with respect to the axis $y = \frac{1}{2}$ of the unit square $\mathbb{I}^2 \subset \mathbb{R}^2$ maps the subset

$$\mathcal{E}(f) = \left\{\left(\frac{x \bmod p^n}{p^n}; \frac{f(x) \bmod p^n}{p^n}\right) : x \in \mathbb{Z}_p, n \in \mathbb{N}\right\} \subset \mathbb{I}^2$$

onto the subset $\mathcal{E}(-f)$ and vice versa; so $\alpha(f(x)) = \alpha(-f(x))$. A similar argument proves that $\alpha(f(x)) = \alpha(f(-x))$.

Theorem 13.13, which we are going to prove, yields:

Automata whose automata functions are polynomials of degree $\geqslant 2$ over $\mathbb{Z}$ are not only completely transitive (as this immediately follows from Theorem 13.11); they are absolutely transitive.

Theorem 13.13 *If automaton function $f = f_{\mathfrak{A}}$ of an automaton $\mathfrak{A}$ is a univariate polynomial of degree $\geqslant 2$ with rational integer coefficients, then the automaton $\mathfrak{A}$ is absolutely transitive.*

Proof As f is a polynomial, f has not more than a finite number of zeros in $\mathbb{R}$, so there exists $d \in \mathbb{N}_0$ such that for all $b \geqslant d$ either values $f(b)$ are all positive or they are all negative. It suffices to consider only the case when all $f(b) > 0$. Indeed, let $\mathfrak{A} = \langle \mathbb{F}_p, \mathcal{S}, \mathbb{F}_p, S, O, s_0\rangle$; consider the automaton $\mathfrak{A}' = \langle \mathbb{F}_p, \mathcal{S}, \mathbb{F}_p, S, \sigma(O), s_0\rangle$, where for $\chi \in \mathbb{F}_p$

$$\sigma(\chi) = \begin{cases} \chi, & \text{if } \chi \notin \{0, p-1\}; \\ 0, & \text{if } \chi = p-1; \\ p-1, & \text{if } \chi = 0. \end{cases}$$

As the map σ is just a permutation on the alphabet $\mathbb{F}_p$, it is clear that the automaton $\mathfrak{A}$ is absolutely transitive if and only if the automaton $\mathfrak{A}'$ is absolutely transitive (cf. equivalent definition of absolute transitivity in terms of words). However, under the alphabet substitution σ, negative rational integers (the ones in whose canonical p-adic representations there are only finitely many coefficients distinct from $(p - 1)$) will change to non-negative rational integers (the ones in whose canonic p-adic representations there are only finitely many non-zero coefficients) and vice versa.

Given a finite non-empty word $\tilde{g}$ over alphabet $\mathbb{F}_p$, take a finite word $\tilde{v}$ whose prefix is $\tilde{g}$ and such that the corresponding non-negative rational integer v (the one whose base p-expansion is $\tilde{v}$) is greater than d, $v > d$, and $f''(v) \neq 0$ (recall that according to our conventions words are read from right to left, that is the rightmost letters of $\tilde{v}$ correspond to low order digits in the base-p expansion of v). The word $\tilde{v}$ that satisfies these three conditions simultaneously exists as f'' is a polynomial (and thus has not more than a finite number of zeros); fixing arbitrary $\tilde{g}$ means that only some less significant digits in the base-p expansion of v are fixed; so the condition $v > d$ can be also satisfied just by taking v whose base-p expansion is sufficiently long. Therefore both f and the so constructed v satisfy conditions of Theorem 13.11.

Now note that the claim stated at the very beginning of the proof of Theorem 13.11 is just a restatement of item 2 from Definition 13.5: Indeed, in the notation of Definition 13.5 and the one from the beginning of the proof of Theorem 13.11, the concatenation $w \circ y$ corresponds to a, w corresponds to u, w' corresponds to z, and w is a k-letter suffix of the output word which is the base-p expansion for $f(a) \bmod p^M$, whereas M is the length of the word $w \circ y$. Up to these correspondences, condition (13.6) is equivalent to the statement of item 2 from Definition 13.5. Furthermore, as the word $\tilde{v}$ has an arbitrarily chosen prefix $\tilde{g}$, and as the condition (13.6) holds for $a = v + p^{\ell}t$ from the proof of Theorem 13.11 (as the whole Theorem 13.11 holds for f and v), this means that statement of item 2 from Definition 13.5 holds for input word with arbitrarily chosen prefix $\tilde{g}$, up to all mentioned correspondences. However, this means that the statement of item 3 from Definition 13.5 also holds for $x = \tilde{g}$ in the case under consideration. This finally proves Theorem 13.13. □

Combining Theorem 13.11 with results of Sects. 6.3.3 and 9.5.1, it is not difficult to construct examples like that one:

Example 13.14 The automata whose automata functions are as follows are completely transitive:

- $f(x) = cx + c^x$ if $c \in \{2, 3, 4, \ldots,\}$ and $c \equiv 1 \pmod{p}$;
- $f(x) = (x \text{ AND } c) + ((x^2) \text{ OR } c)$ if $c \in \mathbb{Z}$ and $p = 2$.

We note that the following properties of automata (and of their automata functions) functions are independent:

- to be ergodic/non-ergodic;
- to be of measure 0/of measure 1.

Indeed, by Theorem 13.13, a polynomial of degree not less than 2 with rational integer coefficients is of measure 1; however, it may be ergodic or non-ergodic depending on its coefficients (cf. Theorem 10.9). A polynomial of degree 1 over $\mathbb{Z}$ is of measure 0; it may be ergodic or non-ergodic depending on its coefficients (cf. Theorem 9.1). The function $f\colon \mathbb{Z}_2 \to \mathbb{Z}_2$ such that $f(z) = z + z^2 \mathsf{OR}\, C$ and $C = -131065$ is ergodic since $-131065 \equiv 7 \pmod 2$ (cf. Example 9.21), but the function is of measure 0 since its limit plot is a straight line $y = x$; see the example in Sect. 12.1.

With the use of Theorem 13.11, new types of absolutely transitive automata can be constructed as well. The following corollary from Theorems 13.11 is a key tool in the construction of these.

Corollary 13.15 *Let an automaton function $f = f_{\mathfrak{A}(s_0)}$ map $\mathbb{N}_0$ into $\mathbb{N}_0$, let f be two times differentiable on $\mathbb{Z}_p$, and let $f''(x)$ have no more than a finite number of zeros in $\mathbb{N}_0$. Then the automaton $\mathfrak{A}(s_0)$ is absolutely transitive.*

Proof The proof goes along the lines similar to that of the proof of Theorem 13.13. Given a finite non-empty word $\tilde{g}$ (say, of length $m > 0$) over the alphabet $\mathbb{F}_p$, take a finite word $\tilde{v}$ whose prefix is $\tilde{g}$ and such that the corresponding non-negative rational integer v is a non-zero of f'': $f''(v) \neq 0$. The word $\tilde{v}$ which satisfies these conditions simultaneously exists as f'' has not more than a finite number of zeros in $\mathbb{N}_0$ (fixing arbitrary $\tilde{g}$ means that only some less significant digits in the base-p expansion of v are fixed); so by taking v whose base-p expansion is sufficiently long (thus making v large enough), we find $v \in \mathbb{N}_0$ such that $f''(v) \neq 0$ and the m-letter prefix of the word $\tilde{v}$ is $\tilde{g}$.

To put it in other words, given arbitrary finite word $\tilde{g}$ over the alphabet $\mathbb{F}_p$, by properly choosing $r \in \mathbb{N}_0$, we find a positive rational integer $v = g + p^m r$ (where $g \in \{0, 1, \ldots, p^m - 1\}$, $\tilde{g}$ is the base-p expansion of g) such that $f''(v) \neq 0$. This is possible due to the finiteness of a number of zeros of f'' in $\mathbb{N}_0$. We see that both f and the so constructed v satisfy conditions of Theorem 13.11: Just assume that the ball $\mathbf{B}$ from the conditions of the mentioned theorem is the whole space $\mathbb{Z}_p$. Then we finish the proof of the corollary exactly as this is done in the proof of Theorem 13.13. □

We remark that Note 13.12 can be applied to Corollary 13.15 as well. Note also that Theorem 13.13 can be easily deduced from Corollary 13.15. A number of automata can be proved to be absolutely transitive by using Corollary 13.15. For instance,

Example 13.16 An automaton whose both input and output alphabets are $\mathbb{F}_2$ and whose automaton function is $f(x) = a + bx + ((x^2)\ \mathsf{OR}\ c)$, where $a, b, c \in \mathbb{N}_0$, is absolutely transitive.

The assertion easily follows from Corollary 13.15 as $f(\mathbb{N}_0) \subset \mathbb{N}_0$ and $f''(x) = 2$ for all $x \in \mathbb{Z}_2$ (see Sect. 9.5.1 on how to calculate the derivative f'').

13.4.3 Uniformity of Distribution Produced by the "Nonlinear" Automata Functions

Theorems 13.8, 13.11, and 13.13 provoke the following natural question: Given a 1-Lipschitz map $f: \mathbb{Z}_p \to \mathbb{Z}_p$ of measure 1, how the points of the set

$$\mathcal{E}_n^s(f) = \left\{ \left(\frac{x \bmod p^n}{p^n}, \frac{f(x) \bmod p^n}{p^n}, \ldots, \frac{f^{s-1}(x) \bmod p^n}{p^n} \right), x \in \mathbb{Z}_p \right\}$$

are distributed in the unit s-dimensional hypercube? The following strong result in that direction was obtained in [136, 137]:

Theorem 13.17 *For any polynomial f with integer coefficients whose degree is greater than 1 and for any positive integer s the sequence of random vectors*

$$\left(\frac{\kappa_n}{p^n}, \frac{f(\kappa_n) \bmod p^n}{p^n}, \ldots, \frac{f^{s-1}(\kappa_n) \bmod p^n}{p^n} \right)$$

where κ_n are discrete uniform distributions on the finite set $\{0, 1, \ldots, p^n - 1\}$, weakly converges (as n tends to infinity) to the continuous uniform distribution on $[0, 1)^s$.

This result was extended to a wider class of automata functions in [135]. We will not prove Theorem 13.17 here since the proof is lengthy and involved; we only note that the idea underlying the proof is further development of the approach from the proof of Theorem 13.11. It is worth noticing also that Theorem 13.17 is true for arbitrary integer $p > 1$, and not necessarily a prime.

13.5 Automata Finiteness Criterion

Given an automaton function, currently no criterion is still known to determine whether the respective automaton is of measure 1 or 0. In the foregoing section, we already have proved some sufficient conditions for an automaton to be of measure 1. On the other hand, from Theorem 13.10 we already know that a finite automaton is always of measure 0. Therefore if, given an automaton function, one can determine whether the function corresponds to a finite automaton, this will serve as a sufficient condition for an automaton to be of measure 0. Moreover, the problem to determine whether a given 1-Lipschitz function is an automaton function has its own value since it may be infeasible to prove finiteness of the automaton by constructing its reduced state transition diagram.

We are going to prove automata finiteness criterion assuming that the automaton function is represented via van der Put series (cf. Theorem 5.14). We first remind some notions and facts from the theory of automata sequences following [3].

An infinite sequence $\mathbf{a} = (a_i)_{i=0}^{\infty}$ over a finite alphabet $\mathcal{A}$, $\#\mathcal{A} = L < \infty$, is called *p-automatic* if there exists a finite transducer $\mathfrak{T} = \langle \mathbb{F}_p, \mathcal{S}, \mathcal{A}, S, O, s_0\rangle$ such that for all $n = 0, 1, 2, \ldots$, if $\mathfrak{T}$ is fed by the word $\chi_k\chi_{k-1}\cdots\chi_0$ which is the base-p expansion of $n = \chi_0 + \chi_1 p + \cdots \chi_k p^k$, $\chi_k \neq 0$ if $n \neq 0$, then the k-th output symbol of $\mathfrak{T}$ is a_n; or, in other words, such that $\delta_k^{\mathcal{A}}(f_{\mathfrak{T}}(n)) = a_n$ for all $n \in \mathbb{N}_0$, where $k = \lfloor \log_p n \rfloor$ and $\delta_k^{\mathcal{A}}(r)$ stands for the k-th digit in the base-L expansion of r.

Note that definition of p-automatic sequence from [3] reads at this place "DFAO," a discrete finite automaton with output (which is also known under the name of Moore automaton) rather than "transducer." However in automata theory, it is well-known that any Moore automaton is equivalent to a suitable Mealy automaton in the following meaning: Given a Mealy automaton, there exists a Moore automaton whose automaton function is the same as the one of the Mealy automaton (and vice versa); see, e.g., [73, Section 1.9]. As in the meaning of our book finite transducers are Mealy automata, therefore, we may speak of transducers rather than of DFAO in the definition.

Further, the *p-kernel* of the sequence $\mathbf{a}$ is, by the definition, a set $\ker_p(\mathbf{a})$ of all subsequences $(a_{jp^m+t})_{j=0}^{\infty}$, $m = 0, 1, 2, \ldots; 0 \leqslant t < p^m$. We will need the following automaticity criterion (cf. [3, Theorem 6.6.2]):

Theorem 13.18 *Let $p \geqslant 2$; then the sequence* $\mathbf{a}$ *is p-automatic if and only if its p-kernel is finite.*

Now we are able to state the main result of the section (cf. [15]):

Theorem 13.19 *Given a 1-Lipschitz function $f : \mathbb{Z}_p \to \mathbb{Z}_p$ represented by van der Put series* (5.32)

$$f(x) = \sum_{m=0}^{\infty} b_m p^{\lfloor \log_p m \rfloor} \chi(m, x),$$

the function f is the automaton function of a finite automaton if and only if the following conditions hold simultaneously:

1. *All coefficients b_m, $m = 0, 1, 2, \ldots$, constitute a finite subset $B_f \subset \mathbb{Q} \cap \mathbb{Z}_p$.*
2. *The p-kernel of the sequence $(b_m)_{m=0}^{\infty}$ is finite.*

Note 13.20 Condition 2 of the theorem is equivalent to the condition that the sequence $(b_m)_{m=0}^{\infty}$ is p-automatic (cf. Theorem 13.18).

The assertion of Theorem 13.19 can be stated in equivalent form, in terms of formal power series. To do this, we need first to recall some notions and facts from the theory. Given a q-element field $\mathbb{F}_q$, denote via $\mathbb{F}_q[[X]]$ the ring of formal power series in variable X over $\mathbb{F}_q$

$$\mathbb{F}_q[[X]] = \left\{\sum_{i=0}^{\infty} a_i X^i : a_i \in \mathbb{F}_q\right\};$$

denote via $\mathbb{F}_q((X))$ the ring of formal Laurent series over $\mathbb{F}_q$

$$\mathbb{F}_q((X)) = \left\{\sum_{i=-n_0}^{\infty} a_i X^i : n_0 \in \mathbb{N}_0, a_i \in \mathbb{F}_q\right\}.$$

Denote via $F_q(X)$ the field of (univariate) rational functions over $\mathbb{F}_q$

$$\mathbb{F}_q(X) = \left\{\frac{u(X)}{v(X)} : u(X), v(X) \in \mathbb{F}_p[X], u(X) \neq 0\right\},$$

where $\mathbb{F}_p[X]$ is the ring of polynomials in variable X over $\mathbb{F}_q$. As the field $\mathbb{F}_q((X))$ contains a subfield $\mathbb{F}_q(X)$, it is possible to define algebraicity over $\mathbb{F}_q(X)$ in a standard way: A formal Laurent series $F(X) = \sum_{i=-n_0}^{\infty} a_i X^i$ is *algebraic over* $\mathbb{F}_p(X)$ if and only if there exist $d \in \mathbb{N}$ and polynomials $u_0(X), \ldots, u_d(X) \in \mathbb{F}_p[X]$, not all zero, such that in the field $\mathbb{F}_q((X))$ the following identity holds:

$$u_0(X) + u_1(X)F(X) + \cdots + u_d(X)(F(X))^d = 0.$$

Now we remind Christol's theorem (cf., e.g., [3, Theorem 12.2.5]) which is a criterion of automaticity of a sequence in terms of algebraicity over the field of rational functions over a finite field:

Theorem 13.21 *Let p be a prime, and let $\mathbf{a} = (a_i)_{i=0}^{\infty}$ be infinite sequence over a finite non-empty alphabet $\mathcal{A}$. The sequence $\mathbf{a}$ is p-automatic if and only if there exists an integer $\ell \in \mathbb{N}$ and an injection $\tau : \mathcal{A} \rightarrowtail \mathbb{F}_{p^\ell}$ such that the formal power series $\sum_{i=0}^{\infty} \tau(a_i) X^i$ is algebraic over $\mathbb{F}_{p^\ell}(X)$.*

By Christol's theorem, we now may replace condition 2 from the statement of Theorem 13.19 by equivalent one, thus getting the following *equivalent statement of automata finiteness criterion*:

Theorem 13.22 *Given a 1-Lipschitz function $f : \mathbb{Z}_p \to \mathbb{Z}_p$ represented by van der Put series* (5.32)

$$f(x) = \sum_{m=0}^{\infty} b_m p^{\lfloor \log_p m \rfloor} \chi(m, x),$$

the function f is the automaton function of a finite automaton if and only if the following conditions hold simultaneously:

1. *All coefficients* b_m, $m = 0, 1, 2, \dots$, *constitute a finite subset* $B_f \subset \mathbb{Q} \cap \mathbb{Z}_p$.
2. *Under a suitable injection* $\tau\colon B_f \to \mathbb{F}_{p^\ell}$, *the formal power series*

$$\sum_{m=0}^{\infty} \tau(b_m) X^m$$

over $\mathbb{F}_{p^\ell}$ *is algebraic over* $\mathbb{F}_{p^\ell}(X)$.

Proof of Theorem 13.19 Given a 1-Lipschitz function f, for $n \in \mathbb{N}_0$, $k \geqslant \lfloor \log_p n \rfloor + 1$, consider functions $f_{n,k}\colon \mathbb{Z}_p \to \mathbb{Z}_p$ defined as follows:

$$f_{n,k}(z) = \frac{1}{p^k}\Big(f(n + p^k z) - (f(n) \bmod p^k)\Big); \; z \in \mathbb{Z}_p.$$

The function f is an automaton function of a finite automaton if and only if in the collection $\mathcal{F}$ of functions $f_{n,k}$ ($n \in \mathbb{N}_0$, $k \in \mathbb{N}$, $k \geqslant \lfloor \log_p n \rfloor + 1$) contains only finitely many pairwise distinct functions: Note that $f_{n,k}$ is the automaton function that corresponds to the automaton $\mathfrak{A}(s(n_k)) = \langle \mathbb{F}_p, \mathcal{S}, \mathbb{F}_p, S, O, s(n_k)\rangle$, where $s(n_k) \in \mathcal{S}$ is the state the automaton $\mathfrak{A} = \mathfrak{A}_f = \langle \mathbb{F}_p, \mathcal{S}, \mathbb{F}_p, S, O, s_0\rangle$ reaches after it has been fed with the input word n_k (of length p^k) that corresponds to the base-p expansion of n (so the word n_k may contain some leading zeros that correspond to higher-order digits of the expansion).

Note that by Theorem 5.14,

$$\begin{aligned} b_{n+p^k s} &= \frac{1}{p^k}(f(n + p^k s) - f(n)) = \\ &= \frac{1}{p^k}(f(n + p^k s) - (f(n) \bmod p^k)) - \frac{1}{p^k}(f(n) - (f(n) \bmod p^k)) = \\ &\qquad = f_{n,k}(s) - f_{n,k}(0) \end{aligned}$$

if $n \leqslant p^k - 1$ and $s \in \{1, 2, \dots, p-1\}$ (cf. (4.13)); so the finiteness of $\mathcal{F}$ implies that in the sequence $(b_m)_{m=0}^{\infty}$ there are only finitely many pairwise distinct terms. We proceed with this in mind.

Take $n \in \mathbb{N}_0$ and $k \geqslant \lfloor \log_p n \rfloor + 1$. By (5.32), the value $f(n + p^k z)$ can be represented as $f(n + p^k z) = A_{n,k}(z) + B_{n,k}(z)$, where

$$A_{n,k}(z) = \sum_{m=0}^{p^k-1} b_m p^{\lfloor \log_p m \rfloor} \chi(m, n + p^k z); \tag{13.15}$$

$$B_{n,k}(z) = \sum_{m=p^k}^{\infty} b_m p^{\lfloor \log_p m \rfloor} \chi(m, n + p^k z). \tag{13.16}$$

By (4.11) and (4.13), once $m \leqslant p^k - 1$, the equality $\chi(m, n + p^k z) = 0$ holds if and only if $m \not\equiv n \pmod{p^{\lfloor \log_p m \rfloor + 1}}$; and once $m \geqslant p^k$, the equality $\chi(m, n+p^k z) = 0$ holds if and only if $m \not\equiv n \pmod{p^{\lfloor \log_p m \rfloor + 1}}$ (note that $\lfloor \log_p m \rfloor + 1 \geqslant k+1$ under conditions of the latter case); thus

$$A_{n,k}(z) = \sum_{m=0}^{p^k-1} b_m p^{\lfloor \log_p m \rfloor} \chi(m, n); \tag{13.17}$$

$$\begin{aligned} B_{n,k}(z) &= \sum_{t=1}^{\infty} b_{n+p^k t} p^{k+\lfloor \log_p t \rfloor} \chi(n + p^k t, n + p^k z) \\ &= \sum_{t=1}^{\infty} b_{n+p^k t} p^{k+\lfloor \log_p t \rfloor} \chi(t, z). \end{aligned} \tag{13.18}$$

From here in particular, it follows that $A_{n,k}(z)$ does not depend on z and that $B_{n,k}(z) = 0 \pmod{p^k}$ for all $z \in \mathbb{Z}_p$; consequently,

$$f_{n,k}(z) = \frac{1}{p^k}\left(A_{n,k}(0) + B_{n,k}(z) - (A_{n,k}(0) \bmod p^k)\right) = C_{n,k} + D_{n,k}(z), \tag{13.19}$$

where

$$C_{n,k} = \frac{1}{p^k}\left(A_{n,k}(0) - (A_{n,k}(0) \bmod p^k)\right); \tag{13.20}$$

$$D_{n,k}(z) = \sum_{t=1}^{\infty} b_{n+p^k t} p^{\lfloor \log_p t \rfloor} \chi(t, z). \tag{13.21}$$

Note that from (13.21) it follows that $D_{n,k}(0) = 0$ (cf. (4.11) and (4.13)), so from (13.19) we deduce that $f_{n,k}(0) = C_{n,k}$.

Thus, we have arrived to the following criterion of the finiteness of the number of distinct functions in the collection $\mathcal{F}$: There are only finite number of pairwise distinct functions $f_{n,k} \in \mathcal{F}$, $n \in \mathbb{N}_0$, $k \geqslant \lfloor \log_p n \rfloor + 1$, if and only if the following two conditions hold simultaneously:

1′. There are only finitely many pairwise distinct constants $C_{n,k} \in \mathbb{Z}_p$, $n \in \mathbb{N}_0$, $k \geqslant \lfloor \log_p n \rfloor + 1$.
2′. There are only finitely many pairwise distinct functions $D_{n,k} \colon \mathbb{Z}_p \to \mathbb{Z}_p$, $n \in \mathbb{N}_0$, $k \geqslant \lfloor \log_p n \rfloor + 1$.

However, since representation (13.21) is a (unique) van der Put expansion of the function $D_{n,k}$, condition 2′ is equivalent to the condition that in the sequence $(b_n)_{n=0}^{\infty}$ there are only finitely many pairwise distinct subsequences $(b_{n+p^k t})_{t=1}^{\infty}$, where $k \geqslant \lfloor \log_p n \rfloor + 1$, $n \in \mathbb{N}_0$. In turn, the latter condition is equivalent to the condition that there are only finitely many pairwise distinct subsequences $(b_{n+p^k t})_{t=0}^{\infty}$, $n \in \mathbb{N}_0$, $k \in \mathbb{N}$. Note that if the condition holds, there are only finitely many pairwise distinct terms b_m in the sequence $(b_m)_{m=0}^{\infty}$.

Consider condition $1'$. Note that for $k \geqslant \lfloor \log_p n \rfloor + 1$, we have that

$$\sum_{m=0}^{p^k-1} b_m p^{\lfloor \log_p m \rfloor} \chi(m,n) = \sum_{m=0}^{p^{\lfloor \log_p n \rfloor + 1}-1} b_m p^{\lfloor \log_p m \rfloor} \chi(m,n) \tag{13.22}$$

since $\chi(m,n) = 0$ once $\lfloor \log_p m \rfloor > \lfloor \log_p n \rfloor$ (cf. (4.11) and (4.13)). Thus, $A_{n,k}(z)$ does not depend on k (and on z as we have already shown); so denoting the right hand side in (13.22) via $A(n)$, we have that $C_{n,k} = p^{-k}(A(n) - ((A(n)) \bmod p^k))$ (cf. (13.20) and (13.17)).

From here by (4.11) and (4.13), we get that $C_{n,k} = p^{-k}(b_n - (b_n \bmod p^k))$ once n is such that $\lfloor \log_p n \rfloor = 0$. Consequently, given $n \in \{0, 1, \ldots, p-1\}$, the finiteness of the number of pairwise distinct $C_{n,k}$, where $k \geqslant \lfloor \log_p 0 \rfloor + 1 = 1$ is equivalent to the condition that the sequence $(\delta_i(b_n))_{i=0}^{\infty}$ is eventually periodic, that is, to the condition that $b_n \in \mathbb{Q} \cap \mathbb{Z}_p$. Using this as a base for induction on $\lfloor \log_p n \rfloor$, assuming that all $b_n \in \mathbb{Q} \cap \mathbb{Z}_p$ for n such that $\lfloor \log_p n \rfloor < N$, we see that, given n such that $\lfloor \log_p n \rfloor = N$, the finiteness of the number of pairwise distinct $C_{n,k} = p^{-k}(A(n) - ((A(n)) \bmod p^k))$ for $k \geqslant \lfloor \log_p n \rfloor + 1 = N + 1$ is equivalent to the condition that $A(n) \in \mathbb{Q} \cap \mathbb{Z}_p$. However, in view of (4.11) and (4.13) from the definition of $A(n)$, it follows that

$$A(n) = b_n p^{\lfloor \log_p n \rfloor} + \sum_{m=0}^{p^{\lfloor \log_p n \rfloor}-1} b_m p^{\lfloor \log_p m \rfloor} \chi(m,n).$$

The left-hand side sum is in $\mathbb{Q} \cap \mathbb{Z}_p$ by induction hypothesis; so $b_n p^{\lfloor \log_p n \rfloor} \in \mathbb{Q} \cap \mathbb{Z}_p$ and whence $b_n \in \mathbb{Q} \cap \mathbb{Z}_p$ since $b_n \in \mathbb{Z}_p$.

We finally have proved that conditions $1'$–$2'$ hold simultaneously if and only if the following conditions hold simultaneously:

$1''$. All coefficients b_m, $m = 0, 1, 2, \ldots$, constitute a non-empty finite subset (denoted as B_f) in $\mathbb{Q} \cap \mathbb{Z}_p$.

$2''$. There are only finitely many pairwise distinct subsequences $(b_{n+p^k t})_{t=0}^{\infty}$, $n \in \mathbb{N}_0$, $k \in \mathbb{N}$, $k \geqslant \lfloor \log_p n \rfloor + 1$; that is, the p-kernel of the sequence $(b_m)_{m=0}^{\infty}$ is finite.

This ends the proof. □

Note 13.23 Theorem 13.19 was generalized for arbitrary $p > 1$, not necessarily a prime; see [83].

Criteria to determine if an automaton function is finite which are based on expansions other than van der Put are also known (cf. [173, 184]).

It is worth recalling here a well-known fact (which also can be proved by using Theorem 13.19) that *addition of two p-adic integers can be performed by a finite automaton with two inputs and one output*: Actually the automaton just finds successively (digit after digit) the sum by a standard addition-with-carry algorithm which is used to find a sum of two non-negative integers represented by base-p expansions, thus calculating the sum with arbitrarily high accuracy with respect to the p-adic metric. On the contrary, *no finite automaton can perform multiplication of two arbitrary p-adic integers* since it is well-known that no finite automaton can calculate the base-p expansion of a square of an arbitrary non-negative integer given the base-p expansion of the latter (cf., e.g., [31, Theorem 2.2.3]). From these remarks combined with Theorem 13.19 the following properties of finite automata functions can be deduced:

Proposition 13.24 *Let $\mathfrak{A}$, $\mathfrak{B}$ be finite automata; let $a, b \in \mathbb{Z}_p \cap \mathbb{Q}$ be p-adic rational integers. Then the following is true:*

1. *The mapping $z \mapsto f_{\mathfrak{A}}(z) + f_{\mathfrak{B}}(z)$ of $\mathbb{Z}_p$ into $\mathbb{Z}_p$ is a finite automaton function;*
2. *The composite function $f(z) = af_{\mathfrak{A}}(z) + b$, $(z \in \mathbb{Z}_p)$, is a finite automaton function;*
3. *The constant function $f(z) = c$ is a finite automaton function if and only if $c \in \mathbb{Z}_p \cap \mathbb{Q}$;*
4. *The affine mapping $f(z) = cz + d$ is a finite automaton function if and only if $c, d \in \mathbb{Z}_p \cap \mathbb{Q}$.*

Proof Note that the van der Put expansion of the constant function $z \mapsto c$ is

$$c = c\chi(0, z) + c\chi(1, z) + \cdots + c\chi(p-1, z) + 0\chi(p, z) + 0\chi(p+1, z) + \cdots, \tag{13.23}$$

while the van der Put expansion for the identity function $z \mapsto z$ is

$$z = \sum_{i=0}^{\infty} \delta_{\lfloor \log_p i \rfloor}(i) p^{\lfloor \log_p i \rfloor} \chi(i, z), \tag{13.24}$$

where $\delta_j(i)$ stands for the j-th digit in the base-p expansion of i. Now all statements of the proposition follow immediately from Theorem 13.19 and facts from automata theory mentioned in Sect. 1.6. □

Note that assertions of Proposition 13.24 are known: For instance, they can be deduced from the old work [144]. To the best of our knowledge, A. G. Lunts was the first who revealed connections between automata theory and p-adic analysis. It is worth noticing that Lunts defines automata functions in a slightly different way than we do: In his work, an automaton function is a 1-Lipschitz function $F: \mathbb{Q}_p \to \mathbb{Q}_p$ such that $F(pz) = pF(z)$ for all $z \in \mathbb{Q}_p$. Also, Lunts' methods of proofs are completely different form the ones of Proposition 13.24. Unfortunately, most automata theorists seem to be unaware of the paper [144] since it was never translated into English and even was never reviewed by Mathematical Reviews. Concluding the subsection, we remark that in literature (finite) automata functions are also known under names of *(bounded) determinate functions* or *(bounded) deterministic functions* (cf., e.g., [37, 38, 173, 190]).

13.6 The Affinity of Smooth Real Functions Which Are Computable on Finite Automata

From Theorems 13.10 and 13.8, we know that limit plots of measure-0 automata (in particular, of finite automata) cannot contain "figures"; but they *may* contain "lines." It turns out that "smooth lines" which belong to limit plots of measure-0 automata can be straight lines only; moreover, if these plots are considered as the sets on the surface of a torus (cf. Sect. 12.1), these lines are necessarily *windings* (or, which is the same, *cables*) of the torus. The central result of this section is rigorous statement of that assertion and proofs of respective theorems. This property of measure-0 automata is extremely important and will serve as a base of reasoning in Chap. 16. We follow [17–19] in the rest of the section.

We first recall some more properties of *finite automata functions*, that is, about functions $f_{\mathfrak{A}}\colon \mathbb{Z}_p \to \mathbb{Z}_p$ where $\mathfrak{A} = \mathfrak{A}(s_0)$ is a finite automaton (i.e., with a finite set $\mathcal{S}$ of states). It is well-known that *the class of finite automata functions is closed with respect to composition of functions and a sum of functions*: Once $f, g\colon \mathbb{Z}_p \to \mathbb{Z}_p$ are finite automata functions, either of mappings $x \mapsto f(g(x))$ and $x \mapsto f(x)+g(x)$ $(x \in \mathbb{Z}_p)$ is a finite automaton function. Another important property of finite automata functions is that *any finite automaton function maps $\mathbb{Z}_p \cap \mathbb{Q}$ into itself.* We already have mentioned that property in preceding sections, but it is worth noticing once again that by (2.8), the property is just a restatement of a well-known property of finite automata which yields that any finite automaton, being fed by an eventually periodic sequence, produces an eventually periodic sequence as output (cf., e.g., [31, Corollary 2.6.9]). Since further we often use that property of finite automata, we state it as a lemma for future references:

Lemma 13.25 *If a finite automaton $\mathfrak{A}$ is being fed by a left-infinite periodic word w^{∞}, where $w \in \mathcal{W}$ is a finite non-empty word, then the corresponding output left-infinite word is eventually periodic; i.e., it is of the form $u^{\infty}v$, where $u \in \mathcal{W}$, $v \in \mathcal{W}_{\varphi}$. To put it in other words, if a finite automaton is being fed by an eventually periodic finite word $(w)^k t$, where $w \in \mathcal{W}$, $t \in \mathcal{W}_{\varphi}$, and $k \in \mathbb{N}$ is sufficiently large, then the output word is of the form $r(u)^{\ell} v$, where $\ell \in \mathbb{N}$, $u \in \mathcal{W}$, $r, v \in \mathcal{W}_{\varphi}$ and r is either empty or a prefix of u: $u = hr$ for a suitable $h \in \mathcal{W}_{\varphi}$. Therefore the output word is of the form $(\bar{u})^{\ell} v'$, where $\bar{u}$ is a cyclically shifted word u.*

Then we recall some conventions which have been made earlier in Sect. 12.1 and introduce new ones. Given an m-letter non-empty word $v = \gamma_{m-1}\gamma_{m-2}\ldots\gamma_0$ over the alphabet $\mathbb{F}_p = \{0, 1, \ldots, p-1\}$, we put into the correspondence the number $0.v \in \mathbb{R}$ whose base-p expansion is

$$0.\gamma_{m-1}\gamma_{m-2}\ldots\gamma_0 = \sum_{j=0}^{m-1} \gamma_j p^{-m+j}$$

so to every m-letter input word $w = \alpha_{m-1}\alpha_{m-2}\cdots\alpha_0$ of the $\mathfrak{A}$ whose automaton function is $f = f_{\mathfrak{A}}\colon \mathbb{Z}_p \to \mathbb{Z}_p$ and to respective m-letter output word $\mathfrak{a}(w) = \beta_{m-1}\beta_{m-2}\cdots\beta_0$ of $\mathfrak{A}$, there corresponds the point $(0.w; 0.\mathfrak{a}(w)) \in \mathbb{R}^2$.

Obviously, Definition 12.1 implies that the *limit plot* $\mathcal{P}'(\mathfrak{A})$ of the automaton $\mathfrak{A}$ is the point set which also can be defined as follows:

Definition 13.26 A point $(x; y) \in \mathbb{R}^2$ lies in the limit plot $\mathcal{P}'(\mathfrak{A})$ if and only if there exist $z \in \mathbb{Z}_p$ and a strictly increasing infinite sequence $k_1 < k_2 < \ldots$ of numbers from $\mathbb{N}$ such that the following equation holds simultaneously:

$$\lim_{i\to\infty} \frac{z \bmod p^{k_i}}{p^{k_i}} = x; \quad \lim_{i\to\infty} \frac{f_{\mathfrak{A}}(z) \bmod p^{k_i}}{p^{k_i}} = y \tag{13.25}$$

Note 13.27 Further in the book we consider the limit plot $\mathcal{P}'(\mathfrak{A})$, as well as the plots $\mathcal{P}(\mathfrak{A})$ and $\mathcal{P}(f)$) of the automaton $\mathfrak{A}$ and of the 1-Lipschitz function $f\colon \mathbb{Z}_p \to \mathbb{Z}_p$, either as a subset of the unit square $\mathbb{I}^2 \subset \mathbb{R}^2$ or as a subset of the unit torus $\mathbb{T}^2 = \mathbb{R}^2/\mathbb{Z}^2$ when appropriate. Note that when considering the plot on the unit torus, we reduce coordinates of the points modulo 1, that is, we just "glue together" 0 and 1 of the unit segment $\mathbb{I}$ thus transforming it into the unit circle $\mathbb{S}$ (whose points we usually identify with the points of the half-open segment $[0, 1)$ via a natural one-to-one correspondence, say, $\omega \leftrightarrow \sin^2(\omega/2)$). Also, sometimes we consider $\mathcal{P}'(\mathfrak{A})$ (as well as $\mathbf{P}(\mathfrak{A})$ and $\mathbf{P}(f)$) as a subset of the cylinder $\mathbb{I} \times \mathbb{S}$ or of the cylinder $\mathbb{S} \times \mathbb{I}$ by reducing modulo 1 either y- or x-coordinate respectively. We denote the corresponding plot via $\mathcal{P}'_{\mathbb{M}}(\mathfrak{A})$ by using the subscript $\mathbb{M} \in \{\mathbb{I}^2, \mathbb{T}^2, \mathbb{I} \times \mathbb{S}, \mathbb{S} \times \mathbb{I}\}$, and we omit the subscript when it is clear (or when it is no difference) on which of the surfaces the plot is considered.

13.6.1 Topological Properties of Automata Plots

In this subsection we recall some well-known topological notions and prove some related facts about automata plots which we will need further. In the sequel, given a subset S of a topological (in particular, metric) space $\mathbb{M}$ which satisfies the Hausdorff axiom, we denote via $\mathbf{AP}_{\mathbb{M}}(S)$ the set of all accumulation points of S. Recall that the point $x \in \mathbb{M}$ is called an *accumulation point of* $S \subset \mathbb{M}$ once every neighborhood of x contains infinitely many points from S; and a point $y \in \mathbb{M}$ is called *isolated point of* S (or, the *point isolated from* S; or, the *point isolated with respect to* S) once there exists a neighborhood $U \ni y$ such that U contains no points from S other than (maybe) y. We may omit the subscript and use notation $\mathbf{AP}(S)$ when it is clear from the context what metric space is meant.

We also write $\mathbf{AP}((a_i)_{i=0}^{\infty})$ (or briefly $\mathbf{AP}(a_i)$, or $\mathbf{AP}(C)$) for the set of all limit points of the sequence $C = (a_i)_{i=0}^{\infty}$ over $\mathbb{M}$. Recall that a point $x \in \mathbb{M}$ is called a *limit (or, cluster) point of the sequence* $(a_i)_{i=0}^{\infty}$ if every neighborhood of x contains infinitely many terms of the sequence $(a_i)_{i=0}^{\infty}$; that is, given any neighborhood U of x, the number of i such that $a_i \in U$ is infinite (note that the very $a_i \in U$ are not assumed to be pairwise distinct points of $\mathbb{M}$; some, or even all of them may be identical). Note that in topology the terms "accumulation point of a set" and "limit point of a set" are used as synonyms; however to avoid possible misunderstanding, we reserve the term "limit point" only for sequences while for sets we use the term "accumulation point." Though topological properties of plots/limits plots of automata as point sets in Euclidean space are quite obvious, we make some clarifications in order to avoid misunderstanding in the rest of the book:

Note 13.28 Definition 12.1 of the plot $\mathcal{P}(\mathfrak{A})$ of the automaton $\mathfrak{A}$ immediately implies that $(x; y) \in \mathcal{P}(\mathfrak{A})$ if and only if there exists a sequence $(w_i)_{i=0}^{\infty}$ of finite non-empty words $w_i \in \mathcal{W}$ such that $\Lambda(w_i) = k_i$ for all $i = 0, 1, 2, \dots$ and $\lim_{i\to\infty} \rho(w_i) = x$, $\lim_{i\to\infty} \rho(\mathfrak{a}(w_i)) = y$. Note that once $(x; y) \in \mathcal{P}'(\mathfrak{A})$ then there exists a sequence $(w_i)_{i=0}^{\infty}$ of words such that the sequence $(\Lambda(w_i) = k_i)_{i=0}^{\infty}$ of their lengths is strictly increasing: One just may take $w_i = \mathsf{wrd}_{k_i}(z \bmod p^{k_i})$ (cf. (1.4) and Sect. 1.4.1). Therefore $\mathcal{P}'(\mathfrak{A}) \subset \mathbf{AP}(\mathcal{P}(f_{\mathfrak{A}}))$. Moreover, from Definition 13.26 it readily follows that $\mathbf{AP}(\mathcal{P}(f_{\mathfrak{A}})) = \mathbf{AP}(\mathcal{E}(f_{\mathfrak{A}})) = \mathbf{AP}(\mathcal{P}(\mathfrak{A}))$ since given a finite non-empty word w and taking any $z \in \mathbb{Z}_p$ such that the prefix of the corresponding infinite word is w (i.e., such that $w = \mathsf{wrd}_{\Lambda(w)}(z \bmod p^{\Lambda(w)})$), we see that $\rho(\mathfrak{a}(w)) = ((f_{\mathfrak{A}}(z)) \bmod p^{\Lambda(w)})/p^{\Lambda(w)}$. This implies that $\mathcal{P}(\mathfrak{A}) = \mathcal{P}(f_{\mathfrak{A}})$ since $\mathcal{P}(f_{\mathfrak{A}}) = \mathcal{E}(f_{\mathfrak{A}}) \cup \mathbf{AP}(\mathcal{E}(f_{\mathfrak{A}})) = \mathcal{P}(\mathfrak{A})$; so in the sequel we do not differ automata plots from the plots of automata functions and use both $\mathcal{P}(\mathfrak{A})$ and $\mathcal{P}(f_{\mathfrak{A}})$ as notation for the plot of the automaton $\mathfrak{A}$. Also we may use notation $\mathcal{P}'(f_{\mathfrak{A}})$ along with $\mathcal{P}'(\mathfrak{A})$ to denote the limit plot of the automaton $\mathfrak{A}$.

It is worth stressing once again here a crucial difference in the construction of plots and of Monna graphs of automata (cf. Sect. 12.1): Given a canonical expansion of the p-adic integer $z = \sum_{i=0}^{\infty} \gamma_i p^i$, we put into correspondence to z a *single* real number $\mathsf{Mon}(z) = \sum_{i=0}^{\infty} \gamma_i p^{-i-1}$ (cf. Sect. 7.2.1) while constructing the graph of Monna maps; whereas while constructing plots we associate to z a *whole set of all limit points* of the sequence $(p^{-m}(z \bmod p^m))_{m=1}^{\infty}$, and the latter set may not consist of a single point; moreover, "usually" the set never consists of a single point since with a probability 1 the set is a whole segment $[0, 1]$. Therefore to study structure of plots, we need to work with sets of all limit points of (usually non-convergent) sequences rather than with limits of convergent sequences as in the case of Monna maps.

Proposition 13.29 *Let $\mathfrak{A}$ be an arbitrary automaton; then $\mathcal{P}'(\mathfrak{A})$ contains no points isolated with respect to $\mathcal{E}(f_{\mathfrak{A}})$.*

Proof of Proposition 13.29 Let $(x; y) \in \mathcal{P}'(\mathfrak{A})$ be a point isolated w.r.t. $\mathcal{E}(f_{\mathfrak{A}})$. As $(x; y) \in \mathcal{P}'(\mathfrak{A})$, let $z = \sum_{j=0}^{\infty} \zeta_j p^j$ be p-adic canonical representation of the p-adic integer $z \in \mathbb{Z}_p$ and let $f_{\mathfrak{A}}(z) = \sum_{j=0}^{\infty} \gamma_j p^j$ be p-adic canonical expansion of the p-adic integer $f_{\mathfrak{A}}(z)$. Then, as the point $(x; y)$ is isolated, there exists $I \in \mathbb{N}$ such that $z \bmod p^{k_i}/p^{k_i} = x$ and $f_{\mathfrak{A}}(z) \bmod p^{k_i}/p^{k_i} = y$ for all $i \geqslant I$ (cf. (13.25)) (if otherwise, the point (x, y) is not isolated with respect to $\mathcal{E}(f_{\mathfrak{A}})$). Put $w_i = \mathsf{wrd}_{k_i}\left(z \bmod p^{k_i}/p^{k_i}\right) = \zeta_{k_i-1}\zeta_{k_i-2}\ldots\zeta_0$, $u_i = \mathsf{wrd}_{k_i}\left(f_{\mathfrak{A}}(z) \bmod p^{k_i}/p^{k_i}\right) = \gamma_{k_i-1}\gamma_{k_i-2}\ldots\gamma_0$; then

$$0.\zeta_{k_i-1}\ldots\zeta_0 = x; \tag{13.26}$$

$$0.\gamma_{k_i-1}\ldots\gamma_0 = y, \tag{13.27}$$

for all $i \geqslant I$. We claim that then necessarily both $z = 0$ and $f_{\mathfrak{A}}(z) = 0$ (whence both $x = 0$ and $y = 0$).

Indeed, as the sequence $\mathcal{K} = (k_i)_{i=0}^{\infty}$ is infinite and strictly increasing, then taking $i = I$ in (13.26) we conclude that necessarily $\zeta_0 = \zeta_1 = \cdots = \zeta_{k_{I+M}-k_I-1} = 0$ for all $M \in \mathbb{N}$. Therefore, taking M large enough so that $k_{I+M} - k_I \geqslant k_I$ (which is always possible since the sequence $\mathcal{K}$ is strictly increasing) we see that $\zeta_0 = \zeta_1 = \cdots = \zeta_{k_I-1} = 0$ and thus $\zeta_i = 0$ for all $i \in \mathbb{N}_0$ since $0.\zeta_{k_i-1}\ldots\zeta_0 = 0.\zeta_{k_I-1}\ldots\zeta_0$ for all $i \geqslant I$ by (13.26). But this implies that $z = 0$ (whence $x = 0$). The same argument combined with (13.27) shows that $f_{\mathfrak{A}}(0) = 0$ and $y = 0$.

Consider now an automaton $\mathfrak{B}$ whose automaton function $f_{\mathfrak{B}}$ is defined as follows: Given $z \in \mathbb{Z}_p$, let its p-adic canonical representation be $z = \sum_{j=0}^{\infty} \zeta_j p^j$ (where $\delta_j(z) = \zeta_j$, $j = 0, 1, 2, \ldots$) let $\delta_0(f_{\mathfrak{B}}(z)) = 1$ and $\delta_j(f_{\mathfrak{B}}(z)) = \delta_j(f_{\mathfrak{A}}(z))$ for $j > 0$. Such an automaton $\mathfrak{B}$ exists since the so defined function $f_{\mathfrak{B}}$ satisfies (13.1) and thus is 1-Lipschitz. Actually the automaton being fed by the input word $\ldots\zeta_2\zeta_1\zeta_0$ just produces 1 as the first letter of the output and produces γ_j as the j-th letter of the output if $j > 0$, where $\ldots\gamma_2\gamma_1\gamma_0$ is the output word of the automaton $\mathfrak{A}$ fed by the input word $\ldots\zeta_2\zeta_1\zeta_0$; that is, $\mathfrak{B}$ outputs $\ldots\gamma_2\gamma_1 1$ being fed by $\ldots\zeta_2\zeta_1\zeta_0$.

From Definition 12.1, it immediately follows that $\mathcal{P}'(\mathfrak{A}) = \mathcal{P}'(\mathfrak{B})$ and that a point $(x; y) \in \mathbb{R}^2$ is an isolated point of $\mathcal{E}(f_{\mathfrak{B}})$ if and only if it is an isolated point of $\mathcal{E}(f_{\mathfrak{A}})$. But by the claim we have proved above, once $(x; y)$ is an isolated point of $\mathcal{E}(f_{\mathfrak{B}})$, then necessarily $f_{\mathfrak{B}}(0) = 0$. But the first letter of any output word of automaton $\mathfrak{B}$ is 1 by the construction of $f_{\mathfrak{B}}$; thus $\delta_0(f_{\mathfrak{B}}(0)) = 1$ and so $f_{\mathfrak{B}}(0) \neq 0$. From the claim we have proved at the beginning of the proof, it follows now that $\mathcal{P}'(\mathfrak{B})$ cannot contain isolated points of $\mathcal{E}(f_{\mathfrak{B}})$; thus $\mathcal{P}'(\mathfrak{A})$ cannot contain isolated points of $\mathcal{E}(f_{\mathfrak{A}})$. □

Note that Proposition 13.29 only states that $\mathcal{P}'(\mathfrak{A})$ contains no points isolated from $E(f_{\mathfrak{A}})$, but of course $\mathcal{P}'(\mathfrak{A})$ may contain isolated points w.r.t. itself. For instance, let $\mathfrak{A}$ be a p-adic odometer; that is, $f_{\mathfrak{A}}(z) = z + 1$ (the automaton $\mathfrak{A}$ may be taken a finite then). Then the point $(1; 0) \in \mathbb{I}^2$ is an isolated point of $\mathcal{P}'_{\mathbb{I}^2}(\mathfrak{A})$ w.r.t. $\mathcal{P}'_{\mathbb{I}^2}(\mathfrak{A})$; however $\mathcal{P}'_{\mathbb{T}^2}(\mathfrak{A})$ contains no points isolated with respect to $\mathcal{P}'_{\mathbb{T}^2}(\mathfrak{A})$.

Theorem 13.30 *If an automaton $\mathfrak{A}$ is finite and minimal* (cf. Definition 1.22), *then* $\mathbf{AP}(\mathcal{E}(f_{\mathfrak{A}})) = \mathcal{P}'(\mathfrak{A})$.

Proof of Theorem 13.30 By Proposition 13.29, $\mathbf{AP}(\mathcal{E}(f_{\mathfrak{A}})) \supset \mathcal{P}'(\mathfrak{A})$; we need to prove that the inverse inclusion also holds. Let $(x; y) \in \mathbf{AP}(\mathcal{E}(f_{\mathfrak{A}}))$; then there exists a sequence $(z_i)_{i=0}^{\infty}$ of p-adic integers and a sequence $(k_i)_{i=0}^{\infty}$ of integers from $\mathbb{N}$ such that all the points

$$\mathbf{p}_i = \left(\frac{z_i \bmod p^{k_i}}{p^{k_i}}; \frac{f_{\mathfrak{A}}(z_i) \bmod p^{k_i}}{p^{k_i}}\right) \in \mathbb{R}^2$$

are pairwise distinct and

$$\lim_{i\to\infty} \frac{z_i \bmod p^{k_i}}{p^{k_i}} = x;$$

$$\lim_{i\to\infty} \frac{f_{\mathfrak{A}}(z_i) \bmod p^{k_i}}{p^{k_i}} = y.$$

We may assume that the sequence $(k_i)_{i=0}^{\infty}$ is increasing since otherwise in the point sequence $(\mathbf{p}_i)_{i=0}^{\infty}$ there are only finitely many pairwise distinct points. Moreover, we may assume that the sequence $(k_i)_{i=0}^{\infty}$ is strictly increasing; we consider corresponding infinite point subsequence of $(\mathbf{p}_i)_{i=0}^{\infty}$ if otherwise. So we see that there exists an infinite sequence of words $h_i = \mathsf{wrd}_{k_i}(z_i \bmod p^{k_i})$ of strictly increasing lengths k_i such that

$$\lim_{i\to\infty} 0.h_i = x; \tag{13.28}$$

$$\lim_{i\to\infty} 0.\mathfrak{a}(h_i) = y. \tag{13.29}$$

That is, there exists a sequence $(h_i)_{i=0}^{\infty}$ of words $h_i = \alpha_{k_i-1}^{(i)} \cdots \alpha_0^{(i)}$ of strictly increasing lengths $1 \leqslant k_0 < k_1 < k_2 < \ldots$ such that $\lim_{i\to\infty} 0.\alpha_{k_i-1}^{(i)} \cdots \alpha_0^{(i)} = x$. From here it follows that once i is sufficiently large (say, once $i \geqslant M_0 \in \mathbb{N}_0$), then $\alpha_{k_i-1}^{(i)} = \zeta_0$ for a suitable $\zeta_0 \in \mathbb{F}_p$. By the same reason, $\alpha_{k_i-2}^{(i)} = \zeta_1$ for a suitable $\zeta_1 \in \mathbb{F}_p$ once i is large enough (say, once $i \geqslant M_1 \in \mathbb{N}_0$), etc. Moreover, we may assume that the sequence $(M_\ell)_{\ell=0}^{\infty}$ is strictly increasing. Therefore, $x = 0.\zeta_0\zeta_1\ldots$. Applying a similar argument to the sequence $\beta_{k_i-1}^{(i)}\beta_{k_i-2}^{(i)} \cdots \beta_0^{(i)} = \mathfrak{a}(\alpha_{k_i-1}^{(i)}\alpha_{k_i-2}^{(i)} \cdots \alpha_0^{(i)})$ $(i = 0, 1, 2, \ldots)$ we conclude that there exists a strictly increasing sequence $(N_\ell)_{\ell=0}^{\infty}$ such that $\beta^{(i)}_{k_i-\ell-1} = \gamma_\ell \in \mathbb{F}_p$ once $i \geqslant N_\ell$ and therefore $y = 0.\gamma_0\gamma_1\ldots$. Moreover, by the construction of the sequences $(M_\ell)_{\ell=0}^{\infty}$ and $(N_\ell)_{\ell=0}^{\infty}$ we may assume that $M_\ell = N_\ell$ for all $\ell \in \mathbb{N}_0$. Thus we have shown that

$$\alpha^{(i)}_{k_i-1}\ldots\alpha^{(i)}_0 = \zeta_0\zeta_1\ldots\zeta_\ell w^{(i)}_\ell \text{ if } i \geqslant M_\ell \ (\ell = 0, 1, 2, \ldots);$$

$$\beta^{(i)}_{k_i-1}\ldots\beta^{(i)}_0 = \gamma_0\gamma_1\ldots\gamma_\ell u^{(i)}_\ell \text{ if } i \geqslant M_\ell \ (\ell = 0, 1, 2, \ldots),$$

where $w^{(i)}_\ell, u^{(i)}_\ell \in \mathcal{W}_\varnothing$ and $\gamma_0\gamma_1\ldots\gamma_\ell u^{(i)}_\ell = \mathfrak{a}(\zeta_0\zeta_1\ldots\zeta_\ell w^{(i)}_\ell)$, $(\ell = 0, 1, 2, \ldots)$. Let $s^{(i)}_\ell$ be a state the automaton $\mathfrak{A}$ reaches after being fed by the input word $w^{(i)}_\ell$ (note that $s^{(i)}_\ell = s_0$, the initial state, once $w^{(i)}_\ell$ is empty word). As the number of states of $\mathfrak{A}$ is finite, at least one state $s \in \mathcal{S}$ repeats in the sequence $\left(s^{(M_\ell)}_\ell\right)_{\ell=0}^{\infty}$ infinitely many times. Therefore

$$\lim_{\ell\to\infty} 0.\zeta_0\ldots\zeta_\ell = x; \quad \text{whence } x = 0.\zeta_0\zeta_1\zeta_2\ldots \tag{13.30}$$

$$\lim_{\ell\to\infty} 0.\mathfrak{a}_s(\zeta_0\ldots\zeta_\ell) = y; \text{ whence } y = 0.\gamma_0\gamma_1\gamma_2\ldots. \tag{13.31}$$

Denote $w_\ell = \zeta_0\ldots\zeta_\ell$, $v_\ell = \gamma_0\ldots\gamma_\ell$, $(\ell = 0, 1, 2, \ldots)$. As every state of the automaton $\mathfrak{A}$ is reachable from the initial state s_0, there exists a word $t_0 \in \mathcal{W}_\varphi$ such that if the automaton $\mathfrak{A}$ (which is initially at the state s_0) has been fed by the word t_0, then $\mathfrak{A}$ outputs the word $\bar{t}_0 = \mathfrak{a}(t_0)$ and reaches the state s. Thus $\mathfrak{a}(w_0t_0) = \gamma_0\bar{t}_0$, and the automaton $\mathfrak{A}$ after being fed by the word w_0t_0 reaches the state $r^{(0)}$. As the automaton $\mathfrak{A}$ is minimal, there exists a word $t_1 \in \mathcal{W}_\varnothing$ such that once the automaton $\mathfrak{A}_1 = \mathfrak{A}(r^{(0)})$ has been fed by the word t_1, the automaton reaches the state s. Now being fed by the word w_1, the automaton $\mathfrak{A}_s = \mathfrak{A}(s)$ outputs the word v_1 and reaches a state $r^{(1)}$. By the minimality of $\mathfrak{A}$, there exists a word $t_2 \in \mathcal{W}_\varphi$ such that after $\mathfrak{A}(r^{(1)})$ has been fed by the word t_2, the automaton reaches the state s. Now after $\mathfrak{A}_s$ has been fed by the word w_2, the automaton $\mathfrak{A}_s$ reaches the state $r^{(2)}$, and we can find a word t_3 in a manner similar to that of described. Now being fed by the so constructed left-infinite word $\ldots w_2t_2w_1t_1w_0t_0$, the automaton $\mathfrak{A}$ outputs the left-infinite word $\ldots v_2\bar{t}_2v_1\bar{t}_1v_0\bar{t}_0$ where $\bar{t}_j = \mathfrak{a}_{r^{(j-1)}}(t_j)$, $j = 1, 2, 3, \ldots$. Now consider p-adic integers $z = \sum_{i=0}^{\infty}\chi_i p^i$ and $\bar{z} = \sum_{i=0}^{\infty}\xi_i p^i$ which correspond to infinite words $\ldots w_2t_2w_1t_1w_0t_0$ and $\ldots v_2\bar{t}_2v_1\bar{t}_1v_0\bar{t}_0$ accordingly; that is, $\ldots\chi_2\chi_1\chi_0 = \ldots w_2t_2w_1t_1w_0t_0$ and $\ldots\xi_2\xi_1\xi_0 = \ldots v_2\bar{t}_2v_1\bar{t}_1v_0\bar{t}_0$. Then, by the construction, we have that $\bar{z} = f_{\mathfrak{A}}(z)$, and from (13.30)–(13.31) it follows that

$$\lim_{j\to\infty} \frac{z \bmod p^{K_j}}{p^{K_j}} = x; \tag{13.32}$$

$$\lim_{j\to\infty} \frac{f_{\mathfrak{A}}(z) \bmod p^{K_j}}{p^{K_j}} = y, \tag{13.33}$$

where $K_j = \sum_{i=0}^{j} \Lambda(w_i) + \sum_{i=0}^{j} \Lambda(t_i)$. As the sequence $(K_j)_{j=0}^{\infty}$ is strictly increasing, from (13.32)–(13.33), it follows now that $(x; y) \in \mathcal{P}'(\mathfrak{A})$ in view of Definition 12.1. □

It is well-known (see, e.g., [2, Ch.2, Exercise 2]) that the set of all accumulation points of a Hausdorff topological space (the *derived set* of the space) is a closed subset of the space. From Theorem 13.30 it follows that once a finite automaton is minimal then its limit plot is a derived set of its plot (whence, closed):

Corollary 13.31 *Let an automaton $\mathfrak{A}$ be finite and minimal; then the set $\mathcal{P}'(\mathfrak{A})$ is a derived set of $\mathcal{P}(\mathfrak{A})$ and therefore is closed in $\mathbb{R}^2$. A point $(x; y) \in \mathbb{R}^2$ belongs to $\mathcal{P}'(\mathfrak{A})$ if and only if there exists a sequence $\left(\alpha_{k_i-1}^{(i)} \ldots \alpha_0^{(i)}\right)_{i=0}^{\infty}$ of finite non-empty words of strictly increasing lengths $k_0 < k_1 < k_2 < \cdots$ such that the sequence $\left(0.\alpha_{k_i-1}^{(i)}\alpha_{k_i-2}^{(i)} \ldots \alpha_0^{(i)}\right)_{i=0}^{\infty}$ tends to x and the corresponding sequence $\left(0.\beta_{k_i-1}^{(i)}\beta_{k_i-2}^{(i)} \ldots \beta_0^{(i)}\right)_{i=0}^{\infty}$ tends to y as $i \to \infty$, where $\beta_{k_i-1}^{(i)} \ldots \beta_0^{(i)}$ are respective output words of the automaton $\mathfrak{A}$ that correspond to input words $\alpha_{k_i-1}^{(i)} \ldots \alpha_0^{(i)}$* (i.e., $\beta_{k_i-1}^{(i)}\beta_{k_i-2}^{(i)} \ldots \beta_0^{(i)} = \mathfrak{a}(\alpha_{k_i-1}^{(i)}\alpha_{k_i-2}^{(i)} \ldots \alpha_0^{(i)})$, $i = 0, 1, 2, \ldots$).

We stress once again that we send words $\alpha_{k_i-1} \ldots \alpha_0$ to the automaton $\mathfrak{A}$ from right to left; i.e., we first sent the letter α_0 to $\mathfrak{A}$, then we sent the letter α_1 to $\mathfrak{A}$, etc.

Proof of Corollary 13.31 By the definition, the set $\mathbf{AP}(\mathcal{E}(f_{\mathfrak{A}})) = \mathbf{AP}(\mathcal{P}(\mathfrak{A}))$ is a derived set of $\mathcal{P}(\mathfrak{A})$; whence by Theorem 13.30 the set $\mathcal{P}'(\mathfrak{A})$ is a derived (thus, closed) set of $\mathcal{P}(\mathfrak{A})$.

The necessity of conditions of the corollary follows immediately from Definition 12.1 since once $(x; y) \in \mathcal{P}'(\mathfrak{A})$ then there exist a p-adic integer $z = \sum_{i=0}^{\infty} \alpha_i p^i$ and a strictly increasing sequence $1 \leqslant k_1 < k_2 < \ldots$ over $\mathbb{N}$ such that (13.25) holds; that is, we just put $\alpha_{k_i-1}^{(i)} \ldots \alpha_0^{(i)} = \mathsf{wrd}_{k_i}(z \bmod p^{k_i})$ and $\beta_{k_i-1}^{(i)} \ldots \beta_0^{(i)} = \mathsf{wrd}_{k_i}(f(z) \bmod p^{k_i})$, where f is an automaton function of $\mathfrak{A}$ (cf. Note 13.28)‘.

To prove sufficiency of the conditions note that the conditions just yield that there exists an infinite sequence of words $h_i = \alpha_{k_i-1}^{(i)} \ldots \alpha_0^{(i)}$ of strictly increasing lengths k_i such that (13.28)–(13.29) hold. The argument that follows (13.28)–(13.29) of the proof of Theorem 13.30 now proves the sufficiency. □

It is worth noticing here that the limit plot of a finite minimal automaton does not depend on what state of the automaton is taken as initial:

Note 13.32 If s, t are states of a finite minimal automaton $\mathfrak{A}$, $s \neq t$, then $\mathcal{P}'(\mathfrak{A}(s)) = \mathcal{P}'(\mathfrak{A}(t))$.

Indeed, due to the minimality, every state of $\mathfrak{A}$ is reachable from any other state of $\mathfrak{A}$. Therefore if $(x; y) \in \mathcal{P}'(\mathfrak{A}(t))$ then by Definition 13.26 there exist $z \in \mathbb{Z}_p$ and a strictly increasing infinite sequence $k_1 < k_2 < \ldots$ of numbers from $\mathbb{N}$ such that (13.25) holds. By the minimality of $\mathfrak{A}$, there exists a finite word w of length $K > 0$ such that after the automaton $\mathfrak{A}(s)$ has been fed by w, it reaches the state t. Now by substituting $p^K z + \mathsf{num}(w)$ for z and $k_1 + K < k_2 + K < \ldots$ for $k_1 < k_2 < \ldots$ in Definition 13.26, we see that (13.25) holds and therefore $(x; y) \in \mathcal{P}'(\mathfrak{A}(s))$.

Using ideas similar to that of Note 13.32, it can be easily demonstrated that if $\mathfrak{B}$ is a sub-automaton of $\mathfrak{A}$ then $\mathcal{P}(\mathfrak{B}) \subset \mathcal{P}(\mathfrak{A})$ since every state of the automaton $\mathfrak{A}$ is reachable from its initial state:

Note 13.33 Let $\mathfrak{B} = \mathfrak{B}(s)$ be a sub-automaton of the automaton $\mathfrak{A}$. As the initial state s of the automaton $\mathfrak{B}$ is reachable from the initial state s_0 of the automaton $\mathfrak{A}$, from the definition of the respective sets it immediately follows that $\mathcal{P}(\mathfrak{B}) \subset \mathcal{P}(\mathfrak{A})$, $\mathcal{P}'(\mathfrak{B}) \subset \mathcal{P}'(\mathfrak{A})$, and $\mathbf{AP}(\mathfrak{B}) \subset \mathbf{AP}(\mathfrak{A})$.

The following useful lemma is a sort of a counter-part of Lemma 13.25 in terms of points from $\mathcal{P}'(\mathfrak{A})$ rather than in terms of words.

Lemma 13.34 *Given a finite automaton $\mathfrak{A}$ and a point $x \in \mathbb{Z}_p \cap \mathbb{Q}$, if $(x; y) \in \mathcal{P}'(\mathfrak{A})$ for some $y \in \mathbb{R}$ then $y \in \mathbb{Z}_p \cap \mathbb{Q}$.*

Proof of Lemma 13.34 As $(x; y) \in \mathcal{P}'(\mathfrak{A})$ then there exist $z \in \mathbb{Z}_p$ and a strictly increasing sequence $k_0 < k_1 < \ldots$ over $\mathbb{N}$ such that (13.25) holds. Therefore there exists an infinite sequence of words $h_i = \mathsf{wrd}_{k_i}(z \bmod p^{k_i})$ of strictly increasing lengths k_i such that (13.28)–(13.29) hold simultaneously. Now repeating for the case $z_i = z$ the argument that follows (13.28)–(13.29) in the proof of Theorem 13.30, we conclude that (13.30)–(13.31) hold in our case as well (note that nowhere in the mentioned argument from the proof of Theorem 13.30 we used that $\mathfrak{A}$ is minimal). Moreover, in the notation of the argument, there exists a strictly increasing sequence $(M_\ell)_{\ell=0}^{\infty}$ over $\mathbb{N}$ such that

$$\alpha_{k_i-1}^{(i)} \ldots \alpha_0^{(i)} = \zeta_0\zeta_1 \ldots \zeta_\ell w_\ell^{(i)} \text{ if } i \geqslant M_\ell \ (\ell = 0, 1, 2, \ldots); \tag{13.34}$$

$$\beta_{k_i-1}^{(i)} \ldots \beta_0^{(i)} = \gamma_0\gamma_1 \ldots \gamma_\ell u_\ell^{(i)} \text{ if } i \geqslant M_\ell \ (\ell = 0, 1, 2, \ldots); \tag{13.35}$$

But $\alpha_j^{(i)}, \beta_j^{(i)}$ do not depend on i since in our case $z_i = z = \sum_{n=0}^{\infty} \alpha_n p^n$ (where $\alpha_0, \alpha_1, \ldots \in \mathbb{F}_p$) for all i; therefore $\alpha_n^{(i)} = \alpha_n$ for all $n, i \in \mathbb{N}_0$. As $x = 0.\zeta_0\zeta_1 \ldots$ (cf. (13.30)) and $x \in \mathbb{Z}_p \cap \mathbb{Q}$, then the right-infinite word $\zeta_0\zeta_1 \ldots$ must be purely periodic (cf. Corollary 2.15) with a period $\chi_0 \ldots \chi_{t-1}$ of length $t > 0$: that is, $\zeta_0\zeta_1 \ldots = (\chi_0 \ldots \chi_{t-1})^\infty$. Now in (13.34) put $\ell = mt - 1$; then for every $m \in \mathbb{N}$, we have that $\alpha_{k_i-1} \ldots \alpha_{k_i-mt} = (\chi_0 \ldots \chi_{t-1})^m$ for all $i \geqslant M_{mt}$. Now denote via $s_m^{(i)}$ the state the automaton $\mathfrak{A}$ reaches after being fed by the word $w_{mt}^{(i)}$. Fix $m \in \mathbb{N}$ and denote s_m a state which occurs in the sequence $(w_{mt}^{(i)})_{i=M_{mt}}^{\infty}$ infinitely often; due to the finiteness of the automaton $\mathfrak{A}$, such state exists. Denote K_m the smallest $i \geqslant M_{mt}$ such that $s_m = s_m^{(i)}$ (therefore $K_m \geqslant M_{mt}$). And again due to the finiteness of the automaton $\mathfrak{A}$ in the sequence $(s_m)_{m=1}^{\infty}$ some state (say, s) occurs infinitely often. Let $(m_j)_{j=0}^{\infty}$ be the corresponding infinite (thus, strictly increasing) subsequence, i.e., $s_{m_j} = s$; then as the sequence $(M_\ell)_{\ell=0}^{\infty}$ is strictly increasing and as $K_m \geqslant M_{mt}$, in the sequence $(K_{m_j})_{j=0}^{\infty}$ there exists a strictly increasing subsequence, say $(K_{m_{j_r}})_{r=0}^{\infty}$ (note that the sequence $(m_{j_r})_{r=0}^{\infty}$ is also strictly increasing). Now from (13.34)–(13.35) it follows that once being fed successfully by purely periodic words $w_r = \alpha_{k_i-1}\ldots\alpha_{k_i-m_{j_r}t} = (\chi_0\ldots\chi_{t-1})^{m_{j_r}}$ for $i = K_{m_{j_r}}$, $r = 0, 1, 2, \ldots$, the automaton $\mathfrak{A}(s)$ outputs the words $v_r = \gamma_0\gamma_1\ldots\gamma_{tm_{j_r}-1}$. Now by combining Lemma 13.25 with Corollary 2.15, we conclude that if $z' \in \mathbb{Z}_p$ is such that $\mathsf{wrd}\, z' = (\chi_0\ldots\chi_{t-1})^{\infty}$ then $\lim_{r\to\infty}(z' \bmod p^{m_{j_r}})/p^{m_{j_r}} = x$ and $\lim_{r\to\infty}((\mathfrak{a}_s(z')) \bmod p^{m_{j_r}})/p^{m_{j_r}} = y \in \mathbb{Z}_p \cap \mathbb{Q}$. □

Yet one more property of automata plots is their invariance with respect to p-shifts. That is, *given a point* $(x; y) \in \mathbf{P}(\mathfrak{A})$, *take base-p expansions* $x = 0.\chi_1\chi_2\chi_3\ldots$, $y = 0.\xi_1\xi_2\xi_3\ldots$ *of coordinates* x, y; *then* $(0.\chi_2\chi_3\ldots; 0.\xi_2\xi_3\ldots) \in \mathbf{P}(\mathfrak{A})$. To put it in other words, the following proposition is true:

Proposition 13.35 *For arbitrary automaton* $\mathfrak{A}$, *if* $(x; y) \in \mathcal{P}(\mathfrak{A}) \subset \mathbb{T}^2$ *(resp.,* $(x; y) \in \mathcal{P}'(\mathfrak{A}) \subset \mathbb{T}^2$*) then* $((px) \bmod 1; (py) \bmod 1) \in \mathcal{P}(\mathfrak{A})$ *(resp.,* $((px) \bmod 1; (py) \bmod 1) \in \mathcal{P}'(\mathfrak{A})$*).*

Proof of Proposition 13.35 The first assertion follows immediately from Note 13.28 since once

$$(0.\alpha_{k_i}\ldots\alpha_0; 0.\beta_{k_i}\ldots\beta_0) \to (0.\chi_1\chi_2\ldots; 0.\xi_1\xi_2\ldots)$$

as $i \to \infty$ then necessarily

$$(0.\alpha_{k_i-1}\ldots\alpha_0; 0.\beta_{k_i-1}\ldots\beta_0) \to (0.\chi_2\chi_3\ldots; 0.\xi_2\xi_3\ldots)$$

as $i \to \infty$.

To prove the second assertion, let $f = f_{\mathfrak{A}}\colon \mathbb{Z}_p \to \mathbb{Z}_p$ be an automaton function of the automaton $\mathfrak{A}$. As $(x; y) \in \mathcal{P}'(\mathfrak{A})$, there exists $z \in \mathbb{Z}_p$ and a strictly increasing sequence $(k_i)_{i=0}^{\infty}$ over $\mathbb{N}$ such that $x = \lim_{i\to\infty}(z \bmod p^{k_i})/p^{k_i}$ and $y = \lim_{i\to\infty}(f(z) \bmod p^{k_i})/p^{k_i}$, cf. Definition 13.26. Therefore $(px) \bmod 1 = (p\lim_{i\to\infty}(z \bmod p^{k_i})/p^{k_i}) \bmod 1 = \lim_{i\to\infty}(p(z \bmod p^{k_i})/p^{k_i}) \bmod 1 = \lim_{i\to\infty}(z \bmod p^{k_i-1})/p^{k_i-1}$ as $(z \bmod p^{k_i})/p^{k_i} = \zeta_{k_i-1}p^{-1} + \zeta_{k_i-2}p^{-2} + \cdots + \zeta_0 p^{-k_i}$ once $z = \zeta_0 + \zeta_1 p + \cdots + \zeta_{k_i-1}p^{k_i-1} + \cdots$ is a p-adic canonical representation for $z \in \mathbb{Z}_p$. By the same reason, $(py) \bmod 1 = (p\lim_{i\to\infty}(f(z) \bmod p^{k_i})/p^{k_i}) \bmod 1$ $= \lim_{i\to\infty}(p(f(z) \bmod p^{k_i})/p^{k_i}) \bmod 1 = \lim_{i\to\infty}(f(z) \bmod p^{k_i-1})/p^{k_i-1}$. Therefore $((px) \bmod 1; (py) \bmod 1) \in \mathcal{P}'(\mathfrak{A})$ by Definition 13.26. □

The goal we pursue in this section is to prove that if $\mathfrak{A}$ is a finite automaton then smooth curves which lies completely in $\mathcal{P}(\mathfrak{A})$ (thus in $\mathcal{P}'(\mathfrak{A})$) can only be straight lines, and, *if finite automata plots are considered as subsets of the unit torus* $\mathbb{T}^2$ *in* $\mathbb{R}^3$, *then the smooth curves lying in the plots can only be torus windings*; cf., Theorem 13.70). For this purpose, we need some extra information (which follows) about torus knots.

13.6.2 Torus Knots, Torus Links, and Linear Flows on Torus

We will need only a few concepts from the torus knots theory; details may be found in numerous books on knot theory; see, e.g., [44, 147]. For our purposes it is enough to recall only two notions, the knot and the link. Recall that a knot is a smooth embedding of a circle $\mathbb{S}$ into $\mathbb{R}^3$ and a link is a smooth embedding of several disjoint circles in $\mathbb{R}^3$ (cf. [147]). We will consider only special types of knots and links, namely, torus knots and torus links. Informally, a torus knot is a smooth closed curve without intersections which lies completely in the surface of a torus $\mathbb{T}^2 \subset \mathbb{R}^3$, and a link (of torus knots) is a collection of (possibly knotted) torus knots; see, e.g., [55, Section 26] for formal definitions.

Definition 13.36 (Cable of the Torus) A *cable of the torus* (also, *winding of the torus*) is the image of a straight line in $\mathbb{R}^2$ under the map $\mathsf{mod}\,1\colon (x; y) \mapsto (x \,\mathsf{mod}\, 1; y \,\mathsf{mod}\, 1)$ of the Euclidean plain $\mathbb{R}^2$ onto the two-dimensional real torus $\mathbb{T}^2 = \mathbb{R}^2/\mathbb{Z} \times \mathbb{Z} = \mathbb{S} \times \mathbb{S} \subset \mathbb{R}^3$. If the line is defined by the equation $y = ax + b$ we say that a is a *slope* of the cable $\mathbf{C}(a, b)$. We denote via $\mathbf{C}(\infty, b)$ a cable which corresponds to the line $x = b$, the *meridian*, and say that the slope is ∞ in this case. Cables $\mathbf{C}(0, b)$ of slope 0 (i.e., the ones that correspond to straight lines $y = b$) are called *parallels*.

In dynamics, cables of torus $\mathbb{T}^2$ are viewed as orbits of *linear flows on torus*, that is, of dynamical systems on $\mathbb{T}^2$ defined by a pair of differential equations of the form $\frac{dx}{dt} = \beta$; $\frac{dy}{dt} = \alpha$ on $\mathbb{T}^2$, whence, by a pair of parametric equations $x = (\beta t + \tau) \,\mathsf{mod}\, 1$; $y = (\alpha t + \sigma) \,\mathsf{mod}\, 1$ in Cartesian coordinates (cf., e.g., [96, Subsection 4.2.3]).

Note 13.37 It is well-known that a cable defined by the straight line $y = ax + b$ is dense in $\mathbb{T}^2$ if and only if $-\infty < a < +\infty$ and the slope $a = \frac{\alpha}{\beta}$ is irrational; see, e.g., [96, Proposition 4.2.8] or [153, Section 5.4].

Given a Cartesian coordinate system XYZ of $\mathbb{R}^3$, a torus can be obtained by rotation around Z-axis of a circle which lies in the plain XZ. If a radius of the circle is r and the circle is centered at a point lying in X-axis at a distance R from the origin, then in cylindrical coordinates (r_0, θ, z) of $\mathbb{R}^3$ (where r_0 is a radius-vector in Cartesian coordinate system XY, θ is an angle of the radius-vector in coordinates XY, z is a Z-coordinate in Cartesian coordinate system XYZ), the torus is defined by the equation $(r_0 - R)^2 + z^2 = r^2$, and a cable (with a rational slope $\frac{\alpha}{\beta}$ where $\alpha \in \mathbb{Z}$ and $\beta \in \mathbb{N}$) of the torus is defined by the system of parametric equations (with parameter $t \in \mathbb{R}$) of the form

$$\begin{bmatrix} r_0 \\ \theta \\ z \end{bmatrix} = \begin{bmatrix} R + r\cos\left(\frac{\alpha}{\beta}t + \omega\right) \\ t \\ r\sin\left(\frac{\alpha}{\beta}t + \omega\right) \end{bmatrix}, \quad t \in \mathbb{R}. \tag{13.36}$$

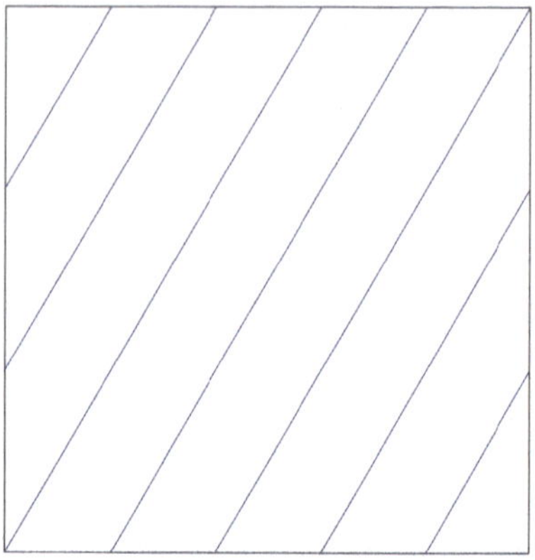

Fig. 13.12 Limit plot of the function $f(z) = \frac{5}{3}z$, $z \in \mathbb{Z}_2$, in $\mathbb{R}^2$

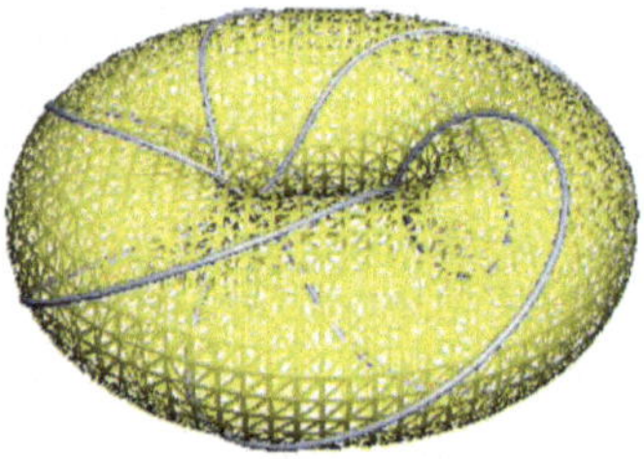

Fig. 13.13 Limit plot of the same function on the torus $\mathbb{T}^2$

The cable defined by the above equations winds β times around Z-axis and $|\alpha|$ times around a circle in the interior of the torus (the sign of α determines whether the rotation is clockwise or counter-clockwise); see, for an example of the corresponding torus knot, Figs. 13.12 and 13.13 where $\alpha = 5$ and $\beta = 3$. Letting ω in the above equations take a finite number of values, we get an example of torus link; see, e.g., Figs. 13.17 and 13.18 which illustrate a link consisting of a pair of torus knots whose slopes are $\frac{3}{5}$. Note that Figs. 13.19 and 13.20 illustrate a union of two distinct torus links (of two and of three knots respectively) rather than a single torus link of five knots. Finally, due to the above representation of a torus link in the form of equations in cylindrical coordinates, we naturally associate the torus link consisting of N cables with slopes $\frac{\alpha}{\beta}$ to a family of complex-valued functions $\psi_k : \mathbb{R} \to \mathbb{C}$ of real variable $t \in \mathbb{R}$

$$\left\{\psi_j(t) = \mathrm{e}^{\mathrm{i}(\frac{\alpha}{\beta}t+\omega_j)} \colon j = 0, 1, 2, \ldots, N-1\right\},$$

where i stands for imaginary unit $\mathrm{i} \in \mathbb{C}$: $\mathrm{i}^2 = -1$.

13.6.3 *Plots of Affine Functions*

In this subsection we completely describe limit plots of finite automata maps $z \mapsto c$ (constant maps), $z \mapsto az$ (linear maps), and $z \mapsto az + b$ (affine maps), where a, b, c are some (suitable) p-adic integers and the variable z takes values in $\mathbb{Z}_p$. Recall that an automaton $\mathfrak{A}(s_0) = \langle \mathcal{I}, \mathcal{S}, \mathcal{O}, S, O, s_0\rangle$ is called *autonomous* once neither its state update function S nor its output function O depend on input, i.e., when $s_{i+1} = S(s_i)$, $\xi_i = O(\chi_i, s_i) = O(s_i)$ $(i = 0, 1, 2, \ldots)$ (cf. Fig. 1.1).

13.6.3.1 Limit Plots of Constants

It is clear that an autonomous automaton function is a constant; however a limit plot of this function is not necessarily a straight line. For instance, the limit plot of a constant $c \in \mathbb{Z}_p$ is the whole unit square $\mathbb{I}^2$ once $c = \sum_{i=0}^{\infty} \alpha_i p^i$ where the infinite word $u = \ldots\alpha_2\alpha_1\alpha_0$ over $\mathbb{F}_p$ is such that every non-empty finite word $w = \gamma_{k-1}\gamma_{k-2}\ldots\gamma_0$ over $\mathbb{F}_p$ occurs as a subword in u, that is, if there exist a finite word v and an infinite word s over $\mathbb{F}_p$ such that u is a concatenation of v, w, and s: $u = swv$. These words exist, e.g., the base-p Champernowne word, that is, the infinite word obtained by successively concatenating finite words which are base-p expansions of numbers $1, 2, 3, \ldots$. It is obvious that respective p-adic integer is measure-1 (cf. Sect. 13.3).

On the other hand, once an autonomous automaton $\mathfrak{A}$ is finite, the corresponding infinite output word must necessarily be eventually periodic. That is, $c = \alpha_0 + \alpha_1 p + \cdots + \alpha_{r-1}p^{r-1} + (\beta_0 + \beta_1 p + \cdots + \beta_{t-1}p^{t-1})\sum_{j=0}^{\infty} p^{r+tj}$ for suitable $\alpha_i, \beta_j \in \mathbb{F}_p$; therefore a finite autonomous automaton function is a rational constant, i.e., $c \in \mathbb{Z}_p \cap \mathbb{Q}$ (cf. Propositions 2.10 and 13.24).

Furthermore, the numbers that correspond to (sufficiently long) finite output words are then all the form

$$0.\beta_k\beta_{k-1}\ldots\beta_0\beta_{t-1}\beta_{t-2}\ldots \\ \ldots\beta_0\beta_{t-1}\beta_{t-2}\ldots\beta_0\ldots\beta_{t-1}\beta_{t-2}\ldots\beta_0\alpha_{r-1}\alpha_{r-2}\ldots\alpha_0$$

for $k = 0, 1, \ldots, t-1$. Consequently, the limit plot of the automaton (in $\mathbb{R}^2$) consists of t pairwise parallel straight lines which correspond to the numbers

$$0.\beta_k\beta_{k-1}\ldots\beta_0\beta_{t-1}\beta_{t-2}\ldots\beta_0\beta_{t-1}\beta_{t-2}\ldots\beta_0\ldots = \\ = 0.\beta_k\beta_{k-1}\ldots\beta_0(\beta_{t-1}\beta_{t-2}\ldots\beta_0)^{\infty}$$

where $k = 0, 1, \ldots, t-1$ (cf. Sect. 12.1), or (which is the same) to the numbers $0.(\beta_k\beta_{k-1}\ldots\beta_0\beta_{t-1}\beta_{t-2}\ldots\beta_{k+1})^{\infty}$. That is, all the lines from the limit plot are $y = p^{\ell}h \bmod 1$, $\ell \in \mathbb{N}_0$, for any line $y = h$ belonging to the limit plot; thus the number of lines in the limit plot does not exceed t. Respectively, being considered as a point set on the torus $\mathbb{T}^2$, the limit plot consists of not more than t parallels (cf., e.g., Figs. 13.14 and 13.15).

Now we present a more formal argument and derive a little bit more information about the number of lines in the limit plot. Given $q \in \mathbb{Z}_p \cap \mathbb{Q}$, represent q as an irreducible fraction $q = a/b$ for suitable $a \in \mathbb{Z}$, $b \in \mathbb{N}$. Note that $p \nmid b$ since $q \in \mathbb{Z}_p$. Denote

$$\mathbf{C}(a/b) = \text{limit points of}\left\{\left(p^\ell\left(1-\frac{a}{b}\right)\right) \operatorname{mod} 1 \colon \ell = 0, 1, 2, \ldots\right\} = \\ = \text{limit points of}\left\{\left(-p^\ell\frac{a}{b}\right) \operatorname{mod} 1 \colon \ell = 0, 1, 2, \ldots\right\}. \qquad (13.37)$$

Since $a/b \in \mathbb{Z}_p \cap \mathbb{Q}$, by Proposition 2.10 a p-adic canonical form of a/b is

$$a/b = \alpha_0 + \alpha_1 p + \cdots + \alpha_{r-1}p^{r-1} + (\beta_0 + \beta_1 p + \cdots + \beta_{t-1}p^{t-1})\sum_{j=0}^{\infty} p^{r+tj} \qquad (13.38)$$

for suitable $\alpha_i, \beta_m \in \{0, 1, \ldots, p-1\}$, or, in other words, the infinite word that corresponds to a/b is $(\beta_{t-1}\ldots\beta_0)^\infty\alpha_{r-1}\ldots\alpha_0$. Then from Proposition 2.14 it follows that

$$(a/b) \operatorname{mod} 1 = (p^r \times 0.(\hat\beta_{t-1}\ldots\hat\beta_0)^\infty) \operatorname{mod} 1 = \\ = 0.(\hat\beta_{t-1-\bar r}\hat\beta_{t-2-\bar r}\ldots\hat\beta_0\hat\beta_{t-1}\hat\beta_{t-2}\ldots\hat\beta_{t-\bar r})^\infty \operatorname{mod} 1,$$

where $\hat\beta_i = p-1-\beta_i$, $i = 0, 1, 2, \ldots, t-1$, and $\bar r$ is the least non-negative residue of r modulo t if $t > 1$ or $\bar r = 0$ if otherwise. From here in view of (2.12) we deduce that

$$(-a/b) \operatorname{mod} 1 = 0.(\beta_{t-1-\bar r}\beta_{t-2-\bar r}\ldots\beta_0\beta_{t-1}\beta_{t-2}\ldots\beta_{t-\bar r})^\infty \operatorname{mod} 1$$

and thus

$$\mathbf{C}(a/b) = \\ = \left\{0.(\beta_{t-1-\ell}\beta_{t-2-\ell}\ldots\beta_0\beta_{t-1}\beta_{t-2}\ldots\beta_{t-\ell})^\infty \operatorname{mod} 1 \colon \ell = 0, 1, 2, \ldots, t-1\right\} = \\ \left\{\frac{\operatorname{num}(\upsilon)}{p^t-1} \colon \upsilon \in \{\hat\zeta_{t-1}\hat\zeta_{t-2}\ldots\hat\zeta_0, \hat\zeta_{t-2}\hat\zeta_{t-3}\ldots\hat\zeta_0\hat\zeta_{t-1}, \hat\zeta_{t-3}\hat\zeta_{t-4}\ldots\hat\beta_0\hat\zeta_{t-1}\hat\zeta_{t-2}, \ldots\}\right\}$$

where $(a/b) \operatorname{mod} 1 = (\zeta_0 + \zeta_1 p + \cdots + \zeta_{t-1}p^{t-1})(p^t-1)^{-1}$ (cf. Proposition 2.11 and Corollary 2.16). Now we can suppose that t is a period length of the rational

p-adic integer $a/b \in \mathbb{Z}_p \cap \mathbb{Q}$ (cf. Sect. 2.3.2); then in view of Proposition 2.17 we conclude that

$$\mathbf{C}(a/b) = \left\{(-p^{\ell}(a/b)) \bmod 1 \colon \ell = 0, 1, \ldots, (\mathsf{mult}_b\, p) - 1\right\} =$$
$$= \{0.(w)^{\infty} \bmod 1 \colon w \text{ runs through all cyclic shifts of the word } \beta_{(\mathsf{mult}_b\, p)-1} \cdots \beta_0\} =$$
$$\left\{0.(v)^{\infty} \bmod 1 \colon v \text{ runs through all cyclic shifts of the word } \hat{\zeta}_{(\mathsf{mult}_b\, p)-1} \cdots \hat{\zeta}_0\right\} =$$
$$= \left\{\left(-p^{\ell}\frac{d}{p^{\mathsf{mult}_b\, p} - 1}\right) \bmod 1 \colon \ell = 0, 1, \ldots, (\mathsf{mult}_b\, p) - 1\right\} \tag{13.39}$$

since

$$1 - \frac{\zeta_0 + \zeta_1 p + \cdots + \zeta_{t-1} p^{t-1}}{p^t - 1} = \frac{\hat{\zeta}_{t-1} + \hat{\zeta}_0 p + \hat{\zeta}_1 p^2 + \cdots + \hat{\zeta}_{t-2} p^{t-1}}{p^t - 1} \text{ and}$$
$$p\frac{\hat{\zeta}_0 + \hat{\zeta}_1 p + \cdots + \hat{\zeta}_{t-1} p^{t-1}}{p^t - 1} = \hat{\zeta}_{t-1} + \frac{\hat{\zeta}_{t-1} + \hat{\zeta}_0 p + \hat{\zeta}_1 p^2 + \cdots + \hat{\zeta}_{t-2} p^{t-1}}{p^t - 1}.$$

Note that $0.(w)^{\infty} \bmod 1 = 0.(w)^{\infty}$ except of the case when $t = 1$ and w is a single-letter word that consists of the only letter $p - 1$ (in the latter case $0.(w)^{\infty} = 1$ and thus $0.(w)^{\infty} \bmod 1 = 0$). Similarly, $0.(v)^{\infty} \bmod 1 = 0.(v)^{\infty}$ except of the case when $a/b \in \mathbb{Z}$ and thus $\zeta_0 = \ldots = \zeta_{t-1} = 0$ (so $\hat{\zeta}_0 = \ldots = \hat{\zeta}_{t-1} = p - 1$ and $0.(v)^{\infty} = 1$). But this case happens if and only if $a/b \in \mathbb{Z}$, i.e., when $\mathbf{C}(a/b) = \{0\}$.

We now summarize all these considerations in the proposition:

Proposition 13.38 *Let $f_{\mathfrak{A}} \colon z \mapsto q$ be an automaton function of a finite automaton* $\mathfrak{A}$ (therefore $q \in \mathbb{Z}_p \cap \mathbb{Q}$ by Proposition 13.24); *then $\mathcal{P}'(\mathfrak{A}) \subset \mathbb{T}^2$ is a disjoint union of t parallels $\mathbf{C}(0, e)$, $e \in \mathbf{C}(q)$, and t is a period length of q* (cf. (13.37) and (13.39)).

Note 13.39 In conditions of Proposition 13.38 the constant $q \in \mathbb{Z}_p \cap \mathbb{Q}$ can be represented as an irreducible fraction $q = a/b$ where $a \in \mathbb{Z}$, $b \in \mathbb{N}$, $p \nmid b$ (we put $b = 1$ and $a = 0$ if $q = 0$). Then the limit plot $\mathcal{P}'(\mathfrak{A}) \subset \mathbb{T}^2$ is a torus link that consists of $t = \mathsf{mult}_b\, p$ trivial torus cables (parallels) with slopes 0; to the link there corresponds a collection of t complex constants (which are b-th roots of 1)

$$\left\{\psi_{\ell} = \mathrm{e}^{-2\pi \mathrm{i} p^{\ell} q} \colon \ell = 0, 1, \ldots, (\mathsf{mult}_b\, p) - 1\right\}$$

where i stands for imaginary unit $i \in \mathbb{C}$: $i^2 = -1$ (cf. Sect. 13.6.2).

Being considered in the unit real square $\mathbb{I}^2$, the limit plot $\mathcal{P}'(\mathfrak{A})$ is a collection of $t = \mathsf{mult}_b\, p$ segments of straight lines $y = c(t, k, u)$ that cross $\mathbb{I}^2$, where

$$c(t,k,u) = \left(-p^k \frac{u}{p^t - 1}\right) \bmod 1 =$$
$$= 0.(\hat{\zeta}_{t-1-k}\hat{\zeta}_{t-2-k}\ldots\hat{\zeta}_0\hat{\zeta}_{t-1}\hat{\zeta}_{t-2}\ldots\hat{\zeta}_{t-k})^\infty \bmod 1;\ k = 0, 1, \ldots, t-1. \tag{13.40}$$

Here $q \bmod 1 = u(p^t - 1)^{-1}$, $0 \leqslant u \leqslant p^t - 2$, and a base p-expansion of u is $u = \zeta_0 + \zeta_1 p + \cdots + \zeta_{t-1}p^{t-1}$ (cf. Proposition 2.11); $\hat{\zeta} = p - 1 - \zeta$ for $\zeta \in \{0, 1, \ldots, p-1\}$. In other words, all the constants $c(t, k, u)$ are of the form

$$c(t,k,u) = 0.v^\infty \bmod 1 = \frac{\mathsf{num}(v)}{p^t - 1} \bmod 1, \tag{13.41}$$

where v runs through all cyclic shifts of the word $\hat{\zeta}_{t-1}\hat{\zeta}_{t-2}\ldots\hat{\zeta}_0$; that is, $v \in \{\hat{\zeta}_{t-1}\hat{\zeta}_{t-2}\ldots\hat{\zeta}_0, \hat{\zeta}_{t-2}\hat{\zeta}_{t-3}\ldots\hat{\zeta}_0\hat{\zeta}_{t-1}, \ldots\}$.

If q is represented in the p-adic canonical form (13.38) rather than in a form of Proposition 2.11, then all the lines of the limit plot can be represented as

$$y = 0.(\beta_{t-1-\ell}\beta_{t-2-\ell}\ldots\beta_0\beta_{t-1}\beta_{t-2}\ldots\beta_{t-\ell})^\infty \bmod 1;\ \ell = 0, 1, 2, \ldots, t-1. \tag{13.42}$$

Note that we may omit $\bmod 1$ in (13.41) and in (13.42) in all cases but the case when simultaneously the length t of the period is 1 and $\hat{\zeta}_0 = p - 1$ (respectively, $\beta_0 = p - 1$); but $q \in \mathbb{Z}$ in that case and therefore $\mathbf{C}(q) = \{0\}$.

The following property of the set $\mathbf{C}(q)$ will be used in further proofs:

Corollary 13.40 *Given $q_1, q_2 \in \mathbb{Z}_p \cap \mathbb{Q} \cap [0, 1)$, the following alternative holds: Either $\mathbf{C}(q_1) = \mathbf{C}(q_2)$ or $\mathbf{C}(q_1) \cap \mathbf{C}(q_2) = \emptyset$.*

Proof of Corollary 13.40 The result is clear enough since the numbers that constitute $\mathbf{C}(q)$ are exactly all numbers whose base-p expansions are of the form $0.(u)^\infty$ where u runs through all cyclic shifts of the finite word w which is the (shortest) period of $q \bmod 1$, cf. Note 13.39; nonetheless we give a formal proof which follows.

Given $q_i \in \mathbb{Z}_p \cap \mathbb{Q} \cap [0, 1)$, $i = 1, 2$, represented as irreducible fractions $q_i = a_i/b_i$ whose denominators b_i are co-prime to p, let $\mathbf{C}(q_1) \cap \mathbf{C}(q_2) \neq \emptyset$; then $p^{\ell_1}(a_1/b_1) = p^{\ell_2}(a_2/b_2)$ for suitable $\ell_1, \ell_2 \in \mathbb{N}_0$. If $\ell_1 = \ell_2$ then $a_1/b_1 = a_2/b_2$ and thus $\mathbf{C}(q_1) = \mathbf{C}(q_2)$. Let $\ell_1 > \ell_2$, then $p^{\ell_1 - \ell_2}a_1 b_2 = a_2 b_1$; so since $\gcd(b_1, p) = 1$ we conclude that $a_2 = p^{\ell_1 - \ell_2 + s}a_2'$ for a suitable $s \in \mathbb{N}_0$ and $a_2' \in \mathbb{Z}$ such that $\gcd(a_2', p) = 1$. Therefore necessarily $a_1 = p^s a_1'$ where $\gcd(a_1', p) = 1$ since $\gcd(b_1, p) = \gcd(b_2, p) = 1$. But then we conclude that $p^{\ell_1 - \ell_2 + s}a_1' b_2 = p^{\ell_1 - \ell_2}a_1 b_2 = a_2 b_1 = p^{\ell_1 - \ell_2 + s}a_2' b_1$ and therefore $q_1 = p^s q$, $q_2 = p^{\ell_1 - \ell_2 + s}q$ where $q = a_1'/b_1 = a_2'/b_2$. Hence $\mathbf{C}(q_1), \mathbf{C}(q_2) \subset \mathbf{C}(q)$; the

Fig. 13.14 Limit plot of the constant function $f(z) = \frac{2}{7}$ ($z \in \mathbb{Z}_2$), in $\mathbb{R}^2$

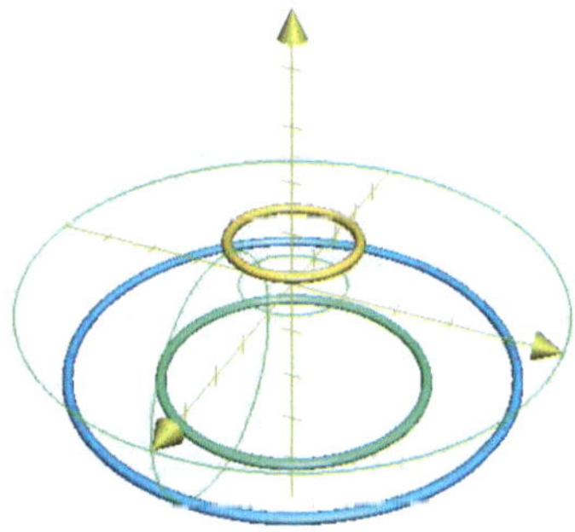

Fig. 13.15 Limit plot of the same function on the torus $\mathbb{T}^2$

inverse inclusion also holds since $\mathbf{C}(p^\ell q) = \mathbf{C}(q)$ for any $q \in \mathbb{Z}_p \cap \mathbb{Q}$ by, e.g., (13.42). □

Example 13.41 Let $p = 2$ and $q = 2/7$. Then $\mathsf{mult}_7 2 = 3$ and the limit plot consists of three lines. The binary infinite word that corresponds to the 2-adic canonical representation of $2/7$ is $(011)^\infty 10$, so the period of $2/7$ is 011, the pre-period is 01, and $u = 2 = 0+1\times 2+0\times 2^2$. Therefore the tree lines of the limit plot are $y = 0.(101)^\infty = 5/7 = (-2/7) \bmod 1 = c(3, 0, 2)$, $y = 0.(011)^\infty = 6/7 = (-1/7) \bmod 1 = c(3, 2, 2)$, $y = 0.(110)^\infty = 3/7 = (-4/7) \bmod 1 = c(3, 1, 2)$. The limit plot (on the unit square and on the torus) is illustrated by Figs. 13.14 and 13.15 accordingly (Fig. 13.16).

13.6.3.2 Limit Plots of Linear Maps

We examine now limit plots of linear maps $z \mapsto cz$ ($z \in \mathbb{Z}_p$) which are finite automaton functions. By Proposition 13.24, the finiteness takes place if and only if $c \in \mathbb{Z}_p \cap \mathbb{Q}$.

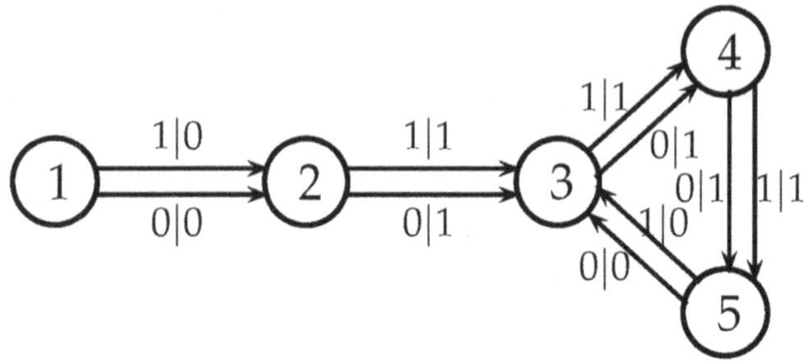

Fig. 13.16 State transition diagram of autonomous automaton whose automaton function $f\colon \mathbb{Z}_2 \to \mathbb{Z}_2$ is a constant: $f(z) = 2/7$, $(z \in \mathbb{Z}_2)$. State 1 is initial

Proposition 13.42 *Given $c \in \mathbb{Z}_p \cap \mathbb{Q}$, represent $c = a/b$, where $a \in \mathbb{Z}$, $b \in \mathbb{N}$, a, b are coprime, $p \nmid b$. If $\mathfrak{A}$ is an automaton such that $f_{\mathfrak{A}}(z) = cz$ $(z \in \mathbb{Z}_p)$ then $\mathcal{P}'(\mathfrak{A}) = \{(x \bmod 1; (cx) \bmod 1)\colon x \in \mathbb{R}\} = \mathbf{C}(c, 0)$ is a cable (with a slope c) of the unit two-dimensional real torus $\mathbb{T}^2$. For every $c \in \mathbb{Z}_p \cap \mathbb{Q}$ the automaton $\mathfrak{A}$ may be taken a finite.*

Proof of Proposition 13.42 By Proposition 13.24, the map $z \mapsto cz$ on $\mathbb{Z}_p$ is an automaton function of a finite automaton if and only if $c \in \mathbb{Z}_p \cap \mathbb{Q}$.

Given $x \in [0, 1)$, take $z \in \mathbb{Z}_p$ such that $\lim_{i\to\infty} z \bmod p^{k_i}/p^{k_i} = x$ for a suitable strictly increasing sequence $k_1, k_2, \ldots \in \mathbb{N}$. As $c \in \mathbb{Z}_p \cap \mathbb{Q}$, then $c = u + v/(p^t - 1)$ for suitable $u \in \mathbb{Z}$, $t \in \mathbb{N}$, $v \in \{0, 1, 2, \ldots, p^t - 2\}$, by Proposition 2.11. If $t > 1$, then considering residues of k_i modulo t, we see that at least one residue (say, $\ell \in \{0, 1, \ldots, t-1\}$) occurs in the sequence $k_1, k_2, \ldots$ infinitely many times. Therefore $\lim_{j\to\infty} z \bmod p^{r_j t+\ell}/p^{r_j t+\ell} = x$ for a respective strictly increasing sequence $r_1, r_2, \ldots \in \mathbb{N}$. The latter equality trivially holds when $t = 1$: one just takes $r_j = k_j$ and $\ell = 0$. So further we assume that $k_i = r_i t + \ell$, $i = 1, 2, \ldots$.

For $i = 1, 2, \ldots$ we have that

$$\frac{(cz) \bmod p^{k_i}}{p^{k_i}} = \frac{1}{p^{k_i}}((c \bmod p^{k_i})(z \bmod p^{k_i})) \bmod p^{k_i} =$$

$$= \left(c \bmod p^{k_i} \frac{z \bmod p^{k_i}}{p^{k_i}}\right) \bmod 1 \tag{13.43}$$

As $0 \leqslant \ell < t$ and $k_i = r_i t + \ell$, we have that

$$c \bmod p^{k_i} = \left(u + \frac{v}{p^t - 1}\right) \bmod p^{k_i} = \left(u - v\frac{p^{(r_i+1)t} - 1}{p^t - 1}\right) \bmod p^{k_i} \tag{13.44}$$

Note that the argument of mod in the right-hand side of (13.44) is negative once i is sufficiently large; therefore once i is large enough then

$$\left(u - v\frac{p^{(r_i+1)t} - 1}{p^t - 1}\right) \bmod p^{k_i} = Lp^{k_i} + u - v\frac{p^{(r_i+1)t} - 1}{p^t - 1}$$

for a suitable $L \in \mathbb{N}$ which does not depend on i (actually it is not difficult to see that $L = \lceil vp^{t-\ell}(p^t - 1)^{-1}\rceil$). Thus,

$$\left(c \bmod p^{k_i} \frac{z \bmod p^{k_i}}{p^{k_i}}\right) \bmod 1 =$$

$$= \left(\left(Lp^{k_i} + u - v\frac{p^{(r_i+1)t} - 1}{p^t - 1}\right) \frac{z \bmod p^{k_i}}{p^{k_i}}\right) \bmod 1 =$$

$$= \left(L(z \bmod p^{k_i}) + u\frac{z \bmod p^{k_i}}{p^{k_i}} + \frac{v}{p^t - 1} \times \frac{z \bmod p^{k_i}}{p^{k_i}} - \frac{vp^{t-\ell}}{p^t - 1}(z \bmod p^{k_i})\right)$$
$$\times \bmod 1 =$$

$$= \left(c\frac{z \bmod p^{k_i}}{p^{k_i}} - \frac{vp^{t-\ell}}{p^t - 1} \times (z \bmod p^{k_i})\right) \bmod 1, \qquad (13.45)$$

$$\frac{wp^{rt}}{p^t - 1} \bmod 1 = \left(w\frac{p^{rt} - 1}{p^t - 1} + \frac{w}{p^t - 1}\right) \bmod 1 = \left(\frac{w}{p^t - 1}\right) \bmod 1 \qquad (13.46)$$

as $p^t - 1$ is a factor of $p^{rt} - 1$.

Secondly, put $\bar{z} = p^{t-\ell}z$, then $p^{t-\ell}(z \bmod p^{r_it+\ell}) = \bar{z} \bmod p^{(r_i+1)t}$ and

$$x = \lim_{i\to\infty} \frac{z \bmod p^{k_i}}{p^{k_i}} = \lim_{i\to\infty} \frac{\bar{z} \bmod p^{r_it}}{p^{r_it}},$$

so in (13.45)

$$\left(\frac{vp^{t-\ell}}{p^t - 1}\right) \times (z \bmod p^{r_it+\ell}) = \left(\frac{v}{p^t - 1}\right) \times (\bar{z} \bmod p^{(r_i+1)t})$$

and $\bar{z} = zp^{t-\ell} \in \mathbb{Z}_p$ (recall that $k_i = r_it + \ell$ where $\ell \in \{0, 1, \ldots, t - 1\}$).

Let $\bar{z} = \zeta_0 + \zeta_1 p^t + \zeta_2 p^{2t} + \cdots$ be the base-p^t representation of $\bar{z}$ (i.e., $\zeta_j \in \{0, 1, \ldots, p^t - 1\}$); then by combining (13.43) and (13.45) with (13.46) we get

$$\frac{(cz) \bmod p^{k_i}}{p^{k_i}} = \left(c\frac{z \bmod p^{k_i}}{p^{k_i}} - \frac{v}{p^t - 1}\operatorname{wt}_{p^t}(\bar{z} \bmod p^{(r_i+1)t})\right) \bmod 1 =$$
$$= \left(c\frac{z \bmod p^{k_i}}{p^{k_i}} - \frac{v}{p^t - 1} \times ((\operatorname{wt}_{p^t}(\bar{z} \bmod p^{(r_i+1)t})) \bmod (p^t - 1))\right) \bmod 1 \tag{13.47}$$

where wt_{p^t} stands for a p^t-weight of a natural number, that is, the sum of digits of the number in its base-p^t representation; i.e., $\operatorname{wt}_{p^t}(\bar{z} \bmod p^{(r_i+1)t}) = \zeta_0 + \zeta_1 + \cdots + \zeta_i \in \mathbb{N}_0$. Therefore every limit point of the sequence

$$\left(\frac{(cz) \bmod p^{k_i}}{p^{k_i}}\right)_{i=1}^{\infty}$$

is of the form

$$\left(cx + \frac{vw}{p^t - 1}\right) \bmod 1 \tag{13.48}$$

for a suitable $w \in \{0, 1, \ldots, p^t - 2\}$.

We claim that, on the other hand, given $x \in [0, 1)$ and $h \in \{0, v, 2v \bmod (p^t - 1), \ldots, ((p^t - 2)v) \bmod (p^t - 1)\}$ (i.e., h lying in the ideal $\langle v \rangle$ of the residue ring $\mathbb{Z}/(p^t - 1)\mathbb{Z}$ generated by v), there exists $z \in \mathbb{Z}_p$ and a strictly increasing sequence $k_1, k_2, \ldots, \in \mathbb{N}$ such that

$$x = \lim_{i \to \infty} \frac{z \bmod p^{k_i}}{p^{k_i}}, \text{ and}$$
$$\frac{(cz) \bmod p^{k_i}}{p^{k_i}} = \left(c\frac{z \bmod p^{k_i}}{p^{k_i}} + \frac{h}{p^t - 1}\right) \bmod 1.$$

Indeed, take $z \in \mathbb{Z}_p$ and $k_i = r_i t + \ell$ as above; then all limit points of the sequence $((cz) \bmod p^{k_i}/p^{k_i})_{i=1}^{\infty}$ are of the form (13.48) for, say, $w = w_1, \ldots, w_s \in \{0, 1, \ldots, p^t - 2\}$. If $h \equiv vw_j \pmod{(p^t - 1)}$ for some $j = 1, 2, \ldots, s$, then there is nothing to prove; if $h \not\equiv vw_j \pmod{(p^t - 1)}$ for all $j = 1, 2, \ldots, s$, then we tweak z as follows. As the point of the form (13.48) for $w = w_1$ is a limit point of the sequence $((cz) \bmod p^{k_i}/p^{k_i})_{i=1}^{\infty}$, then $(-vw_1) \bmod (p^t - 1)$ occurs in the sequence $((-v((\operatorname{wt}_{p^t} \bar{z}) \bmod p^{(r_i+1)t})) \bmod (p^t - 1))_{i=1}^{\infty}$ infinitely many times (cf. (13.47)); so some $\bar{w} \in \{0, 1 \ldots, p^t - 2\}$ such that $v\bar{w} \equiv vw_1 \pmod{(p^t - 1)}$ occurs in the sequence $((\mathrm{wt}_{p^t}\bar{z}) \bmod p^{(r_i+1)t}) \bmod (p^t-1)$ infinitely many times:

$$\bar{w} = ((\mathrm{wt}_{p^t}\bar{z}) \bmod p^{(r_i+1)t}) \bmod (p^t-1) = (\zeta_0+\zeta_1+\cdots+\zeta_i) \bmod (p^t-1)$$

for $i = i_1, i_2, \ldots$ $(1 < i_1 < i_2 < \ldots)$.

As $h \in \langle v\rangle$, then $h \equiv -v\tilde{w} \pmod{(p^t-1)}$ for a suitable $\tilde{w} \in \{0, 1, \ldots, p^t-2\}$. Now put $\tilde{z} = \zeta_0+\tilde{\zeta}_1 p^t+\zeta_2 p^{2t}+\zeta_3 p^{3t}+\cdots$, where $\tilde{\zeta}_1 \equiv \zeta_1-\bar{w}+\tilde{w} \pmod{(p^t-1)}$; then $\tilde{w} = (\mathrm{wt}_{p^t}\tilde{z}) \bmod p^{(r_i+1)t}) \bmod (p^t-1)$. But $\lim_{i\to\infty}(z \bmod p^{k_i}/p^{k_i}) = \lim_{i\to\infty}(\tilde{z} \bmod p^{k_i}/p^{k_i}) = x$; so finally we conclude by (13.47) that

$$\lim_{j\to\infty}\frac{(c\tilde{z}) \bmod p^{r_{i_j}t+\ell}}{p^{r_{i_j}t+\ell}} = \left(cx+\frac{h}{p^t-1}\right) \bmod 1.$$

Thus we have shown that

$$\mathcal{P}'(\mathfrak{A}) = \left\{\left(x; \left(cx+\frac{e}{p^t-1}\right) \bmod 1\right) : x \in [0,1), e \in \langle v\rangle\right\} \tag{13.49}$$

But the right-hand side in (13.49) is a cable of torus with slope c since

$$\left\{\left(x; \left(cx+\frac{h}{p^t-1}\right) \bmod 1\right) : x \in [0,1), h \in \langle v\rangle\right\} =$$
$$= \{(y \bmod 1; (cy) \bmod 1) : y \in \mathbb{R}\} \tag{13.50}$$

Indeed, if $y_1 = y+n$ for some $n \in \mathbb{Z}$, then $y_1 \bmod 1 = y \bmod 1$ and $(cy_1) \bmod 1 = (c(y+n)) \bmod 1 = (((u+v(p^t-1)^{-1})(y+n)) \bmod 1 = (cy+vn(p^t-1)^{-1}) \bmod 1 = (cy+((vn) \bmod (p^t-1))(p^t-1)^{-1}) \bmod 1$, and (13.50) follows. This concludes the proof. □

Example 13.43 Take $p = 2$ and $c = 5/3$. Figures 13.12 and 13.13 illustrate the limit plot of the function $f(z) = (5/3)z$ in $\mathbb{I}^2$ and in $\mathbb{T}^2$, respectively.

13.6.3.3 Limit Plots of Affine Maps

We finally combine the above two cases (constant maps and linear maps) into a single one to describe limit plots of finite automata whose functions are affine, i.e., of the form $z \mapsto cz+q$ $(z \in \mathbb{Z}_p)$. It is evident that the limit plot should be a torus link consisting of several disjoint cables with slopes c since the limit plot of the constant q is a collection of parallels (cf. Propositions 13.42 and 13.38). We will give a formal proof of this claim and find the number of knots in the link.

Recall that by Proposition 13.24 the map $z \mapsto cz+q$ of $\mathbb{Z}_p$ into itself is an automaton function of some finite automaton if and only if $c, q \in \mathbb{Z}_p \cap \mathbb{Q}$. The following proposition shows that we do not alter the limit plot of the map once we replace q by $q+n$ for arbitrary $n \in \mathbb{Z}$.

Proposition 13.44 *Let $f\colon z \mapsto cz+q$ where $c, q \in \mathbb{Z}_p \cap \mathbb{Q}$, $z \in \mathbb{Z}_p$; let $\bar{q} = q \bmod 1$, $\bar{f}\colon z \mapsto cz+\bar{q}$. Then $\mathcal{P}'(f) = \mathcal{P}'(\bar{f})$.*

Proof of Proposition 13.44 Indeed, once $n \in \mathbb{Z}$ then $\lim_{k\to\infty} n \bmod p^k/p^k \in \{0,1\}$; the limit is 1 if and only if n is negative since given canonical p-adic representation $n = \alpha_0 + \alpha_1 p + \cdots$ of negative $n \in \mathbb{Z}$, all $\alpha_i = p-1$ if i is large enough (cf. Sect. 2.3.2). Therefore $(\lim_{k\to\infty}(z+n) \bmod p^k/p^k) \bmod 1 = (\lim_{k\to\infty}(z \bmod p^k + n \bmod p^k) \bmod p^k/p^k) \bmod 1 = \lim_{k\to\infty}(z \bmod p^k/p^k + n \bmod p^k/p^k) \bmod 1 = (\lim_{k\to\infty} z \bmod p^k/p^k) \bmod 1$ for all $z \in \mathbb{Z}$. □

Note that the map $z \mapsto cz+\bar{q}$ from the statement of Proposition 13.44 is an automaton function for a suitable finite automaton $\mathfrak{B}$, and $\mathcal{P}'(\mathfrak{A}) = \mathcal{P}'(\mathfrak{B})$, where $\mathfrak{A}$ is a finite automaton whose automaton function is f.

Now we describe the limit plot of a special affine map with $c=1, q \neq 0$.

Lemma 13.45 *Given a finite automaton $\mathfrak{A}$ whose automaton function is $f(z) = z+q$ (whence, $q \in \mathbb{Z}_p \cap \mathbb{Q}$), the limit plot $\mathcal{P}'(\mathfrak{A}) \subset \mathbb{T}^2$ is a link of a finite number of torus knots which are cables $\mathbf{C}(1,e)$ where e is running over $\mathbf{C}(q)$.*

Proof of Lemma 13.45 We will prove that once $\mathfrak{A}$ is a finite automaton such that $f_{\mathfrak{A}}(z) = f(z) = z+q$ then

$$\mathcal{P}'(\mathfrak{A}) = \bigcup_{e \in \mathbf{C}(q)} \mathbf{C}(1,e). \tag{13.51}$$

Note that if $e=0$ and $e \in \mathbf{C}(q)$ then $\mathbf{C}(q) = \{0\}$ by Proposition 13.38 and there is nothing to prove. So further we assume that $e \in \mathbf{C}(q)$ and $e \neq 0$.

By Proposition 13.44 we may assume that $q \in \mathbb{Z}_p \cap \mathbb{Q} \cap (-1,0]$ then $q = d(p^t-1)^{-1} - 1$ for suitable $d \in \{0,1\ldots,p^t-2\}$, cf. Proposition 2.11; that is, $d = \zeta_{t-1} + \zeta_{t-2}p + \cdots + \zeta_0 p^{t-1}$, where $\zeta_0,\ldots,\zeta_{t-1} \in \{0,1,\ldots,p-1\}$ and therefore

$$q = -(\zeta_{t-1} + \zeta_{t-2}p + \cdots + \zeta_0 p^{t-1})(1+p^t+p^{2t}+\cdots) - 1 =$$
$$= ((p-1-\zeta_{t-1}) + (p-1-\zeta_{t-2})p + \cdots + (p-1-\zeta_0)p^{t-1})(1+p^t+p^{2t}+\cdots)$$

as $(p^t-1)^{-1} = -(1+p^t+p^{2t}+\cdots)$ in $\mathbb{Z}_p$ by Note 2.12. Therefore, in $\mathbb{Z}_p$ the rational number q can be represented as

$$q = (\eta_0 + \eta_1 p + \cdots + \eta_{t-1}p^{t-1})(1+p^t+p^{2t}+\cdots) \tag{13.52}$$

where $\eta_j = p-1-\zeta_{t-1-j}$, $j = 0,1,\ldots,t-1$.

Given $x \in [0, 1)$ take a sequence $n_i \in \mathbb{N}_0$, $i = 1, 2, \ldots$, and a strictly increasing sequence $k_i \in \mathbb{N}$, $i = 1, 2, \ldots$, such that $k_i \geqslant \lfloor \log_p n_i \rfloor + 1$, $\lim_{i \to \infty} n_i / p^{k_i} = x$, and $k_i \bmod t = s \in \{0, 1, \ldots, t-1\}$ for all $i = 1, 2, \ldots$. This is always possible since if, e.g., $x = \xi_1 p^{-1} + \xi_2 p^{-2} + \cdots$ for suitable $\xi_1, \xi_2, \ldots \in \{0, 1, \ldots, p-1\}$, then one takes $n_i = \xi_1 p^{j-1} + \xi_2 p^{j-2} + \cdots + \xi_{j-1}$ where $j = it + s$ and put $k_i = it + s$ for $i = 1, 2, \ldots$.

Considering a sequence $(n_i + q)_{i=0}^{\infty}$ in $\mathbb{Z}_p$, we see that

$$\frac{(n_i + q) \bmod p^{k_i}}{p^{k_i}} = \left(\frac{n_i}{p^{k_i}} + \frac{q \bmod p^{k_i}}{p^{k_i}} \right) \bmod 1 \tag{13.53}$$

But $k_i = it + s$, $s \in \{0, 1 \ldots, t-1\}$; thus

$$\left(\lim_{i \to \infty} \left(\frac{q \bmod p^{k_i}}{p^{k_i}} \right) \right) \bmod 1 = (\lim_{i \to \infty} ((\eta_0 + \eta_1 p + \cdots + \eta_{s-1} p^{s-1}) p^{-s} +$$

$$(\eta_0 + \eta_1 p + \cdots + \eta_{t-1} p^{t-1})(p^{-it-s} + p^{(-i+1)t-s} + \cdots + p^{-t-s}))) \bmod 1 =$$

$$= 0.(\eta_{s-1} \eta_{s-2} \ldots \eta_0 \eta_{t-1} \eta_{t-2} \ldots \eta_s)^{\infty} \bmod 1 \tag{13.54}$$

if $s \neq 0$, or

$$\left(\lim_{i \to \infty} \left(\frac{q \bmod p^{k_i}}{p^{k_i}} \right) \right) \bmod 1 =$$

$$= (\lim_{i \to \infty} ((\eta_0 + \eta_1 p + \cdots + \eta_{t-1} p^{t-1})(p^{-it} + p^{(-i+1)t} + \cdots + p^{-t}))) \bmod 1 =$$

$$= 0.(\eta_{t-1} \eta_{t-2} \ldots \eta_0)^{\infty} \bmod 1 \tag{13.55}$$

if $s = 0$. From (13.55) and (13.54), it follows that $\lim_{i \to \infty} (q \bmod p^{k_i} / p^{k_i}) \bmod 1 \in \mathbf{C}(q)$ by (13.40) of Note 13.39. Thus we have proved that given $x \in [0, 1)$ and $e \in \mathbf{C}(q)$, necessarily $(x, (x + e) \bmod 1) \in \mathcal{P}'(\mathfrak{A})$; so $\mathcal{P}'(\mathfrak{A}) \supset \mathbf{C}(1, e)$ for every $e \in \mathbf{C}(b)$.

On the other hand, given $z \in \mathbb{Z}_p$ and a strictly increasing sequence $k_1, k_2, \ldots \in \mathbb{N}$, by combining (13.55) and (13.54) with (13.40) of Note 13.39 we conclude that all limit points of the sequence $q \bmod p^{k_i} / p^{k_i}$, $i = 1, 2, \ldots$, are in $\mathbf{C}(q)$ by an argument similar to the above one. Therefore, limit points of the sequence $(z \bmod p^{k_i} / p^{k_i} + q \bmod p^{k_i} / p^{k_i}) \bmod 1$, $i = 1, 2, \ldots$, are all of the form $(x + e) \bmod 1$, where x is an limit point of the sequence $z \bmod p^{k_i} / p^{k_i}$ and $e \in \mathbf{C}(q)$. This proves that $\mathcal{P}'(\mathfrak{A}) \subset \cup_{e \in \mathbf{C}(q)} \mathbf{C}(1, e)$ and that (13.51) is true. □

Now we are ready to prove the main result of this subsection.

Theorem 13.46 *Given $c, q \in \mathbb{Z}_p$, a map $z \mapsto cz + q$ of $\mathbb{Z}_p$ into itself is an automaton function of a finite automaton if and only if $c, q \in \mathbb{Z}_p \cap \mathbb{Q}$. Given a* *finite automaton* $\mathfrak{A}$ *whose automaton function is* $f(z) = cz + q$ *for* $c, q \in \mathbb{Z}_p \cap \mathbb{Q}$, *represent* c, q *as irreducible fractions* $c = a/b$, $q = a'/b'$, *where* $a, a' \in \mathbb{Z}$, $b, b' \in \mathbb{N}$ *and* $\gcd(a, b) = \gcd(a', b') = \gcd(b, p) = \gcd(b', p) = 1$; *then the limit plot* $\mathcal{P}'(\mathfrak{A}) \subset \mathbb{T}^2$ *is a link of* $\mathsf{mult}_m\, p$ *torus knots, where* $m = b'/\gcd(b, b')$, *and every knot of the link is a cable* $\mathbf{C}(c, e)$ *for* $e \in \mathbf{C}(q)$:

$$\mathcal{P}'(\mathfrak{A}) = \{(y \bmod 1; (cy + e) \bmod 1) : y \in \mathbb{R}, e \in \mathbf{C}(q)\}. \tag{13.56}$$

Moreover, $\mathbf{C}(c, e_1) = \mathbf{C}(c, e_2)$ *for* $e_1, e_2 \in \mathbf{C}(q)$ *if and only if* $r_1 \equiv r_2 \pmod m$ *where* $e_i = (-p^{r_i} q) \bmod 1$, $i = 1, 2$ *(cf.* (13.40)*).*

Note 13.47 Once $m = 1$, i.e., once $b' \mid b$, the congruence $r_1 \equiv r_2 \pmod m$ holds trivially, $\mathsf{mult}_1\, p = 1$ and the link consists of a single knot; so in that case $\mathbf{C}(c, e_1) = \mathbf{C}(c, e_2)$ for all $e_1, e_2 \in \mathbf{C}(q)$.

Proof of Theorem 13.46 The first assertion of the theorem is already proved; see Proposition 13.24. Let us prove the rest one.

Given $q, c \in \mathbb{Z}_p \cap \mathbb{Q}$, we have that

$$c = u + \frac{v}{p^t - 1}, \tag{13.57}$$

$$q = \frac{w}{p^T - 1} \tag{13.58}$$

for suitable $u \in \mathbb{Z}$, $t, T \in \mathbb{N}$, $v \in \{0, 1, 2, \ldots, p^t - 2\}$, $w \in \{0, 1 \ldots, p^T - 2\}$, by Proposition 2.11. Note that we may assume that $0 < q < 1$ since the set of all limit points of the sequence $((z + q) \bmod p^k / p^k)_{k=1}^{\infty}$ is the same as that of the sequence $((z + q \bmod 1) \bmod p^k / p^k)_{k=1}^{\infty}$ by Proposition 13.44 and the case $q = 0$ is already considered (cf. Proposition 13.42).

Now we will prove that $\mathcal{P}'(\mathfrak{A}) \supset \mathbf{C}(c, e)$ for $e \in \mathbf{C}(q)$. As $q \in \mathbb{Z}_p \cap \mathbb{Q} \cap (0, 1)$, the canonical p-adic representation of q is eventually periodic and the period length of q is T (cf. Sect. 2.3.2). Now fix $e \in \mathbf{C}(q)$, take corresponding $d \in \{0, 1, \ldots, T - 1\}$ and consider a sequence $n_j = d + iT \in \mathbb{N}$ $(j = 1, 2, \ldots)$; then $\lim_{j \to \infty} q \bmod p^{n_j} / p^{n_j} = e$ (cf. the proof of Lemma 13.45). Given $x \in [0, 1)$ take $z \in \mathbb{Z}_p$ and a sequence $\mathcal{K} = (k_i = \ell + r_i t)_{i=1}^{\infty}$ as in the proof of Proposition 13.42; so $x = \lim_{i \to \infty} z \bmod p^{k_i} / p^{k_i}$. Note that if $\bar{z} = p^m z$ for some $m \in \mathbb{N}_0$ then $x = \lim_{i \to \infty} \bar{z} \bmod p^{k_i + m} / p^{k_i + m}$; so the proof of Proposition 13.42 remains valid if one substitutes $\bar{z}$ for z and any strictly increasing subsequence $(\check{k}_i)$ of the sequence $\bar{\mathcal{K}} = (\bar{k}_i = m + \ell + r_i t)$ for the sequence $\mathcal{K}$.

We claim that for some $m \in \mathbb{N}_0$ there exist an increasing sequence $j_s \in \mathbb{N}$ and a subsequence $(\bar{r}_s)_{s=1}^{\infty}$ of the sequence $(r_i)_{i=1}^{\infty}$ such that

$$m + \ell + \bar{r}_s t = d + j_s T \text{ for all } s = 1, 2, 3, \ldots. \tag{13.59}$$

Indeed, let $D = \gcd(T, t)$ be the greatest common divisor of T and t; then $T = \check{T}D$, $t = \check{t}D$, $\check{t}$ and $\check{T}$ are co-prime. As the infinite sequence $(r_i)_{i=1}^{\infty}$ is strictly increasing, there exists $\check{n} \in \{0, 1, \dots, \check{t} - 1\}$ such that $r_i + \check{n} \equiv 0 \pmod{\check{T}}$ for infinitely many $i \in \mathbb{N}$, say, for $i = i_1, i_2, \dots$. Put $\bar{r}_s = r_{i_s}$; $s = 1, 2, 3, \dots$.

Take the smallest $\bar{n} = \check{n} + n\check{T}$, $n \in \mathbb{N}_0$, such that $d - \ell + \bar{n}\check{t}D \geqslant 0$, then put $m = d - \ell + \bar{n}\check{t}D$ and find j_s from the Eq. (13.59) which now is equivalent to the equation $(\bar{n} + \bar{r}_s)\check{t} = j_s\check{T}$: As $\bar{n} + \bar{r}_s = h_s\check{T}$ for a suitable $s \in \mathbb{N}$ by the definition of $\bar{n}$, one sees that $j_s = \check{t}h_s$ for $s = 1, 2, 3, \dots$ thus proving our claim.

We conclude now that given arbitrary $y \in \mathbb{R}$ and $e \in \mathbf{C}(q)$ there exist $\bar{z} \in \mathbb{Z}_p$ and a sequence $\check{\mathcal{K}} = (\check{k}_s = \bar{r}_s t + \ell + m = d + j_s T)$ such that

$$y \bmod 1 = x = \lim_{s\to\infty} \frac{\bar{z} \bmod p^{\check{k}_s}}{p^{\check{k}_s}}, \tag{13.60}$$

$$(cy) \bmod 1 = \lim_{s\to\infty} \frac{(c\bar{z}) \bmod p^{\check{k}_s}}{p^{\check{k}_s}}, \tag{13.61}$$

$$e = \lim_{s\to\infty} \frac{q \bmod p^{\check{k}_s}}{p^{\check{k}_s}}; \tag{13.62}$$

cf. (13.49), (13.50) and Proposition 13.38. Therefore, $\lim_{s\to\infty}(c\bar{z} + q) \bmod p^{\check{k}_s} / p^{\check{k}_s} = \lim_{s\to\infty}((c\bar{z}) \bmod p^{\check{k}_s} / p^{\check{k}_s} + q \bmod p^{\check{k}_s} / p^{\check{k}_s}) \bmod 1 = (cy + e) \bmod 1$ and so the point $(y \bmod 1, (cy + e) \bmod 1) \in \mathbf{C}(c, e)$ is in $CalP'(\mathfrak{A})$. Thus we have proved that $\mathcal{P}'(\mathfrak{A}) \supset \mathbf{C}(c, e)$ for every $e \in \mathbf{C}(q)$.

On the other hand, given arbitrary $z \in \mathbb{Z}_p$ and arbitrary strictly increasing sequence $k_1, k_2, \dots \in \mathbb{N}$, limit points of the point sequence $(z \bmod p^{k_i} / p^{k_i}; (cz) \bmod p^{k_i} / p^{k_i})$ are all in $\mathbf{C}(c, 0)$ by Proposition 13.42 whereas limit points of the sequence $q \bmod p^{k_i} / p^{k_i}$ are all in $\mathbf{C}(q)$ by Proposition 13.38. Therefore limit points of the point sequence $((z \bmod p^{k_i} / p^{k_i}; (cz + q) \bmod p^{k_i} / p^{k_i}))_{i=1}^{\infty} = ((z \bmod p^{k_i} / p^{k_i}, ((cz) \bmod p^{k_i} / p^{k_i} + q \bmod p^{k_i} / p^{k_i}) \bmod 1))_{i=1}^{\infty}$ are all in $\cup_{e\in\mathbf{C}(q)} \mathbf{C}(c, e)$. Finally we conclude that $\mathcal{P}'(\mathfrak{A}) = \cup_{e\in\mathbf{C}(q)}\mathbf{C}(c, e)$; or (which is the same) that

$$\mathcal{P}'(\mathfrak{A}) = \{(y \bmod 1; (cy + e) \bmod 1) : y \in \mathbb{R}, e \in \mathbf{C}(q)\} \tag{13.63}$$

Note that it may happen that $\mathbf{C}(c, q) = \mathbf{C}(c, q_1)$ even if $q \neq q_1$ (and even $q \notin \mathbf{C}(q_1)$): For instance, (13.49) shows that $\mathbf{C}(c, q) = \mathbf{C}(c, 0)$ for some $q \neq 0$. Therefore to finish the proof we must now calculate the number of pairwise distinct cables $\mathbf{C}(c, e)$ when $e \in \mathbf{C}(q)$.

During the proof of Proposition 13.42, we have shown that (in the notation of the proposition under the proof)

$$\{(y \bmod 1; (cy) \bmod 1) : y \in \mathbb{R}\} = \left\{\left(y \bmod 1; \left(cy + \frac{j}{b}\right) \bmod 1\right) : y \in \mathbb{R}\right\}$$

for every $j \in \mathbb{Z}$ (cf. Eq. (13.50) and the text which follows it). Therefore $\mathbf{C}(c, e_1) = \mathbf{C}(c, e_2)$ if $e_1 - e_2 \equiv (j/b) \bmod 1$ for some $j \in \mathbb{Z}$. The converse statement is also true: If $\mathbf{C}(c, e_1) = \mathbf{C}(c, e_2)$, then $e_1 - e_2 \equiv (j/b) \bmod 1$ for some $j \in \mathbb{Z}$.

To prove this, for $h \in \mathbf{C}(q)$ let $A(c, h)$ be a set of all points where the cable $\mathbf{C}(c, h)$ crosses zero meridian of the torus $\mathbb{T}^2$; that is,

$$A(c, h) = \mathbf{AP}\left(\left\{\left(0; \left(\frac{(cz) \bmod p^{s_r}}{p^{s_r}} + h\right) \bmod 1\right) \right.\right.$$
$$\left.\left. : z \in \mathbb{Z}_p, \lim_{r\to\infty} \frac{z \bmod p^{s_r}}{p^{s_r}} = 0\right\}\right),$$

where $s_1, s_2, \ldots \in \mathbb{N}$, $s_1 < s_2 < \ldots$; therefore by (13.48)

$$\mathbf{AP}\left(\left(\left(\frac{(cz) \bmod p^{s_r}}{p^{s_r}} + h\right) \bmod 1\right)_{r=0}^{\infty}\right) =$$
$$= \left\{\left(\frac{j}{b} + h\right) \bmod 1 : j = 0, 1, 2, \ldots\right\}. \quad (13.64)$$

Finally, as $\mathbf{C}(c, e_1) = \mathbf{C}(c, e_2)$ if and only if $A(c, e_1) = A(c, e_2)$ since the both cables cross zero meridian at a same angle (which is equal to $\arctan c$), this means that $\mathbf{C}(c, e_1) = \mathbf{C}(c, e_2)$ if and only if $e_1 - e_2 \equiv jb^{-1} \pmod 1$ for some $j \in \mathbb{N}_0$, as claimed.

Now we are able to calculate the number of torus knots (cables) which constitutes the link $\mathcal{P}'(\mathfrak{A})$. Let for some $j_1, j_2 \in \{0, 1, \ldots, b-1\}$, $(j_1 \neq j_2)$ and $e_1, e_2 \in \mathbf{C}(q)$ the following equality holds:

$$\left(\frac{j_1}{b} + e_1\right) \bmod 1 = \left(\frac{j_2}{b} + e_2\right) \bmod 1. \quad (13.65)$$

We see that $e_i = -p^{r_i}(\frac{a'}{b'}) \bmod 1$ for suitable $r_i \in \{0, 1, \ldots, (\operatorname{mult}_{b'} p) - 1\}$ by Note 13.39 ($i = 1, 2$). Therefore (13.65) is equivalent to the congruence

$$p^{r_1}\frac{a'}{b'} - p^{r_2}\frac{a'}{b'} \equiv \frac{j}{b} \pmod 1$$

for a suitable $j \in \{0, 1, \ldots, b-1\}$; but the latter congruence in turn is equivalent to the congruence

$$p^{r_2}\left(p^{r_1-r_2}-1\right)a'n \equiv jm \pmod{nmd}, \qquad (13.66)$$

where $d = \gcd(b', b)$, $m = b'/d$, $n = b/d$ (we assume that $r_1 > r_2$ since the case $r_1 = r_2$ is trivial). From here it follows that $p^{r_2}\left(p^{r_1-r_2}-1\right)a'n \equiv 0 \pmod{m}$ once $m \neq 1$; therefore necessarily $r_1 \equiv r_2 \pmod{\mathsf{mult}_m\, p}$ since $\gcd(b', b) = \gcd(p, b) = \gcd(p, b') = 1$. So $\left(p^{r_1-r_2}-1\right) = mh$ for a suitable $h \in \mathbb{N}$ and thus (13.66) is equivalent to the congruence $p^{r_2}ha'n \equiv j \pmod{nd}$, and the latter congruence gives the value of j (modulo $b = nd$) so that (13.65) is satisfied. This means that when $m \neq 1$, (13.65) holds if and only if $r_1 \equiv r_2 \pmod{\mathsf{mult}_m\, p}$. Thus, if $m \neq 1$ (i.e., if b' is not a factor of b), then the number of pairwise distinct torus knots in the link is $\mathsf{mult}_m\, p$.

In the remaining case when $m = 1$ (i.e., when b' divides b), (13.66) always holds: If $p^{r_1-r_2} \equiv 1 \pmod{d}$, then we can take $j = 0$ to satisfy (13.66); otherwise the left-hand side of (13.66) just gives an expression for a unique residue j modulo $b = nd$ (which thus satisfies (13.66)). Therefore the link consist of a unique cable; so the number of pairwise distinct cables in the link is $1 = \mathsf{mult}_1\, p$ in this case as well. This concludes the proof. □

Note 13.48 In conditions of Theorem 13.46, note that $b'|b$ is the only case when the link $\mathcal{P}'(\mathfrak{A})$ consists of a single cable. Note also that from the proof of Theorem 13.46, it is clear that if the number $\#\mathbf{C}(q)$ of points in $\mathbf{C}(q)$ is 1, then the link necessarily consists of a single cable. By Note 13.39, $\#\mathbf{C}(q) = 1$ if and only if the period length of q is 1 and therefore $q \bmod 1 = 0.(\xi)^{\infty} \bmod 1$ for some $\xi \in \{0, 1, \ldots, p-1\}$).

Example 13.49 Let $p = 2$ and $f(z) = (3/5)z + (1/3)$. Then in conditions of Theorem 13.46 we have that $m = 3$ and therefore the link consists of $\mathsf{mult}_3\, 2 = 2$ cables with slopes 3/5 (cf. Figs. 13.17 and 13.18).

Fig. 13.17 Limit plot of the function $f(z) = \frac{3}{5}z + \frac{1}{3}$, $z \in \mathbb{Z}_2$, in $\mathbb{R}^2$

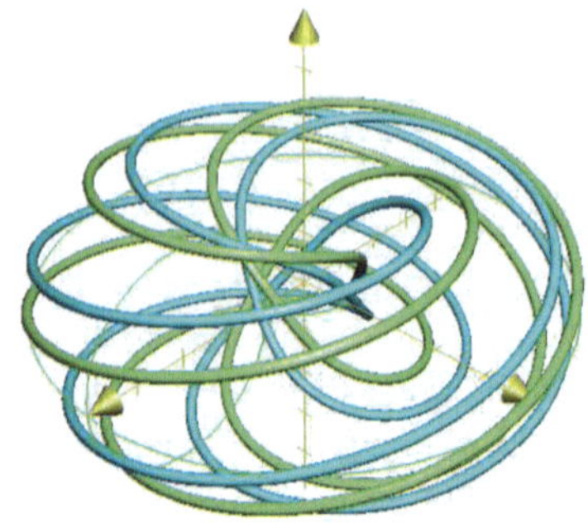

Fig. 13.18 Limit plot of the same function on the torus $\mathbb{T}^2$

Corollary 13.50 *There is a one-to-one correspondence between maps of the form* $f\colon z \mapsto \frac{a}{b}z + \frac{a'}{b'}$ *on* $\mathbb{Z}_p$ *(where* $\frac{a}{b}, \frac{a'}{b'} \in \mathbb{Z}_p \cap \mathbb{Q}$*;* $a, a' \in \mathbb{Z}$*;* $b, b' \in \mathbb{N}$*) and collections of* $\mathsf{mult}_m\, p$ *complex-valued exponential functions* $\psi_k\colon \mathbb{R} \to \mathbb{C}$ *of real variable* $y \in \mathbb{R}$

$$\left\{\psi_k(y) = \mathrm{e}^{\mathrm{i}(\frac{a}{b}y - 2\pi p^k \frac{a'}{b'})} : k = 0, 1, 2, \ldots, (\mathsf{mult}_m\, p) - 1\right\}$$

Here $\mathrm{i} \in \mathbb{C}$ *is imaginary unit and* $m = b'/\gcd(b, b')$.

Proof of Corollary 13.50 Indeed, embedding the unit torus $\mathbb{T}^2$ into a three-dimensional Euclidean space $\mathbb{R}^3$ and using cylindrical coordinates as in Note 13.37, in view of Theorem 13.46, every knot from the link can be expressed in the form (13.36) with $\omega = 2\pi e$ for $e \in \mathbf{C}(q)$ since $\cos\omega$ and $\sin\omega$ specifies position of the point where the knot crosses zero meridian of the torus (i.e., when $\theta \equiv 0 \pmod{2\pi}$ in (13.36)). But $q = a'/b'$ and thus $\mathbf{C}(q) = \left\{(-p^\ell(a'/b')) \bmod 1 : \ell = 0, 1, \ldots, (\mathsf{mult}_{b'}\, p) - 1\right\}$ by (13.39). As two such knots (with accordingly $\omega_i = 2\pi e_i$, $i = 1, 2$) coincide if an only if $\omega_1 \equiv \omega_2 \pmod{2\pi(a/b)}$ by (13.36), i.e., if and only if $e_1 \equiv e_2 \pmod{a/b}$. But the latter congruence is equivalent to (13.65); so finally the assertion follows from Theorem 13.46. □

13.6.4 Finite Computability

In this subsection we introduce central notion of the section, the *finite computability*, and prove some technical results which will be needed further during the proof of main result of the section on the affinity of finitely computable smooth functions.

Definition 13.51 A non-empty point set $S \subset \mathbb{I}^2$ ($S \subset \mathbb{T}^2$, $S \subset \mathbb{I} \times \mathbb{S}$, $S \subset \mathbb{S} \times \mathbb{I}$) is called *(ultimately) finitely computable* (or, (ultimately) computable by a finite automaton) if there exists a finite automaton $\mathfrak{A}$ such that S is a subset of $\mathcal{P}(\mathfrak{A})$ (of

$\mathcal{P}'(\mathfrak{A})$). We say that the automaton $\mathfrak{A}$ (*ultimately*) *computes* the set S; and $\mathfrak{A}$ is called an (*ultimate*) *computing automaton* of the set S.

In the most cases which follow, given a real function $g\colon D \to \mathbb{R}$ with the domain $D \subset \mathbb{R}$, by the *graph of the function* (on the torus $\mathbb{T}^2$) we mean the point subset $\mathbf{G}_D(g) = \{(x \bmod 1; g(x) \bmod 1)\colon x \in D\} \subset \mathbb{T}^2$. However, given a function $g\colon D \to T$ where either $D \subset [0,1]$ or $D \subset \mathbb{S}$ and T is either $[0,1]$ or $\mathbb{S}$, we call a graph $\mathbf{G}_D$ of the function g the set $\{(\bar{x}; \overline{g(x)})\colon x \in D\}$ where either $\bar{x} = x$ if $D \subset [0,1]$ or $\bar{x} = x \bmod 1$ if $D \subset \mathbb{S}$ and accordingly either $\overline{g(x)} = g(x)$ if $T = [0,1]$ or $\overline{g(x)} = (g(x)) \bmod 1$ if $T = \mathbb{S}$. In the sequel, we always explain what is meant by $\mathbf{G}_D(g)$ if this is not clear from the context. Also, we may omit the subscript D when it is clear what is the domain.

Definition 13.52 Given a real function $g\colon D \to \mathbb{R}$ with domain $D \subset \mathbb{R}$ and an automaton $\mathfrak{A}$, the function g is called (*ultimately*) *computable by* $\mathfrak{A}$ *at the point* $x \in D$ if $(x \bmod 1; g(x) \bmod 1) \in \mathcal{P}'(\mathfrak{A}) \subset \mathbb{T}^2$ ($(x \bmod 1; g(x) \bmod 1) \in \mathcal{P}'(\mathfrak{A}) \subset \mathbb{T}^2$). Also, if either $D \subset [0,1]$ or $D \subset \mathbb{S}$ and $g\colon D \to T$ where either $T = [0,1]$ or $T = \mathbb{S}$ we will say that $\mathfrak{A}$ (*ultimately*) *computes* g *at the point* $x \in D$ if $(\bar{x}; \overline{g(x)}) \in \mathcal{P}'(\mathfrak{A})$ where either $x = x$ if $D \subset [0,1]$ or $\bar{x} = x \bmod 1$ if $D \subset \mathbb{S}$ and accordingly either $\overline{g(x)} = g(x)$ if $T = [0,1]$ or $\overline{g(x)} = (g(x)) \bmod 1$ if $T = \mathbb{S}$ (cf. Note 13.27)

Given a real function $g\colon D \to \mathbb{R}$ with domain $D \subset \mathbb{R}$, the function g is called (*ultimately*) *finitely computable* (or, (*ultimately*) *computable by a finite automaton*) if there exists a finite automaton $\mathfrak{A}$ such that $\mathbf{G}(g) \subset \mathcal{P}(\mathfrak{A}) \subset \mathbb{T}^2$ ($\mathbf{G}(g) \subset \mathcal{P}'(\mathfrak{A}) \subset \mathbb{T}^2$). The automaton $\mathfrak{A}$ which (*ultimately*) *computes* the function g is called the (*ultimate*) *computing automaton* of the function g. In a similar manner, we define these notions for the cases when $g\colon D \to T$ and D, T are as above.

13.6.4.1 The Mark-Ups

To move ahead we need a yet one more auxiliary technical notion, the *mark-up*. In loose terms, when assigning a real-valued function $f^{\mathfrak{A}}\colon [0,1] \to [0,1]$ to automaton $\mathfrak{A}$ via Monna map $\mathsf{Mon}\colon \mathbb{Z}_p \to \mathbb{R}$ (cf. Sect. 7.2.1), one feeds the automaton by the base-p-expansion of argument $x \in [0,1]$ and get the output as the base-p expansion of $f^{\mathfrak{A}}(x)$: The base-p expansion specifies a unique right-infinite word in the alphabet $\mathbb{F}_p$ and the automaton "reads the word from head to tail," i.e., the automaton is fed by digits of the base-p expansion from left to right (i.e., digits on more significant positions are fed prior to digits on less significant positions); the output word specifies the base-p expansion of a unique real number from $[0,1]$.

To examine functions computed by automata in the meaning of Definition 13.52, it would also be convenient to work with base-p expansions of real numbers; but the problem is that we need feed the automaton by a right-infinite word in the inverse order "from tail (which is at infinity) to head": Digits on less significant positions (the rightmost ones) should be fed prior to digits on more significant positions (the leftmost ones). That "straightforward" inversion is impossible since it is unclear which letter should be the first when feeding the automaton this way; thus output word is undefined and so is the real number whose base-p expansion is the output word. We are going to specify rigorously this inversion and develop some techniques needed in further proofs.

Let a function $g\colon D \to \mathbb{S}$ (or $g\colon D \to [0, 1]$) whose domain D is either a subset of a real unit circle $\mathbb{S}$ or a subset of a unit segment $[0, 1]$ be ultimately computable by a finite automaton $\mathfrak{A} = \mathfrak{A}(s_0)$; that is, for any $x \in D$ there exists $x \in \mathbb{Z}_p$ such that x is a limit point of the sequence $(z \operatorname{mod} p^k/p^k)_{k=1}^{\infty}$ and $g(x)$ is a limit point of the sequence $((f_{\mathfrak{A}}(z)) \operatorname{mod} p^k/p^k)_{k=1}^{\infty}$, where $f_{\mathfrak{A}}\colon \mathbb{Z}_p \to \mathbb{Z}_p$ is the automaton function of the automaton $\mathfrak{A}$ (cf. Definitions 13.52 and 12.1). As said, in further examination of finitely computable real functions, it will be more convenient to work with automata maps as of maps of reals into reals rather than to consider automata functions on p-adic integers and then represent $x \in \mathbb{R}$ and $g(x) \in \mathbb{R}$ as limit points of the sequences $(z \operatorname{mod} p^k/p^k)_{k=1}^{\infty}$ and $((f_{\mathfrak{A}}(z)) \operatorname{mod} p^k/p^k)_{k=1}^{\infty}$, respectively.

Recall that given $x \in (0, 1)$, there exists a (right-)infinite word $w = \gamma_0\gamma_1\ldots$ over $\{0, 1, \ldots, p-1\}$ such that

$$x = 0.\gamma_0\gamma_1\ldots = 0.w = \sum_{i=0}^{\infty} \gamma_i p^{-i-1}, \tag{13.67}$$

the *base-p expansion* of x. If x is not of the form $x = n/p^k$ for some $n = \alpha_0 + \alpha_1 p + \cdots + \alpha_\ell p^\ell \in \{0, 1, \ldots, p^k - 1\}$, where $\ell = \mathbf{le}(n) = \lfloor \log_p n \rfloor + 1$ is the length of the base-p expansion of $n \in \mathbb{N}_0$ (recall that we put $\lfloor \log_p 0 \rfloor = 0$; see the reasoning which follows (4.12)), $\alpha_0, \alpha_1, \ldots \alpha_\ell \in \{0, 1, \ldots, p-1\}$, then the right-infinite word $\mathsf{wrd}(x) = \gamma_0\gamma_1\ldots$ over $\{0, 1, \ldots, p-1\}$ is uniquely defined (and the corresponding x is said to have a *unique* base-p expansion); else there are exactly two infinite words $\mathsf{wrd}^r(x)$ $\mathsf{wrd}^l(x)$ such that $x = 0.\mathsf{wrd}_r(x) = 0.\mathsf{wrd}_l(x)$:

$$\mathsf{wrd}^r(x) = \alpha_0\alpha_1\ldots\alpha_{\ell-1}\alpha_\ell 00\ldots = \alpha_0\alpha_1\ldots\alpha_{\ell-1}\alpha_\ell(0)^\infty \tag{13.68}$$

$$\mathsf{wrd}^l(x) = \alpha_0\alpha_1\ldots\alpha_{\ell-1}(\alpha_\ell - 1)(p-1)\ldots = \alpha_0\alpha_1\ldots\alpha_{\ell-1}(\alpha_\ell - 1)(p-1)^\infty \tag{13.69}$$

Here $\alpha_\ell \neq 0$. In that case x is said to have a *non-unique* base-p expansion; the corresponding base-p expansions are called *right* and *left*, respectively. Both 0 and 1 are assumed to have unique base-p expansions since $0 = 0.00\ldots$, $1 = 0.(p-1)(p-1)\ldots$; so $\mathsf{wrd}(0) = 00\ldots$, $\mathsf{wrd}(1) = (p-1)(p-1)\ldots$. This way we define $\mathsf{wrd}(x)$ for all $x \in [0, 1]$; and to $x = n/p^k$ we will usually put into the correspondence the both infinite words $\mathsf{wrd}^l(x)$ and $\mathsf{wrd}^r(x)$ if converse is not stated explicitly. The only difference in considering a unit circle $\mathbb{S}$ rather than the unit segment $\mathbb{I} = [0, 1]$ is that we identify 0 and 1 and thus have two representations for 0, namely, $0.(0)^\infty$ and $0 = 1 \operatorname{mod} 1 = 0.(p-1)^\infty$.

Given a finite word $w = \alpha_{m-1}\alpha_{m-2}\cdots\alpha_0$, we denote via $\overrightarrow{w}$ the (right-)infinite word $\overrightarrow{w} = \alpha_{m-1}\alpha_{m-2}\cdots\alpha_0(0)^\infty$ and we put

$$0.\overrightarrow{w} = 0.\alpha_{m-1}\alpha_{m-2}\cdots\alpha_0(0)^\infty\ldots$$

(note that then $0.\overrightarrow{w} = \rho(w)$). Of course, $0.\overrightarrow{w} = 0.w = \sum_{i=0}^{m-1}\alpha_i p^{-m+i}$; but we use notation $0.\overrightarrow{w}$ if we want to stress that we deal with infinite base-p expansion. To unify our notation, we also may write $\overrightarrow{w} = \zeta_1\zeta_2\ldots$ for a (right-)infinite word $w = \zeta_1\zeta_2\ldots$; then $0.\overrightarrow{w} = 0.w = 0.\zeta_1\zeta_2\ldots$.

Let $\overrightarrow{w} = \gamma_0\gamma_1\ldots$ be a (right-)infinite word over $\mathbb{F}_p = \{0, 1, \ldots, p-1\}$. Given an automaton $\mathfrak{A}$ with the initial state s, we further denote via $\mathfrak{a}_s(\overrightarrow{w})$ the set of all limit points of the sequence $(\rho(\mathfrak{a}_s(\gamma_0\gamma_1\ldots\gamma_k)))_{k=0}^\infty$. We may omit the subscript s if it is clear from the context what is the initial state of the automaton.

Given $x \in [0, 1]$, further $\mathfrak{a}_s(x)$ stands for $\mathfrak{a}_s(\overrightarrow{w}(x))$ if x admits a unique base-p expansion, and $\mathfrak{a}_s(x) = \mathfrak{a}_s(\overrightarrow{w}(x)_l) \cup \mathfrak{a}_s(\overrightarrow{w}(x)_r)$ if the expansion is non-unique (thus, if x admits both left and right base-p expansions). We also consider $\mathfrak{a}_s(x)$ for $x \in \mathbb{S}$ rather that for $x \in [0, 1]$; in that case we take for 0 both base-p expansions $0.(0)^\infty$ and $0.(p-1)^\infty$ (since $0 = (0.(p-1)^\infty) \bmod 1$) and reduce modulo 1 all limit points of all sequences $(\rho(\mathfrak{a}_s(\gamma_0\gamma_1\ldots\gamma_k)))_{k=0}^\infty$. We further use the same symbol $\mathfrak{a}_s(x)$ independently of whether we consider $x \in [0, 1]$ or $x \in \mathbb{S}$; we make special remarks when this may cause a confusion.

We stress that $\mathfrak{a}_s(w)$ is a uniquely defined finite word whenever $w \in \mathcal{W}$ is a finite word (and therefore $\rho(\mathfrak{a}_s(w))$ consists of a single number), but in the case when w is an infinite word or w is a real number from $[0, 1]$ (or $w \in \mathbb{S}$), the set $\mathfrak{a}_s(w)$ may contain more than one element.

Given $x \in \mathbb{Q} \cap [0, 1]$, in view of Lemma 13.25 it is clear that if the automaton $\mathfrak{A}$ is finite then $\mathfrak{a}(x) \in \mathbb{Q} \cap [0, 1]$ since a real number is rational if and only if its base-p expansion is eventually periodic. The following propositions reveals some more details about $\mathfrak{a}(x)$ for a rational x; and especially for $x = 0$.

Proposition 13.53 *If $\mathfrak{A}$ is finite, $x \in \mathbb{Q} \cap [0, 1]$ then $\mathfrak{a}(x) \subset \mathbb{Q} \cap [0, 1]$ and $\mathfrak{a}(x)$ is a finite set. Moreover, if $x \in \mathbb{Z}_p \cap \mathbb{Q} \cap [0, 1]$ then $\mathfrak{a}(x) \subset \mathbb{Z}_p \cap \mathbb{Q} \cap [0, 1]$. In particular, if $x = 0 \in \mathbb{S}$ then $\mathfrak{a}(x) = \mathbf{C}(q_1) \cup \mathbf{C}(q_2)$ for suitable $q_1, q_2 \in \mathbb{Z}_p \cap \mathbb{Q} \cap [0, 1)$* (cf. Sect. 13.6.3). *Let a $\mathfrak{A}$-computable function $g\colon D \to \mathbb{S}$ be defined on the domain $D \subset \mathbb{S}$ and continuous at $0 \in D$. If the domain D is open, then there exists $q \in \mathbb{Z}_p \cap \mathbb{Q} \cap [0, 1)$ such that $\mathfrak{a}(0.(0)^\infty) = \mathfrak{a}(0.(p-1)^\infty) \in \mathbf{C}(q)$ and either $\mathfrak{a}(0.(0)^\infty) \in \mathbf{C}(q)$ or $\mathfrak{a}(0.(p-1)^\infty) \in \mathbf{C}(q)$ if the domain D is half-open and x is a boundary of D.*

Proof of Proposition 13.53 Follows from Lemma 13.25 and Proposition 13.38 (see the proof of the latter).

Given $x \in \mathbb{Q} \cap [0, 1]$, the base-$p$ expansion of x is eventually periodic (cf. (2.11)): $x = 0.\chi_0\ldots\chi_{k-1}(\xi_0\ldots\xi_{n-1})^\infty$. Given x, take n, k the smallest possible (note that then the word $v = \chi_0\ldots\chi_{k-1}$ may be empty). From the definition of $\mathfrak{a}(x)$, it follows that $\mathfrak{a}(x)$ consists of all limit points of the sequences $\mathcal{K}(r) =$

$(\rho((v(u)^k r))_{k=K}^{\infty}$, where K is large enough and $r \in \{\xi_0, \xi_0\xi_1, \ldots, \xi_0\xi_1\ldots\xi_{n-1}\}$ are all suffixes of the word $u = \xi_0 \ldots \xi_{n-1}$, for every right-infinite word w which corresponds to the base-p expansion of $x = 0.w$. As the automaton $\mathfrak{A} = \mathfrak{A}(s_0)$ is finite, the number of states it reaches after being fed by either of words r, is finite; say, these states are $s_1, \ldots, s_N \in \mathcal{S}$. By the same reason, being fed by the words $(u)^K$ where K is large enough, the automata $\mathfrak{A}_1 = \mathfrak{A}(s_1), \ldots, \mathfrak{A}_N = \mathfrak{A}(s_N)$ output respectively words $v_1(u_1)^{K_1}t_1, \ldots, v_N(u_N)^{K_N}t_N$, where the words $u_1, \ldots, u_N, t_1, \ldots, t_N$ do not depend on K, each of the words $v_1, \ldots, v_N$ is either empty or a prefix of the respective word $w_1, \ldots, w_N$, and all the output words $v_1(u_1)^{K_1}t_1, \ldots, v_N(u_N)^{K_N}t_N$ have the same length as the one of the input word $(u)^K$ (cf. Lemma 13.25). That is, the output words of automata $\mathfrak{A}_i$ are all of the form $(\bar{u})^L t$, where $\bar{u}$ stands for a cyclically shifted word u, and after outputting the word $(\bar{u})^L t$, each automaton $\mathfrak{A}_i$, $i = 1, 2, \ldots, N$, reaches some of finitely many its states, say, $s'_{i1}, \ldots, s'_{iM(i)}$. After reaching respective state, the automaton is fed by the word v and outputs the corresponding output word $v'_{i,j}$, $j = 1, 2, \ldots, M(i)$. Therefore, all limit points of the sequences $\mathcal{K}(r)$ are of the form $0.v'_{i,j}(\bar{u}_i)^{\infty}$ where $\bar{u}_i$ runs through a (sub)set of all cyclic shifts of the word u_i, $i = 1, 2, \ldots, N$, $j = 1, 2, \ldots, M(i)$. But there are only finite number of points of that form; therefore, given the base-p expansion of $x = 0.\overrightarrow{w}$, the set $\mathfrak{a}(0.\overrightarrow{w})$ is a union of a finite number points of the said form. As every $x \in \mathbb{Q}$ has at most two base-p expansions, this proves the first claim of the proposition.

If $x \in \mathbb{Z}_p \cap \mathbb{Q} \cap [0, 1]$, then the base-$p$ expansion of x is purely periodic by Proposition 2.14: $x = 0.(\chi_0 \ldots \chi_{n-1})^{\infty}$. Therefore once the automaton $\mathfrak{A}$ is being fed by finite words of the form $w^L t$, where $w = \chi_0 \ldots \chi_{n-1}$, t is a suffix of w or empty, and L is large enough, by Lemma 13.25 the corresponding output word will be of the form $(u_t)^{N(t)} v_t$, and the number of different words u_t is finite since the number of different words t is finite. Applying the same argument as above, we conclude that all limit points of corresponding sequences are of the form $0.(u)^{\infty}$ where u runs through a finite number of finite words. But by Corollary 2.15, all these points are in $\mathbb{Z}_p \cap \mathbb{Q} \cap [0, 1]$. This proves the second claim of the proposition.

To prove final assertion of the proposition, we must consider both base-p representations of zero point of the unit circle $\mathbb{S}$: $0 = 0.(0)^{\infty}$ and $0 = 0.(p-1)^{\infty}$. Sending a left-infinite zero sequence to the automaton A, the output sequence will be of the form $w^{\infty} t$ by suitable finite words w, t by Lemma 13.25; so $\mathfrak{a}(0^{\infty})$ consists of all points of the form $0.(u)^{\infty}$, where u runs through all cyclic shifts of the word $w = \chi_{n-1} \ldots \chi_0$; therefore $\mathfrak{a}(0^{\infty}) = \mathbf{C}(q_1)$ for a suitable $q_1 \in \mathbb{Z}_p \cap \mathbb{Q} \cap [0, 1)$ (cf. Note 13.39). By the same reason, $\mathfrak{a}((p-1)^{\infty}) = \mathbf{C}(q_2)$ for a suitable $q_2 \in \mathbb{Z}_p \cap \mathbb{Q} \cap [0, 1)$. In the case when the $\mathfrak{A}$-computable function $g \colon D \to \mathbb{S}$ is continuous at 0 and there exists an open neighborhood U of 0 such that $U \subset D$ then necessarily $g(0) \in \mathfrak{a}(0.0^{\infty})$ and $g(0) \in \mathfrak{a}(0.(p-1)^{\infty})$; so $\mathbf{C}(q_1) \cap \mathbf{C}(q_2) \neq \emptyset$ and therefore $\mathbf{C}(q_1) = \mathbf{C}(q_2)$ by Corollary 13.40. If no such neighborhood U exists, then the domain is half-open, and 0 is a boundary point; thus due to the continuity of g at 0, we see that either $g(0) \in \mathfrak{a}(0.0^{\infty})$ or $g(0) \in \mathfrak{a}(0.(p-1)^{\infty})$, and the conclusion follows. □

Corollary 13.54 *Let $\mathfrak{A}$ be a finite automaton, let $(x; y) \in \mathcal{P}(\mathfrak{A}) \subset \mathbb{T}^2$, and let $x \in \mathbb{Z}_p \cap \mathbb{Q} \setminus \{0\}$; then $y \in \mathbb{Z}_p \cap \mathbb{Q}$. If $x = 0$ then $y \in [0, 1) \cap \mathbb{Q}$; moreover, there exists $y \in \mathbb{Z}_p \cap \mathbb{Q}$ such that $(0; y) \in \mathcal{P}(\mathfrak{A})$.*

Proof of Corollary 13.54 As $x \in \mathbb{Z}_p \cap \mathbb{Q}$ and $x \in [0, 1)$ then by Corollary 2.16 the base-p expansion is purely periodic; that is, $x = 0.w$, where $w \in \mathcal{W}^\infty$ is a right-infinite periodic word: $w = (v)^\infty$ for a suitable finite non-empty word $v \in \mathcal{W}$. As $(x, y) \in \mathcal{P}(\mathfrak{A})$ then by Note 13.28, there exists a sequence $(u_i)_{i=0}^\infty$ of finite non-empty words such that $\lim_{i\to\infty} 0.u_i = x$ and simultaneously $\lim_{i\to\infty} 0.\mathfrak{a}(u_i) = y$. Let $x \neq 0$; then v is not a 1-letter zero word: $v \neq 0$. Therefore since $\lim_{i\to\infty} 0.u_i = x$, for all sufficiently large i the words u_i must be of the form $u_i = (v)^{L_i}\bar{u}_i$ where L_i increases unboundedly while $i \to \infty$. Therefore we may assume that the sequence L_i is strictly increasing (we consider a strictly increasing subsequence of the sequence (L_i) if otherwise) and that all u_i are of the form $u_i = (v)^{L_i}\bar{u}_i$. Let $s(i)$ be a state the automaton $\mathfrak{A}$ reaches after being fed by the word $\bar{u}_i$ ($s(i) = s_0$ if $\bar{u}_i = \varphi$ is empty). As the automaton $\mathfrak{A}$ is finite, then there are only finitely many pairwise distinct $s(i)$, say these are $s'(1), \dots, s'(n)$. Now we consider automata $\mathfrak{A}(s'(1)), \dots, \mathfrak{A}(s'(n))$ and apply the same argument as in the proof of the second statement of Proposition 13.53 thus proving that $y \in \mathbb{Z}_p \cap \mathbb{Q}$. The same argument can be applied for the case when $x = 0$, but there exists an infinite sequence of words u_i whose lengths are increasing unboundedly while $i \to \infty$. Therefore the only rest case is now $x = 0$ and once the word sequence $(u_i)_{i=0}^\infty$ is such that $\lim_{i\to\infty} 0.u_i = x$ and lengths of u_i are bounded; $\Lambda(u_i) \leqslant K$ for all $i \in \mathbb{N}_0$. But this just means that for all sufficiently large i all the words u_i are K-letter zero words: $u_i = (0)^K$ for a suitable $K \in \mathbb{N}$. But then $y = 0.\mathfrak{a}((0)^K)$; thus $y \in [0, 1) \cap \mathbb{Q}$. The last claim of the corollary trivially follows from Proposition 13.53 since once $y \in \mathfrak{a}(0)$ then $(0; y) \in \mathcal{P}(\mathfrak{A})$ by the definition of $\mathcal{P}(\mathfrak{A})$. □

Now everything is ready to introduce an important technical notion, the *mark-up*, which will be used in further proofs:

Definition 13.55 Given a function $g : D \to [0, 1]$ defined on the domain $D \subset [0, 1]$, an automaton $\mathfrak{A}$, and a point $x \in D$, consider a right-infinite word w such that $x = 0.w$ (cf. (13.67), (13.68), (13.69)). An infinite strictly increasing sequence $i_0, i_1, i_2, \dots$ over $\mathbb{N}_0$ is called an *s-mark-up of the* (right-infinite) *word* $w = \gamma_0\gamma_1\dots$ *with respect to g and $\mathfrak{A}$* (or briefly a *mark-up* when it is clear what g, $\mathfrak{A}$ and s are meant) if there exists a state $s \in \mathcal{S}$ of the automaton $\mathfrak{A}$ such that $\lim_{k\to\infty} \rho(\mathfrak{a}_s(\gamma_0\gamma_1\dots\gamma_{i_k})) = g(x)$.

Given $x \in D \subset \mathbb{S}$, the mark-ups of x are defined exactly in the same way. Note only that the point $x = 0$ of $\mathbb{S}$ coincides with the point $x = 1$ and $1 = 0 = 0.000\ldots = 0.(p-1)(p-1)(p-1)\ldots \in \mathbb{S}$ as $\mathbb{S} = \mathbb{R} \bmod 1$. In a similar way, we define the mark-up when $g : D \to \mathbb{S}$. We will also speak about *mark-up of the point x* meaning the mark-up of the word which corresponds to the base-p expansion of x.

In simple words, the following proposition shows that if a continuous real function is finitely computable, then all base-p expansions of all its arguments can be marked up.

Proposition 13.56 *Given a continuous function* $g\colon (a,b) \to [0,1]$ (or, which makes no difference, $g\colon (a,b) \to \mathbb{S}$ where $(a,b) \subset \mathbb{S}$ is an arc of the unit circle $\mathbb{S}$), *let* $\mathbf{G}(g) \subset \mathcal{P}(\mathfrak{A})$ *for a suitable finite automaton* $\mathfrak{A}$. *Then for every* $x \in (a,b)$ *and every infinite word* w *such that* $x = 0.w$, *there exists an* s*-mark-up, for a suitable state* $s = s(w) \in \mathcal{S}$ *of the automaton* $\mathfrak{A}$.

Proof of Proposition 13.56 The idea of the proof is as follows: Once feeding a finite automaton $\mathfrak{A}$ by infinite sequence of finite words $\gamma_0, \gamma_0\gamma_1, \gamma_0\gamma_1\gamma_2, \ldots$ over $\mathbb{Z}_p$, the automaton reaches some of its states infinitely many times; this state s specifies a mark-up $(i(s))$ since corresponding sequences of approximations $0.\gamma_0 \ldots \gamma_{i(s)}$ and $\mathfrak{a}_s(0.\gamma_0 \ldots \gamma_{i(s)})$ tend accordingly to $x = 0.\gamma_0\gamma_1\gamma_2 \ldots$ and to $g(x)$ due to the continuity of g at x. Now we prove the proposition rigorously.

Firstly consider the case when $x \in (a,b)$ has a unique base-p expansion, say, $x = 0.\gamma_0\gamma_1 \ldots = 0.w$, where $w = \gamma_0\gamma_1 \ldots$. As $g(x)$ is $\mathfrak{A}$-computable, there exists a sequence $w_0, w_1, w_2, \ldots \in \mathcal{W}$ of finite non-empty words such that $\rho(w_0), \rho(w_1), \rho(w_2), \ldots \in (a,b)$, $\lim_{i\to\infty} \rho(w_i) = x$ and $\lim_{j\to\infty} \rho(\mathfrak{a}(w_i)) = g(x)$. Note that the sequence $(\Lambda(w_i))_{i=0}^{\infty}$ is increasing since otherwise $x = n/p^r$ for suitable $n, r \in \mathbb{N}_0$ as $\lim_{i\to\infty} \rho(w_i) = x$ and thus x has a non-unique base-p expansion. We may assume that $(\Lambda(\omega_i))_{i=0}^{\infty}$ is a strictly increasing sequence since otherwise we just take a suitable subsequence of $(w_i)_{i=0}^{\infty}$. Moreover, by the same reason, we may assume that $\Lambda(w_i) > i$.

Consider a word sequence $\bar{w}_0 = \gamma_0, \bar{w}_1 = \gamma_0\gamma_1, \bar{w}_2 = \gamma_0\gamma_1\gamma_2, \ldots$. As $x \in (a,b)$, there exists $N \in \mathbb{N}_0$ such that $0.\bar{w}_i \in (a,b)$ once $i \geqslant N$. Without loss of generality we may assume that $N = 1$. As $\lim_{i\to\infty} 0.w_i = 0.\gamma_0\gamma_1 \ldots = x$, for every $n \in \mathbb{N}_0$ there exists $M(n) \in \mathbb{N}_0$, $M(n) \geqslant n$, such that $|0.w_i - 0.\bar{w}_j| < p^{-n}$ provided $i, j \geqslant M(n)$; therefore as $\Lambda(w_i) > i \geqslant M(n) \geqslant n$, we conclude that $w_i = \gamma_0 \ldots \gamma_n v_i$ for $i \geqslant M(n)$ and suitable finite word v_i. Given $n \in \mathbb{N}_0$, let $M(n)$ be the smallest with the said property; this way we obtain an increasing sequence $M(0) < M(1) < M(2) < \ldots$. Considering the subsequence $(w_{M(j)})_{j=0}^{\infty}$ of the word sequence $(w_i)_{i=0}^{\infty}$, we see that $w_{M(j)} = \gamma_0 \ldots \gamma_j r_j$ for $j = 0, 1, 2, \ldots$ and suitable finite non-empty words r_j, that $\lim_{j\to\infty} \rho(w_{M(j)}) = x$ and that $\lim_{j\to\infty} \rho(\mathfrak{a}(w_{M(j)})) = g(x)$. Now to the word sequence $(r_j)_{j=0}^{\infty}$ we put into the correspondence the sequence $(s(r_j)))_{j=0}^{\infty}$ of states of the automaton $\mathfrak{A}$, where $s(r_j)$ is the state the automaton $\mathfrak{A}$ reaches after being fed by the input word r_j. As the number of different states of $\mathfrak{A}$ is finite, in the sequence $(s(r_j))_{j=0}^{\infty}$ at least one state, say s, occurs in the sequence $(s(r_j))_{j=0}^{\infty}$ infinitely many times; say, for $j = j_0, j_1, j_2, \ldots$ $(j_0 < j_1 < j_2 < \ldots)$. Therefore

$$\lim_{k\to\infty} \rho(\gamma_0 \ldots \gamma_{j_k}) = 0.\gamma_0\gamma_1 \ldots = x \tag{13.70}$$

$$\lim_{k\to\infty} \rho(\mathfrak{a}_s(\gamma_0 \ldots \gamma_{j_k})) = \lim_{k\to\infty} \rho(\mathfrak{a}(w_{M(i_k)})) = g(x) \tag{13.71}$$

Recall that $\mathfrak{a} = \mathfrak{a}_{s_0}$ where s_0 is the initial state of the automaton $\mathfrak{A} = \mathfrak{A}(s_0)$. Note that nowhere in the argument above we have used that g is continuous.

Now consider the case when $x = n/p^r$ for suitable $n, r \in \mathbb{N}_0$, $n \in \{0, 1, \dots, p^r - 1\}$. In this case, let $0.\gamma_0\gamma_1\dots$ be either of base-p expansions $\mathsf{wrd}^r(x)$, $\mathsf{wrd}^l(x)$. We will show that similarly to the case when the base-p expansion is unique, in the case under consideration there also exist $s \in \mathcal{S}$ and sequences $i_0 < i_1 < i_2 < \dots$, $M(i_0) < M(i_1) < M(i_2) < \dots$ over $\mathbb{N}_0$ such that $M(i_k) \geqslant i_k$ for all $k \in \mathbb{N}_0$ and both (13.70) and (13.71) hold.

For this purpose, consider arbitrary sequence $w_0, w_1, w_2, \dots$ of right-infinite words over $\{0, 1, \dots, p-1\}$ which are not eventually periodic and such that $\gamma_0 \dots \gamma_i$ is a prefix of w_i for all $i = 0, 1, 2, \dots$. Then $\lim_{i\to\infty} 0.w_i = x$ and therefore all $0.w_i \in (a, b)$ once $i \geqslant I$ where I is large enough (we may assume that $I = 0$; otherwise we consider a subsequence $(w_i)_{i=I}^{\infty}$ rather than the whole sequence $(w_i)_{i=0}^{\infty}$). Note that then $\lim_{i\to\infty} g(0.w_i) = g(x)$ as g is continuous on (a, b); therefore there exists a sequence $(S(i))_{i=0}^{\infty}$ over $\mathbb{N}$ such that for all $i = 0, 1, 2, \dots$ the following inequality holds:

$$|g(0.w_i) - g(x)| < p^{-S(i)} \tag{13.72}$$

Moreover, we may assume that the sequence $(S(i))_{i=0}^{\infty}$ is strictly increasing (if not, we consider a corresponding infinite subsequence of the sequence $(w_i)_{i=0}^{\infty}$ rather than the whole sequence).

Consider now a word $w_i = \gamma_{i0}\gamma_{i1}\dots$ from the above word sequence (note that $\gamma_{i\ell} = \gamma_\ell$ for $\ell = 1, 2, \dots, i$). As every $0.w_i$ is a unique base-p expansion of the corresponding real number from (a, b), there exists a state $s(i)$ of the automaton $\mathfrak{A}$ and a strictly increasing sequence $\mathcal{J}(s(i)) = (j_{ik})_{k=0}^{\infty}$ of numbers from $\mathbb{N}_0$ such that

$$\lim_{k\to\infty} \rho(\mathfrak{a}_{s(i)}(\gamma_{i0}\gamma_{i1}\dots\gamma_{ij_{ik}})) = g(0.w_i), \tag{13.73}$$

compare the case we just have considered above at the beginning of the proof of the proposition. Therefore, for any $k \in \mathbb{N}_0$ there exists $K(j_{ik}) \in \mathbb{N}$ such that

$$|\rho(\mathfrak{a}_{s(i)}(\gamma_{i0}\gamma_{i1}\dots\gamma_{ij_{ik}})) - g(0.w_i)| < p^{-K(j_{ik})}, \tag{13.74}$$

and there exists a strictly increasing sequence of k such that the corresponding sequence of $K(j_{ik})$ is also strictly increasing in force of (13.73). Without loss of generality, we may assume that the sequence $(K(j_{ik}))_{k=0}^{\infty}$ is strictly increasing (otherwise we consider a corresponding subsequence of the sequence $\mathcal{J}(s(i))$).

As a total number of states of $\mathfrak{A}$ is finite, in the infinite sequence $(s(i))_{i=0}^{\infty}$, at least one state, say s, occurs infinitely many times. We may assume that $s(i) = s$ for all $i \in \mathbb{N}_0$; otherwise we just consider respective subsequence of $(w_i)_{i=0}^{\infty}$ rather than the whole sequence. As the sequence $(j_{ik})_{k=0}^{\infty}$ is strictly increasing, all $j_{ik} > i$ once k is large enough. Given $i \in \mathbb{N}_0$, denote via $N(i) \in \mathbb{N}$ the smallest number such that $j_{ik} > i$ once $k \geqslant N(i)$. We again may assume that $N(i) = 0$; if otherwise we will just consider the subsequence $(j_{ik})_{k=N(i)}^{\infty}$ rather than the whole sequence $(j_{ik})_{k=0}^{\infty}$. Then $\gamma_{i0}\gamma_{i1}\dots\gamma_{ij_{ik}} = \gamma_0\dots\gamma_i r_{ik}$ for all k where r_{ik} is a non-empty finite word.

Let $s'(i,k) \in \mathcal{S}$ be a state the automaton $\mathfrak{A}(s)$ reaches after being fed by the input word r_{ik} (the latter state s is defined above). As the total number of states of $\mathfrak{A}(s)$ is finite, in the sequence $(s'(i,k))_{k=0}^{\infty}$, at least one state, say $s'(i)$, occurs infinitely many times. Moreover, by the same reason, at least one state, say s', occurs in the sequence $(s'(i))_{i=0}^{\infty}$ infinitely many times. And again, without loss of generality we may assume that $s'(i,k) = s'$ for all i,k; otherwise we consider corresponding subsequences of the sequences $(w_i)_{i=0}^{\infty}$ and $(j_{ik})_{k=0}^{\infty}$. Note that being fed by the input word $\gamma_{i0}\gamma_{i1}\dots\gamma_{ij_{ik}} = \gamma_0\dots\gamma_i r_{ik}$, the automaton $\mathfrak{A}(s)$ (rightmost letters are fed prior to leftmost ones), the automaton outputs a word of length $j_{ik}+1$ whose (left) suffix of length $i+1$ is the word which outputs the automaton $\mathfrak{A}' = \mathfrak{A}(s')$ if being fed by the word $\gamma_0\dots\gamma_i$. Therefore

$$|\rho(\mathfrak{a}'(\gamma_0\dots\gamma_i)) - \rho(\mathfrak{a}_s(\gamma_{i0}\gamma_{i1}\dots\gamma_{ij_{ik}}))| < p^{-i+1} \tag{13.75}$$

Now combining (13.72), (13.74), and (13.75), we conclude that

$$\lim_{i\to\infty} \rho(\mathfrak{a}'(\gamma_0\dots\gamma_i)) = g(x)$$

However, $\lim_{i\to\infty}\rho(\gamma_0\dots\gamma_i) = x$ since $0.\gamma_0\gamma_1\dots$ is the base-p expansion of x. This finally proves the proposition. □

Note 13.57 Actually during the proof of Proposition 13.56, we have shown that the following claim is true: *Let the function* $g\colon U \to \mathbb{S}$ (or, $g\colon U \to \mathbb{I}$) *be defined on an open neighborhood* $U \subset \mathbb{S}$ (or, $U \subset \mathbb{I}$) *of a point* x, *let* g *be continuous at* x, *and let* $\mathbf{G}(g) \subset \mathcal{P}(\mathfrak{A})$ *for a suitable finite automaton* $\mathfrak{A}$; *then there exists a mark-up for every base-*p *expansion of* x. *Moreover, if* $g\colon [a,b] \to [0,1]$ *is a continuous function on the closed segment* $[a,b]$, *then there exists an* s*-mark-up for right base-*p *expansion of* a *and an* s*-mark-up for left base-*p *expansion of* b.

Corollary 13.58 *In conditions of Proposition 13.56, if the automaton* $\mathfrak{A}$ *is minimal, then* $\mathbf{G}(g) \subset \mathcal{P}'(\mathfrak{A})$.

Proof of Corollary 13.58 Follows immediately from Corollary 13.31 by the definition of mark-up. □

The following proposition reduces examination of continuous functions computable by a finite automaton $\mathfrak{A}$ for the case when the function is defined on a segment for which there exists a state s of the automaton such that the set of all points from the segment which have s-mark-ups is dense in the segment; thus, values at these points completely specify the function on the segment.

Proposition 13.59 *Let* $g\colon [a,b] \to [0,1]$, *where* $[a,b] \subset [0,1]$, *be a continuous function; let* $\mathbf{G}(g) \subset \mathcal{P}(\mathfrak{A})$ *for a suitable automaton* $\mathfrak{A}$ *whose set of states* $\mathcal{S}$ *is finite. Then* $[a,b]$ *is a union of a countably many sub-segments* $[a'_j, b'_j] \subset [a,b]$, $a'_j < b'_j$,

$j = 1, 2, \dots$ having the following property: For every $j = 1, 2, \dots$ there exists a state $s_q \in \mathcal{S}$, $q = q(j)$, such that the set $M_q([a'_j, b'_j])$ of all points from $[a'_j, b'_j]$ which have s_q-mark-ups is dense in $[a'_j, b'_j]$.

Proof of Proposition 13.59 For $s \in \mathcal{S}$ denote via $\mathcal{W}_\infty(s)$ the set of all right-infinite words $w \in \mathcal{W}_\infty$ such that $0.w \in (a, b)$ and an s-mark-up for w exists; put $0.\mathcal{W}_\infty(s) = \{0.w : w \in \mathcal{W}_\infty(s)\}$.

Given $s \in \mathcal{S}$ such that $\mathcal{W}_\infty(s) \neq \emptyset$, let $\mathbf{W}(s)$ be intersection of the closure $\overline{\mathbf{W}}(s)$ of $0.\mathcal{W}^\infty(s)$ with (a, b); so $\mathbf{W}(s)$ is closed in (a, b) w.r.t. the induced topology on (a, b), and there are only finitely many pairwise distinct $\mathbf{W}(s)$; say, these are $\mathbf{W}(s_1), \dots, \mathbf{W}(s_k)$. Proposition 13.56 implies that $\mathbf{W}(s_1) \cup \dots \cup \mathbf{W}(s_k) = (a, b)$. We argue that for some $\mathbf{W}(s_1), \dots, \mathbf{W}(s_k)$, their interiors $\mathbf{W}(s_1)^o, \dots, \mathbf{W}(s_k)^o$ are not empty. Indeed, $\mathbf{W}(s_i)^o = \overline{\mathbf{W}}(s_i)^o \cap (a, b)$ for all $i = 1, 2, \dots, k$; but from Proposition 13.56, it follows that $\overline{\mathbf{W}}(s_1) \cup \dots \cup \overline{\mathbf{W}}(s_k) = [a, b]$ and therefore $\overline{\mathbf{W}}(s_1)^o \cup \dots \cup \overline{\mathbf{W}}(s_k)^o$ is dense in $[a, b]$ as $[a, b]$ is Baire (cf., e.g., [2, Theorems 6.16–6.17]).

As some (without loss of generality we may assume that all) of the interiors $\overline{\mathbf{W}}(s_1)^o, \dots, \overline{\mathbf{W}}(s_k)^o$ are non-empty, the interiors are countable unions of open intervals: $\overline{\mathbf{W}}(s_i)^o = \cup_{\ell=1}^\infty (a'_{i\ell}, b'_{i\ell})$, $a'_{i\ell} < b'_{i\ell}$, $(i = 1, 2, \dots, k)$. Therefore $\overline{\mathbf{W}}(s_i) = \cup_{\ell=1}^\infty [a'_{i\ell}, b'_{i\ell}]$. This completes the proof as $\overline{\mathbf{W}}(s_1) \cup \dots \cup \overline{\mathbf{W}}(s_k) = [a, b]$. □

Corollary 13.60 *In conditions of Proposition 13.59, the segment $[a, b]$ admits a countable covering by closed sub-segments $[a'_j, b'_j] \subset [a, b]$ such that the graph $\mathbf{G}(g_j)$ of the restriction of the function g to the sub-segment $[a'_j, b'_j]$ lies in $\mathcal{P}(\mathfrak{A}(s)) \subset \mathcal{P}(\mathfrak{A})$ for a suitable sub-automaton $\mathfrak{A}(s)$ of the automaton $\mathfrak{A}$, $s = s(j) \in \mathcal{S}$.*

Proof of Corollary 13.60 Indeed, a closure in $\mathbb{R}^2$ of the point set $\{(x; g(x)) : x \in M_q([a'_j, b'_j])\}$ is a graph of the restriction g_j of the function g to the segment $[a'_j, b'_j]$ as g_j is continuous on $[a'_j, b'_j]$. On the other hand, the closure must lie in $\mathcal{P}(\mathfrak{A}(s))$ for $s = s_q$ as $g(x) = \lim_{k\to\infty} \rho(\mathfrak{a}_s(\gamma_0 \dots \gamma_{i_k}))$ where $x = 0.\gamma_0\gamma_1 \dots$ and $i_0, i_1, \dots$ is an s-mark-up. Note also that $\mathcal{P}(\mathfrak{A}(s)) \subset \mathcal{P}(\mathfrak{A})$ as every state of the automaton $\mathfrak{A}$ is reachable from its initial state s_0 (since we consider reachable automata only; cf., Sect. 1.6). □

Note 13.61 From the respective proofs, it follows that both Proposition 13.59 and Corollary 13.60 remain true for a continuous function $g : [a, b] \to \mathbb{S}$ as well as for the case when $[a, b] \subset \mathbb{S}$.

The following theorem shows that we may restrict our considerations of finitely computable continuous functions to the case when computing automata are minimal.

Theorem 13.62 *Given a continuous function $g : [a, b] \to [0, 1]$ (where $[a, b] \subset [0, 1]$) such that $\mathbf{G}(g) \subset \mathcal{P}(\mathfrak{A})$ for a finite automaton $\mathfrak{A}$, there exists a countable covering $\{[a'_j, b'_j] \subset [a, b] : j = 1, 2, \dots ; a'_j < b'_j\}$ of the segment $[a, b]$ such that* *for every j the graph $\mathbf{G}(g_j)$ of the restriction g_j of the function g to the segment $[a'_j, b'_j]$ lies in $\mathcal{P}'(\mathfrak{A}_n)$ for a suitable minimal sub-automaton $\mathfrak{A}_n$ of $\mathfrak{A}$, $n = n(j)$.*

Proof of Theorem 13.62 The state s_q from Proposition 13.59 is either ergodic or transient, see Sect. 1.6. We consider these two cases separately.
Case 1: The state s_q is ergodic. As the set $M_q([a'_j, b'_j])$ from Proposition 13.59 is dense in $[a'_j, b'_j]$ and g_j is continuous, every point $g_j(x)$ for $x \in [a'_j, b'_j]$ is a limit of a sequence $(g(x_i))_{i=0}^{\infty}$ where $(x_i)_{i=0}^{\infty}$ is a sequence of points from $M_q([a'_j, b'_j])$ and $(x_i)_{i=0}^{\infty}$ tends to x as i tends to infinity:

$$x = \lim_{i\to\infty} x_i; \tag{13.76}$$

$$g_j(x) = \lim_{i\to\infty} g_j(x_i). \tag{13.77}$$

But $x_i = 0.w_i$ where w_i is a right-infinite word for which there exists an s_q-mark-up (cf. the construction of the set $M_q([a'_j, b'_j])$); therefore from (13.76)–(13.77) it follows now that there exists a sequence $(h_\ell)_{\ell=0}^{\infty}$ of finite words h_ℓ of strictly increasing lengths such that

$$x = \lim_{\ell\to\infty} 0.h_\ell; \tag{13.78}$$

$$g_j(x) = \lim_{\ell\to\infty} \rho(\mathfrak{a}_{s_q}(h_\ell)). \tag{13.79}$$

Indeed, the words h_ℓ are (left) prefixes of words $w_q = \omega_0^{(q)}\omega_1^{(q)}\ldots$ that correspond to s_q-mark-up; that is, h_ℓ are of the form $\omega_0^{(q_\ell)}\omega_1^{(q_\ell)}\ldots\omega_{r_{q_\ell,k_\ell}}^{(q_\ell)}$ where the sequence $(r_{q_\ell,k})_{k=0}^{\infty}$ is the s_q-mark-up of the word w_{q_ℓ}. Now, as the state s_q is ergodic (i.e., s_q a state of a certain minimal sub-automaton, say $\mathfrak{A}_q = \mathfrak{A}(s_q)$, of the automaton $\mathfrak{A}$; cf. Sect. 1.6), then we just mimic the proof of Theorem 13.30 starting with (13.28)–(13.29) and show that $(x, g_j(x)) \in \mathcal{P}'(\mathfrak{A}_q)$.
Case 2: Now let the state s_q from Proposition 13.59 be not ergodic (whence transient). Thus there exists a finite word $u = \alpha_0\ldots\alpha_{k-1}$ such that after the automaton $\tilde{\mathfrak{A}} = \mathfrak{A}(s_q)$ has been fed by the word u (rightmost letters are fed to the automaton prior to leftmost ones), the automaton reaches some ergodic state (say t) which is a state of a minimal sub-automaton $\mathfrak{A}' = \mathfrak{A}(t)$; see Sect. 1.6. Note that then all words of the form vu have the same property, for all $v \in \mathcal{W}_\varnothing$: After being fed by vu, the automaton reaches some state from the set of states of $\mathfrak{A}'$ due to the minimality of $\mathfrak{A}'$. Therefore, the set $B_q \subset \mathbb{Z}_p$ of all p-adic integers whose p-adic canonical representation correspond to left-infinite words $w \in \mathcal{W}_\infty$ such that if the automaton $\mathfrak{A}(s_q)$ while being fed by the word w reaches at a finite step some ergodic state (i.e., reaches a state which is a state of some minimal sub-automaton of $\mathfrak{A}$) is a union of balls of non-zero radii in $\mathbb{Z}_p$; thus, the set B_q is an open subset in $\mathbb{Z}_p$ since every ball of a non-zero radius is open in $\mathbb{Z}_p$ w.r.t. the p-adic topology (cf. Sect. 2.2.1). Hence the set $A_q = \mathbb{Z}_p \setminus B_q$ is a closed subset of $\mathbb{Z}_p$; and the set A_q consists of all p-adic integers such that if the automaton $\mathfrak{A}(s_q)$ is being fed by a

left-infinite word that is the base-p expansion of some p-adic integer from A_q, the automaton $\mathfrak{A}$ never reaches an ergodic state. Let P_q be the set of all finite prefixes of words from A_q; denote $\mathbf{P}_q$ a closure of the set $0.P_q = \{0.w \colon w \in P_q\}$ in $\mathbb{R}$. *Claim:The interior of* $\mathbf{P}_q$ *is empty* (therefore $\mathbf{P}_q$ is nowhere dense in $[0, 1]$).

Indeed, if not then $\mathbf{P}_q$ contains an open interval (a_q, b_q). Take a finite non-empty word u such that $a_q < 0.u < b_q$. As $\mathbf{P}_q \supset (a_q, b_q)$ then, given an arbitrary finite non-empty word $v = \alpha_1 \ldots \alpha_k$ where $\alpha_k \neq 0$, there exists a sequence $\mathcal{W}(v) = (w_i)_{i=1}^{\infty}$ of finite non-empty words $w_i \in P_q$ such that $\lim_{i\to\infty} 0.w_i = 0.uv \in (a_q, b_q)$ (recall that uv is a concatenation of words u and v). Therefore either $uv \in P_q$ (thus $v \in P_q$ by the construction of P_q) or $\mathcal{W}(v)$ contains an infinite subsequence of words of the form $w_i' = u\alpha_1 \ldots \alpha_k'(p-1)^{r_i}$ where $\alpha_k' = \alpha_k - 1$, $r_1 < r_2 < \ldots$ (recall that $(p-1)^{r_i}$ is a word of length r_i all whose letters are $p-1$); hence by the construction of P_q there exists an infinite sequence $w_i'' = \alpha_1 \ldots \alpha_k'(p-1)^{r_i}$ over P_q. Thus we conclude that once $n \in \mathbb{N}$, the closure of A_q in $\mathbb{Z}_p$ (thus, the very set A_q itself as it is closed) must either contain n or $-n$ (recall that negative rational integers in $\mathbb{Z}_p$ are exactly that ones whose canonical p-adic expansions have only a finitely many terms with coefficients other than $p-1$; cf. Sect. 2.3.2). But this implies that $A_q = \mathbb{Z}_p$ as the set $\{\pm n \colon n \in \mathbb{N}\}$ (where $+$ or $-$ are taken in arbitrary order) is dense in $\mathbb{Z}_p$. On the other hand, by the construction the set A_q consists of all p-adic integers such that if the automaton $\mathfrak{A}(s_q)$ is being fed by a left-infinite word that is the base-p expansion of some p-adic integer from A_q, the automaton $\mathfrak{A}$ never reaches an ergodic state; therefore the equality $A_q = \mathbb{Z}_p$ contradicts our assumption that s_q is transient (since then there must exist a left-infinite word w such that at a finite step the automaton $\tilde{\mathfrak{A}} = \mathfrak{A}(s_q)$ reaches an ergodic state if being fed by w). This proves our claim.

Denote now via $\tilde{f}$ an automaton function of the automaton $\tilde{\mathfrak{A}} = \mathfrak{A}(s_q)$; and for $k = 1, 2, \ldots$ put

$$E_k'(\tilde{f}) = \left\{ \left(\frac{z \bmod p^k}{p^k}; \frac{\tilde{f}(z) \bmod p^k}{p^k} \right) \in \mathbb{I}^2 \colon z \in \mathbb{Z}_p \setminus A_q = B_q \right\} \tag{13.80}$$

a point set in the unit real square $\mathbb{I}^2 = [0, 1] \times [0, 1]$; then take a union $E'(\tilde{f}) = \cup_{k=1}^{\infty} E_k'(\tilde{f})$; denote via $\mathbf{P}'(\tilde{\mathfrak{A}}) = \mathbf{P}'(\tilde{f})$ a closure (in topology of $\mathbb{R}^2$) of the set $E'(\tilde{f})$ (cf. (12.6)). Denote via g_j a restriction of the function g to $[a_j', b_j']$. As the function g_j is continuous on $[a_j', b_j']$ (cf. Corollary 13.60) and $\mathbf{G}(g) \subset \mathcal{P}(\mathfrak{A})$, then necessarily $\mathbf{G}(g_j) \subset \mathbf{P}'(\tilde{\mathfrak{A}})$ since the set $\mathbf{P}_q$ is nowhere dense in $[a_j', b_j']$ by Claim 1.

As the set $\mathcal{S}$ of all states of the automaton $\mathfrak{A}$ is finite, there are only finitely many ergodic components in $\mathcal{S}$; say they are $\mathcal{S}_1, \ldots, \mathcal{S}_m \subset \mathcal{S}$. Given an ergodic component $\mathcal{S}_n$ $(n = 1, 2, \ldots, m)$ denote

$$E_n = \left\{ \left(\frac{z \bmod p^k}{p^k}; \frac{\tilde{f}(z) \bmod p^k}{p^k} \right) \in \mathbb{I}^2 \colon z \in B_q, k > k_n(z) \right\}$$

where $k_n(z)$ is the smallest $k \in \mathbb{N}$ such that after the automaton $\tilde{\mathfrak{A}} = \mathfrak{A}(s_q)$ has been fed by the word $\mathsf{wrd}(z \bmod p^{k_n(z)})$, the automaton reaches a state from $\mathcal{S}_n$. Then the union $E = \cup_{n=1}^{m} E_n$ is disjoint and $\mathbf{P}'(\tilde{\mathfrak{A}})$ is a closure of E by (13.80). Therefore $\mathbf{P}'(\tilde{\mathfrak{A}}) = \cup_{n=1}^{m} \mathbf{E}_n$ where $\mathbf{E}_n$ is a closure of E_n in $\mathbb{I}^2$; hence $\mathbf{G}(g_j) = \cup_{n=1}^{m}(\mathbf{G}(g_j) \cap \mathbf{E}_n)$. Note that from the definition of $\mathcal{P}(\mathfrak{A})$ (cf. Sect. 12.1), it follows that $\mathbf{E}_n = \mathcal{P}(\mathfrak{A}_n)$ where $\mathfrak{A}_n$ is a minimal sub-automaton (of the automaton $\mathfrak{A}$) whose set of states is $\mathcal{S}_n$.

Further, as the function g_j is continuous on $[a'_j, b'_j]$, the set $\mathbf{G}(g_j)$ is closed in $\mathbb{I}^2$; therefore the set $\mathbf{G}_n = \mathbf{G}(g_j) \cap \mathbf{E}_n$ is closed in $\mathbb{R}^2$. Hence, the set $\mathbf{R}_n = \{x \in [a'_j, b'_j]\colon (x, g(x)) \in \mathbf{E}_n\}$ is closed in $\mathbb{R}$ and $[a'_j, b'_j] = \cup_{n=1}^{m} \mathbf{R}_n$. Now by argument similar to that from the proof of Proposition 13.59, we conclude that some of the interiors $\mathbf{R}_n^o$ must be non-empty and hence either of the non-empty interiors is a union of a countably many open intervals. By taking closures of the intervals, we see that $[a'_j, b'_j]$ is a union of the closures, that is, $[a'_j, b'_j]$ is a union of a countably many its closed sub-segments $[a'_{j.i}, b'_{j.i}]$ $(i \in \mathbb{N}_0)$ of non-zero lengths, and the graph of the restriction $g_{j.i}$ of g_j to either of the sub-segments lies in $\mathbf{E}_n = \mathcal{P}(\mathfrak{A}_n)$ for a suitable $n \in \{1, 2, \ldots, m\}$. Now we apply Proposition 13.59 substituting $g_{j.i}$ for g and $[a'_{j.i}, b'_{j.i}]$ for $[a, b]$; but as every s_q from the statement of Proposition 13.59 is now a state of the minimal sub-automaton $\mathfrak{A}_n$, we now are in conditions of Case 1. Therefore $\mathbf{G}(g_{j.i}) \subset \mathcal{P}'(\mathfrak{A}_n(s_q))$; but $\mathcal{P}'(\mathfrak{A}_n(s)) = \mathcal{P}'(\mathfrak{A}_n(t))$ for all states s, t of the automaton $\mathfrak{A}_n(s_q)$ due to the minimality of the automaton (cf. Note 13.32). This finally proves the theorem. □

Note 13.63 From the proof of Theorem 13.62 it follows that the theorem remains true for a continuous function $g\colon [a, b] \to \mathbb{S}$ as well as for the case when $[a, b] \subset \mathbb{S}$.

The following proposition shows that we may if necessary consider only finitely computable continuous functions defined everywhere on the unit segment $[0, 1]$ rather than on sub-segments of $[0, 1]$.

Proposition 13.64 *If a continuous function* $g\colon [a, b] \to \mathbb{S}$, $[a, b] \subset [0, 1]$, *is such that* $\mathbf{G}_{[a,b]}(g) \subset \mathcal{P}(\mathfrak{A})$ *for a suitable finite automaton* $\mathfrak{A} = \mathfrak{A}(s_0)$, *then for every* $n, m \in \mathbb{N}_0$ *such that* $m \geqslant \lfloor \log_p n \rfloor + 1$ *and* $n/p^m, (n+1)/p^m \in [a, b]$, *the function* $g_d(x) = (p^m g(d + p^{-m}x)) \bmod 1$, *where* $d = np^{-m}$, *is continuous on* $[0, 1]$, *and* $\mathbf{G}_{[0,1]}(g_d) \subset \mathcal{P}(\mathfrak{A})$.

Proof of Proposition 13.64 As the base-p expansion of d is $d = 0.\chi_0 \ldots \chi_{m-1}00\ldots$ then, given the base-p expansion for $x = 0.\zeta_0\zeta_1 \ldots \in [0, 1]$, the base-$p$ expansion for $d + xp^{-m}$ is $d + xp^{-m} = 0.\chi_0 \ldots \chi_{m-1}\zeta_0\zeta_1 \ldots$ and $d + xp^{-m} \in [a, b]$ for all right-infinite words $\zeta_0\zeta_1 \ldots$ (thus, for all $x \in [0, 1]$). Therefore if $i_0 < i_1 < i_2 < \ldots$ is a mark-up for $\chi_0 \ldots \chi_{m-1}\zeta_0\zeta_1 \ldots$ (cf. Proposition 13.56) then $(j_m = i_{r+m} - m)_{m=0}^{\infty}$, where $r = \min\{\ell\colon i_\ell > m\}$, is an s-mark-up of the infinite word $\zeta_0\zeta_1 \ldots$ for a suitable state $s \in \mathcal{S}$ of the automaton $\mathfrak{A} = \mathfrak{A}(s_0)$ with respect to the function g. Hence,

$$\lim_{k\to\infty} \rho(\mathfrak{a}_s(\zeta_0\zeta_1\ldots\zeta_{j_k})) \equiv \left(p^m\left(g\left(d+\frac{x}{p^m}\right)\right)\right) \pmod 1 \tag{13.81}$$

as $\rho(\mathfrak{a}_s(\zeta_0\zeta_1\ldots\zeta_{j_k})) = (p^m(\rho(\mathfrak{a}_s(\chi_0\ldots\chi_{m-1}\zeta_0\zeta_1\ldots\zeta_{j_k})))) \bmod 1$. By our assumption on reachability of the automaton $\mathfrak{A}$ (cf. Sect. 1.6), there exists a finite word $u = u(s)$ such that the automaton $\mathfrak{A}$ being fed by u reaches the state s and outputs the corresponding finite word $u' = \mathfrak{a}_{s_0}(u)$; therefore the automaton $\mathfrak{A} = \mathfrak{A}(s_0)$ being fed by a concatenated finite word $\zeta_0\zeta_1\ldots\zeta_{j_k}u$ outputs the concatenated finite word $\mathfrak{a}_s(\zeta_0\zeta_1\ldots\zeta_{j_k})u'$. But

$$\lim_{k\to\infty} \rho(\mathfrak{a}_{s_0}(\zeta_0\zeta_1\ldots\zeta_{j_k}u)) = \lim_{k\to\infty} \rho(\mathfrak{a}_s(\zeta_0\zeta_1\ldots\zeta_{j_k})u') =$$

$$= \lim_{k\to\infty} \rho(\mathfrak{a}_s(\zeta_0\zeta_1\ldots\zeta_{j_k}))$$

and simultaneously $x = \lim_{k\to\infty} \rho(\zeta_0\zeta_1\ldots\zeta_{j_k}u) = \lim_{k\to\infty} \rho(\zeta_0\zeta_1\ldots\zeta_{j_k})$ since the words u, u' are finite and fixed; therefore $(x; g_d(x) \bmod 1) \in \mathcal{P}(\mathfrak{A})$ for all $x \in [0,1]$ in view of (13.81).

The function g_d is conjugated to a continuous function by a continuous map and therefore is also continuous: Once $e, h \in [0,1]$ are such that $|e-h| < p^{-K(L)}$ to ensure that $|g(d+p^{-m}e) - g(d+p^{-m})h| < p^{-L}$ for a sufficiently large $L \in \mathbb{N}$ then $|g_d(e) - g_d(h)| < p^{-L+m}$. □

Corollary 13.65 *If a continuous function $g\colon [a,b] \to \mathbb{S}$, $[a,b] \subset [0,1]$, is such that $\mathbf{G}_{(a,b)}(g) \subset \mathcal{P}(\mathfrak{A})$ for a suitable finite automaton $\mathfrak{A} = \mathfrak{A}(s_0)$, then for every $n, m \in \mathbb{N}_0$ such that $m \geqslant \lfloor\log_p n\rfloor + 1$ and $d = n/p^m \in [a,b)$ the following is true:*

- *The function $g_{d,M}(x) = (p^M g(d+p^{-M}x)) \bmod 1$ is continuous on $[0,1]$ for all sufficiently large $M \geqslant m$, and*
- $\mathbf{G}_{[0,1]}(g_{d,M}) \subset \mathcal{P}(\mathfrak{A})$.

Proof of Corollary 13.65 In the proof of Proposition 13.64, take $0.\chi_0\ldots\chi_{m-1}(0)^{M-m}$ as the base-p expansion of $d = np^{-m}$ where $M \geqslant m$ is large enough to provide $0.\chi_0\ldots\chi_{m-1}(0)^{M-m-1}1 \in [a,b]$. Note that nowhere in the proof of the proposition we used that some of $\chi_0,\ldots,\chi_{m-1}$ are not zero. □

Note 13.66 Corollary 13.65 shows that given any point $d' \in [a,b)$ and a rational approximation $d = np^{-m}$ of d', the graph of the function g on a sufficiently small closed neighborhood $[a',b']$ of the point $d' \neq b'$ is similar to the graph of the function $g_{d,M}$ on $[0,1]$ where $d = np^{-m}$ and M is large enough.

We summarize the above results as follows:

While examining a continuous function $g\colon [a,b] \to \mathbb{S}$ (where $[a,b] \subset [0,1]$ or $[a,b] \subset \mathbb{S}$) whose graph $\mathbf{G}(g)$ lies in $\mathcal{P}(\mathfrak{A})$ for some finite automaton $\mathfrak{A}$ one can if necessary assume that the function is defined and continuous on $[0,1]$ (or on $\mathbb{S}$ except for maybe a single point), the automaton $\mathfrak{A}$ is minimal, the function g is ultimately computable by $\mathfrak{A}$ and that for some state s of $\mathfrak{A}$ the set of all points from $[0,1]$ which have base-p expansions admitting s-mark-ups is dense in $[0,1]$ (respectively, in $\mathbb{S}$).

13.6.4.2 Finite Computability of Compositions

It is quite clear by intuition that a composition of finitely computable continuous functions should be a finitely computable continuous function. The following proposition states this formally and gives some extra information about the graph of a composite finitely computable function.

Proposition 13.67 *Let* $[a,b],[c,d] \subset [0,1]$ *and let* $g\colon [a,b] \to [0,1]$, $f\colon [c,d] \to [0,1]$ *be two continuous functions such that* $g([a,b]) \subset [c,d]$ *and there exist finite automata* $\mathfrak{A}$ *and* $\mathfrak{B}$ *such that* $\mathbf{G}_{[a,b]}(g) \subset \mathcal{P}(\mathfrak{A})$, $\mathbf{G}_{[c,d]}(f) \subset \mathcal{P}(\mathfrak{B})$. *Then there exists a covering* $\{[a'_j,b'_j] \subset [a,b]\colon j \in J\}$ *such that if* h_j *is a restriction of the composite function* $f(g)$ *to the sub-interval* $[a'_j,b'_j]$ *then* $\mathbf{G}_{[a'_j,b'_j]}(h_j) \subset \mathcal{P}(\mathfrak{C}_j)$ *for every* $j \in J$, *where* $\mathfrak{C}_j$ *is a sequential composition of the automaton* $\mathfrak{A}(s_j)$ *with the automaton* $\mathfrak{B}(t_j)$ *and* s_j, t_j *are suitable* (depending on j) *states of the automata* $\mathfrak{A}$, $\mathfrak{B}$ *accordingly.*

Proof of Proposition 13.67 By Note 13.57, for every right-infinite word $w = \gamma_0\gamma_1\ldots \in \mathcal{W}_\infty$ such that $x = 0.w \in (a,b)$ there exists a mark-up (with respect to some state s of the finite automaton $\mathfrak{A}$) $i_0, i_1, i_2, \ldots$; i.e., $\lim_{k\to\infty}\rho(\mathfrak{a}(w_k)) = g(x)$, where $w_k = \gamma_0\gamma_1\ldots\gamma_{i_k} \in \mathcal{W}$; and if $x = a$ (respectively, $x = b$), then the mark-up exists at least for right (respectively, left) base-p expansion. By the same reason, for $y = g(x) = 0.v$, where $v = \nu_0\nu_1\ldots \in \mathcal{W}_\infty$, there exists a mark-up $j_0, j_1, \ldots$ (with respect to some state t of the finite automaton $\mathfrak{B}$) such that $\lim_{n\to\infty}\rho(\mathfrak{b}(v_n)) = g(y)$, where $v_n = \nu_0\nu_1\ldots\nu_{i_n} \in \mathcal{W}$. Now for $m \in \mathbb{N}_0$ denote $N(m) = \min\{k\colon i_k \geqslant j_m\}$, consider the sequence $(N(m))_{m=0}^\infty$ and let $q(0) = N(m_0), q(1) = N(m_1), \ldots$ be a strictly increasing subsequence of $(N(m))_{m=0}^\infty$. Denote $s(\ell)$ the state the automaton $\mathfrak{A}$ reaches after being fed by the word $\gamma_{j_{m_\ell}+1}\gamma_{j_{m_\ell}+2}\cdots\gamma_{i_{q(\ell)}}$; put $s(\ell) = s$ if the latter word is empty. As the automaton $\mathfrak{A}$ is finite, there is a state, say s', that occurs in the sequence $(s(\ell))$ infinitely often. Then the sequence $(j_{m(\ell)}\colon s(\ell) = s')$ is a mark-up of the word w with respect to the automaton $\mathfrak{A}(s')$, and simultaneously the same sequence is a mark-up of the word v with respect to the automaton $\mathfrak{B}(t)$. Therefore

$\lim_{\ell\to\infty} \rho(\mathfrak{c}'(\gamma_0\gamma_1\ldots\gamma_{j_{m(\ell)}}))) = f(y) = f(g(x))$, where $\mathfrak{C}'$ is a sequential composition of automata $\mathfrak{A}(s')$ and $\mathfrak{B}(t)$.

By Corollary 13.60, the segment $g([a,b])$ can be covered by a countably many segments $[c_k, d_k]$, $k \in \mathbb{N}$ where for every k there exists a state t_k of the automaton $\mathfrak{B}$ such that the set of all points from $[c_k, d_k]$ whose base-p expansions (with respect to the function f) admit t_k-mark-ups is dense in $[c_k, d_k]$. Given a real number $y \in [c_k, d_k]$ and its base-p expansion, in view of Proposition 13.56, there exists a t_k-mark-up of the base-p-expansion. Having this mark-up and by acting as above, we, given $x \in g^{-1}(y)$, find corresponding state s'_k of the automaton $\mathfrak{A}$ and construct a strictly increasing sequence over $\mathbb{N}$ such that the sequence is simultaneously a mark-up for y (with respect to t_k and the function f) and for x (with respect to s'_k and the function g).

Let $s'_1, \ldots, s'_r$ be all pairwise distinct states of the automaton $\mathfrak{A}$ that satisfy the following condition: For every $w, v \in \mathcal{W}_\infty$ such that $0.v \in [a,b]$, $g(0.w) = 0.v$ there exists an s'_i-mark-up (for suitable $i \in \{1, 2, \ldots, r\}$) such that the mark-up is a mark-up both for w (with respect to s'_i and g) and for v (with respect to t_k and f) simultaneously. For $i \in \{1, 2, \ldots, r\}$ denote via $\mathcal{W}_\infty(s'_i)$ the set of all infinite words $w \in \mathcal{W}_\infty$ such that there exists an s-mark-up which is a mark-up both for w and for v simultaneously; then proceeding in the same way as in the proof of Proposition 13.59 we conclude that there exists $s' = s'_i$ and a closed subinterval $[a', b']$ such that $W = \mathcal{W}_\infty(s'_i) \cap [a', b']$ is dense in $[a', b']$. But then $g(W)$ is dense in $g([a', b'])$ and $f(g(W))$ is dense in $[f(g([a', b']))$ as g is continuous on $[a', b']$ and f is continuous on $g([a', b'])$. Therefore for a finite automaton $\mathfrak{C}'_{ik}$ which is a sequential composition of the automata $\mathfrak{A}(s'_i)$ and $\mathfrak{B}(t_k)$, we have that the graph of the restriction h of the function $f(g)$ to $[a', b']$ lies in $\mathcal{P}(\mathfrak{C}'_{ik})$. □

Note 13.68 By arguing as in the proof of Proposition 13.67, the following can be shown: *Let* $[a,b] \subset [0,1]$, *let* $g\colon [a,b] \to \mathbb{S}$, $f\colon [a,b] \to \mathbb{S}$ *be two continuous functions, and let there exist finite automata* $\mathfrak{A}$ *and* $\mathfrak{B}$ *such that* $\mathbf{G}_{[a,b]}(g) \subset \mathcal{P}(\mathfrak{A})$, $\mathbf{G}_{[a,b]}(f) \subset \mathcal{P}(\mathfrak{B})$. *Then there exists a covering* $\{[a'_j, b'_j] \subset [a,b] \colon j \in J\}$ *such that if* h_j *is a restriction of the function* $(f+g) \bmod 1$ *to the sub-interval* $[a'_j, b'_j]$ *then* $\mathbf{G}_{[a'_j, b'_j]}(h_j) \subset \mathcal{P}(\mathfrak{C}_j)$ *for every* $j \in J$, *where* $\mathfrak{C}_j$ *is a sum of the automaton* $\mathfrak{A}(s_j)$ *with the automaton* $\mathfrak{B}(t_j)$ *and* s_j, t_j *are suitable* (depending on j) *states of the automata* $\mathfrak{A}$, $\mathfrak{B}$ *accordingly.* Here by the sum of automata $\mathfrak{A}$ and $\mathfrak{B}$ we mean a sequential composition of the automata by automaton which has two inputs and a single output and performs addition of p-adic integers. The latter automaton is finite; see Proposition 13.24. Note also that we may assume that both f and g are defined on an arc of $\mathbb{S}$ rather than on $[a,b]$.

Corollary 13.69 *Given* $A, B \in \mathbb{Z}_p \cap \mathbb{Q}$ *and continuous finitely computable functions* $f, g\colon [a,b] \to \mathbb{S}$, *there exists a covering* $\{[a'_j, b'_j] \subset [a,b] \colon j \in J\}$ *such that the function* $Af + Bg$ *is finitely computable on every* $[a'_j, b'_j]$.

13.6.5 Main Theorems on Finite Computability

In this subsection we prove the following:

A graph of any C^2-smooth finitely computable function $g\colon [a,b] \to \mathbb{S}$, where $[a,b] \subset [0,1)$, lies (under a natural association of the half-open interval $[0,1)$ with the unit circle $\mathbb{S}$) on a torus winding with a p-adic rational slope; and if $\mathfrak{A}$ is a finite automaton that computes g, then necessarily the plot of the automaton contains the whole winding. Moreover, we prove a generalization of this theorem for multivariate functions.

To make further proofs (which are somewhat involved) more transparent, we begin with a brief (and no too rigorous) outline of their general underlying idea.

Given g as above, fix $x = np^{-m} \in [a,b]$; then for $h \in [0,1]$ and all sufficiently large ℓ from the differentiability of g it follows that $g(x+p^{-m-\ell}h) = g(x) + g'(x)p^{-m-\ell}h + p^{-m-t(\ell)}\theta(\ell,h)$, where $|\theta(\ell,h)| \leqslant 1$ and t is a map from $\mathbb{N}_0$ to $\mathbb{N}_0$ such that $p^{t(\ell)} \to \infty$ faster than $p^\ell \to \infty$ while $\ell \to \infty$. Once h is fixed (say, $h = p^{-1}$) then the above equality for large ℓ implies (in view of Proposition 13.67 and Corollary 13.69) that there exists a finite automaton $\mathfrak{B}_x$ which computes $(g'(x)) \bmod 1 = ((g(x+p^{-m-\ell-1}) - g(x))p^{m+\ell+1} - p^{1+\ell-t(\ell)}\theta(\ell,p^{-1})) \bmod 1$ being fed by an infinite sequence of zero words whose lengths increase unboundedly, i.e., $g'(x) \in \mathcal{P}(\mathfrak{B}_x)$: This is because, speaking loosely, the error term $p^{1+\ell-t(\ell)}\theta(\ell,h)$ makes no perturbations of the infinite output sequence due to the fast growth of $t(\ell)$. But then necessarily $g'(x) \in \mathbb{Z}_p \cap \mathbb{Q}$ by Proposition 13.53. Lemma 13.71 which follows proves this fact rigorously.

We then (see Lemma 13.72 below) play a similar trick with the second derivative $g''(x)$: As g is two times differentiable and $g'(x) \in \mathbb{Z}_p \cap \mathbb{Q}$, the function $g_1(u) = g(u) - g'(x)u + c$ of argument $u \in [a,b]$ is also a C^2-smooth finitely computable function for every $c \in \mathbb{Q} \cap \mathbb{Z}_p$. As $g_1'(x) = 0$, $g_1''(u) = g''(u)$, we have (for all sufficiently large ℓ) that $g_1(x+p^{-m-\ell}h) = g_1(x) + \frac{g''(x)}{2}p^{-2m-2\ell}h^2 + p^{-2m-t_1(\ell)}\theta_1(\ell,h)$ where $|\theta_1(\ell,h)| \leqslant 1$, $t(\ell) = 2\ell + w(\ell)$, and w is a map from $\mathbb{N}_0$ to $\mathbb{N}_0$ such that $w(\ell) \to \infty$ as $\ell \to \infty$. From here in a way similar to that of above, we deduce that $\frac{g''(x)}{2} \in \mathbb{Z}_p \cap \mathbb{Q}$. But then, if $g''(x) \neq 0$, the argument means that there exists a finite automaton which performs squaring $h \mapsto h^2$ of every $h \in [0,1]$ with arbitrarily high accuracy. However it is well-known (cf., e.g., [31, Theorem 2.2.3]) that no finite automaton can perform squaring of natural numbers; so necessarily $g''(x) = 0$ for all $x = np^{-m} \in [a,b]$. But the set of these x is dense in $[a,b]$; therefore $g''(x) = 0$ for all $x \in [a,b]$ as g'' is continuous on $[a,b]$. Hence g must be affine: $g(u) = g'(x)u + q$ for all $u \in [a,b]$. Note that then necessarily $q \in \mathbb{Z}_p \cap \mathbb{Q}$ since $q = g(0)$ and g is finitely computable (cf. Proposition 13.53). After that by Proposition 13.64, we can "stretch" the graph of

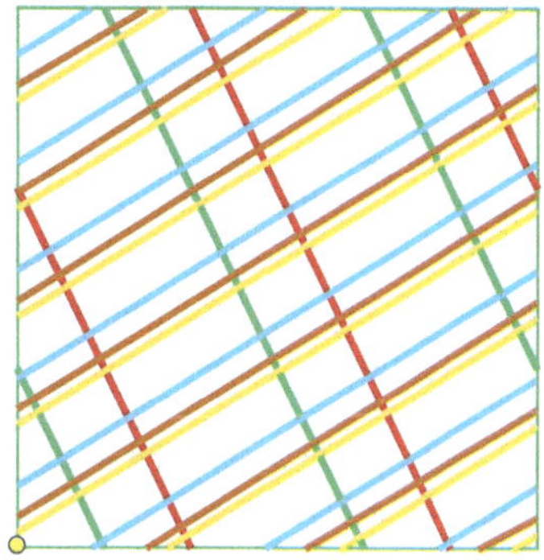

Fig. 13.19 Limit plot in $\mathbb{R}^2$ of an automaton that has two affine sub-automata $\mathfrak{A}$ and $\mathfrak{B}$; $f_{\mathfrak{A}}(z) = -2z + \frac{1}{3}$ and $f_{\mathfrak{B}}(z) = \frac{3}{5}z + \frac{2}{7}$, where $z \in \mathbb{Z}_2$

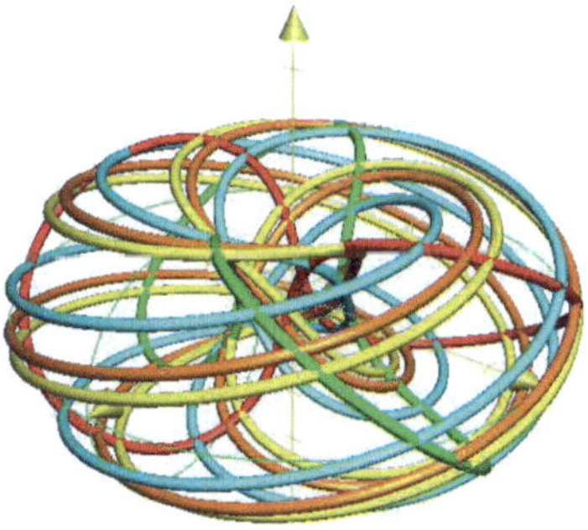

Fig. 13.20 Limit plot of the same automaton on the torus $\mathbb{T}^2$ in $\mathbb{R}^3$. The plot consists of two torus links; the links consist of two and of three knots

the function g from $[a, b]$ to the whole unit circle $\mathbb{S}$ and thus finally obtain a whole cable which lies in the plot of the finite automaton which calculates g. But then by Theorem 13.46, the plot must contain the whole link of torus windings; and the graph $\mathbf{G}_{[a,b]}(g)$ must lie completely on some of these windings. The number of links is finite since every link corresponds to some minimal sub-automaton (see Sect. 1.6 and Theorem 13.62) of the automaton which computes g; and the number of minimal sub-automata of a finite automaton is clearly a finite. Finally, every such link corresponds to a finite family of complex-valued exponential functions of the form $\psi_k(y) = \mathrm{e}^{\mathrm{i}(Ay-2\pi p^k B)}$, $k = 0, 1, 2, \ldots$, for suitable $A, B \in \mathbb{Z}_p \cap \mathbb{Q}$ as shown in Corollary 13.50. Figures 13.19 and 13.20 illustrate how the graphs of C^2-functions from the plots of finite automata look like. Now we proceed with rigorous assertions and proofs.

13.6.5.1 The Univariate Case

We start with a univariate case: That is, we show that C^2-smooth finitely computable functions defined on $[a, b] \subset [0, 1)$ and valuated in $[0, 1)$ are only affine ones. Once we associate the half-open interval $[0, 1)$ with a unit circle $\mathbb{S}$ under a natural bijection, we may consider graphs of the functions as subsets on a surface of the unit torus $\mathbb{T}^2 = \mathbb{S} \times \mathbb{S}$. We show that then the graphs lie only on cables of the torus

$\mathbb{T}^2$, and the slopes of the cables must be p-adic rational integers (i.e., must lie in $\mathbb{Z}_p \cap \mathbb{Q}$); see Sect. 13.6.2 for definitions of torus knots, cables of torus, and links of knots.

Theorem 13.70 *Consider a finite automaton $\mathfrak{A}$ and a continuous function g with domain $[a, b] \subset [0, 1)$, valuated in $[0, 1)$. Let $\mathbf{G}(g) \subset \mathcal{P}(\mathfrak{A})$, let g be two times differentiable on $[a, b]$, and let the second derivative g'' of g be continuous on $[a, b]$. Then there exist $A, B \in \mathbb{Q} \cap \mathbb{Z}_p$ such that $g(x) = (Ax + B) \bmod 1$ for all $x \in [a, b]$; moreover, the graph $\mathbf{G}_{[a,b]}(g)$ of the function g lies completely in the cable $\mathbf{C}(A, B) \subset \mathcal{P}'(\mathfrak{A})$ and $\mathbf{C}(A, \bar{B}) \subset \mathcal{P}'(\mathfrak{A})$ for all $\bar{B} \in \mathbf{C}(B \bmod 1)$.*

Given a finite automaton $\mathfrak{A}$, there are not more than a finite number of pairwise distinct cables $\mathbf{C}(A, B)$ of the unit torus $\mathbb{T}^2$ such that $\mathbf{C}(A, B) \subset \mathcal{P}(\mathfrak{A})$ (note that then $A, B \in \mathbb{Z}_p \cap \mathbb{Q}$).

Lemma 13.71 *Consider a finite automaton $\mathfrak{A}$ and a continuous function g with domain $[a, b] \subset [0, 1)$ valuated in $[0, 1)$. Let $\mathbf{G}(g) \subset \mathcal{P}(\mathfrak{A})$ and let g be differentiable at the point $x = np^{-m} \in [a, b)$ where $n \in \mathbb{N}_0$. Then $g'(x) \in \mathbb{Z}_p \cap \mathbb{Q}$.*

Lemma 13.72 *Under conditions of Theorem 8.2, let x be the same as in the statement of Lemma 13.71; then $g''(x) = 0$.*

Proof of Lemma 13.71 Under conditions of the lemma, the right base-p expansion of x is $x = 0.\gamma_0 \ldots \gamma_{m-1}00\ldots$, for suitable $\gamma_0, \ldots, \gamma_{m-1} \in \{0, 1, \ldots, p-1\}$. We claim that $(p^m g(x)) \bmod 1 \in \mathbb{Z}_p \cap \mathbb{Q}$. Indeed, as g is continuous and as $x = 0.v0^\infty$ where $v = \gamma_0 \ldots \gamma_{m-1}$, there exists an s-mark-up of the right-infinite word $v0^\infty$ with respect to some state $s \in \mathcal{S}$ (cf. Note 13.57). That is, there exists a strictly increasing sequence $k_0 < k_1 < \ldots$ over $\mathbb{N}$ such that for the infinite sequence of words $w_i = v0^{k_i - m}$ of strictly increasing lengths k_i (where $i \geqslant K$ and K is large enough so that $k_i - m > 0$), the following is true:

$$\lim_{i\to\infty} 0.w_i = x;$$

$$\lim_{i\to\infty} 0.\mathfrak{a}_s(w_i) = g(x).$$

Therefore, the mark-up $(\bar{k}_j = k_{K+j} - m)_{j=0}^{\infty}$ of the zero right-infinite word 0^∞ is such that (in the notation of Proposition 13.64) the following equalities hold simultaneously:

$$\lim_{j\to\infty} 0.0^{\bar{k}_j} = 0;$$

$$\lim_{j\to\infty} 0.\mathfrak{a}_s(0^{\bar{k}_j}) = g_d(0).$$

Now by combining Proposition 13.64 (or Corollary 13.65 if necessary) and Proposition 13.53, we conclude that $g_d(0) \in \mathbb{Z}_p \cap \mathbb{Q}$ where $d = x = 0.\gamma_0 \ldots \gamma_{m-1}$; therefore $(p^m g(x)) \bmod 1 \in \mathbb{Z}_p \cap \mathbb{Q}$ as $g_d(0) = (p^m g(d)) \bmod 1$. Note that if

$(n+1)p^{-m} \notin [a,b]$ then we apply Corollary 13.65 rather than Proposition 13.64 and use $g_{d,M}$ instead of g_d and M instead of m here and after.

Take $\ell \in \mathbb{N}$; then by differentiability of g, for all $0 \leqslant h < 1$ and all sufficiently large $\ell \in \mathbb{N}_0$ we can represent $g(x + p^{-m-\ell}h)$ as

$$g(x + p^{-m-\ell}h) = g(x) + c(x)p^{-m-\ell}h + p^{-m-t(\ell)}\theta(\ell, h), \tag{13.82}$$

where $c(x) = g'(x)$, $|\theta(\ell, h)| \leqslant 1$ and t is a map from $\mathbb{N}_0$ to $\mathbb{N}_0$ such that $p^{t(\ell)} \to \infty$ faster than $p^{\ell} \to \infty$ while $\ell \to \infty$. That is, for all sufficiently large ℓ we may represent $t(\ell)$ as $t(\ell) = \ell + w(\ell)$, where w is a map from $\mathbb{N}_0$ to $\mathbb{N}_0$ such that $w(\ell) \to \infty$ as $\ell \to \infty$.

Further, by Proposition 13.64, the function $\tilde{g}(y) = (p^m g(x + p^{-m}y)) \bmod 1$ is continuous on $[0, 1]$ and $\mathbf{G}_{[0,1]}(\tilde{g}) \subset \mathcal{P}(\mathfrak{A})$. From here by combining Proposition 13.67 and Theorem 13.46, we conclude that there exists a finite automaton $\mathfrak{C}$

$$\begin{aligned}\bar{g}(y) = (p^m g(x + p^{-m}y) - p^m g(x)) \bmod 1 = \\ = ((p^m g(x + p^{-m}y)) \bmod 1 - (p^m g(x)) \bmod 1) \bmod 1\end{aligned}$$

lies completely in $\mathcal{P}(\mathfrak{C})$.

Indeed, as $(p^m g(x)) \bmod 1 \in \mathbb{Z}_p \cap \mathbb{Q}$, then by (13.56) the graph of the continuous function $y \mapsto y - (p^m g(x)) \bmod 1$ on $[0, 1]$ lies completely in $\mathcal{P}'(\mathfrak{B}) \subset \mathcal{P}(\mathfrak{B})$ for a finite automaton $\mathfrak{B}$ whose automaton function is $f_{\mathfrak{B}}(z) = z - (p^m g(x)) \bmod 1$, $(z \in \mathbb{Z}_p)$, and therefore the composite function $\bar{g}(y)$ is finitely computable on $[0, 1]$ (cf. Proposition 13.67). We proceed with this in mind.

We see from (13.82) that for all sufficiently large ℓ and all h the following is true:

$$\bar{g}(p^{-\ell}h) = (p^m g(x + p^{-m-\ell}h) - p^m g(x)) \bmod 1 = c(x)p^{-\ell}h + p^{-t(\ell)}\theta(\ell, h). \tag{13.83}$$

Now for the rest of the proof, we take (and fix) $h = p^{-1} = 0.1$. Let $0.\alpha_s\alpha_{s+1}\ldots$ be the base-p expansion of $(c(x)) \bmod 1$; so $c(x) = \alpha_0 \ldots \alpha_{s-1}.\alpha_s\alpha_{s+1}\ldots$ is the base-p expansion of $c(x)$. We may assume that $c(x) \geqslant 0$ since if otherwise we consider the function $(-g) \bmod 1$ which satisfies conditions of the lemma as g satisfies these conditions. If there exists two different base-p expansions for $(c(x)) \bmod 1$, we will consider only one of these. Recall that these expansions are of the form $0.\zeta_1 \ldots \zeta_n 0^\infty$ and $0.\zeta_1 \ldots \zeta_{n-1}\zeta_n'(p-1)^\infty$ where $\zeta_1, \ldots, \zeta_n \in \{0, 1, \ldots, p-1\}$, $\zeta_n \neq 0$ and $\zeta_n' = \zeta_n - 1$. Now, if the function $\theta(\ell, p^{-1})$ is non-negative for an infinite number of $\ell \in \mathbb{N}$, then we take the first of the base-p expansions; and we take the second one in the opposite case.

We claim that in all cases mentioned above, there exists a strictly increasing sequence $\mathcal{L}$ of $\ell \in \mathbb{N}$ such that, speaking loosely, the term $p^{-t(\ell)}\theta(\ell, h)$ has no affect on higher order digits of the base-p expansion of the right-hand part of (13.83). In the case when $(c(x)) \bmod 1$ admits only one base-p expansion, this follows from the fact that $p^{-t(\ell)}$ tends to 0 faster than $p^{-\ell}$ as we may take for $\mathcal{L}$ all sufficiently large

ℓ. In the case when $(c(x)) \bmod 1$ admits two base-p expansions, the claim is also true since we consider right base-p expansion $(c(x)) \bmod 1 = 0.\zeta_1 \ldots \zeta_n 0^\infty$ and assume that the function $\theta(\ell, p^{-1})$ in (13.83) is non-negative for an infinite number of $\ell \in \mathbb{N}$: In that case we take for $\mathcal{L}$ all sufficiently large ℓ such that $\theta(\ell, p^{-1}) \geqslant 0$. When $(c(x)) \bmod 1$ admits two base-p expansions and the function $\theta(\ell, p^{-1})$ is non-negative only for a finite number of $\ell \in \mathbb{N}$, we consider left base-p expansion $(c(x)) \bmod 1 = 0.\zeta_1 \ldots \zeta_{n-1}\zeta'_n(p-1)^\infty$ where $\zeta_1, \ldots, \zeta_n \in \{0, 1, \ldots, p-1\}$, $\zeta_n \neq 0$ and $\zeta'_n = \zeta_n - 1$. Then there exists infinitely many $\ell \in \mathbb{N}_0$ such that $\theta(\ell, p^-1) \leqslant 0$, and we take for $\mathcal{L}$ all these sufficiently large ℓ.

In other words, if we take $\ell \in \mathcal{L}$, substitute $y = p^{-\ell-1}$ to $\bar{g}(y)$, and apply (13.83), then we get

$$\bar{g}(0.(0)^\ell 1(0)^\infty) = 0.\underbrace{0\ldots0}_{\ell-s+1}\alpha_0 \ldots \alpha_{t_1(\ell)}\delta_{t_1(\ell)+1}\delta_{t_1(\ell)+2}\ldots, \tag{13.84}$$

where $\delta_j \in \{0, 1, \ldots, p-1\}$ for $j \geqslant t_1(\ell)+1$, $t_1(\ell) = -\ell + s + t(\ell) = s + w(\ell)$ (note that δ_j depends on ℓ). Further, by Note 13.57 we conclude now that given a right-infinite word u and $\ell \in \mathbb{N}$, there exists an s-mark-up of the word $0^\ell u$ with respect to the function $\bar{g}$ where s is a suitable (depending on u and ℓ) state of a finite automaton $\mathfrak{C}$ which is a sequential composition of the automaton $\mathfrak{A}$ with the automaton $\mathfrak{B}$. This means in particular that given an infinite word $v(\ell) = 0^\ell 100\ldots$, for any $\ell \in \mathcal{L}$, there exists a mark-up (with respect to a suitable state $s = s(v(\ell))$ of the automaton $\mathfrak{C}$; cf. Proposition 13.56) $i_0(\ell), i_1(\ell), i_2(\ell), \ldots$ of the word $v(\ell)$. As a total number of states of the automaton $\mathfrak{C}$ is finite, at least one state, say s', in the sequence $(s(v(\ell)) \colon \ell \in \mathcal{L})$ occurs infinitely many times. Denote $\mathfrak{C}' = \mathfrak{C}(s')$ (then $\mathfrak{C}'$ is a finite automaton as well), and consider an infinite strictly increasing sequence $\mathcal{L}' = (\ell' \colon s(v(\ell')) = s'; \ell' \in \mathcal{L})$.

Given a term ℓ' of the latter sequence $\mathcal{L}'$ take the smallest $k \in \mathbb{N}_0$ such that $i_k(\ell') > t(\ell')$; denote via $s(\ell')$ the state the automaton $\mathfrak{C}'$ reaches after being fed by the word $0^{i_k(\ell')-t(\ell')}$. As the number of states of the automaton $\mathfrak{C}'$ is finite, in the infinite sequence $(s(\ell'))$ at least one term, say $\bar{s}$, occurs infinitely many times. Consider an infinite sequence $(\ell'_j)_{j=0}^\infty$ such that $s(\ell'_j) = \bar{s}$ and consider an automaton $\mathfrak{C}'(\bar{s})$ (whence the latter automaton is finite also). If the automaton $\mathfrak{C}'(\bar{s})$ is being fed by the word $0^{\ell'_j}10^{w(\ell'_j)}$ then the automaton outputs the word $q_j = \sigma_0^{(j)} \ldots \sigma_{t(\ell'_j)}^{(j)}$; therefore being fed by the word $0^{w(\ell'_j)}$ the automaton outputs the word $q'_j = \sigma_{\ell'_j-s+1}^{(j)} \ldots \sigma_{t(\ell'_j)}^{(j)}$ of length $L_j = w(\ell'_j) - s$. As $w(\ell) \to \infty$ while $\ell \to \infty$ we may assume without loss of generality that the sequence (L_j) is strictly increasing (since otherwise we consider a subsequence $(j_i)_{i=0}^\infty$ of the sequence (j) such that the sequence $(L_{j_i})_{i=0}^\infty$ is strictly increasing). Now we just mimic the proof of Proposition 13.53 and thus show that the sequence $(0.q'_j)$ has only finitely many limit points and all these limit points are in $\mathbb{Z}_p \cap \mathbb{Q}$. But from (13.84), it follows that $\lim_{j\to\infty} 0.q'_j = (c(x)) \bmod 1$; therefore $(c(x)) \bmod 1 \in \mathbb{Z}_p \cap \mathbb{Q}$ and thus $c(x) = g'(x) \in \mathbb{Z}_p \cap \mathbb{Q}$. □

Proof of Lemma 13.72 Let x be as in the statement of Lemma 13.71; i.e., let the right base-p expansion of x be as in the proof of Lemma 13.71. Then $g'(x) \in \mathbb{Q} \cap \mathbb{Z}_p$ by Lemma 13.71.

Consider the function $g_1(u) = g(u) - g'(x)u + c$ of argument $u \in [a, b]$ where $c \in \mathbb{Q} \cap \mathbb{Z}_p$; then g_1 is two times differentiable on $[a, b]$ and $g_1'(x) = 0$, $g_1''(u) = g''(u)$ for all $u \in [a, b]$. As g_1 is continuous on $[a, b]$, the constant c may be taken so that $g_1(u) \in [0, 1]$ for all u from a sufficiently small closed neighborhood $[a_1, b_1]$ of x.

We are going to prove that $g_1''(x) = 0$ for all $x \in [a, b]$. Note that if $g_1''(x) \neq 0$ for some $x \in [a, b]$ then g_1'' does not change its sign on a sufficiently small neighborhood $(a_2, b_2) \subset [a_1, b_1]$ of x. Indeed, if not, then there exist two infinite sequences, $(\check{x}_i)_{i=0}^{\infty}$ and $(\hat{x}_i)_{i=0}^{\infty}$ such that all the terms of either sequence are pairwise distinct, $\lim_{i\to\infty} \check{x}_i = \lim_{i\to\infty} \hat{x}_i = x$, and $g_1''(\check{x}_i) \geqslant 0$, $g_1''(\hat{x}_i) \leqslant 0$ for all $i \in \mathbb{N}_0$. But as $\lim_{i\to\infty} g_1''(\check{x}_i) = \lim_{i\to\infty} g_1''(\hat{x}_i) = g_1''(x)$ (since g_1'' is continuous at x) then necessarily $g_1''(x) = 0$; but this contradicts our assumption that $g_1''(x) \neq 0$. We therefore may assume that $g_1''(u) \geqslant 0$ for all $u \in [a_2, b_2]$; otherwise consider the function $1 - g_1$ rather than g_1.

Finally by Corollary 13.69 we conclude that g_1 is finitely computable on a sufficiently small closed neighborhood $U \subset [a, b]$ of x. Further we use g for g_1 and $[a, b]$ for U without risk of misunderstanding. Thus we have:

1. g is finitely computable on $[a, b] \ni x$;
2. g is two times differentiable on $[a, b]$;
3. g'' is continuous on $[a, b]$;
4. $g'' \geqslant 0$ on $[a, b]$;
5. $g'(x) = 0$.

Now, since g is two times differentiable on $[a, b]$, for all $0 \leqslant h \leqslant 1$ and all sufficiently large $\ell \in \mathbb{N}_0$, we can represent $g(x + p^{-m-\ell-1}h)$ as

$$g(x + p^{-m-\ell}h) = g(x) + C(x)p^{-2m-2\ell}h^2 + p^{-2m-t(\ell)}\theta(\ell, h), \tag{13.85}$$

where $C(x)$ stands for $\frac{g''(x)}{2}$, $|\theta(\ell, h)| \leqslant 1$, $t(\ell) = 2\ell + w(\ell)$, and w is a map from $\mathbb{N}_0$ to $\mathbb{N}_0$ such that $w(\ell) \to \infty$ as $\ell \to \infty$.

Claim 1: $C(x) \in \mathbb{Z}_p \cap \mathbb{Q}$. To prove that we mimic the respective part of the proof of Lemma 13.71. Firstly we show that $(p^m g(x)) \bmod 1 \in \mathbb{Z}_p \cap \mathbb{Q}$ as in the proof of Lemma 13.71; thus considering a function $\bar{g}(y) = (p^m g(x + p^{-m}y) - p^m g(x)) \bmod 1$, we see that

$$\begin{aligned} \bar{g}(p^{-\ell}h) &= (p^m g(x + p^{-m-\ell}h) - p^m g(x)) \bmod 1 \\ &= C(x)p^{-m-2\ell}h^2 + p^{-m-t(\ell)}\theta(\ell, h) \end{aligned} \tag{13.86}$$

for all sufficiently large ℓ.

Let $0.\alpha_s\alpha_{s+1}\ldots$ be the base-p expansion of $(C(x)) \bmod 1$ (we may take either of the expansions if there exist two different ones); so $C(x) = \alpha_0 \ldots \alpha_{s-1}.\alpha_s\alpha_{s+1}\ldots$

is the base-p expansion of $C(x)$. Take $h = 1$; then for all sufficiently large ℓ the base-p expansion of the right-hand part of (13.86) is of the form

$$0.\underbrace{0\ldots0}_{m+2\ell-s}\alpha_0\ldots\alpha_{t_1(\ell)}\delta_{t_1(\ell)+1}\delta_{t_1(\ell)+2}\ldots \tag{13.87}$$

where $\delta_j \in \{0, 1, \ldots, p-1\}$ for $j \geqslant t_1(\ell)+1 = t(\ell)-m-2\ell+s = w(\ell)-m+s$ depend on ℓ.

The function $(p^m g(x + p^{-m}z) - p^m g(x)) \bmod 1$ of argument z is continuous and finitely computable on $[0, 1]$ by a finite automaton $\mathfrak{C}$. Now considering an infinite word $0^\ell 10^\infty$ with the corresponding mark-up, we prove in the same way as in Lemma 13.71 that the corresponding sequence of finite output words of the automaton $\mathfrak{C}$ is a sequence of initial finite sub-words of the infinite word $0^{m+2\ell-s}\alpha_0\alpha_1\ldots$ and then deduce as in the proof of Lemma 13.71 that $C(x) \in \mathbb{Z}_p \cap \mathbb{Q}$ (note that given $x = np^{-m}$ we may always take m so that $m + 2\ell - s > \ell$ without altering the value of x just by multiplying both numerator and denominator by a suitable power of p).

Claim 2: Now we prove that $C(x) = 0$. Assume that $C(x) \neq 0$; that is, that $g''(x) \neq 0$. Let $p^c = |C(x)|_p$, $c \in \mathbb{Z}$, be a p-adic absolute value of $C(x)$; therefore $C(x) = qp^{-c}$, where $q \in \mathbb{Z}_p$, q is a unit of $\mathbb{Z}_p$, i.e., its multiplicative inverse is in $\mathbb{Z}_p$ (cf. Sect. 2.2.2). By Claim 1, $C(x) \in \mathbb{Z}_p \cap \mathbb{Q}$, so necessarily $c \leqslant 0$ and $q \in \mathbb{Q}$; whence $q \in \mathbb{Z}_p \cap \mathbb{Q}$ and thus $q^{-1} \in \mathbb{Z}_p \cap \mathbb{Q}$ as q is a unit. Therefore the function $\breve{g}(y) = (q^{-1}(p^m g(x + p^{-m}y) + p^m g(x)) \bmod 1) \bmod 1$ of argument $y \in [0, 1]$ is finitely computable on $[0, 1]$ (say, by the automaton $\mathfrak{B}$): This follows from Propositions 13.64 and 13.67 and Corollary 13.69. Now by (13.86) we conclude that

$$\begin{aligned}(q^{-1}(p^m g(x + p^{-m-\ell}h) + p^m g(x)) \bmod 1) \bmod 1 \\ = p^{-k-2\ell}h^2 + p^{-m-t_1(\ell)}\theta_1(\ell, h),\end{aligned} \tag{13.88}$$

where $k = c + m \geqslant 0$ (we may assume that the inequality is true just by taking m sufficiently large by multiplying both numerator and denominator by a suitable power of p and thus without altering the value of x), $|\theta_1(\ell, h)| \leqslant 1$, $t_1(\ell) = 2\ell + w_1(\ell)$, and w_1 is a map from $\mathbb{N}_0$ to $\mathbb{N}_0$ such that $w_1(\ell) \to \infty$ as $\ell \to \infty$.

Further, given $n \in \mathbb{N}$ and a word $v_n = \chi_0 \ldots \chi_{n-1} \in \mathcal{W}$ and taking $y = p^{-\ell}h$ with $h = 0.v_n0^\infty$, we have that $(p^{-\ell}h; \breve{g}(p^{-\ell}h)) \in \mathcal{P}(\mathfrak{B})$. Let $i_0(\ell, v_n) < i_1(\ell, v_n) < i_2(\ell, v_n) < \ldots$ be corresponding mark-up of the infinite word $0^\ell v_n 0^\infty$ (the latter infinite word corresponds to $y = p^{-\ell}h$ once $h = 0.v_n0^\infty$). Take r the smallest so that $i_r = i_r(\ell, v_n) > t_1(\ell) + m$; denote $s(\ell, v_n)$ the state the automaton $\mathfrak{B}$ reaches after being fed by the word $0^{i_r(\ell,v_n)-k-2\ell-2n}$. As the number of states of the automaton $\mathfrak{B}$ is finite, in the sequence $(s(\ell, v_n))_{\ell=1}^\infty$ at least one state, say $\bar{s}(v_n)$, occurs infinitely many times. Consider a strictly increasing sequence $\mathcal{L} = (\ell_j \in \mathbb{N}\colon s(\ell_j, v_n) = \bar{s}(v_n))_{j=0}^\infty$; then once $w_1(\ell_j) > 2n + r$, the automaton $\mathfrak{B}(s(\ell_j, v_n)) = \mathfrak{B}(\bar{s}(v_n))$, being fed by the word $0^{\ell_j} v_n 0^{k+\ell_j+n}$, outputs the word $\zeta_0^{(j)} \dots \zeta_{k+2\ell_j-1}^{(j)} \xi_0^{(j)} \dots \xi_{2n-1}^{(j)}$. From (13.88) it follows that $(\lim_{j\to\infty}(p^{k+2\ell_j} \times 0.\zeta_0^{(j)} \dots \zeta_{k+2\ell_j-1}^{(j)} \xi_0^{(j)} \dots \xi_{2n-1}^{(j)})) \bmod 1 = h^2$. Therefore $(\lim_{j\to\infty} 0.\xi_0^{(j)} \dots \xi_{2n-1}^{(j)}) \bmod 1 = h^2 = 0.\xi_0 \dots \xi_{2n-1}$ where

$$\xi_0 p^{2n-1} + \xi_1 p^{2n-2} + \dots + \xi_{2n-1} = \left(\chi_0 p^{n-1} + \chi_1 p^{n-2} + \dots + \chi_{n-1}\right)^2$$

(in other words, $\xi_0\xi_1 \dots \xi_{2n-1}$ is the base p expansion of the square of the number whose base-p expansion is $v_n = \chi_0 \dots \chi_{n-1}$). Thus necessarily $\xi_0^{(j)} = \xi_0, \dots, \xi_{2n-1}^{(j)} = \xi_{2n-1}$ for all sufficiently large j. But $(p^{-\ell}h; \breve{g}(p^{-\ell}h)) \in \mathbf{P}(\mathfrak{B})$ for all ℓ; thus by Proposition 13.35, $((p^{k+\ell_j}h) \bmod 1; (p^{k+2\ell_j}\breve{g}(p^{-\ell_j}h)) \bmod 1 \in \mathcal{P}(\mathfrak{B})$ for all $j \in \mathbb{N}_0$. Therefore for all sufficiently large j (such that $k + \ell_j > n$) we have that $(0; 0.\xi_0 \dots \xi_{2n-1}) \in \mathcal{P}(\mathfrak{B})$ since $h = 0.v_n 0^\infty$. In other words, as the automaton $\mathfrak{B}(s(\ell_j, v_n)) = \mathfrak{B}(\bar{s}(v_n))$, being fed by the word $0^{\ell_j} v_n 0^{k+\ell_j+n}$, outputs the word $\zeta_0^{(j)} \dots \zeta_{k+2\ell_j-1}^{(j)} \xi_0 \dots \xi_{2n-1}$ once j is sufficiently large, $k + \ell_j > n$, the automaton $\mathfrak{B}(\bar{s}(v_n))$, being fed by the zero word 0^{2n}, outputs the word $\xi_0\xi_1 \dots \xi_{2n-1}$. This means that, given an arbitrary number $N \in \mathbb{N}_0$ whose base-p expansion (where higher order digits might be 0) is of length n, n being sufficiently large, there exists a state s of the finite automaton $\mathfrak{B}$ such that the automaton $\mathfrak{B}(s)$, being fed by a zero sequence of length $2n$, outputs a word (of length $2n$) that is the base-p expansion of N^2. But this is not possible since it is well-known that squaring is not possible by a finite automaton (cf., e.g., [31, Theorem 2.2.3] or [143, Proposition 7.1.6]); however, a short proof follows.

As the automaton $\mathfrak{B}$ is finite, then there are only finitely many sub-automata $\mathfrak{B}(s(\ell, v_n))$. But any finite automaton, being fed by a sufficiently long zero word 0^L outputs the word of the form $u_1(u_2)^M u_3$, where $M = M(L) \in \mathbb{N}$, $u_2 \in \mathcal{W}$, $u_1, u_3 \in \mathcal{W}_\varphi$, and the words u_1, u_2, u_3 are completely determined by the finite automaton, u_1 is a right prefix of u_2, cf. Lemma 13.25. But given finitely many words $u_{1.i}, u_{2.i}, u_{3.i}$ of that sort, $i = 0, 1, 2, \dots, K$, there exist infinitely many words $\xi_0\xi_1 \dots \xi_{2n-1}$ which are base-p expansions of squares of numbers from $\mathbb{N}_0$ and which are not of the form $u_{1.i}u_{2.i}^M u_{3.i}$, $i = 1, 2, \dots, K$, $M = 1, 2, \dots$.

The contradiction proves that $C(x) = 0$; therefore, $g''(x) = 0$. □

Proof of Theorem 13.70 We already have proved that $g''(x) = 0$ if $x = np^{-m} \in [a, b)$, where $n \in \mathbb{N}_0$, $m \geqslant \lfloor \log_p n \rfloor + 1$. But the set of all these x is dense in $[a, b]$, so, as the second derivative g'' is continuous on $[a, b]$ by the condition of the theorem under the proof, g'' must vanish everywhere on $[a, b]$; therefore, $g' = const$. But this implies that there exist $A, B \in \mathbb{R}$ such that $g(x) = Ax + B$ for all $x \in [a, b]$ and $g'(x) = A$ for all $x \in [a, b]$. From Lemma 13.71 it follows now that $A \in \mathbb{Z}_p \cap \mathbb{Q}$. Now taking an arbitrary number $y \in (a, b) \cap \mathbb{Z}_p \cap \mathbb{Q}$, we see that $g(y) \in \mathbb{Z}_p \cap \mathbb{Q}$ by Proposition 13.53; hence $g(y) - Ay = B$ must be also in $\mathbb{Z}_p \cap \mathbb{Q}$ as $Ay \in \mathbb{Z}_p \cap \mathbb{Q}$.

Now we will prove that $\mathbf{C}(A, B) \subset \mathcal{P}'(\mathfrak{A})$. To begin with, we note that by Theorem 13.62 there exists a minimal sub-automaton $\mathfrak{A}'$ and a segment $[a', b'] \subset [a, b]$ such that $\mathbf{G}_{[a',b']}(g) \subset \mathcal{P}'(\mathfrak{A}')$. Taking $d \in (a', b')$ as in the statement of Proposition 13.64, we conclude that the graph $\mathbf{G}_{[0,1]}(g_d)$ of the function $g_d(x) = (Ax + An + p^m B) \bmod 1$ on $[0, 1]$ lies completely in $\mathcal{P}(\mathfrak{A}')$; thus $\mathbf{G}_{[0,1]}(g_d) = \{(x; g_d(x)) \colon x \in [0, 1]\} \subset \mathcal{P}'(\mathfrak{A}')$ by Corollary 13.58. $An \in \mathbb{Z}$; that is, $p^m a \leqslant n < p^m b$. As $A \in \mathbb{Z}_p \cap \mathbb{Q}$ then $A = P/Q$ for suitable $P \in \mathbb{Z}$, $Q \in \mathbb{N}$. Now given arbitrary $R \in \{0, 1, \dots, Q-1\}$ we take n and m so that $d = p^{-m} n$ satisfies conditions of Proposition 13.64 (i.e., $p^m a \leqslant n < p^m b$) and $n = LQ + R \in \{0, 1, \dots, p^m - 1\}$ for a suitable $L \in \mathbb{N}_0$ and conclude that

$$\{(x; g_d(x)) \colon x \in [0, 1]\} = \{(x; (Ax + AR + p^m B) \bmod 1) \colon x \in [0, 1]\} \subset \mathcal{P}'(\mathfrak{A}')$$

Given arbitrary $R \in \{0, 1, \dots, Q-1\}$, the above inclusion holds for all sufficiently large m; therefore due the structure of $\mathbf{C}(B)$ (cf. Sect. 13.6.3), the following inclusion holds for every $R \in \{0, 1, \dots, Q-1\}$ and every $B' \in \mathbf{C}(B)$:

$$\{(x; (Ax + AR + B') \bmod 1) \colon x \in [0, 1]\} \subset \mathcal{P}'(\mathfrak{A}').$$

But $\cup_{R=0}^{Q-1} \{(x; (Ax + AR + B') \bmod 1) \colon x \in [0, 1]\} = \{((x \bmod 1; (Ax + B') \bmod 1) \colon x \in \mathbb{R}\} = \mathbf{C}(A, B')$; therefore we have shown that $\mathbf{C}(A, B') \subset \mathcal{P}'(\mathfrak{A}')$ for all $B' \in \mathbf{C}(B \bmod 1)$. That is, $\mathcal{P}'(\mathfrak{A}')$ contains the whole link of cables $\mathbf{C}(A, B')$ for all $B' \in \mathbf{C}(B)$ (i.e., contains $\mathcal{P}'(F)$ where $F \colon z \mapsto Az + B$, $z \in \mathbb{Z}_p$; cf. Theorem 13.46), and $\mathbf{G}_{[a,b]}(g)$ lies completely in a suitable cable of the link. This proves the first claim of Theorem 8.2 since $\mathcal{P}'(\mathfrak{A}') \subset \mathcal{P}'(\mathfrak{A})$, cf. Note 13.33.

To prove the second claim, given a finite automaton $\mathfrak{A}$, consider all cables $\mathbf{C}(A, B)$ such that $\mathbf{C}(A, B) = \{(y \bmod 1; (Ay + B) \bmod 1) \colon y \in \mathbb{R}\} \subset \mathcal{P}(\mathfrak{A})$; whence by the first claim of the theorem, all these cables lie in $\mathcal{P}'(\mathfrak{A})$. Moreover, as we have shown during the proof of the first claim of the theorem, for either of the cables $\mathbf{C}(A, B)$, there exists a minimal sub-automaton $\mathfrak{A}'_{A,B}$ of the automaton $\mathfrak{A}$ such that $\mathbf{C}(A, B) \subset \mathfrak{A}'_{A,B}$. The cables cross zero meridian $\mathbf{O} = \{(0; t \bmod 1) \colon t \in \mathbb{R}\} \subset \mathbb{T}^2$ of the torus $\mathbb{T}^2$ only when $y \bmod 1 = 0$; therefore the point set $\mathbf{S}$ of all the points where the cables cross zero meridian consists of the points of the form $(0; e)$ where $e \in \mathfrak{a}(0)$ and $\mathbf{S}$ contains all the points of the form $(0; B \bmod 1)$ where B are constant terms of the cables. As $\mathfrak{a}(0)$ is a finite set (cf. Proposition 13.53), there are no more than a finite number of pairwise distinct numbers $B \bmod 1$ (note that cables with equal slopes whose constant terms are congruent modulo 1 coincide). Now taking $y \in \mathbb{Z}$, we see that all the points of the form $(0; (Ay+B) \bmod 1)$ of the cables belong to zero meridian and therefore to the finite set $\{(0; r) \colon r \in \mathfrak{a}(0)\}$; hence, there exist no more than a finite number of pairwise distinct numbers $Ay \bmod 1$ where y ranges over rational integers $\mathbb{Z}$ and A are slopes of the cables from $\mathcal{P}(\mathfrak{A})$. Thus if there exists an infinite number of cables in $\mathcal{P}(\mathfrak{A})$, then there exists a minimal sub-automaton $\mathfrak{A}'$ of the automaton $\mathfrak{A}$ such that $\mathcal{P}'(\mathfrak{A}')$ contains an infinite number of cables of the form $\mathbf{C}(AC, B)$ with A, B fixed and C ranging through an infinite subset $\mathbf{C}$ of $\mathbb{Z}$ so that $AC \bmod 1$ are all equal one to another. Therefore for the rest of the proof, we may (and will) assume that the automaton $\mathfrak{A}$ is minimal.

By the first claim of the theorem, $A \in \mathbb{Z}_p \cap \mathbb{Q}$; so there exists a unique representation of A in the form

$$A = c + \frac{d}{p^t - 1}$$

where $t \in \mathbb{N}$ is a period length of A, $c \in \mathbb{Z}$, and $d \in \{0, 1, \ldots, p^t - 2\}$ (cf. Proposition 2.11 and Note 2.13). Therefore $\mathbf{C}$ must contain an infinite subset of numbers from the coset $q + (p^t - 1)\mathbb{Z}$ for a suitable $q \in \{0, 1, \ldots, p^t - 2\}$ since $AC_1 \equiv AC_2 \pmod 1$ implies $A(C_1 - C_2) \equiv 0 \pmod 1$, i.e., $A(C_1 - C_2) \in \mathbb{Z}$. Thus from the assumption that there are infinitely many pairwise distinct cables in $\mathcal{P}(\mathfrak{A})$, it follows that then in $\mathcal{P}'(\mathfrak{A})$ there exist infinitely many cables of the form $\mathbf{C}(D + E, B)$ with B, E fixed ($B \in \mathbb{Z}_p \cap \mathbb{Q}$, $E \in \mathbb{Z}_p \cap \mathbb{Q} \cap [0, 1)$) and D running through an infinite subset $\mathbf{D} \subset \mathbb{Z}$. By considering $-f_{\mathfrak{A}}$ (and corresponding finite automaton) if necessary, we may assume that $\mathbf{D}$ is an infinite subset of $\mathbb{N}$. Therefore, $\mathbf{D}$ constitutes a strictly increasing sequence $(D_i)_{i=0}^{\infty}$ of natural numbers. Now take arbitrary $u \in [0, 1)$ and consider a sequence $x_i = uD_i^{-1}$. As the sequence (D_i) is strictly increasing, $\lim_{i\to\infty} x_i = 0$; therefore $\lim_{i\to\infty}(x_i; (D_i x_i + Ex_i + B) \bmod 1) = (0; (u + B) \bmod 1) \in \mathcal{P}'(\mathfrak{A})$ as $(x_i; (u + Ex_i + B) \bmod 1) \in \mathbf{C}(D_i + E, B) \subset \mathcal{P}'(\mathfrak{A})$ and $\mathcal{P}'(\mathfrak{A})$ is closed in $\mathbb{T}^2$ (cf. Corollary 13.31). Thus we have proved that zero meridian $\mathbf{O} = \{(0; y) : y \in [0, 1)\}$ of the torus $\mathbb{T}^2$ lies completely in $\mathcal{P}'(\mathfrak{A})$.

On the other hand, if $(0; y) \subset \mathcal{P}'(\mathfrak{A})$ then $y \subset \mathfrak{a}(0)$ by definitions of $\mathcal{P}'(\mathfrak{A})$ and $\mathfrak{a}(0)$ (see Sect. 12.1, Definition 13.55 and the results on mark-ups); but there are only finitely many points in $\mathfrak{a}(0)$ by Proposition 13.53. The contradiction proves the second assertion of the theorem. □

13.6.5.2 The Multivariate Case

Now we are going to extend Theorem 13.70 for the case of finite automata with multiply input/output channels (cf. Fig. 13.2 and the text from Sect. 13.1.2 on automata of this kind): Actually an automaton over alphabet $\mathbb{F}_p = \{0, 1, \ldots, p-1\}$ with m input channels and n output channels can be considered as a letter-to-letter transducer with a single input channel over the alphabet $\{0, 1, \ldots, p^m - 1\}$ and a single output channel over the alphabet $\{0, 1, \ldots, p^n - 1\}$; therefore the plot of that automaton is a closed subset of the unit square $\mathbb{I}^2$. We however are going to consider plots of automata of that sort as subsets of multidimensional unit hypercube $\mathbb{I}^{m+n}$. Therefore automata functions of such automata are 1-Lipschitz mappings from $\mathbb{Z}_p^m$ to $\mathbb{Z}_p^n$; and vice versa, every 1-Lipschitz mapping from $F : \mathbb{Z}_p^m \to \mathbb{Z}_p^n$ is an automaton function of a suitable automaton $\mathfrak{A}$ with m inputs and n outputs over the alphabet $\mathbb{F}_p$. Note that $F = (F_1; \ldots; F_m)$ where $F_k : \mathbb{Z}_p^m \to \mathbb{Z}_p$ $(k = 1, 2, \ldots, m)$ is 1-Lipschitz and therefore is an automaton function of an automaton with m input channels and a single output channel. Now we restate the definition of a (limit) plot for that case of automata with m input channels and n output channels (cf. Sect. 12.1 and especially (12.7) there).

Definition 13.73 (Automata Plots, the Multivariate Case) Given an automaton function $F = F_{\mathfrak{A}}\colon \mathbb{Z}_p^m \to \mathbb{Z}_p^n$, define a set $\mathcal{P}(F_{\mathfrak{A}})$ of points of $\mathbb{R}^{n+m}$ as follows: For $k = 1, 2, \ldots$ denote

$$\mathcal{E}_k(F) = \left\{ \left(\frac{\mathbf{z} \bmod p^k}{p^k}; \frac{F(\mathbf{z}) \bmod p^k}{p^k} \right) \in \mathbb{I}^{m+n} : \mathbf{z} \in \mathbb{Z}_p^m \right\} \tag{13.89}$$

a point set in a unit real hypercube $\mathbb{I}^{m+n}$; here given $\mathbf{y} = (y_1; \ldots; y_q) \in \mathbb{Z}_p^q$, we put

$$\frac{\mathbf{y} \bmod p^k}{p^k} = \left(\frac{y_1 \bmod p^k}{p^k}; \ldots; \frac{y_q \bmod p^k}{p^k} \right) \in (\mathbb{Z}/p^k\mathbb{Z})^q.$$

Then take a union $\mathcal{E}(F) = \cup_{k=1}^{\infty} \mathcal{E}_k(f)$ and denote via $\mathcal{P}(F) = \mathcal{P}(\mathfrak{A})$ a closure (in topology of $\mathbb{R}^{m+n}$) of the set $\mathcal{E}(F)$.

Given an automaton $\mathfrak{A}$, we call a *plot of the automaton* $\mathfrak{A}$ the set $\mathcal{P}(\mathfrak{A})$. We call a *limit plot* of the automaton $\mathfrak{A}$ the point set $\mathcal{P}'(\mathfrak{A})$ which is defined as follows: A point $(\mathbf{x}; \mathbf{y}) \in \mathbb{R}^{m+n}$ lies in $\mathcal{P}'(\mathfrak{A})$ if and only if there exist $\mathbf{z} \in \mathbb{Z}_p^m$ and a strictly increasing infinite sequence $k_1 < k_2 < \ldots$ of numbers from $\mathbb{N}$ such that simultaneously

$$\lim_{i\to\infty} \frac{\mathbf{z} \bmod p^{k_i}}{p^{k_i}} = \mathbf{x}; \ \lim_{i\to\infty} \frac{F_{\mathfrak{A}}(\mathbf{z}) \bmod p^{k_i}}{p^{k_i}} = \mathbf{y}. \tag{13.90}$$

To put it in other words, at every step, a letter-to-letter transducer $\mathfrak{A}$ (which has m input channels and n output channels over a p-symbol alphabet $\mathbb{F}_p$)

- obtains a vector $\mathbf{a} = (\alpha^{(1)}; \ldots, \alpha^{(m)}) \in \mathbb{F}_p^m$ (each i-th letter $\alpha^{(i)}$ is sent accordingly to the i-th input channel of the automaton, $i = 1, 2, \ldots, m$),
- produces output vector $\mathbf{b} = (\beta^{(1)}; \ldots, \beta^{(n)}) \in \mathbb{F}_p^n$ (each j-th output channel of the automaton produces accordingly the letter $\beta^{(j)}$, $i = 1, 2, \ldots, n$) which depends both on the current state and on the input vector $\mathbf{a}$,
- reaches the next state (which depends both on $\mathbf{a}$ and on the current state).

Then the routine repeats. Therefore after k steps the automaton $\mathfrak{A}$ transforms the input m-tuple $\mathbf{w} = (w_1; \ldots; w_m)$ of k-letter words $w_i = \alpha_k^{(i)} \ldots \alpha_1^{(i)}$ $(i = 1, 2, \ldots, m)$ into the output n-tuple $\mathbf{v} = \mathfrak{a}(\mathbf{w}) = (v_1; \ldots; v_n)$ of k-letter words $v_j = \mathfrak{a}^{(j)}(\mathbf{w}) = \beta_k^{(j)} \ldots \beta_1^{(j)}$ $(j = 1, 2, \ldots, n)$. For $\mathbf{w}$ running over all m-tuples of k-letter words, $k = 1, 2, \ldots$ we consider the set $\mathcal{E}(\mathfrak{A})$ of all points $(0.\mathbf{w}; 0.\mathfrak{a}(\mathbf{w})) \in \mathbb{R}^{m+n}$; here $0.\mathbf{u}$ stands for $(0.u_1; \ldots; 0.u_\ell)$ where $u_1, \ldots, u_\ell$ are k-letter words. Then we define $\mathcal{P}(\mathfrak{A})$ as a closure in $\mathbb{R}^{m+n}$ of the set $\mathcal{E}(\mathfrak{A})$. Following the lines of Note 13.28 it can be shown that $\mathbf{P}(\mathfrak{A}) = \mathcal{P}(F_{\mathfrak{A}})$. We stress that $\mathfrak{A}$ is a synchronous letter-to-letter transducer; that is why in the definition of the plot, all

m input words as well as corresponding n output words of the automaton must have pairwise equal lengths.

Given a real function $G: D \to \mathbb{R}^n$ with the domain $D \subset \mathbb{R}^m$, by the *graph of the function* (on the torus $\mathbb{T}^{m+n}$) we mean the point subset $\mathbf{G}_D(g) = \{(\mathbf{x} \bmod 1; G(\mathbf{x}) \bmod 1) : \mathbf{x} \in D\} \subset \mathbb{T}^{m+n}$. Note that if $\mathbf{y} = (y_1; \ldots; y_k) \in \mathbb{R}^k$ then $\mathbf{y} \bmod 1$ stands for $(y_1 \bmod 1; \ldots; y_k \bmod 1)$.

Theorem 13.74 *Let $\mathfrak{A}$ be a finite automaton over the alphabet $\{0, 1, \ldots, p-1\}$, let $\mathfrak{A}$ have m inputs and n outputs, and let $G = (G_1; \ldots; G_n): [\mathbf{a}, \mathbf{b}] = [a_1, b_1] \times \cdots \times [a_m, b_m] \to [0, 1)^n$ (where $[a_i, b_i] \subset [0, 1)$, $G_i : [\mathbf{a}, \mathbf{b}] \to [0, 1)$, $i = 1, 2, \ldots, m$) be a two times differentiable function such that all its second partial derivatives are continuous on $[\mathbf{a}, \mathbf{b}]$. If $\mathbf{G}(G) \subset \mathcal{P}(\mathfrak{A}) \subset \mathbb{T}^{m+n}$ then there exist an $m \times n$ matrix $\mathbf{A} = (A_{ij})$ and a vector $\mathbf{B} = (B_1; \ldots; B_n)$ such that $A_{ij} \in \mathbb{Q} \cap \mathbb{Z}_p$, $B_j \in \mathbb{Q} \cap \mathbb{Z}_p \cap [0, 1)$ $(i = 1, 2, \ldots, m;\ j = 1, 2, \ldots, n)$ and $G(\mathbf{x}) = (\mathbf{x}\mathbf{A} + \mathbf{B}) \bmod 1$ for all $\mathbf{x} \in [\mathbf{a}, \mathbf{b}]$. There are not more than a finitely many $\mathbf{A}$ and $\mathbf{B}$ such that $A_{ij} \in \mathbb{Q} \cap \mathbb{Z}_p$, $B_j \in \mathbb{Q} \cap \mathbb{Z}_p \cap [0, 1)$ $(i = 1, 2, \ldots, m;\ j = 1, 2, \ldots, n)$ and $\mathbf{G}_{[\mathbf{a},\mathbf{b}]}((\mathbf{x}\mathbf{A} + \mathbf{B}) \bmod 1) \subset \mathcal{P}(\mathfrak{A})$ for some $[\mathbf{a}, \mathbf{b}] \subset [0, 1)^m$; moreover, if $\mathbf{G}_{[\mathbf{a},\mathbf{b}]}(\mathbf{x}\mathbf{A} + \mathbf{B}) \subset \mathcal{P}(\mathfrak{A})$ for some $[\mathbf{a}, \mathbf{b}] \subset [0, 1)^m$ then $\mathbf{G}_{\mathbb{R}^m}((\mathbf{x}\mathbf{A} + \mathbf{B}) \bmod 1) \subset \mathcal{P}(\mathfrak{A}) \subset \mathbb{T}^{n+m}$.*

Proof of Theorem 13.74 Let $F_{\mathfrak{A}} = (F_1; \ldots; F_n): \mathbb{Z}_p^m \to \mathbb{Z}_p^n$ be automaton function of the automaton $\mathfrak{A}$. Having $i \in \{1, 2, \ldots, m\}$ and $j \in \{1, 2, \ldots, n\}$ fixed, take arbitrary numbers $z_k \in \mathbb{Z}_p \cap \mathbb{Q} \cap [a_k, b_k]$, $k = 1, 2, \ldots, i-1, i+1, \ldots, m$, consider the map $\bar{F}_{ij}(z) = F_j(z_1; \ldots; z_{i-1}; z; z_{i+1}; \ldots; z_m)$ and the function $\bar{G}_{ij}(x) = G(z_1; \ldots; z_{i-1}; x; z_{i+1}; \ldots; z_m)$.

As $z_k \in \mathbb{Z}_p \cap \mathbb{Q} \cap [a_k, b_k]$ and $[a_k, b_k] \subset [0, 1)$, then the map $\bar{F}_{ij} : \mathbb{Z}_p \to \mathbb{Z}_p$ is a finite automaton function: Actually the corresponding automaton $\mathfrak{A}_{ij}$ is a sequential composition of the automaton $\mathfrak{A}$ with autonomous automata $\mathfrak{B}_k$ which produce accordingly purely periodic output words $\mathsf{wrd}(z_k) \in \mathcal{W}_\infty$ (cf. Corollary 2.16) and feed accordingly k-th input channels $(k = 1, \ldots, i-1, i+1, \ldots, m)$ of the automaton $\mathfrak{A}$ while the output of the automaton $\bar{\mathfrak{A}}_{ij}$ is the j-th output of the automaton $\mathfrak{A}$.

Claim: We assert that $\mathbf{G}_{[a_i,b_i]}(\bar{G}_{ij}) \subset \mathcal{P}(\bar{\mathfrak{A}}_{ij})$.

To prove the claim, we first note that by Corollary 2.16, we have for every $k = 1, \ldots, i-1, i+1, \ldots, m$ that $z_k = 0.(\zeta^{(k)}_{T_k-1}\zeta^{(k)}_{T_k-2} \cdots \zeta^{(k)}_0)^\infty$ where T_k is a period length of z_k. Let T be the least common multiple of all t_i; then $z_k = 0.(\eta^{(k)}_{T-1}\eta^{(k)}_{T-2} \cdots \eta^{(k)}_0)^\infty$ for all $k = 1, \ldots, i-1, i+1, \ldots, m$. Denote the right-infinite purely periodic word $(\eta^{(k)}_{T-1}\eta^{(k)}_{T-2} \cdots \eta^{(k)}_0)^\infty$ via $u(z_k) = \tau^{(k)}_1 \tau^{(k)}_2 \ldots$ for suitable $\tau^{(\ell)}_q \in \mathbb{F}_p$.

Take arbitrary $x \in [a_i, b_i]$ and put $\mathbf{x} = (z_1; \ldots; z_{i-1}; x; z_{i+1}; \ldots; z_m) \in [\mathbf{a}, \mathbf{b}]$; then $(\mathbf{x}; G(\mathbf{x})) \in \mathcal{P}(\mathfrak{A})$. Let $x = 0.\chi_1\chi_2 \ldots$ be the base-p expansion of x (the word $u(x) = \chi_1\chi_2 \ldots$ is right-infinite); then from the definition of the plot, it follows that there exists a strictly increasing sequence $\bar{r}_1 < \bar{r}_2 < \ldots$ over $\mathbb{N}$ such that

$$\lim_{\ell\to\infty} 0.\bar{\chi}_1\bar{\chi}_2\ldots\bar{\chi}_{\bar{r}_\ell} = x; \tag{13.91}$$

$$\lim_{\ell\to\infty} 0.\bar{\tau}_1^{(k)}\bar{\tau}_2^{(k)}\ldots\bar{\tau}_{\bar{r}_\ell}^{(k)} = z_k \ \ (k = 1,\ldots,i-1,i+1,\ldots,m); \tag{13.92}$$

$$\lim_{\ell\to\infty} 0.\mathfrak{a}(\mathbf{u}_{\bar{r}_\ell}(\bar{x})) = G(\mathbf{x}), \tag{13.93}$$

where

$$\mathbf{u}_{\bar{r}_\ell}(\bar{x}) = (u_{\bar{r}_\ell}(\bar{z}_1);\ldots;u_{\bar{r}_\ell}(\bar{z}_{i-1});u_{\bar{r}_\ell}(\bar{x});u_{\bar{r}_\ell}(\bar{z}_{i+1});\ldots;u_{\bar{r}_\ell}(\bar{z}_m));$$
$$u_{\bar{r}_\ell}(\bar{z}_k) = \bar{\tau}_1^{(k)}\bar{\tau}_2^{(k)}\ldots\bar{\tau}_{r_\ell}^{(k)} \ \ (k = 1,\ldots,i-1,i+1,\ldots,m);$$
$$u_{\bar{r}_\ell}(\bar{x}) = \bar{\chi}_1\bar{\chi}_2\ldots\bar{\chi}_{\bar{r}_\ell};$$

see remarks which follow Definition 13.73 above. Moreover, since base-p expansions of all z_k are unique, then arguing like in the first part of the proof of Proposition 13.56, we conclude that there exists a state s of the automaton $\mathfrak{A}$ and a strictly increasing sequence $r_1 < r_2 < \ldots$ over $\mathbb{N}$ such that

$$\lim_{\ell\to\infty} 0.\chi_1\chi_2\ldots\chi_{r_\ell} = x; \tag{13.94}$$

$$\lim_{\ell\to\infty} 0.\tau_1^{(k)}\tau_2^{(k)}\ldots\tau_{r_\ell}^{(k)} = z_k \ \ (k = 1,\ldots,i-1,i+1,\ldots,m); \tag{13.95}$$

$$\lim_{\ell\to\infty} 0.\mathfrak{a}_s(\mathbf{u}_{r_\ell}(x)) = G(\mathbf{x}), \tag{13.96}$$

where $\mathfrak{A}_s$ is the automaton which differs from $\mathfrak{A}$ only maybe by the initial state (which is s rather than s_0). Now recall that

$$\tau_1^{(k)}\tau_2^{(k)}\ldots = (\eta_{T-1}^{(k)}\eta_{T-2}^{(k)}\cdots\eta_0^{(k)})^\infty$$

for all $k = 1,\ldots,i-1,i+1,\ldots,m$; so given $\ell \in \mathbb{N}$ let $q(\ell) \in \mathbb{N}$ be the largest such that $q_\ell < r_\ell$ and $\tau_{q_\ell}^{(k)} = \eta_0^{(k)}$ for some (thus, for all) $k = 1,\ldots,i-1,i+1,\ldots,m$. Since all the words $\tau_1^{(k)}\tau_2^{(k)}\ldots$ are periodic with a period of length T such q_ℓ exists for all sufficiently large $\ell \geqslant N$. Denote via s_ℓ the state the automaton $\mathfrak{A}(s)$ reaches after being fed (via respective inputs) by words $\tau_{q_\ell+1}^{(k)}\cdots\tau_{r_\ell}^{(k)}$ $(k = 1,\ldots,i-1,i+1,\ldots,m)$ and $\chi_{q_\ell+1}^{(k)}\cdots\chi_{r_\ell}^{(k)}$. By the finiteness of the automaton, in the sequence $(s_\ell)_{\ell=N}^\infty$, at least one state, say $\hat{s}$, occurs infinitely many times; therefore from (13.94)–(13.96) it follows that

$$\lim_{\ell\to\infty} 0.\chi_1\chi_2\ldots\chi_{q_\ell} = x; \tag{13.97}$$

$$\lim_{\ell\to\infty} 0.(\eta_{T-1}^{(k)}\eta_{T-2}^{(k)}\cdots\eta_0^{(k)})^{q_\ell/T} = z_k \ (k = 1,\ldots,i-1,i+1,\ldots,m); \tag{13.98}$$

$$\lim_{\ell\to\infty} 0.\mathfrak{a}_{\hat{s}}(\mathbf{u}_{q_\ell}(x)) = G(\mathbf{x}), \tag{13.99}$$

where $\mathfrak{A}_{\hat{s}}$ is the automaton which differs from $\mathfrak{A}$ only maybe by the initial state (which is $\hat{s}$ rather than s_0). Note that T is a divisor of q_ℓ by the construction of q_ℓ since all the words $\tau_1^{(k)}\tau_2^{(k)}\ldots$ are periodic with a period of length T. By the definition of the plot we conclude that (13.97)–(13.99) prove our claim.

Thus the function $\bar{G}_{ij}$ satisfies all conditions of Theorem 13.70; therefore the second derivative of $\bar{G}_{ij}$ is zero. But this means by the construction of G_{ij} that every second partial derivative $\partial^2 G_j/\partial^2 x_i$ is zero for all $z_k \in \mathbb{Z}_p \cap \mathbb{Q} \cap [a_k, b_k]$ $(k = 1,\ldots,i-1,i+1,\ldots,m)$ and all $x \in [a_i, b_i]$. As $\mathbb{Z}_p \cap \mathbb{Q} \cap [a_k, b_k]$ is dense in $[a_k, b_k]$ for all $k = 1,\ldots,i-1,i+1,\ldots,m$ we conclude that $\partial^2 G_j/\partial x_i^2 = 0$ everywhere on $[\mathbf{a}, \mathbf{b}]$ and for all $j = 1, 2, \ldots, n$, $i = 1, 2, \ldots, m$.

Now we are going to prove that $\partial^2 G_j/\partial x_i \partial x_t$ vanishes everywhere on $[\mathbf{a}, \mathbf{b}]$ and for all $j = 1, 2, \ldots, n$, $i, t = 1, 2, \ldots, m$, $i \neq t$ (without loss of generality, let $t > i$ in what follows). Assume that the opposite is true, that is, that there exist i, j, t and a point $\mathbf{x} \in [\mathbf{a}, \mathbf{b}]$ such that $\partial^2 G_j(\mathbf{x})/\partial x_i \partial x_t \neq 0$. Then due to the continuity of second partial derivatives of the function G, by using the argument similar to that from the beginning of the proof of Lemma 13.72, we conclude that there exist a point (which without risk of misunderstanding we denote by the same symbol $\mathbf{x}$) in $(\mathbf{a}, \mathbf{b}) = (a_1, b_1) \times \cdots \times (a_m, b_m)$ and a neighborhood U of that point such that $\partial^2 G_j/\partial x_i \partial x_t > 0$ everywhere on U. Therefore we always may take $z_k \in \mathbb{Z}_p \cap \mathbb{Q} \cap [a_k, b_k]$; $k \neq i, t$; $M \in \mathbb{N}$ and $c, d \in \{0, 1, \ldots, p^M - 1\}$ such that the point $\mathbf{z}(x, y) = (z_1; \ldots; z_{i-1}; x; z_{i+1}; \ldots; z_{t-1}; y; z_{t+1}; \ldots; z_m)$ lies in U for all $x = p^{-M}(c + e)$, $y = p^{-M}(d + h)$ and all $e, h \in [0, 1)$. Arguing like in the proof of Proposition 13.64 we see that the following inclusion holds:

$$\left\{(\bar{\mathbf{z}}(e, h); (p^M G(\bar{\mathbf{z}}(e, h))) \bmod 1) \colon e, h \in [0, 1]\right\} \subset \mathbf{P}(\mathfrak{A}),$$

where $\bar{\mathbf{z}}(e, h) = (\mathbf{z}(x, y)) \bmod 1$ (we reduce all coordinates modulo 1).

Consider a finite automaton $\tilde{\mathfrak{A}}$ which is obtained by "gluing together" the i-th and the t-th inputs of the automaton $\mathfrak{A}$ while feeding the rest k-th inputs with infinite words $\mathsf{wrd}((p^M z_k) \bmod 1)$; that is, the automaton function of the automaton $\tilde{\mathfrak{A}}$ is

$$f_{\tilde{\mathfrak{A}}}(v) = f_{\mathfrak{A}}(w_1; \ldots; w_{i-1}; v; w_{i+1}; \ldots; w_{t-1}; v; w_{t+1}; \ldots; w_m)$$

where $w_\ell = \mathsf{wrd}((p^M z_k) \bmod 1) \in \mathcal{W}_\infty$, $\ell \in \{1, 2, \ldots, m\} \setminus \{i, t\}$. By argument similar to that for the case $i = t$ (see the proof of the Claim above) we conclude that the automaton $\tilde{\mathfrak{A}}$ is finite and that the graph of the function $\bar{G}_j(h, h) = (p^M G_j(\mathbf{z}(p^{-M}(c+h), p^{-M}(d+h))) \bmod 1 \colon [0, 1]^2 \to [0, 1)$ when h is running through $[0, 1)$ lies in $\mathcal{P}(\tilde{\mathfrak{A}})$. But on the other hand, we have that

$$\frac{\partial^2 \bar{G}_j(h, h)}{\partial h^2} = \left(\frac{\partial}{\partial x_i} + \frac{\partial}{\partial x_t}\right)^2 G_j(\mathbf{z}(x, y)) = 2\frac{\partial^2 G_j(\mathbf{z}(x, y))}{\partial x_i \partial x_t}$$

since $\partial^2 G_j(\mathbf{z}(x, y))/\partial x_i^2 = \partial^2 G_j(\mathbf{z}(x, y))/\partial x_t^2 = 0$ by what we have already proven above. But this is a contradiction to Theorem 13.70 since the function $\bar{G}_j(h, h)$ of argument h satisfies all conditions of the theorem and has a non-zero second derivative. Thus we have proved that under conditions of Theorem 13.74 the function G must be affine: $G(\mathbf{x}) = \mathbf{xA} + \mathbf{B}$ for all $\mathbf{x} \in [\mathbf{a}, \mathbf{b}]$.

Now fix arbitrary $i \in \{1, 2, \ldots, m\}$, $j \in \{1, 2 \ldots, n\}$, and $z_k \in [a_k, b_k] \cap \mathbb{Z}_p \cap \mathbb{Q}$ for $k = 1, 2, \ldots, m$, $k \neq i$; consider the function $\bar{G}_{ij}$ and the automaton $\bar{\mathfrak{A}}_{ij}$ as in the beginning of the proof of Theorem 13.74. Then from the affinity of the function G it follows that $\bar{G}_{ij}(x) = xA_{ij} + B_j$. Since $\mathbf{G}_{[a_i,b_i]}(\bar{G}_{ij}) \subset \mathbf{P}(\bar{\mathfrak{A}}_{ij})$ by the Claim above, Theorem 13.70 implies that $A_{ij}, B_j \in \mathbb{Z}_p \cap \mathbb{Q}$.

Further, arguing like in the proof of Proposition 13.64 we conclude that for suitable $M \in \mathbb{N}$ and $\mathbf{h} \in \{0, 1, \ldots, p^M - 1\}^m$, the graph $\mathbf{G}_{[0,1]^m}(H(\mathbf{v}))$ of the function $H(\mathbf{v}) = H_{\mathbf{h},M}(\mathbf{v}) = (p^M((p^{-M}(\mathbf{h} + \mathbf{v})\mathbf{A} + \mathbf{B}) \bmod 1 = (\mathbf{vA} + (\mathbf{hA} + p^M\mathbf{B})) \bmod 1$ lies completely in $\mathcal{P}(\mathfrak{A})$. Now considering the function H and the corresponding automaton $\bar{\mathfrak{A}}_{ij}$ as above for $z_k = 0$, $k = 1, 2, \ldots, m$, $k \neq i$, $G = H$, we conclude by Theorem 13.70 that there are only finitely many A_{ij}; whence finitely many $\mathbf{A}$. If for some of these $\mathbf{A}$ there were infinitely many $\mathbf{B} \bmod 1$ such that $\mathbf{G}_{[0,1]^m}(H(\mathbf{v})) \subset \mathcal{P}(\mathfrak{A})$, then for some $j \in \{1, 2, \ldots, n\}$, there were infinitely many pairwise distinct $B_j \bmod 1$. But given arbitrary $z_k \in [a_k, b_k] \cap \mathbb{Z}_p \cap \mathbb{Q}$ for $k = 2, 3, \ldots, m$ and considering corresponding automata $\bar{\mathfrak{A}}_{1j}$ for various $(m-1)$-tuples $(z_2, \ldots, z_m)$ (cf. the beginning of the proof of Theorem 13.74), from the construction of $\bar{\mathfrak{A}}_{1j}$, it follows (cf. the proof of the Claim) that there are only finitely many these automata $\bar{\mathfrak{A}}_{1j}$ since the automaton $\mathfrak{A}$ is finite. Therefore applying Theorem 13.70 to every automaton $\bar{\mathfrak{A}}_{1j}$, we finally conclude that there are only finitely many $B_j \bmod 1$, a contradiction to our assumption.

Therefore there are only finitely many pairwise distinct functions $H_{\mathbf{h},M}$ as above. Now by mimicking the respective part of the proof of the first assertion of Theorem 13.70, we conclude that given an $(m \times n)$-matrix $\mathbf{A}$ and a vector $\mathbf{B}_j$ over $\mathbb{Z}_p \cap \mathbb{Q}$ such that the graph of the function $G(\mathbf{x}) = \mathbf{xA} + \mathbf{B}$ on $[\mathbf{a}, \mathbf{b}] \subset [0, 1]^m$ lies completely in $\mathcal{P}(\mathfrak{A})$ then necessarily $\mathbf{G}_{\mathbb{R}^m}((\mathbf{xA} + \mathbf{B}) \bmod 1) \subset \mathcal{P}(\mathfrak{A}) \subset \mathbb{T}^{n+m}$. □

Note 13.75 An automaton with a single input channel and a single output channel over respective alphabets $\{0, 1, \ldots, p^n - 1\}$ and $\{0, 1, \ldots, p^k - 1\}$, where $n, k \geqslant 1$, can be considered as an automaton with n input channels and k output channels over the alphabet $\{0, 1, \ldots, p - 1\}$; therefore Theorem 13.74 can be applied to automata of that sort as well.

It is worth noting that the affinity of finitely computable functions still holds true under the condition of continuity of the first (rather than of the second, as in the statement of Theorem 13.70) derivative: See [98], where the authors by using completely different methods (model-theoretic, to be more exact) prove only the affinity, i.e., the first assertion of Theorem 13.70.

A few important notes in conclusion of the chapter: When studying real functions that can be computed by an automaton $\mathfrak{A}$ whose input/output alphabets are $\mathcal{A} = \{0, 1, \ldots, p-1\}$ (where $p > 1$ is an integer from $\mathbb{N} = \{1, 2, 3, \ldots\}$), most authors follow common approach which is described in, e.g., [60, Chapter XIII, Section 4]. They associate an infinite word $\alpha_1\alpha_2\ldots\alpha_n\ldots$ over $\mathcal{A}$ to a real number whose base-p expansion is $0.\alpha_1\alpha_2\ldots\alpha_n\ldots = \sum_{i=1}^{\infty}\alpha_i p^{-i}$ and consider a real function $f^{\mathfrak{A}}$ defined as follows: Given $x \in [0, 1]$, take its base-p expansion $x = \sum_{i=1}^{\infty}\alpha_i p^{-i}$; then produce an infinite output sequence $\beta_1\beta_2\ldots\beta_n\ldots$ of $\mathfrak{A}$ by successfully feeding the automaton with the letters α_1, α_2, etc.; and put $f^{\mathfrak{A}}(x) = \sum_{i=1}^{\infty}\beta_i p^{-i}$. Being fed by infinite input sequence $\alpha_1\alpha_2\ldots\alpha_n\ldots$, the automaton $\mathfrak{A}$ produces a unique infinite output sequence $\beta_1\beta_2\ldots\beta_n\ldots$; therefore the function $f^{\mathfrak{A}}$ is well defined everywhere on the real closed unit interval (segment) $\mathbb{I} = [0, 1]$ with the exception of maybe a countable set $D \subset [0, 1]$ of points; namely, of those having two different base-p expansions $0.\gamma_1\gamma_2\ldots\gamma_n 0\ldots 0\ldots = 0.\gamma_1\gamma_2\ldots\gamma_{n-1}(\gamma_n - 1)(p-1)\ldots(p-1)\ldots$. The point set $\mathbf{M}(\mathfrak{A}) = \{(x; f^{\mathfrak{A}}(x)) \in \mathbb{R}^2 : x \in [0, 1]\}$ can be considered as a graph of the real function $f^{\mathfrak{A}}$ specified by the automaton $\mathfrak{A}$ (note that every time, before being fed by the very first letter of each infinite input word, the automaton $\mathfrak{A}$ is assumed to be in a fixed state s_0, the *initial state*). Indeed, $f^{\mathfrak{A}}(x)$ is defined uniquely for $x \in [0, 1] \setminus D$ and $f^{\mathfrak{A}}(x)$ can be ascribed to at most two values for $x \in D$; so $f^{\mathfrak{A}}$ can be treated as a real function which is defined on the unit segment $[0, 1]$ and has not more than a countable number points of discontinuity in $[0, 1]$. To put it in other words, $\mathbf{M}(\mathfrak{A})$ is the *image* of the automaton function $f_{\mathfrak{A}}$ with respect to Monna map; see Sects. 7.2.1 and 12.1.

The said common approach (and its various generalizations) is utilized, e.g., in [36–38, 76, 128, 141, 170]. Speaking loosely, the common approach looks as if one feeds the automaton $\mathfrak{A}$ by the base-p expansion of a real number $x \in [0, 1]$ so that *leftmost* (i.e., the most significant) *digits are fed to the automaton prior to rightmost ones* and observes output as real numbers (since the automaton outputs accordingly leftmost digits of the base-p expansion of $f^{\mathfrak{A}}(x) \in [0, 1]$ prior to rightmost ones) thus ascribing to the automaton $\mathfrak{A}$ the real function $f^{\mathfrak{A}}$. We stress that the function $f^{\mathfrak{A}}$ is well defined almost everywhere on $[0, 1]$ due to *namely that order* in which digits of the base-p expansion are fed to (and produce the output word of) the automaton $\mathfrak{A}$.

A crucial difference of the approach we have employed in this chapter from the mentioned one is that in the mentioned papers, *digits are fed to (and are read from) the automaton in another* (i.e., inverse) *order*, namely,

1. Given a real number $x \in [0, 1]$, represent x via the base-p expansion $x = 0.\alpha_1\alpha_2\ldots\alpha_n\ldots$ (take both expansions if x has two different ones).

2. From the base-p expansion $0.\alpha_1\alpha_2\ldots\alpha_n\ldots$ derive corresponding sequence $\alpha_1, \alpha_1\alpha_2, \alpha_1\alpha_2\alpha_3, \ldots$ of words.
3. Then by feeding the automaton $\mathfrak{A}$ successively by the words $\alpha_1, \alpha_1\alpha_2, \alpha_1\alpha_2\alpha_3, \ldots$ so that *rightmost letters are fed to* $\mathfrak{A}$ *prior to leftmost ones*, construct corresponding output word sequence $\zeta_{11}, \zeta_{12}\zeta_{22}, \zeta_{13}\zeta_{23}\zeta_{33}, \ldots$.
4. To the output sequence put into a correspondence the sequence $\mathcal{S}(x)$ of rational numbers whose base-p expansions are $0.\zeta_{11}, 0.\zeta_{12}\zeta_{22}, 0.\zeta_{13}\zeta_{23}\zeta_{33}, \ldots$ thus obtaining a point set $\mathcal{X}(x) = \{(0.\alpha_1\ldots\alpha_i; 0.\zeta_{1i}\zeta_{2i}\ldots\zeta_{ii})\ i = 1, 2, \ldots\}$ in the real unit square $\mathbb{I}^2 = [0, 1] \times [0, 1]$.
5. Then construct the set $\mathcal{F}(x)$ of *all* cluster points of the sequence $\mathcal{S}(x)$.
6. Finally, specify the plot of the automaton $\mathfrak{A}$ as a point set $\cup_{x\in[0,1], y\in\mathcal{F}(x)}((x; y) \cup \mathcal{X}(x))$ in a unit square $\mathbb{I}^2$ (or on the surface of the torus $\mathbb{T}^2$).

The difference between the two point sets, the plot of the automaton and the image of the automaton function with respect to Monna map, is sharp; see Sect. 7.2.1; but, namely, plots represent the long-term behavior of automata. That is why the plots are used in Chap. 16: The plots are capable to catch specific features of automata models of quantum systems.

13.7 Automata Functions over $\mathbb{F}_p[[X]]$

Basically, an automaton whose input/output alphabets are $\mathbb{F}_p = \{0, 1, \ldots, p-1\}$ just maps words from $\mathcal{W}_\infty$ to words from $\mathcal{W}_\infty$ so that each letter of ξ_i of output word does not depend on letters $\chi_{i+1}, \chi_{i+2}, \ldots$ for all $i \in \mathbb{N}_0$; moreover, that any such mapping $\mathcal{W}_\infty \to \mathcal{W}_\infty$ can be produced by a suitable automaton which need not be necessarily finite (cf. the proof of Theorem 13.1 which actually is an argument about words from $\mathcal{W}_\infty$ rather than of p-adic integers).

On the set $\mathcal{W}_\infty$ a structure of commutative integral domain, i.e., of the ring with no zero divisors, can naturally be defined; as a metric space, this ring is complete and compact with respect to the metric d on $\mathcal{W}_\infty$ defined in Sect. 1.6.2, but is not isomorphic to the ring $\mathbb{Z}_p$ of p-adic integers. It is well-known that such alternative ring structure which not ring-isomorphic to $\mathbb{Z}_p$ is unique: This is the ring $\mathbb{F}_p[[X]]$ of formal power series in indeterminate X over a p-element finite field $\mathbb{F}_p$; the one which already has appeared in Sect. 13.5.

Therefore it is quite natural to develop a theory of automata functions as 1-Lipschitz mappings $\mathbb{F}_p[[X]] \to \mathbb{F}_p[[X]]$, not only in view of theoretical, but also of an applied value the theory has. Indeed, if $p = 2$, then the addition in the ring $\mathbb{F}_2[[X]]$ is just a bitwise logical addition modulo 2, the XOR instruction of modern computers (cf. Sect. 9.5.1). Unfortunately, in a contrast to ordinary multiplication of numbers represented via the base-2 expansions, multiplication in the ring $\mathbb{F}_2[[X]]$ (the "multiplication without carry", speaking loosely) is not a standard processor instruction; but nevertheless multiplication in $\mathbb{F}_p[[X]]$ is easily programmable. Therefore it may happen that some useful applications of automata functions as

1-Lipschitz mappings of $\mathbb{F}_p[[X]]$ into itself will be found in the near future, similar to that of the theory of T-functions.

Currently the ergodic theory of 1-Lipschtz mappings $\mathbb{F}_p[[X]] \to \mathbb{F}_p[[X]]$ is under development. The following results within that theory are obtained:

- Criteria of automaticity (i.e., of 1-Lipschizness) of a mapping $\mathbb{F}_2[[X]] \to \mathbb{F}_2[[X]]$, as well as criteria of measure-preservation/ergodicity of a 1-Lipschitz mapping $\mathbb{F}_2[[X]] \to \mathbb{F}_2[[X]]$ are found, in terms of van der Put basis and in terms of Carlitz basis over $\mathbb{F}_2[[X]]$; see [139].
- For general mapping $\mathbb{F}_p[[X]] \to \mathbb{F}_p[[X]]$, criteria of 1-Lipschitzness in terms of van der Put basis, in terms of Carlitz basis, in terms of digit derivatives basis, and in terms of digit shifts basis were found in [103, 104, 191]. The criterion in terms of van der Put basis over $\mathbb{F}_p[[X]]$ is looking similar to the one from Theorem 5.14; namely, the mapping $F\colon \mathbb{F}_p[[X]] \to \mathbb{F}_p[[X]]$ represented via van der Put series $F(x) = \sum_{g\in\mathbb{F}_p[X]} B_g\chi(g,x)$ is an automaton function if and only if $|B_g|_X \leqslant p^{-\deg g}$, where $|\cdot|_X$ is the absolute value on $\mathbb{F}_p[[X]]$ associated with the metric d_p (i.e., $|G|_X = d_p(G,0)$, where 0 is additive neutral of the ring $\mathbb{F}_p[[X]]$, the formal power series in X all whose coefficients are zero polynomials over $\mathbb{F}_p$), $B_g \in \mathbb{F}_p[[X]]$, and $\chi(g,x)$ is a characteristic function of the ball in $\mathbb{F}_p[[X]]$ centered at $g \in \mathbb{F}_p[X]$ of radius $p^{-\deg g-1}$, $g \in \mathbb{F}_p[X]$ are polynomials in indeterminate X over a p-element field $\mathbb{F}_p$.
- Criteria for measure-preservation/ergodicity for 1-Lipschitz mappings of $\mathbb{F}_p[[X]]$ into itself represented in van der Put basis and in Carlitz basis over $\mathbb{F}_p[[X]]$ were found in [139] for $p = 2$; for general case, as well as for representations in other bases, see [103, 104, 191]. We note that the criteria in terms of van der Put basis over $\mathbb{F}_2[[X]]$ are looking very similar to those in terms of van der Put series over $\mathbb{Z}_2$ (cf. Theorems 9.15 and 9.17).

Note that the non-Archimedean theory of automata functions considered as 1-Lipschitz mappings $\mathbb{F}_p[[X]] \to \mathbb{F}_p[[X]]$ is still less developed in comparison to its counterpart over $\mathbb{Z}_p$.

Chapter 14
Applications to Computer Science

Abstract While the preceding chapters are devoted mostly to one type of automata, the transducers, in the current chapter, we focus on another type of automata, the generators. We are mostly interested in pseudorandom number generators, the PRNG, as the latter generators can be used to produce pseudorandom sequences that "look like" random ones. As a rule, uniform distribution is the weakest statistical property the sequence must necessarily satisfy to be considered as a pseudorandom in any reasonable meaning as frequencies of occurrences of terms in a truly random sequence are (approximately) equal. That is why in the current chapter, we are focused on algorithms which produce uniformly distributed sequences out of a given short random string and which are the most suitable for implementation as programs (mostly, as straight line programs, SLP) for contemporary digital computers.

14.1 Pseudorandom Number Generator Is a Dynamical System

Recall that according to general definition of an automaton, the generator is a (non-initial) automaton whose input alphabet is empty while the output alphabet is not empty (however, finite), (cf. Sect. 1.6). Pseudorandom number generators, the PRNG, as the generators which can be used to produce pseudorandom sequences which "look like" random ones.

Of course, to construct a mathematical theory, the ill-defined term "look like" must be defined rigorously. However, currently there are no standard mathematical definition of what are pseudorandom sequences as in various applications a sequence is considered as pseudorandom if it passes a given set of statistical tests which vary depending on the application. Consequently, the definition of what is a pseudorandom sequence (whence, what is a PRNG) depends on the choice of these tests. We stress that the class of tests a PRNG must pass is set beforehand; for instance, if one takes all polynomial-time tests, he obtains a definition of pseudorandomness in the sense of the complexity theory. However, in practice they often use some standard batteries of tests, e.g., NIST, DIEHARD, or some other.

V. Anashin, *Causality: The p-adic Theory*, STEAM-H: Science, Technology, Engineering, Agriculture, Mathematics & Health,
https://doi.org/10.1007/978-3-031-85818-5_14

Pseudorandom generators are widely used in numerous applications, especially in modeling, computer simulation (e.g., in quasi-Monte Carlo methods), in information security, cryptography (e.g., for password generation, in stream ciphers), etc. In most cases, the generators are computer programs; that is why in the current chapter we address a problem of design of these programs which must be based on standard computer instructions to achieve good performance and reduce time/memory consumption. Moreover, in many "sensitive" applications (e.g., in cryptography), it is not sufficient just to demonstrate that the PRNG passes all standard batteries of statistical tests like NIST or DIEHARD; quite sophisticated methods and mathematical theories (e.g., "provable security") were developed in order to *prove mathematically* that the PRNG meets some crucial requirements, e.g., the length of the period of produced sequence, good distribution of k-tuples, etc. In this chapter, we discuss some of these crucial requirements and show how to satisfy them by using the p-adic ergodic theory. We follow mainly [10, 14, 24, 27] in this chapter.

Basically, a real-life PRNG which is implemented as a program for modern computer can be considered as finite non-initial automaton $\mathfrak{A} = \langle \mathcal{N}, \mathcal{M}, f, F\rangle$ *without input* (i.e., with empty input alphabet) where $\mathcal{N}$ is a finite set of states, $f\colon \mathcal{N} \to \mathcal{N}$ is a state transition function, $\mathcal{M}$ is a finite output alphabet, and $F\colon \mathcal{N} \to \mathcal{M}$ is an output function (sometimes in cryptology called a *filter*). That is, mathematical model of a real-life PRNG is (in our terms) an initial automaton *without input*, Sect. 1.6 and Fig. 1.1 there. The initial state (which sometimes is called also a *seed*) is subject to change at every new run of the PRNG to produce a new pseudorandom sequence. Schematics of a typical PRNG is depicted by Fig. 14.1.

Given an initial state $u_0 \in \mathcal{N}$, the PRNG produces a sequence

$$\mathcal{Z} = \{F(u_0), F(f(u_0)), F(f^2(u_0)), \ldots, F(f^j(u_0)), \ldots\}$$

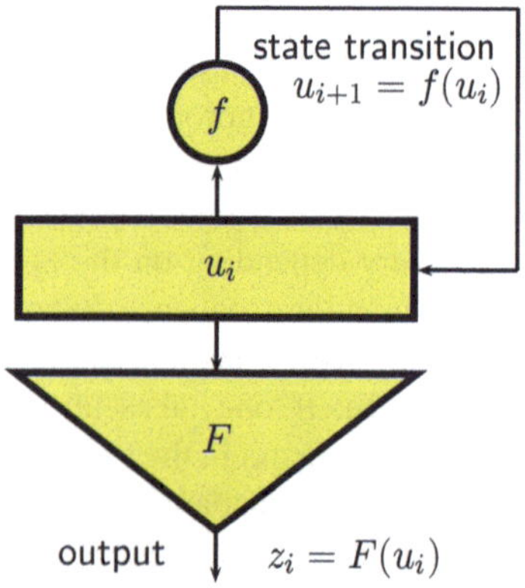

Fig. 14.1 Pseudorandom generator

over the set $\mathcal{M}$, where

$$f^j(u_0) = \underbrace{f(\dots f(}_{j \text{ times}}u_0)\dots) \ (j = 1, 2, \dots); \qquad f^0(u_0) = u_0.$$

As the set of states $\mathcal{N}$ is finite, the output sequence is necessarily eventually periodic. Note that the output sequence depends on the initial state u_0. That is, the PRNG can be considered as a mapping from $\mathcal{N}$ into the set of all (eventually) periodic sequences over $\mathcal{M}$.

The generators may be considered either as pseudorandom generators per se, or as components of more complicated automata, the so-called counter-dependent generators; the latter produce sequences $\{z_0, z_1, z_2, \dots\}$ over $\mathcal{M}$ according to the rule

$$z_0 = F_0(u_0), u_1 = f_0(u_0); \dots z_i = F_i(u_i), u_{i+1} = f_i(u_i); \dots$$

That is, at the $(i+1)$-th step the automaton $\mathfrak{A}_i = \langle \mathcal{N}, \mathcal{M}, f_i, F_i, u_i \rangle$ is applied to the state $u_i \in \mathcal{N}$, producing a new state $u_{i+1} = f_i(u_i) \in \mathcal{N}$, and a new output symbol $z_i = F_i(u_i) \in \mathcal{M}$. It is clear that actually the counter-dependent generator may also be considered as a generator whose set of states is a Cartesian product $\mathbb{N}_0 \times \mathcal{N}$ and whose initial states are pairs $(0, s) \in \mathbb{N}_0 \times \mathcal{N}$.

We will mostly focus on ordinary PRNGs (cf. Fig. 14.1). Note that formally speaking the sequence of states

$$u_0, u_1 = f(u_0), u_2 = f(u_1), \dots, u_{i+1} = f(u_i) = f^{i+1}(u_0), \dots \tag{14.1}$$

can be judged as an orbit of the dynamical system $\langle \mathcal{N}, f \rangle$, whereas the output sequence

$$z_0 = F(u_0), z_1 = F(u_1), \dots, z_i = F(u_i) = F(f^i(u_0)), \dots \tag{14.2}$$

is an observable; see Sect. 1.3. We will show now that such a look at PRNG is not only formal but reveals the essence of the problem how to construct a good PRNG. Moreover, in Chap. 16 such a look leads to a model of evolution of isolated physical system.

14.1.1 What Pseudorandom Number Generators Are Good?

A PRNG which could be considered any good obviously must meet the following conditions:

- The output sequence must be pseudorandom; i.e., must pass certain statistical tests which depend on the requirements for the equipment/application for which the PRNG is designed.

- The PRNG must be suitable for software (or hardware) implementations; the performance must be sufficiently fast; memory consumption must satisfy prescribed conditions.

In the case the PRNG is an automaton represented by Fig. 14.1, we can restate these conditions as follows:

Condition 1: The state transition function f must provide pseudorandomness; in particular, it must guarantee uniform distribution and long period of the sequence of states $(u_i)_{i=0}^{\infty}$.

Condition 2: The output function F must not spoil pseudorandomness; at least, the output sequence $(z_i)_{i=0}^{\infty}$ must be uniformly distributed and must have a long period; also, F must make the output sequence "as unpredictable as possible."

In practice, the PRNG-s are implemented in software or hardware, and it is highly desirable to make respective programs platform-independent to make possible to run the same algorithm on various platforms. Moreover, performance of respective programs must be sufficiently fast on all platforms. This requirement results in the following condition:

Condition 3: To make a PRNG any suitable for software/hardware implementations, and to make it platform-independent, both f and F must be (not too complicated) T-functions which are compositions of basic computer instructions (cf. Sect. 9.5.1).

To satisfy Condition 1, one may take transitive state transition function $f: \mathcal{N} \to \mathcal{N}$; the sequence of states (14.1) will have then the longest possible period (of length $\#\mathcal{N}$) and strict uniform distribution: Every element from $\mathcal{N}$ will occur at the period exactly once; see Sect. 7.1.1.

To satisfy the first part of Condition 2, one may take a balanced output function $F: \mathcal{N} \to \mathcal{M}$ where $\#\mathcal{M}$ is a factor of $\#\mathcal{N}$. Whenever $\#\mathcal{N} = \#\mathcal{M}$, balanced mappings are just invertible (i.e., bijective, one-to-one) mappings. It is easy to see that if a balanced output function is applied to a strictly uniformly distributed sequence of states, the output sequence is also strictly uniformly distributed: It is periodic with a period of length $\#\mathcal{N}$, and every element from $\mathcal{M}$ occurs at the period exactly $\frac{\#\mathcal{N}}{\#\mathcal{M}}$ times. We state this as a proposition for further references.

Proposition 14.1 *If the state transition function f of the automaton $\mathfrak{A}$ is transitive on the state set $\mathcal{N}$, i.e., if f is a permutation with a single cycle of length $N = \#\mathcal{N}$; if, further, $M = \#\mathcal{M}$ is a factor of N, and if the output function $F : \mathcal{N} \to \mathcal{M}$ is balanced* (i.e., $\#F^{-1}(s) = \#F^{-1}(t)$ for all $s, t \in \mathcal{M}$), *then the output sequence $\mathcal{Z}$ of the automaton $\mathfrak{A}$ is purely periodic with a period of length N (i.e., of a maximum possible), and each element of $\mathcal{M}$ occurs at the period the same number of times: $\frac{N}{M}$ exactly. That is, the output sequence $\mathcal{Z}$ is strictly uniformly distributed* (cf. Sect. 7.1.1).

To satisfy Condition 3, one may use only operations that are common to all platforms: These are arithmetical (numerical) instructions, i.e., such operations as addition, multiplication, subtraction, division, and exponentiation of integers. In this case both $\mathcal{N}$ and $\mathcal{M}$ can be associated with the sets of integers $0, 1, 2, \ldots, N-1$ and

$0, 1, 2, \ldots, M-1$ and moreover, to residue rings $\mathbb{Z}/N\mathbb{Z}$ and $\mathbb{Z}/M\mathbb{Z}$, respectively. Moreover, if one takes $N = 2^n$ and $M = 2^m$, then actually both f and F will work with n-bit words to produce output sequence of m-bit words. This case is the most convenient for programming; moreover, in this case one may use along with arithmetical instructions the bitwise logical instructions and other basic instructions from Sect. 9.5.1 to construct f and F.

14.1.2 Why the p-Adic Ergodic Theory?

Now we explain a general way to construct transitive mappings f and balanced mappings F out of arithmetic operations (in the case both N and M are composite numbers) and out of arithmetic and bitwise logical operations (in the case both N and M are powers of 2). The idea is as follows: Let, say, $N = 2^n$ and $M = 2^m$, $m \leqslant n$, $n = kr$, $m = ks$; then using the p-adic ergodic theory developed in Part II, we construct an ergodic mapping $f\colon \mathbb{Z}_2 \to \mathbb{Z}_2$ and a measure-preserving mapping $F\colon \mathbb{Z}_2^r \to \mathbb{Z}_2^s$ out of arithmetic and bitwise logical operations as these operations are 1-Lipschitz functions defined on the space of 2-adic integers $\mathbb{Z}_2$ and valuated in $\mathbb{Z}_2$ (cf. Chap. 2) (note that 2-adic canonical expansions are just representations of integers in the so-called two's complement code). Then according to Theorem 8.2, taking residues of f and of F modulo 2^k and 2^k, respectively, we obtain a transitive transformation $f \bmod 2^n$ of the residue ring $\mathbb{Z}/2^n\mathbb{Z}$ and a balanced mapping $F \bmod 2^k\colon (\mathbb{Z}/2^k\mathbb{Z})^r \to (\mathbb{Z}/2^k\mathbb{Z})^s$. So $f \bmod 2^n$ will serve as a state transition function, whereas $F \bmod 2^k$ will serve as an output function since elements of residue ring $\mathbb{Z}/2^n\mathbb{Z}$ and of Cartesian powers $(\mathbb{Z}/2^k\mathbb{Z})^r$ and $(\mathbb{Z}/2^k\mathbb{Z})^s$ can be treated as n-bit and m-bit words, respectively. Note also that any number whose base-2 representation is longer than a machine word length k is reduced modulo 2^k automatically by a computer.

The case when both N and M are composite numbers can be reduced to the case of prime powers: That is, we will construct ergodic mappings $f\colon \mathbb{Z}_p \to \mathbb{Z}_p$ and measure-preserving mappings $F\colon \mathbb{Z}_p^r \to \mathbb{Z}_p^s$ and then take $f \bmod p^n$ and $F \bmod p^k$, for all all prime factors of N and M (we assume that prime factors of N and of M form the same set). Then with the use of Chinese Reminder Theorem 1.1, we construct mappings modulo N and M which coincides accordingly with $f \bmod p^n$ and $F \bmod p^k$ for all prime factors p of N and of M (this approach in more detail will be discussed further in Chap. 15; see Theorem 15.4 and the example thereafter).

Now we make some conventions on terminology (cf. Sect. 7.1.1):

Definition 14.2 A sequence $(s_i)_{i=0}^{\infty}$ of p-adic integers is called *strictly uniformly distributed modulo* p^k whenever the sequence $(s_i \bmod p^k)_{i=0}^{\infty}$ of residues modulo p^k is strictly uniformly distributed over the residue ring $\mathbb{Z}/p^k\mathbb{Z}$.

Note 14.3 A sequence $(s_i)_{i=0}^{\infty}$ of p-adic integers is uniformly distributed (with respect to the normalized Haar measure μ on $\mathbb{Z}_p$) if and only if it is uniformly distributed modulo p^k for all $k = 1, 2, \ldots$; that is, for every $a \in \mathbb{Z}/p^k\mathbb{Z}$ relative numbers of occurrences of a in the initial segment of length ℓ in the sequence $\{s_i \bmod p^k\}$ of residues modulo p^k are asymptotically equal, i.e., $\lim_{\ell\to\infty} \frac{A(a,\ell)}{\ell} = \frac{1}{p^k}$, where $A(a, \ell) = \#\{s_i \equiv a \pmod{p^k} : i < \ell\}$; see [130] for details. So strictly uniformly distributed sequences are uniformly distributed in usual sense of the theory of distribution of sequences.

Note that in view of Proposition 14.1, one can vary both the state transition and the output function of PRNG *without* affecting uniform distribution of the output sequence, as the only conditions that must be satisfied to make the output uniformly distributed are ergodicity of the state transition function and measure-preservation of the output function.

Of course, to make this reasoning any practical, we must choose these functions f and F from suitably large classes of ergodic and measure-preserving functions. In other words, we must develop certain tools to produce a number of various measure-preserving and/or ergodic mappings out of arithmetical (and of bitwise logical) operations. We consider these methods in the next section.

14.2 Congruential Generators of the Longest Period

In this section we consider the so-called congruential generators, a class of pseudorandom number generators which are widely used in various applications and widely studied in literature. We will show that actually the theory of these generators is a part of the p-adic ergodic theory: Numerous known sporadic results of these generators can be explained in a unified way by p-adic ergodic theory considered in Part II. We will show that all known results about periods of these generators can be deduced from basic theorems of the p-adic ergodic theory; also, we will prove some new general results in this area.

Actually, in this section, we explain how to construct a transformation on a given finite set $\mathcal{N}$ such that this transformation has a prescribed form and the longest possible period. These transformations will be compositions of arithmetic operators and also of bitwise logical operators whenever $\#\mathcal{N}$ is a power of 2. Thus, generators based on so-called T-functions, which are of interest for modern cryptology and which are just triangular functions from Definition 5.3 when $p = 2$, are within the scope of our study as well: Actually, the T-functions are 1-Lipschitz 2-adic functions; see Sect. 5.1.1; so the theory of T-functions is a part of p-adic theory. Now we introduce the main notion of this section.

Definition 14.4 A *congruential generator* is a non-initial automaton $\mathfrak{A} = \langle \mathcal{N}, \mathcal{M}, f, F \rangle$ (cf. the very beginning of Sect. 14.1) where $\mathcal{M} = \mathcal{N} = \mathbb{Z}/N\mathbb{Z}$, $F : \mathcal{M} \to \mathcal{M}$ is the identity mapping, and the state transition function $f : \mathbb{Z}/N\mathbb{Z} \to$

$\mathbb{Z}/N\mathbb{Z}$ preserves all congruences of the residue ring $\mathbb{Z}/N\mathbb{Z}$: $f(a) \equiv f(b) \pmod{L}$ whenever $a \equiv b \pmod{L}$ and $L \neq 1$ is a factor of N. The function f is called the *recurrence law* of the congruential generator.

Note 14.5 In view of Chinese Reminder Theorem 1.12, it is obvious that the output sequence of the congruential generator has the longest possible period (i.e., of length N) if and only if every function $f \bmod p^n$ is transitive modulo p^n, where $n = \operatorname{ord}_p N$, for all prime factors p of N.

In literature, some authors consider one more class of generators, which they call explicit congruential generators.

Definition 14.6 *Explicit congruential generator* corresponds to the case when the state transition function of automaton $\mathfrak{A}$ from Definition 14.4 is a map $f(x) = x + 1 \bmod N$, the *odometer* (which is sometimes called *adding machine*), whereas the output function $F: \mathbb{Z}/N\mathbb{Z} \to \mathbb{Z}/N\mathbb{Z}$ preserves all congruences of the residue ring $\mathbb{Z}/N\mathbb{Z}$.

We stress here that according to Chap. 8 to determine whether a congruential generator (in the meaning of Definition 14.4) attains the longest period (of length N) we should study *ergodicity* of the function f on the space $\mathbb{Z}_p$, for all primes $p \mid N$; whenever in the case of explicit congruential generator, we should study *measure-preservation* of F. This is the leading idea of the current section.

14.2.1 Types of Congruential Generators

Congruential generators from Definition 14.4 (as well as explicit congruential generators from Definition 14.6) were studied in a number of works; see monographs [62, 123, 156] and references therein. In the current subsection, we consider some known and widely used types of congruential generators. We will demonstrate that in all cases the longest possible periods are attained by these generators whenever the corresponding state transition function f is ergodic on certain subspaces of $\mathbb{Z}_p$, for some prime numbers p. This gives a general method to determine lengths of periods of congruential generators with the use of various techniques of the p-adic ergodic theory from Part II. Further we explain how to tweak some of these generators to lengthen their periods if their periods are not the longest possible.

14.2.1.1 Linear, Quadratic, and Cubic Congruential Generators

One of the most widespread types of congruential generators is *linear congruential generators* (which sometimes are also called *Lehmer* generators); they correspond to the case when $f(x) = (ax + b) \bmod N$, where a, b are rational integers and $N > 1$ is a natural number. Note that they speak about *congruential method* of generating pseudorandom numbers whenever $b \equiv 0 \pmod{N}$ and of *mixed congruential method* otherwise; see [123]. Other congruential generators that are often used in applications are quadratic and cubic; they correspond to the cases when $f(x)$ is a polynomial with rational integer coefficients, of degree 2 or 3, respectively. Note that Corollary 10.24 yields necessary and sufficient conditions for transitivity modulo N of a polynomial of *arbitrary* degree, with rational integer coefficients; thus, Corollary 10.24 gives a criterion when a quadratic or cubic congruential generator attains the longest period.

A question when a linear congruential generator has the longest possible period (i.e., of length N) was answered in 1962 by Hull and Dobell (cf. [123]). In view of Note 14.5 and Theorem 8.2, the criterion is actually stated by Theorem 9.1. Note that the longest possible period (of length N) can be achieved only with the use of mixed congruential method, when $b \not\equiv 0 \pmod N$ (actually, only when b and N are coprime, see Theorem 9.1). However, a *multiplicative* generator (with $f(x) = ax \bmod N$) is also often used in applications. In this case every ideal of the residue ring $\mathbb{Z}/N\mathbb{Z}$ is an invariant subset of the mapping $f(x) = ax$, so the longest possible period is achieved whenever f is ergodic on spheres around 0; this holds if and only if a is primitive either modulo p^2 for all prime p such that $p^2|N$, or modulo p, if $p \mid N$ and $p^2 \nmid N$; see Theorem 11.10. Usually a multiplicative generator is assumed to work only on the unit group of the residue ring $\mathbb{Z}/N\mathbb{Z}$, that is, on the multiplicative group $(\mathbb{Z}/N\mathbb{Z})^*$ of all invertible elements of the ring $\mathbb{Z}/N\mathbb{Z}$. In this case (for odd N) the generator is obviously equivalent to a linear congruential generator modulo $\varphi(N)$, the value of Euler's totient function, as the group $(\mathbb{Z}/p^k\mathbb{Z})^*$ is a cyclic group of order $(p-1)p^{k-1}$, for odd prime p; so the longest period of the generator is of length $\varphi(N)$ in this case. Note that for $N = 2^k$, $k \geqslant 2$, the multiplicative group $(\mathbb{Z}/2^k\mathbb{Z})^*$ is a direct product of a group of order 2 by a cyclic group of order 2^{k-2}; so the maximum length of the period of a multiplicative generator is 2^{k-2} in this case.

14.2.1.2 Power Generators

Another type of congruential generators which are used in real life applications are *power generators*, with $f(x) = x^n \bmod N$. They cannot achieve periods of length N since every p-adic sphere centered at 1 is an invariant subset of the transformation $x \mapsto x^n$ on $\mathbb{Z}_p$: They achieve the longest possible period when they are ergodic on p-adic spheres centered at 1; this holds if and only if n is primitive either modulo p^2 for all prime p such that $p^2|N$, or modulo p, if $p \mid N$ and $p^2 \nmid N$; see Theorem 11.10. Note that the maximum length of a period of the power generator can be determined with the use of Lemma 11.8.

14.2.1.3 Inversive Generators

Inversive generators are studied in numerous papers; see, e.g., survey paper [59] and references therein. When N is a prime, $f(x)$ (or $F(x)$, for explicit generators) are of the form $ax^{-1} + b$ or $(a + bx)^{-1}$; here $0^{-1} = 0$ by the definition, $a, b \in \mathbb{Z}$. These functions cannot be expanded directly to residue rings modulo composite N; in the latter case, domains of f and F are assumed to be restricted to the unit group $(\mathbb{Z}/N\mathbb{Z})^*$, which is a Cartesian product of the unit groups $(\mathbb{Z}/p^{\mathrm{ord}_p N}\mathbb{Z})^*$, for all prime $p \mid N$. Now we can study behavior of functions $ax^{-1} + b$ or $(b + ax)^{-1}$ on the unit group $\mathbb{Z}_p^*$ of all invertible p-adic integers to determine periods of these functions modulo N. As the unit group is a p-adic sphere of radius 1 centered at 0, and as both functions are 1-Lipschitz, the problem of maximality of the period length can be reduced to the problem of ergodicity of these functions on the p-adic sphere.

There are known inversive generators of another kind, the ones based on the mapping inv_p, the *generalized multiplicative inverse* (cf. (3.3)). It is clear that the mapping $x \mapsto \mathrm{inv}_p x$ is a 1-Lipschitz bijection of $\mathbb{Z}_p$ onto $\mathbb{Z}_p$; therefore the mapping is measure-preserving (cf. Sect. 8.1).

Proposition 9.22 is the ergodicity criterion of inv_2-based T-functions. For instance, the main result of [58] is that, given $a, b \in \mathbb{Z}$, the function $f(x) = a \times \mathrm{inv}_2(x) + b$ is transitive modulo 2^n, $n \geqslant 2$, if and only if $a \equiv 1 \pmod 4$ and $b \equiv 1 \pmod 2$. By virtue by Theorem 8.2, the transitivity of f modulo 2^n for all is equivalent to ergodicity of this function on $\mathbb{Z}_2$; but from Proposition 9.22 combined with Theorem 9.1, it follows immediately that the ergodicity takes place if and only if $a \equiv 1 \pmod 4$ and $b \equiv 1 \pmod 2$.

14.2.2 Periods of Congruential Generators

In this subsection we discuss various techniques which may be used to construct congruential generators of the longest period or to determine the lengths of periods of congruential generators. We will demonstrate that *the theory of congruential generators is, as a matter of fact, a part of the p-adic ergodic theory.*

14.2.2.1 Techniques Based on Convergent p-Adic Series

The most general characterizations of 1-Lipschitz measure-preserving and/or ergodic transformations on $\mathbb{Z}_p$ are given in terms of Mahler expansions, that is, by representation of the transformation via convergent interpolation series; see Sect. 9.3. This method is the most general as every continuous transformation on $\mathbb{Z}_p$ admits Mahler expansion. Another method is based on van der Put expansion (cf. Sect. 9.4). As every continuous transformation of $\mathbb{Z}_p$ can be represented by convergent van der Put series, the method is as general as the one based on Mahler expansion. In some cases, e.g., for analytic functions, we can also use representations via power series or via falling factorial series to determine whether the function is measure-preserving or ergodic by applying the results of Sect. 10.4.

Also we can use these methods to tweak generators in order to lengthen their periods.

In some cases (e.g., for analytic functions), we can also use representations via power series or via falling factorial series to determine whether the function is measure-preserving or ergodic by applying the results of Sect. 10.4. Also we can use these methods to tweak generators in order to lengthen their periods. We proceed with examples. As said, the exponential generator whose recurrence law is $f(x) = a^x \bmod N$ never attains the longest period, i.e., of length N. However by using the techniques of Mahler series, we immediately can tweak generators of this kind to make lengths of their shortest periods the longest possible, i.e., equal to N, just by adding a linear term to the recurrence law since for every prime p and every $a \equiv 1 \pmod{p}$, the function $f(x) = ax + a^x$ is a 1-Lipschitz ergodic transformation on $\mathbb{Z}_p$; see Example 9.12: Combining Example 9.12 with Theorem 8.2 and with Chinese Reminder Theorem 1.1, we are able to construct exponential generators that attains the longest period (of length N) modulo N for arbitrary composite N. For instance, the function $f(x) = 11x + 11^x$ is transitive modulo 10^n for all $n = 1, 2, \ldots$, as f is ergodic on $\mathbb{Z}_p$ for $p = 2$ and for $p = 5$, thus transitive modulo p^n for all $n = 1, 2, \ldots$ in view of Theorem 8.2; whence, f is transitive modulo 10^n for all $n = 1, 2, \ldots$ in view of Chinese Reminder Theorem 1.12.

In a similar manner, we can make tweaks to inversive generators modulo N to lengthen their periods to the maximum value, N. The idea is to use the mapping $\iota_p \colon x \mapsto (1+pmx)^{-1}$ (for some $m \in \mathbb{Z}_p$) in the composition of $f(x)$ rather than the mapping $x \mapsto x^{-1}$: Although both mappings are 1-Lipschitz p-adic mappings, the first one is defined everywhere on $\mathbb{Z}_p$, whereas the domain of second one is the unit group $\mathbb{Z}_p^*$ (i.e., the p-adic sphere $S_1(0) = \bigcup_{a=1}^{p-1} a + p\mathbb{Z}_p$ of radius 1 centered at 0). Moreover, the function ι_p is a C-function; that is, a p-adic analytic function defined by power series with p-adic integer coefficients which converges everywhere on $\mathbb{Z}_p$; see Sect. 6.2: $(1+pmx)^{-1} = 1 - pmx + p^2m^2x^2 - p^3m^3x^3 + \cdots$. As a C-function is ergodic if and only if it is transitive either modulo p^2 (when $p > 3$), or modulo p^3 (when $p \leqslant 3$) (cf. Corollary 10.23), the function $f(x) = x + (1 + p^3x)^{-1}$ is transitive modulo p^n for all $n = 1, 2, \ldots$ by Theorem 8.2; by the same reason, if $p > 3$, then the function $f(x) = x + (1 + p^2x)^{-1}$ is transitive modulo p^n for all $n = 1, 2, \ldots$. Now using Chinese Reminder Theorem 1.1, we can construct inversive generator modulo N such that the length of its shortest period is N, modulo arbitrary composite N. For instance, taking $f(x) = (x+(1+200x)^{-1}) \bmod 10^n$, we get inversive generator the length of whose period is a maximum, 10^n, whatever $n = 1, 2, 3, \ldots$ is taken: Again, this follows from Theorem 8.2 and Chinese Reminder Theorem 1.12 as this transformation f is ergodic on $\mathbb{Z}_p$ for $p \in \{2, 5\}$.

By exploiting these ideas, we immediately construct a number of different generators of these two kinds (inversive and exponential) that have the longest possible periods modulo p^n, that is, of length p^n:

- $f(x) = 1 + x + \check{p}^2 a^{b^x}$, $a \equiv b \equiv 1 \pmod{p}$, (the doubly exponential generator),

- $f(x) = 1 + x + \frac{p^2}{1+px}$ (the inversive generator),
- $f(x) = 1 + x + p^2(1 + px)^{\frac{1}{1+px}}$ (the exponential-inversive generator).

Now we explain how, given a congruential generator with the recurrence law $f(x) \bmod N$, one can determine the length a period of the generator. In view of Chinese Reminder Theorem 1.12, it suffices to consider only prime power moduli N. For $N = p^k$, p a prime, the idea is to reduce the problem of finding the period length to the problem of finding a closed subset of $\mathbb{Z}_p$ (usually a ball or a sphere), where a certain iterate $f^i(x)$ is ergodic.

For illustration, consider an exponential generator with the law $f(x) = a^x$, where $a \equiv 1 \pmod p$; i.e., $a = 1 + pz$ for some $z \in \mathbb{Z}_p$. It is clear that f maps $\mathbb{Z}_p$ into the ball $\mathbf{B}_{p^{-1}}(1) = 1 + p\mathbb{Z}_p$; so we can write $1 + pg(x) = (1 + pz)^x$ and then examine the function $g(x)$. As $(1 + pz)^x = \sum_{i=0}^{\infty} p^i z^i \binom{x}{i}$ is the Mahler expansion for a^x, we see that $g(x) = zx + pz^2\binom{x}{2} + p^2z^3\binom{x}{3} + \cdots$. Whenever $z \not\equiv 0 \pmod p$, all p-adic spheres around 0 are invariant under action of g, so the period will be the longest possible if g is ergodic on spheres $S_{p^{-r}}(0)$ around 0. Now we can apply Theorems 11.13 and 11.10 on ergodicity on spheres. From these Theorems we deduce that whenever $p \neq 2$, the derivative $g'(0)$ must be primitive modulo p^2; however, as $g'(0) \equiv z - \frac{p}{2}z^2 \pmod{p^2}$, and as $(1 - \frac{p}{2}z)^i \equiv 1 - iz\frac{p}{2} \pmod{p^2}$, the element $z - \frac{p}{2}z^2 = z(1 - \frac{p}{2}z)$ of the residue ring modulo p^n, $n \geqslant 2$, is primitive modulo p^2 whenever z is primitive modulo p^2 (we remind that 2 has a multiplicative inverse in $\mathbb{Z}_p$ whenever $p \neq 2$, so $\frac{p}{2} \in \mathbb{Z}_p$ in this case and least non-negative residue of $\frac{p}{2}$ modulo p^k is well defined). Now easy calculation shows that $g^{p-1}(x) \equiv xz^{p-1} + \frac{p}{2}xz \pmod{p^2}$ or, which is the same, that $g^{p-1}(x) \equiv xz^{p-1}(1+z\frac{p}{2}) \pmod{p^2}$; so $g^{p-1}(x)$ is ergodic on the ball $1+p\mathbb{Z}_p$ by Corollary 10.23 and Theorem 9.1 (if $p > 3$).

Indeed, represent $x \in 1 + p\mathbb{Z}_p$ as $x = 1 + py$. We must show ergodicity of the transformation $w\colon y \mapsto \frac{1}{p}(g^{p-1}(1 + py) - 1)$ on $\mathbb{Z}_p$. As $w(y) = (\frac{z^{p-1}-1}{p} + \frac{z}{2}) + (z^{p-1} + \frac{p}{2})y$ is an affine transformation, in view Theorem 9.1 w is ergodic if and only if simultaneously the following condition hold: $z^{p-1} + \frac{p}{2} \equiv 1 \pmod p$ (this holds as z is primitive modulo p^2; whence, modulo p) and $\frac{z^{p-1}-1}{p} + \frac{z}{2} \not\equiv 0 \pmod p$. Assuming $\frac{z^{p-1}-1}{p} + \frac{z}{2} \equiv 0 \pmod p$, we obtain that $z^{p-1} \equiv 1 - \frac{p}{2}z \pmod{p^2}$, whence, that $z^p \equiv z - \frac{p}{2}z^2 \pmod{p^2}$. As $z - \frac{p}{2}z^2$ is primitive modulo p^2, then z^p must be primitive modulo p^2. However, as z is primitive modulo p^2, the multiplicative order of z^p modulo p^2 is $p - 1$, and not $p(p - 1)$. A contradiction.

Finally by Note 11.9 we conclude that g is ergodic on the sphere $S_1(0)$ of radius 1 around 0. We conclude therefore that *the length of the shortest period of the exponential generator with the law* $f(x) = (1 + pz)^x \bmod p^k$, *where* $p > 3$ *and* z *is primitive modulo* p^2, *is* $(p - 1)p^{k-2}$, *for all* $k = 2, 3, \ldots$.

In a similar manner, one may use van der Put series and Theorem 9.17 (rather than Mahler series and Theorem 9.5) to find the lengths of periods of congruential generators, especially when the recurrence law of the generator is a T-function combined from computer instructions (cf., e.g., Theorem 9.24). The proof of the latter theorem exploits the Mahler series and is quite involved and lengthy. We use an opportunity to give here a much shorter proof of the second claim of the theorem by using van der Put series.

Recall that $\delta_i(x)$ is the value of the i-th digit in the base-2 expansion of $x \in \mathbb{Z}_2$; so $\sum_{k=0}^{\infty} 2^k \delta_k(x) = \sum_{k=0}^{\infty} \chi_k 2^k = x$. where $\chi_k \in \{0, 1\}$. Note that $\delta_k(x) = 1$ if and only if x is congruent modulo 2^{k+1} to either of the numbers $2^k, 2^k+1, \ldots, 2^{k+1}-1$; so

$$\delta_k(x) = \sum_{m=2^k}^{2^{k+1}-1} \chi(m, x), \tag{14.3}$$

as $\chi(m, x)$ is characteristic function of the ball $\mathbf{B}_{2^{-\lfloor \log_2 m \rfloor -1}}(m)$; see Sect. 4.3.

Example 14.7 Given a sequence $c, c_0, c_1, c_2, \ldots$ of 2-adic integers, the series

$$c + \sum_{i=0}^{\infty} c_i \delta_i(x) \tag{14.4}$$

defines an ergodic T-function $f\colon \mathbb{Z}_2 \to \mathbb{Z}_2$ if and only if the following conditions hold simultaneously:

1. $c \equiv 1 \pmod 2$;
2. $c_0 \equiv 1 \pmod 4$;
3. $c_i \equiv 0 \pmod{2^i}$, for $i = 1, 2, 3, \ldots$.

Indeed, substituting (14.3) to (14.4) we obtain the series

$$c + \sum_{i=0}^{\infty} c_i \sum_{m=2^i}^{2^{i+1}-1} \chi(m, x)$$

so the van der Put coefficients are $B_0 = c$, $B_1 = c + c_0$, $B_m = c_{\lfloor \log_2 m \rfloor}$ for $m \geqslant 2$. Now from Condition 1 of Proposition 9.19, we have that $c \equiv 1 \pmod 2$; however, from Condition 2 of Proposition 9.19, we have that $2c + c_0 \equiv 3 \pmod 4$ which gives us that $c_0 \equiv 1 \pmod 4$. Condition 3 of Proposition 9.19 is equivalent to the condition $|c_i|_2 = 2^i$ for $i \geqslant 1$. Due to these three conditions, Condition 4 of Proposition 9.19 is satisfied since $\sum_{m=2^i}^{2^{i+1}-1}(c_i - 2^i) = 2^i c_i - 2^{2i} \equiv 0 \pmod{2^{i+2}}$ for $i \geqslant 1$. This ends the proof.

It is worth noting here that the T-function $f(x) = c + \sum_{i=0}^{\infty} c_i \delta_i(x)$ can be expressed via computer instructions addition, multiplication by constants, and masking, $\mathsf{MASK}(x, 2^i) = x \mathsf{\ AND\ } 2^i = 2^i \delta_i(x)$ (cf. Sect. 9.5.1): As $c_i = 2^i d_i$ for suitable $d_i \in \mathbb{Z}_2$, $i = 0, 1, 2, \ldots$, by Conditions 2–3, we have that $f(x) = c + \sum_{i=0}^{\infty} d_i \mathsf{MASK}(x, 2^i)$; so the respective T-function $\bar{f} = f \bmod 2^k$ on k-bit words is $\bar{f}(x) = c + \sum_{i=0}^{k-1} d_i \,\mathsf{MASK}(x, 2^i)$. Note also that this T-function f is a "knapsack-like" function.

14.2.2.2 Techniques Based on p-Adic Derivations

As it was demonstrated above, the problem to determine whether a congruential generator (or, respectively, an explicit congruential generator) attains the longest period can be reduced to the problem of verifying whether given 1-Lipschitz transformations on $\mathbb{Z}_p$, for some prime p, are ergodic or, respectively, measure-preserving. In a number of practically interesting cases, these transformations are differentiable, so we can apply results of Sects. 10.1 and 10.3 to check measure-preservation and ergodicity. The method is not as general as techniques based on Mahler (or van der Put) expansion since the class of functions it can be applied to is smaller; however, in a number of cases it is easier to calculate derivatives of compositions of functions rather than their Mahler expansions. Moreover, in the case $p = 2$ (which is the most important case for computer applications) it turns out that when we limit our study to differentiable functions only, we actually do not make the class of measure-preserving functions under consideration smaller.

Proposition 14.8 *If a 1-Lipschitz function $f\colon \mathbb{Z}_2 \to \mathbb{Z}_2$ is measure-preserving, then it is uniformly differentiable modulo 2, its derivative modulo 2 is 1 everywhere on $\mathbb{Z}_2$, and $N_1(f) = 1$.*

Proof Indeed, by Corollary 9.9, f is measure-preserving if and only if $f(x) = c + x + 2v(x)$, where $c \in \mathbb{Z}_2$ is a constant and $v\colon \mathbb{Z}_2 \to \mathbb{Z}_2$ is a 1-Lipschitz transformation. Then $f(x + 2^k h) = c + x + 2^k h + 2v(x + 2^k h) \equiv f(x) + 2^k h \pmod{2^{k+1}}$ as $2v(x + 2^k h) \equiv 2v(x) \pmod{2^{k+1}}$ since v is 1-Lipschitz. Thus, f is uniformly differentiable modulo 2, $f_1'(x) \equiv 1 \pmod 2$, and $N_1(f) = 1$ by Definition 3.16. □

Thus, Proposition 14.8 implies that if a recurrence law of a congruential generator is *not* differentiable modulo 2 at some point of $\mathbb{Z}_2$, then the generator is *not* transitive modulo 2^n for all sufficiently large n (actually, it is not even bijective modulo 2^n for these n). This also means that the corresponding explicit congruential generator does not achieve maximum period length on n-bit words, for all sufficiently large n.

So, to determine whether the length of the shortest period of the explicit congruential generator with the law $y_i = f(i) \bmod 2^n$, $i = 1, 2, \ldots$, is 2^n, we just use Theorem 10.1 which states that whenever f is uniformly differentiable modulo 2, then f is measure-preserving if and only if f is bijective modulo $2^{N_1(f)}$ and $f_1'(x) \equiv 1 \pmod 2$ for all $x \in \mathbb{Z}/2^{N_1(f)}\mathbb{Z}$.

Below are examples of congruential generators modulo 2^n, both explicit and non-explicit, which illustrate the approach. Recall that (explicit) congruential generator modulo 2^n attains the longest period if and only if its law is (bijective) transitive modulo 2^n. The first portion of examples are results from [121] to which we give short proofs based on the p-adic ergodic theory:

Example 14.9 An explicit congruential generator, whose output function is one of the following, attains the longest possible period, of length 2^n ($n = 1, 2, 3, \ldots$):

$$x \mapsto (x + 2x^2) \bmod 2^n,$$

$$x \mapsto (x + (x^2 \text{ OR } 1)) \bmod 2^n,$$

$$x \mapsto (x \text{ XOR } (x^2 \text{ OR } 1)) \bmod 2^n$$

Indeed, all three these mappings are uniformly differentiable modulo 2 and $N_1 = 1$ for all of them. So in view of Corollary 10.4, it suffices to prove that all three mappings are bijective modulo 2, i.e., as mappings of the residue ring $\mathbb{Z}/2\mathbb{Z}$ modulo 2 onto itself (this can be checked by direct calculations) and that their derivatives modulo 2 vanish at no point of $\mathbb{Z}/2$. The latter also holds, since the derivatives are, respectively,

$$1 + 4x \equiv 1 \pmod 2,$$

$$1 + 2x \equiv 1 \pmod 2,$$

$$1 + 2x \equiv 1 \pmod 2,$$

Here we have used the "table of derivations of bitwise logical instructions," i.e., Example 9.20, to calculate respective derivatives: $(x^2 \text{ OR } 1)' = 2x \equiv 1 \pmod 2$, and $\frac{\partial_1 (x \text{ XOR } y)}{\partial_1 x} \equiv \frac{\partial_1 (x \text{ XOR } y)}{\partial_1 y} \equiv 1 \pmod 2$.

Further, the next is an example explicit polynomial generator whose output function is $P(x) = (a_0 + a_1 x + \cdots + a_d x^d) \bmod 2^n$, where $n > 1$ and $a_0, a_1, \ldots, a_d \in \mathbb{Z}$. By using Theorem 10.1 it is easy to show that the C-function $\sum_{i=0}^{\infty} c_i x^i$, where $c_i \in \mathbb{Z}_2$, is measure-preserving on $\mathbb{Z}_2$ if and only if the following tree conditions hold simultaneously:

$$c_2 + c_4 + c_6 + \cdots \equiv 0 \pmod 2; \tag{14.5}$$

$$c_3 + c_5 + a_7 + \cdots \equiv 0 \pmod 2; \tag{14.6}$$

$$c_1 \equiv 1 \pmod 2. \tag{14.7}$$

This immediately implies results of [121, 163] on the generator, i.e., the criterion when the generator attains the longest period, of length 2^n.

Moreover, the instruction XOR is uniformly differentiable modulo 2 as a 2-variate function on $\mathbb{Z}_2$ since $u \text{ XOR } v = u + v - 2(u \text{ AND } v)$ for all $u, v \in \mathbb{Z}_2$; see (9.64); therefore the map $z \mapsto f(z) = c_0 \bigcirc_1 c_1 z \bigcirc_2 \cdots \bigcirc_d c_d z^d$, where $c_i \in \mathbb{Z}_2$, $\bigcirc_i \in \{+, \text{XOR}\}$, $i = 0, 1, 2, \ldots, d$, is measure-preserving by Theorem 10.1. This means that respective explicit generator, where instructions XOR and + follow in arbitrary order, attains the period of length 2^n on n-bit words if and only if c_i satisfy conditions (14.5).

Though we already know from Example 9.21 that the congruential generator whose recurrence law is $f(x) = (x + (x^2 \mathsf{OR}\, 5)) \bmod 2^n$ attains the period of length 2^n, for all $n = 1, 2, \ldots$, there is another much shorter proof in this particular case if Theorem 10.9 is used rather that Theorem 9.4 as in general case examined in Example 9.21. Indeed, the function f is uniformly differentiable on $\mathbb{Z}_2$ and $N_2(f) = 3$. Therefore it is necessary and sufficient to check whether this function is transitive modulo 32, and direct calculations modulo 32 show that f is transitive.

By using this argument we immediately prove that congruential generator whose recurrence law is one of the following functions attains the period of length 2^n, for all $n = 1, 2, 3, \ldots$:

- $f(x) = (x + (5x^2 \mathsf{OR}\, 5)) \bmod 2^n$,
- $f(x) = (x + (5^{x^2} \mathsf{OR}\, 5)) \bmod 2^n$,
- $f(x) = (x + (5^{-x^2} \mathsf{OR}\, 5)) \bmod 2^n$,
- $f(x) = (x + (5^{x^2} \mathsf{AND}\, (-5))) \bmod 2^n$,
- $f(x) = (5x + (5^{x^5} \mathsf{AND}\, (-5))) \bmod 2^n$,
- $f(x) = (5x + (5^{5^x} \mathsf{AND}\, (-5))) \bmod 2^n$,
- $f(x) = (x + (((1 + 4(x^2 \mathsf{AND}\, (-5)))^{(1+2(x \,\mathsf{XOR}\, (-5)))^{-5}} \mathsf{OR}\, 5)) \bmod 2^n$

Finally, as Corollary 10.4 gives general criterion of ergodicity of $\mathcal{B}$-functions (whence, of $\mathcal{C}$-functions; whence, of polynomials, rational functions, etc.) which reads that *the $\mathcal{B}$-function is ergodic on $\mathbb{Z}_p$ if and only if it is transitive modulo p^2 when $p > 3$ or modulo p^3 when $p \leqslant 3$*, one immediately obtains a powerful method to design congruential generators of the longest period whose law of recursion is a $\mathcal{B}$-function or to determine whether such a generator achieves the longest period.

14.2.2.3 Techniques Based on Algebraic Normal Forms

In the case when we need to determine whether a given congruential generator with the recursion law $f \bmod 2^n$, where f is a 1-Lipschitz transformation of $\mathbb{Z}_2$, has the longest period, we may use a yet one more method, that of Theorem 9.4. Compared to the two methods we presented above, the method based on Theorem 9.4 can be applied only to relatively simple compositions of arithmetic and bitwise logical instructions; nonetheless some useful results can be obtained by this technique as well. Moreover, namely, this method can be expanded to the design of already-mentioned counter-dependent generators of the longest possible period which are *wreath products* of automata.

We recall the following definition:

Definition 14.10 (Wreath Product of Maps) Given a map $g \colon Y \to Y$ and a family of maps $\mathcal{F} = \{(f_y \colon X \to X) \colon y \in Y\}$, the *wreath product* (or, the *skew* *product*; or, the *skew shift*) of the family $\mathcal{F}$ by the map g is the map $g \wr \mathcal{F}\colon (y, x) \mapsto (g(y), f_y(x))$ of the Cartesian product $Y \times X$ into itself.

It is obvious that the wreath product is bijective if and only if both g and all f_y are bijective.

The following theorem presents the method how to "make a wreath of generators" to obtain a new one of the longest possible period. That new generator is counter-dependent (cf. Sect. 14.1). Actually the theorem can be judged as a generalization of Theorem 9.4.

Theorem 14.11 *Let $g_0, \dots, g_{m-1}$ be a finite sequence of T-functions; let this sequence satisfy the following conditions:*

- $g_j(x) \equiv x + c_j \pmod 2$ *for* $j = 0, 1, \dots, m-1$;
- $\sum_{j=0}^{m-1} c_j \equiv 1 \pmod 2$;
- *the sequence* $C = (c_{i \operatorname{mod} m} \operatorname{mod} 2)_{i=0}^{\infty}$ *is purely periodic, and m is the length of its shortest period;*
- $\delta_k(g_j(z)) \equiv \zeta_k + \varphi_k^j(\zeta_0, \dots, \zeta_{k-1}) \pmod 2$, $k = 1, 2, \dots$, *where* $\zeta_r = \delta_r(z)$, $r = 0, 1, 2, \dots$,
- *for each* $k = 1, 2, \dots$, *the total number of Boolean functions* $\varphi_k^j(\zeta_0, \dots, \zeta_{k-1})$ *that have odd weight is odd.*

Then the recurrence sequence $\mathcal{X} = (x_i)_{i=0}^{\infty}$ defined by the recursion $x_{i+1} = g_{i \operatorname{mod} m}(x_i)$ is a strictly uniformly distributed sequence over $\mathbb{Z}_2$: Namely, the sequence $\mathcal{X} \operatorname{mod} 2^k = (x_i \operatorname{mod} 2^k)_{i=0}^{\infty}$ is purely periodic for all $k = 1, 2, \dots$, the length of its shortest period is $m2^k$, and every element from $\mathbb{Z}/2^k\mathbb{Z}$ occurs at the period exactly m times. Moreover,

1. $m2^{s+1}$ *is the length of a period of the sequence* $\delta_s(\mathcal{X}) = (\delta_s(x_i))_{i=0}^{\infty}$, $s = 0, 1, \dots, k-1$; (i.e., the sequence $\delta_s(\mathcal{X})$ *may have* periods that are shorter than $m2^{s+1}$)
2. $\delta_s(x_{i+2^s m}) \equiv \delta_s(x_i) + 1 \pmod 2$ *for all* $s = 0, 1, \dots, k-1$, $i = 0, 1, 2, \dots$;
3. *for all* $t = 1, 2, \dots, k$, $r = 0, 1, 2, \dots$ *the sequence* $x_r \operatorname{mod} 2^t, x_{r+m} \operatorname{mod} 2^t, x_{r+2m} \operatorname{mod} 2^t, \dots$ *is a purely periodic sequence, the length of its shortest period is 2^t, and every element from $\mathbb{Z}/2^t\mathbb{Z}$ occurs at the period exactly once.*

Note that by Theorem 9.4, Conditions 1 and 4 of the theorem hold if and only if all g_j are measure-preserving. The structure of the wreath product from Theorem 14.11 is depicted by Fig. 14.2.

Proof of Theorem 14.11 As every g_j is bijective modulo 2^n, then from the definition of wreath product it follows that the wreath product $od \wr_{j=0}^{m-1} g_j \operatorname{mod} 2^k$ of the family $(g_j)_{j=0}^{m-1}$ by the odometer transformation $od\colon i \mapsto i+1$ on the residue ring $\mathbb{Z}/m\mathbb{Z}$ is a permutation on the direct product $\mathbb{Z}/m\mathbb{Z} \times \mathbb{Z}/2^k\mathbb{Z}$. Hence, the recurrence sequence $\mathcal{X} \operatorname{mod} 2^k$ defined by the recursion $x_{i+1} = g_{i \operatorname{mod} m}(x_i) \operatorname{mod} 2^k$ is purely periodic. With this in mind, we proceed with induction on k.

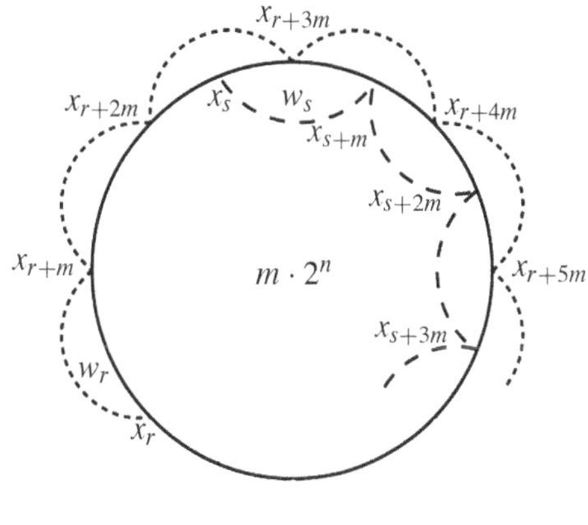

Fig. 14.2 Structure of the sequence produced by a counter-dependent generator

If $k = 1$, we have that $x_{i+1} = (c_{i \bmod m} + x_i) \bmod 2$. Thus,

$$x_i \equiv x_0 + \sum_{j=0}^{i-1} c_{j \bmod m} \pmod 2 \tag{14.8}$$

We need to calculate the length P of the shortest period of the sequence $b_i = (\sum_{j=0}^{i-1} c_{j \bmod m}) \bmod 2$, $i = 0, 1, 2 \ldots$, (i.e., $b_0 = 0, b_1 = c_0, b_2 = c_0 + c_1, \ldots$). Note that the sequence is purely periodic since $b_{2m} = 0$ by the second condition of the theorem.

From (14.8) it is easy to deduce that $0 \equiv \sum_{j=i}^{P+i-1} c_{j \bmod m} \pmod 2$ for all $i = 0, 1, 2, \ldots$; this means that the sequence $C = (c_{j \bmod m} \bmod 2)_{j=0}^{\infty}$ is a linear recurrence sequence over the two-element field $\mathbb{F}_2$ and that the characteristic polynomial of this sequence is $1 + y + \cdots + y^{P-1} \in \mathbb{F}_2[y]$. Since the latter polynomial is a factor of the polynomial $y^P - 1$, P is the length of some period of the sequence C. Therefore, as m is the length of the shortest period of the sequence C, m is a factor of P. However, $x_{i+m} \equiv x_0 + \sum_{j=0}^{m-1} c_{j \bmod m} \equiv x_0 + 1 \pmod 2$, and $x_{i+2m} \equiv x_0 + 2\sum_{j=0}^{m-1} c_{j \bmod m} \equiv x_0 \pmod 2$; thus, $P = 2m$. This proves the theorem if $k = 1$ since $\delta_0(\mathcal{X}) = \mathcal{X} \bmod 2$ in this case.

Let the theorem be true for $k = n$; let us prove it for $k = n + 1$. Denote $\delta_n(x_i) = \chi_n^i$, then

$$\chi_n^i \equiv \chi_n^0 + \sum_{j=0}^{i-1} \varphi_n^j(\chi_0^j, \ldots, \chi_{n-1}^j) \pmod 2 \tag{14.9}$$

Since by induction hypothesis the length of the shortest period of the sequence $\mathcal{X} \bmod 2^n$ is $m2^n$, and since all g_j are T-functions, the length of the shortest period of the sequence $\mathcal{X} \bmod 2^{n+1}$ must be a multiple of $m2^n$. Thus, the only alternative can take place; either the length of the shortest period of the sequence $\mathcal{X} \bmod 2^{n+1}$

is $m2^{n+1}$, or this length is $m2^n$. We will prove that $m2^n$ is not the case. To prove this, we only need to demonstrate that $\chi_n^{m2^n} \not\equiv \chi_n^0 \pmod 2$.

In view of induction hypothesis, the congruence

$$\chi_n^{m2^n+r} \equiv \chi_n^r + \sum_{j=r}^{m2^n-1+r} \varphi_n^j(\chi_0^j, \ldots, \chi_{n-1}^j) \equiv$$

$$\equiv \chi_n^r + \sum_{j=0}^{m-1} \sum_{z\in\mathbb{Z}/2^n\mathbb{Z}} \varphi_n^j(\delta_0(z), \ldots, \delta_{n-1}(z)) \equiv \chi_n^r + 1 \pmod 2 \tag{14.10}$$

holds for all $r = 0, 1, 2, \ldots$, since the total number of Boolean functions $\varphi_n^0, \varphi_n^1, \ldots \varphi_n^{m-1}$ that have odd weight is odd. This proves claim 2 of the theorem.

From (14.10) it follows that $\chi_n^{m2^n} \not\equiv \chi_n^0 \pmod 2$; therefore the length of the shortest period of the sequence $\mathcal{X} \operatorname{mod} 2^{n+1}$ is $m2^{n+1}$ in view of the note we have made above. Congruence (14.10) implies that $\chi_n^{m2^{n+1}+r} \equiv \chi_n^r \pmod 2$, thus proving claim 1 of the theorem.

By claim 3 of induction hypothesis, the following finite sequence of $m2^n$ numbers

$$x_r \operatorname{mod} 2^n, x_{r+m} \operatorname{mod} 2^n, x_{r+2m} \operatorname{mod} 2^n, \ldots, x_{r+(2^n-1)m} \operatorname{mod} 2^n \tag{14.11}$$

is a permutation of $0, 1, 2, \ldots, 2^n - 1$. Hence, all the numbers

$x_r, x_{r+m}, x_{r+2m}, \ldots, x_{r+(2^n-1)m}$ from (14.11) are pairwise distinct modulo 2^{n+1}. Thus, for each $z \in \{0, 1, \ldots, 2^n - 1\}$ among the numbers

$$x_r, x_{r+m}, x_{r+2m}, \ldots, x_{r+(2^{n+1}-1)m} \tag{14.12}$$

there exist exactly two numbers (say, x_u and x_v) such that $u \neq v$ and $z \equiv x_u \equiv x_v \pmod{2^n}$. Therefore $u \equiv v \pmod{m2^n}$ by claim 3 of induction hypothesis; whence $v = u + m2^n$. But then $x_u \not\equiv x_v \pmod{2^{n+1}}$ as $\delta_n(x_v) \equiv \delta_n(x_u) + 1 \pmod 2$ by claim 2 which we already have proved. Thus, all 2^{n+1} numbers from (14.12) are pairwise distinct modulo 2^{n+1}. This proves claim 3 of the theorem.

As we have already have proved that the sequence $\mathcal{X} \operatorname{mod} 2^{n+1}$ is purely periodic, and the length of its shortest period is $m2^{n+1}$, the following finite sequence

$$x_0 \operatorname{mod} 2^{n+1}, x_1 \operatorname{mod} 2^{n+1}, \ldots, x_{m2^{n+1}-1} \operatorname{mod} 2^{n+1}$$

constitute a period of the sequence $\mathcal{X} \operatorname{mod} 2^{n+1}$. But according to already proven claim 3, among these numbers there exist exactly m numbers which are congruent to z modulo 2^{n+1}, for every given $z \in \{0, 1, \ldots, 2^{n+1} - 1\}$. This completes proof of the theorem. □

Note 14.12 From Theorem 14.11 it follows that the function which generates the sequence $w_r = (x_{r+im})_{i=0}^{\infty}$ is a composition

$$g_{r-1}(\dots g_1(g_0(g_{m-1}(\dots g_{r+1}((g_r(\cdot)))\dots)))\dots),$$

cf., Fig. 14.2.

The following examples are wreath product by odometer $od\colon z \mapsto z+1$; that is, these are counter-dependent generators (cf. Sect. 14.1).

Example 14.13 (Wreath Product of Ergodic T-functions by od **mod** m**,** m **Odd)** Let m be odd, let all u_i, $i = 0, 1, \dots, m-1$ be ergodic T-functions, let $f_0 = u_0 \text{ XOR } 1$, $f_1 = u_1 \text{ XOR } 1$, $f_i = u_i$ for $i = 2, 3, \dots, m-1$. Then for the sequence $(x_i)_{i=0}^{\infty}$ over $\mathbb{Z}_2$ whose law of recursion is $x_{i+1} = f_{i \bmod m}(x_i)$ all conclusions of Theorem 14.11 hold.

It suffices to note that conditions of the example imply that $C = 0, 0, 1, 1, \dots, 1$ in conditions of Theorem 14.11; then the conclusion follows by Theorem 9.4.

Example 14.14 (Wreath Product of Ergodic T-functions by LFSR) Let $m = 2^N - 1$ be a Mersenne prime; $(\alpha_i)_{i=0}^{\infty}$ be generated by a N-bit linear feedback shift register (LFSR); let h_i be ergodic T-functions. Then $f_i = \alpha_i \text{ XOR } h_i$ satisfies Theorem 14.11.

Yet one more example:

Proposition 14.15 *Let $c_0, \dots, c_{m-1}$ be arbitrary sequence of length $m - 2^s$ of integers, i.e., $c_0, \dots, c_{m-1}$ need not be necessarily all distinct. Let $h_0, \dots, h_{m-1}$ be arbitrary sequence of T-functions. For $0 \leqslant j \leqslant m-1$ put $g_j(x) = c_j + x + 4h_j(x)$. The mappings g_j satisfy conditions of Theorem 14.11 if and only if $\sum_{j=0}^{m-1} c_j \equiv 1 \pmod 2$.*

Proof of Proposition 14.15 Denote $\delta_i(x) = \chi_i \in \{0, 1\}$, we claim that $\delta_0(c_i + x) \equiv \chi_0 + \delta_0(c_i) \pmod 2$; $\delta_j(c_i + x) \equiv \chi_j + \delta_0(c_i)\chi_0 \cdots \chi_{j-1} + \gamma_{ji}(\chi_0, \dots, \chi_{j-1}) \pmod 2 \quad (j > 0)$, where $\gamma_{ji}(\chi_0, \dots, \chi_{j-1})$ is a Boolean function the degree of whose ANF is less than j. We will prove this by induction on j.

Indeed, if the j-th coordinate function $\delta_j(c_i + x)$ contains the term $\chi_0 \cdots \chi_{j-1}$, then necessarily $\delta_0(c_i) = 1$ since otherwise there is no carry of a bit from 0-th position to higher-order positions. So we may assume that $c_i = 1 + 2c_i'$. Denote $\delta_j(c_i) = \xi_j$; then $\delta_1(1 + 2c_i' + x) \equiv \chi_0 + \chi_1 + \xi_1$. This is the base of induction.

We have $\delta_j(1 + 2c_i' + x) = \delta_j((1 + x) + 2c_i') \equiv \chi_j + \chi_0 \cdots \chi_{j-1} + \langle \text{carry} \rangle \pmod 2$. By induction hypothesis, $\langle \text{carry} \rangle \equiv (\chi_{j-1} + \chi_0 \cdots \chi_{j-2} + \gamma_{(j-1),\, i})\xi_{i-1} \pmod 2$; this prove our claim.

However, $\delta_j(4h_i(x))$ is a Boolean function in Boolean variables $\chi_0, \dots, \chi_{j-2}$ for $j \geqslant 2$, and is 0 otherwise; thus by a similar argument, $\delta_j(g_i(x)) \equiv \chi_j + \delta_0(c_i)\chi_0 \cdots \chi_{j-1} + \lambda_{ji}(\chi_0, \dots, \chi_{j-1}) \pmod 2$, where $\deg \lambda_{ji} < j$, $j = 1, 2, \dots$, and $\delta_0(g_i(x)) \equiv \chi_0 + \delta_0(c_i) \pmod 2$. □

14.3 Independence of Coordinate Sequences

Now we are going to look more attentively at the output sequences produced by generators whose laws of recursion are p-adic ergodic functions $f: \mathbb{Z}_p \to \mathbb{Z}_p$, especially for the case $p = 2$, i.e., when f is a T-function. As the ergodicity is equivalent to transitivity of $f \bmod p^k$ for every $k = 1, 2, \dots$ by Theorem 8.2, we immediately conclude that the length of the period of every k-th sequence $(\delta_k(f^j(x)))_{j=0}^{\infty}$ which is called the k-th *coordinate sequence*, of f and whose terms are k-th digits of the base-p expansions of $f^j(x)$, is exactly p^k. That is, low-order digits of produced sequence have shorter periods, and only the most significant digit of $(f^i(x) \bmod p^k)_{i=0}^{\infty}$ attains the longest period, that is, of the length p^{k+1}. Of course, this is an obvious deficiency of the sequence which must look as random as possible. We will suggest a simple remedy to cure this deficiency, but before we prove that nonetheless the coordinate sequences can be judged as "independent" in some precise meaning. We restrict ourselves to the case of T-functions though the result holds in general case, after a proper re-statement.

Now, let $f: \mathbb{Z}_2 \to \mathbb{Z}_2$ be an ergodic T-function, let $\mathcal{X}_k = (\delta_k(f^j(z_0)))_{j=0}^{\infty}$ be the k-th coordinate sequence of f. *The sequence $\mathcal{X}_k$ is a strictly periodic sequence over the two-element field* $\mathbb{F}_2 = \{0, 1\}$, *the length of its shortest period is exactly* 2^{k+1}; i.e., $\mathcal{X}_k$ is a periodic infinite binary word, $\mathcal{X}_k = (\chi_0 \dots \chi_{2^k-1})^{\infty}$. From the proof of Theorem 9.4 it immediately follows that the subword $\chi_0 \dots \chi_{2^k-1}$, the period of $\mathcal{X}_k$ has a specific property, namely,

The second half of the period of each k-th coordinate sequence $\mathcal{X}_k$, $(k = 0, 1, 2 \dots)$ is a bitwise negation of its first half

$$\delta_k(f^{j+2^k}(z_0)) \equiv \delta_k(f^j(z_0)) + 1 \pmod 2, \; j = 0, 1, 2, \dots \qquad (14.13)$$

In simple words, this means that the k-th coordinate sequence is as "random" as the first half of its period is. A natural question arises:

Given an infinite sequence of binary words $w_0, w_1, w_2, \dots$ such that the length of the k-th word is 2^k, does there exist an ergodic T-function $f: \mathbb{Z}_2 \to \mathbb{Z}_2$ such that every word w_k is the half-period of the k-th coordinate sequence of f, for every $k = 0, 1, 2, \dots$? And the answer is yes.

We will give a proof which is, to some extent, informal, to stress leading idea and not to overload notation. Namely, we fill a table $\mathcal{Z}$ having a countable infinite number of rows and columns in such a way that the first 2^j entries of the j-th column is the word w_j, the next 2^j bits of the column are negations of the first ones, the next 2^j bits are negations of these ones, etc. Then we shall prove that $\mathcal{Z}$ is dense in $\mathbb{Z}_2$, and then we shall define f on $\mathcal{Z}$ in such a way that makes f 1-Lipschitz and ergodic on $\mathcal{Z}$. This will imply the assertion. In what follows, we treat rows z of $\mathcal{Z}$ as 2-adic integers represented by respective 2-adic canonical expansions.

Claim: $\mathcal{Z} \bmod 2^k = \mathbb{Z}/2^k\mathbb{Z}$ for all $k = 1, 2, 3, \ldots$ That is, ring epimorphism $\bmod 2^k \colon z \mapsto z \bmod 2^k$ maps $\mathcal{Z}$ onto the residue ring $\mathbb{Z}/2^k\mathbb{Z}$.

Proof by induction on k: The claim obviously holds for $k = 1$. Assuming the claim holds for $k < m$, we prove that it holds for $k = m$.

Take $t \in \mathbb{Z}_2$. By induction hypothesis, given arbitrary $t \in \{0, 1, \ldots, 2^m - 1\}$, there exists $z_i \in \mathcal{Z}$ such that $z_i \equiv t \pmod{2^{m-1}}$. If $z_i \not\equiv t \pmod{2^m}$ then $\delta_{m-1}(z_i) \equiv \delta_{m-1}(t) + 1 \pmod 2$ and thus $\delta_{m-1}(z_{i+2^{m-1}}) \equiv \delta_{m-1}(t) \pmod 2$. However, $z_{i+2^{m-1}} \equiv z_i \pmod{2^{m-1}}$. Whence $z_{i+2^{m-1}} \equiv t \pmod{2^m}$; this proves the claim.

From the claim it immediately follows that the all rows of $\mathcal{Z}$ constitute a dense subset of $\mathbb{Z}_2$. A similar argument shows that for each $k \in \mathbb{N}$, *the sequence* $(z_i \bmod 2^k)_{i=0}^{\infty}$ *is purely periodic and has a period of length* 2^k, *and each* $t \in \{0, 1, \ldots, 2^k - 1\}$ *occurs at the period exactly once*. In particular, all the rows of $\mathcal{Z}$ are pairwise distinct 2-adic integers and, moreover, that

$$i \equiv i' \pmod{2^k} \text{ if and only if } z_i \equiv z_{i'} \pmod{2^k}$$

Now, if we put $f(z_i) = z_{i+1}$ for all $i = 0, 1, 2, \ldots$, then $|f(z_i) - f(z_{i'})|_2 = |z_{i+1} - z_{i'+1}|_2 = |(i+1) - (i'+1)|_2 = |i - i'|_2 = |z_i - z_{i'}|_2$. Hence, f is well defined on the rows of $\mathcal{Z}$ and is 1-Lipschitz with respect to the 2-adic metric. Thus, the continuation of f to the whole space $\mathbb{Z}_2$ is 1-Lipschitz as well as the rows constitute a dense subset in $\mathbb{Z}_2$, as we have shown. However, f is transitive modulo 2^k for each $k \in \mathbb{N}$ by the construction, so this continuation is ergodic in view of Theorem 8.2. □

And now the remedy, as promised. The idea of the remedy is depicted by Fig. 14.3. This remedy works for counter-depended generators, as depicted, but without loss of generality we may state and prove the result for standard (non-counter-dependent) generators, i.e., the ones where both state update T-function f_i and output T-function F_i do not depend on i.

Proposition 14.16 *Let* $\mathcal{Z}$ *be output sequence of the autonomous automaton* $\mathfrak{A} = \langle \mathbb{Z}/2^n\mathbb{Z}, \mathbb{Z}/2^n\mathbb{Z}, f \bmod 2^n, F, u_0 \rangle$ *where* f *and* h *are ergodic T-functions,* $F(x) = (h(\pi(x))) \bmod 2^n$, $x \in \{0, 1, \ldots, 2^n - 1\}$. *Then the length of the shortest period of the* j*-th coordinate sequence* $\mathcal{Z}_j = \delta_j(\mathcal{Z})$ *of the output sequence* $\mathcal{Z}$ *of the automaton* $\mathfrak{A}$ *is* 2^n *and linear complexity* $\mathrm{L}_{\mathbb{F}_2}(\delta_j(\mathcal{Z}))$ *over* $\mathbb{F}_2$ *exceeds* 2^{n-1}, *for all* $j = 0, 1, 2, \ldots, n-1$.

Recall that *linear complexity* of a periodic sequence over a finite field $\mathbb{F}_2$ is the minimum number of cells of the registry of the LFSR which reproduces

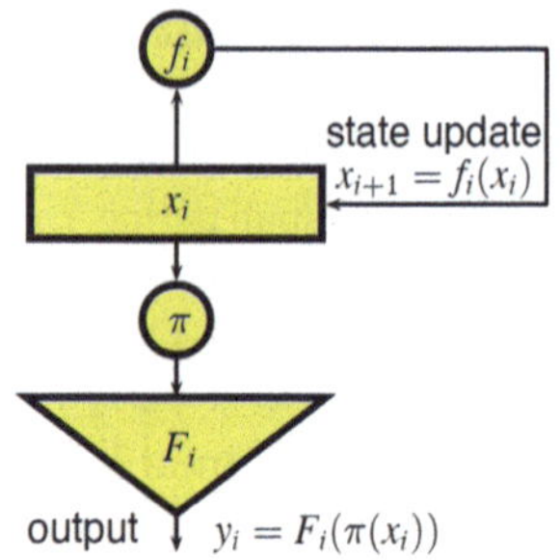

Fig. 14.3 To make the lengths of periods of all coordinate sequences of the output maximum possible, i.e., equal to 2^n, the bit-order reversing permutation π is added which sends higher-order bits of x_i to lower-order bit positions: $\delta_j(\pi(x_i)) = \delta_{n-j-1}(x_i)$, $j = 0, 1, 2 \ldots, n-1$

the sequence, that is, the smallest degree of characteristic polynomials of the sequence; see [138] or Definition 12.2. The characteristic polynomial which has the smallest degree of all characteristic polynomials of a sequence is called the *minimal polynomial of the sequence*.

We start with linear complexity:

Proposition 14.17 *Let $\mathcal{X}$ be a sequence from the from Theorem 14.11. Then linear complexity $\mathrm{L}_{\mathbb{F}_2}(\mathcal{X}_s)$ of the s-th coordinate sequence of $\mathcal{X}_s = \delta_s(\mathcal{X})$ over the field $\mathbb{F}_2$ exceeds 2^s, for all $s = 0, 1, 2, \ldots$.*

Proof As the sequence $\mathcal{X}_s$ has a period of length $m2^{s+1}$ by Theorem 14.11, the polynomial $x^{m2^{s+1}} - 1 = (x^m - 1)^{2^{s+1}}$ is a characteristic polynomial of the sequence $\mathcal{X}_s$. Thus, the minimal polynomial of the sequence $\mathcal{X}_j$ is a factor of $u(x) = x^{m2^{s+1}} - 1$. On the other hand, the minimal polynomial is not a factor of $w(x) = (x^m - 1)^{2^s} = x^{m2^s} - 1$ since otherwise the sequence $\mathcal{X}_j$ would have a period of length $m2^s$; however, the latter is impossible since the second half of the period of length $m2^{s+1}$ of this sequence is a bitwise negation of the first half by Theorem 14.11. □

The proof of this proposition implies that the second assertion of Proposition 14.16 follows from first assertion; so we need to prove only the latter one. As before, in the proofs which follow, we, when convenient, may consider infinite sequence over $\mathbb{F}_2$ as a one-side binary infinite word or as a 2-adic integer from $\mathbb{Z}_2$ considering the sequence as coefficients of the 2-adic canonical form of the integer. Therefore we may apply XOR, AND, and other operations to these sequences since these operations are well defined on $\mathbb{Z}_2$; see Sect. 9.5.1.

To prove the rest assertion of Proposition 14.16, we need an easy lemma.

Lemma 14.18 *Let $\mathcal{X} = (x_i)_{i=0}^{\infty}$ and $\mathcal{Y} = (y_i)_{i=0}^{\infty}$ be purely periodic sequences over the field $\mathbb{F}_2 = \mathbb{Z}/2\mathbb{Z}$; let lengths of their shortest periods be 2^u and 2^v, respectively; let $u > v$. Then the sequence $\mathcal{X}$ XOR $\mathcal{Y} = ((x_i + y_i) \bmod 2)_{i=0}^{\infty}$ is purely periodic, and the length of its shortest period is 2^u. If $x_{i+2^{u-1}} \equiv x_i + 1 \pmod 2$ for all $i = 0, 1, 2, \ldots$, and if $\mathcal{Y} \neq (0)_{i=0}^{\infty}$, then $\mathcal{X}$ AND $\mathcal{Y} = ((x_i y_i) \bmod 2)_{i=0}^{\infty}$ is a purely periodic sequence, and the length of its shortest period is 2^u.*

Proof of Lemma 14.18 The first assertion of the lemma is obvious. To prove the second one assume that P is the length of shortest period of the sequence $(x_i y_i)_{i=0}^{\infty}$.

Then $P = 2^s$ for suitable $s \leqslant u$. However, if $s < u$, then $x_{i+2^{u-1}} y_{i+2^{u-1}} \equiv x_i y_i \pmod 2$ for all $i = 0, 1, 2, \ldots$; thus $(x_i + 1) y_i \equiv x_i y_i \pmod 2$ and hence $y_i \equiv 0 \pmod 2$ for all $i = 0, 1, 2, \ldots$. A contradiction. □

Proof of Proposition 14.16 Denote via C_j the j-th coordinate sequence of the output sequence of the autonomous automaton $\mathfrak{B} = \langle \mathbb{Z}_2, \mathbb{Z}/2^n\mathbb{Z}, f, \operatorname{mod} 2^n, x_r \rangle$; let $\mathcal{E} = 111\ldots$. Then

$$\begin{aligned}
\mathcal{Z}_0 &= C_{n-1} \text{ XOR } \mathcal{E};\\
\mathcal{Z}_1 &= C_{n-2} \text{ XOR } C_{n-1} \text{ XOR } \mathcal{B};\\
\mathcal{Z}_j &= C_{n-j-1} \text{ XOR } C_{n-1} \text{ AND } \alpha_j(C_{n-2}, \ldots, C_{n-j}) \text{ XOR } \beta_j(C_{n-2}, \ldots, C_{n-j})
\end{aligned}$$

Here $j \geqslant 2$ and $\mathcal{B} = \beta_1\beta_1\beta_1\ldots$ is a constant binary sequence. Now by Lemma 14.18 to complete the proof of Proposition 14.16, it suffices to show that the sequence $\alpha_j(C_{n-2}, \ldots, C_{n-j})$, $2 \leqslant j \leq n-1$, is a non-zero binary sequence.

Consider the sequence $\mathcal{G}_j = 2^{n-2}C_{n-2} + \cdots + 2^{n-j}C_{n-j}$ over $\mathbb{Z}/2^{j-1}\mathbb{Z}$. The latter sequence is just an output sequence of the autonomous automaton $\mathfrak{G}_j = \langle \mathbb{Z}/2^{n-1}\mathbb{Z}, \mathbb{Z}/2^{j-1}\mathbb{Z}, f \operatorname{mod} 2^{n-1}, T_{n-j-1} \rangle$, where T_{n-j-1} is a truncation of the first $n-j$ low-order bits: $T_{n-j-1}(z) = \left\lfloor \frac{z}{2^{n-j}} \right\rfloor$. Thus, $\mathcal{G}_j$ is a purely periodic sequence, the length of its shortest period is 2^{n-1}, and each element from $\mathbb{Z}/2^{j-1}\mathbb{Z}$ occurs at the period the same number of times. However, α_j is a non-zero Boolean function; thus it takes value 1 at least at one $(j-1)$-bit word. Consequently, at least one term of the sequence $\alpha_j(C_{n-2}, \ldots, C_{n-j})$ is 1. □

A few remarks in conclusion of the chapter. The goal was to give some evidence that the p-adic theory offers a designer/programmer powerful tools to design pseudorandom number generators and to mathematically prove their main statistical properties, namely,

- Since any 1-Lipschitz map $\mathbb{Z}_p \to \mathbb{Z}_p$ (i.e., an automaton function) transforms input probability distribution into output probability distribution, the output distribution may be of any kind even if the input distribution were uniform (cf. Theorem 7.14); therefore to provide good statistical quality of the output sequence of PRNG, one should use ergodic state update function and measure-preserving output function.
- To obtain longer periods on shorter machine words, one may use a trick described at the end of Sect. 10.2.
- A number of methods on how to construct these functions out of standard computer instructions are given in Sect. 9.5.
- For preliminary "visual" estimates of the quality of output sequence, the methods from Sect. 12.1 usually work fine: The methods may detect bad generators at early stage, before deeper analysis of the algorithm is undertaken.
- For instance, the methods easily catch the PRNG which produce "the points falling mainly in the planes"; see Sect. 12.2. This deficiency of early PRNGs

stayed undetected for more than 20 years which led to incorrect simulations; cf., [148].

- To this end we note that in view of results of Chap. 13, especially of E. Lerner's Theorem 13.17, when polynomials of degree not less than 2 are involved in the composition of pseudorandom generators, the output sequences of these generators are looking "practically indistinguishable" from "truly random" ones, based on physical quantum effects.

Chapter 15
Application to Combinatorics

Abstract This short chapter serves as a yet one more example of how the p-adic ergodic theory is applied to a "non-dynamical" area of mathematics, the combinatorics, namely, to the theory of Latin squares. Latin squares are used in a number of applications, from games to experiment design, password distribution in public networks, etc. In this chapter, we explain how to contract large Latin squares and pairs of orthogonal Latin squares which can be implemented as straight line programs (SLP) even on very weak computers or in constraint environment.

15.1 Latin Squares: Basic Definitions

We recall that a *Latin square of order* P is a $P \times P$ matrix containing P distinct symbols (usually denoted by $0, 1, \dots, P-1$) such that each row and column of the matrix contains each symbol exactly once.

A circulant matrix serves a simple example of a Latin square. Here is a 6×6 one:

$$\begin{matrix} 0&1&2&3&4&5\\ 1&2&3&4&5&0\\ 2&3&4&5&0&1\\ 3&4&5&0&1&2\\ 4&5&0&1&2&3\\ 5&0&1&2&3&4 \end{matrix}$$

In algebra, Latin squares are also known as *binary quasigroups*, an algebraic systems on the set $\mathcal{A} = \{0, 1, \dots, P-1\}$ with the only binary operation $*$ defined by the Cayley table, which is a Latin square. Note that the operation $*$ is invertible with respect to each variable: given $a, b \in \mathcal{A}$, either equation $a * y = b$ or $x * a = b$ has a unique solution. However, the operation $*$ need not be associative. To put it in other words, a Latin square is a two-variate mapping $f : \mathcal{A}^2 \to \mathcal{A}$, where $\mathcal{A} = \{0, 1, \dots, P-1\}$, which is invertible (i.e., bijective) with respect to each variable.

V. Anashin, *Causality: The p-adic Theory*, STEAM-H: Science, Technology, Engineering, Agriculture, Mathematics & Health,
https://doi.org/10.1007/978-3-031-85818-5_15

Latin squares are used widely: For games (recall sudoku), and for more serious applications as, say, experimental design, private communication networks (for password distribution), in coding theory, in some cryptographic algorithms (under the name of multipermutations), etc., see monographs [114, 134]. However, known methods (e.g., the ones from the mentioned books) may not work efficiently in some practical cases.

There is no problem to construct one small Latin square: A real problem is to write a software that produces a number of large Latin squares; however, this is only a part of the problem. Another part of the problem is that in some constraint environments (e.g., in smart cards), it is impossible to store the whole matrix: Given two numbers $a, b \in \{0, 1, \ldots, P-1\}$, the software must calculate the (a, b)-th entry of the matrix on-the-fly. We apply the p-adic ergodic theory to give a solution to this problem. In the next sections we mainly follow [24] improving some flaws in statements and proofs from the latter monograph.

15.2 The p-Adic Ergodic Theory in Design of Latin Squares

According to Theorem 8.2, a bivariate 1-Lipschitz function $f\colon \mathbb{Z}_p^2 \to \mathbb{Z}_p$ is bijective modulo p^k for all $k \in \mathbb{N}$ with respect to either variable if and only if f is measure-preserving with respect to either variable. And Theorem 10.1 actually states that functions that are uniformly differentiable modulo p are bijective modulo p^k for all $k \in \mathbb{N}$ if and only if they are bijective modulo p^k for some (in most cases, small) k. Note that polynomials with integer coefficients are uniformly differentiable functions; whence, they are uniformly differentiable modulo p. Also, polynomials are easily programmable functions as they are just compositions of additions and multiplications. The idea is to use polynomials with integer coefficients to construct easily programmable Latin squares. Moreover, in the case $p = 2$ we can also add to numerical operations (addition and multiplication) some bitwise logical operators (e.g., XOR, AND, etc.) to construct measure-preserving functions; see Chap. 8. So the main tool we use to construct easily programmable Latin squares is the following Corollary 15.1 of Theorem 10.1. We say that a bivariate 1-Lipschitz function $f\colon \mathbb{Z}_p^2 \to \mathbb{Z}_p$ *is a Latin square modulo* p^k whenever the reduced mapping $f \bmod p^k\colon \mathbb{Z}/p^k\mathbb{Z} \times \mathbb{Z}/p^k\mathbb{Z} \to \mathbb{Z}/p^k\mathbb{Z}$ is a Latin square on $\mathcal{A} = \mathbb{Z}/p^k\mathbb{Z} = \{0, 1, \ldots, p^k - 1\}$.

Corollary 15.1 *A uniformly differentiable modulo p 1-Lipschitz function $f\colon \mathbb{Z}_p^2 \to \mathbb{Z}_p$ is a Latin square modulo p^k for all $k = 1, 2, \ldots$ whenever f is a Latin square modulo $p^{N_1(f)}$ and $\frac{\partial_1 f(\mathbf{u})}{\partial_1 x_i} \not\equiv 0 \pmod{p}$ for all $\mathbf{u} \in (\mathbb{Z}/p^{N_1(f)}\mathbb{Z})^2$, $i = 1, 2$* (cf., Definition 3.22). *Equivalent statement: if and only if f is bijective modulo $p^{N_1(f)+1}$ with respect to either variable.*

Proof Indeed, in view of Theorem 10.1, the function f is bijective modulo p^k with respect to either variable if and only if f is bijective modulo $p^{N_1(f)}$ with respect to either variable, and both $\frac{\partial_1 f(x,y)}{\partial_1 x}$ and $\frac{\partial_1 f(x,y)}{\partial_1 y}$ are 0 modulo p nowhere; these conditions are equivalent to the bijectivity modulo $p^{N_1(f)+1}$ of the function f with respect to either variable. □

Example 15.2 (Latin Square on 2^k Symbols) Given arbitrary 1-Lipschitz function $v(x, y)$ (e.g., arbitrary composition of numerical and bitwise logical operators; see Sect. 9.5.1) and arbitrary integer $\gamma \in \mathbb{Z}$, the map

$$f_{2^k}(x, y) = (x + y + \gamma + 2v(x, y)) \bmod 2^k$$

is a Latin square on 2^k symbols for all $k = 1, 2, \ldots$.

Indeed, $f(x, y) \equiv x+y+\gamma \pmod 2$ is a Latin square modulo 2, and $\frac{\partial_1 f(x,y)}{\partial_1 x} \equiv \frac{\partial_1 f(x,y)}{\partial_1 y} \equiv 1 \pmod 2$).

Example 15.3 (Latin Square on $2^k 3^\ell \cdots p^r$ Symbols) Given arbitrary polynomial $v(x, y)$ with integer coefficients and arbitrary rational integer $N > 2$ which has prime decomposition $N = 2^k 3^\ell \cdots p^r$, the function

$$f_N(x, y) = (x + y + 2 \times 3 \times \cdots \times p \times v(x, y)) \bmod N$$

is a Latin square on N symbols.

Indeed, as $f(x, y)$ is a polynomial with integer coefficients, it is compatible with all congruences of the ring $\mathbb{Z}$. So to verify whether f is a Latin square modulo $N = 2^k \cdot 3^\ell \cdots p^r$, in view of compatibility of f, it is sufficient to verify whether f is a Latin square modulo 2^k, modulo 3^ℓ,..., and modulo p^r. We use Corollary 15.1 for this purpose. The conclusion now follows, as f is a Latin square modulo $2, 3, \ldots, p$ and $\frac{\partial f(x,y)}{\partial x} \equiv \frac{\partial f(x,y)}{\partial y} \equiv 1 \pmod q$ for $q = 2, 3, \ldots, p$.

Now we expand the underlying idea of the latter example. Actually, given *arbitrary* Latin squares $f_2, f_3, f_5 \ldots, f_p$ on $2, 3, 5, \ldots, p$ symbols, respectively (some primes may absent), we can construct a bivariate polynomial $f(x, y)$ with integer coefficients so that $f(x, y) \equiv f_2(x, y) \pmod 2$, $f(x, y) \equiv f_3(x, y) \pmod 3$, $f(x, y) \equiv f_5(x, y) \pmod 5$, ..., $f(x, y) \equiv f_p(x, y) \pmod p$ and that $f(x, y) \bmod p^N$ is a Latin square on $N = 2^k \times 3^\ell \times \cdots \times p^r$ symbols, for all $k, \ell, \ldots, r \in N$.

Theorem 15.4 *Let $f_2(x, y)$, $f_3(x, y)$, $f_5(x, y) \ldots, f_p(x, y)$ be Latin squares on $2, 3, 5, \ldots, p$ symbols, respectively* (some primes may absent). *There exists a polynomial over $\mathbb{Z}$ $g(x, y) \in \mathbb{Z}[x, y]$ such that every function* $f(x, y) \bmod p^N$, *where $f_N(x, y) = (g(x, y)+2\times 3\times \cdots \times p \cdot v(x, y)) \bmod N$, is a Latin square on $N = 2^k \times 3^\ell \times \cdots \times p^r$ symbols, for all natural $k, \ell, \ldots, r$, and $f(x, y) \equiv f_q(x, y)$* $\pmod q$ *for all $p = 2, 3, 5, \ldots, p$. Here $v(x, y) \in \mathbb{Z}[x, y]$ is arbitrary polynomial over $\mathbb{Z}$.*

Sketch Proof The key idea of the proof exploits the fact that every bivariate function $f_q\colon (\mathbb{Z}/q\mathbb{Z})^2 \to \mathbb{Z}/q\mathbb{Z}$, q a prime, can be represented by a polynomial over $\mathbb{Z}$ such that a derivative of this polynomial with respect to either variable defines a prescribed mapping of $\mathbb{Z}/q\mathbb{Z}$ into $\mathbb{Z}/q\mathbb{Z}$; see interpolation formula (1.3).

That is, for every $f_q(x, y)$, $q \in \{2, 3, 5, \ldots, p\}$ (some primes may absent), we construct a polynomial $g_q(x, y)$ such that $f_q(x, y) = g_q(x, y)$ for all $(x, y) \in (\mathbb{Z}/q\mathbb{Z})^2$. Then we use a Chinese Reminder Theorem 1.1 to construct a polynomial $\tilde{g}(x, y) \in \mathbb{Z}[x, y]$ such that $\tilde{g}(x, y) \equiv g_q(x, y) \pmod q$ for all $q \in \{2, 3, 5, \ldots, p\}$ (respective primes are absent). Then, with the use of Proposition 1.15, by adding new terms of form

$$\frac{\bar{N}}{q}((x^q - x)u_q(x, y) + (y^q - y)v_q(x, y))$$

to the polynomial $\tilde{g}(x, y)$, where $\bar{N} = 2 \times 3 \times 5 \times \cdots \times p$ (respective primes in the product are absent), we construct a polynomial $g(x, y)$ such that $g(x, y) \equiv \tilde{g}(x, y) \pmod q$, $\frac{\partial g(x,y))}{\partial x} \not\equiv 0 \pmod q$ and $\frac{\partial g(x,y))}{\partial y} \not\equiv 0 \pmod q$ for all corresponding primes q and all $(x, y) \in \mathbb{Z}^2$. Now a combination of Theorem 10.1 with the equivalent form of Chinese Reminder Theorem 1.12 proves Theorem 15.4. □

Note that Theorem 15.4 not only states the existence of the polynomial $g(x, y)$ but gives also a method to construct it explicitly, as both Proposition 1.15 and Chinese Reminder Theorem 1.1 are constructive. For example, let us construct with the use of Theorem 15.4 a Latin square on 10^n symbols.

We skip the first step, the construction of respective interpolation polynomials for Latin squares on 2 and 5 symbols as this procedure is clear from interpolation formula (1.3); we assume that these Latin squares are already represented by bivariate polynomials $f_2(x, y) = x + y$ and $f_5(x, y) = 1 + 3x^3 + y$. The reader by direct calculations may verify that both $f_2(x, y)$ and $f_5(x, y)$ are Latin squares on $\mathbb{Z}/2\mathbb{Z}$ and $\mathbb{Z}/5\mathbb{Z}$, respectively.

We see that $f_5(x, y) \equiv f_2(x, y) + 1 \pmod 2$; so we only must "tweak" constant term (note that in general case we would use Chinese Reminder Theorem 1.1 here): we put $\tilde{g}(x, y) = 6 + 3x^3 + y$ as $6 \equiv 1 \pmod 5$ and $6 \equiv 0 \pmod 2$. Then, as $\frac{\partial \tilde{g}(x,y))}{\partial x} = 9x^2$ and $\frac{\partial \tilde{g}(x,y))}{\partial y} = 1$, we must find a tweak $g(x, y)$ for $\tilde{g}(x, y)$ to make the partial derivative $\frac{\partial g(x,y)}{\partial x}$ non-zero both modulo 2 and modulo 5 everywhere on $\mathbb{Z}/2\mathbb{Z}$ and $\mathbb{Z}/5\mathbb{Z}$, respectively; however, we must not change $\tilde{g}(x, y)$ neither modulo 2 nor modulo 5 by this tweak, that is, $\tilde{g}(x, y) \equiv g(x, y) \pmod 2$ and $\tilde{g}(x, y) \equiv g(x, y) \pmod 5$ must hold for all $(x, y) \in \mathbb{Z}^2$. Let us tweak $\tilde{g}(x, y)$ so that, say, $\frac{\partial \tilde{g}(x,y))}{\partial x} \equiv 1 \pmod 2$ everywhere on $\mathbb{Z}/2\mathbb{Z}$ and $\frac{\partial \tilde{g}(x,y))}{\partial x} \equiv 4 \pmod 5$ everywhere on $\mathbb{Z}/5\mathbb{Z}$. For this purpose, according to formula from Proposition 1.15, we put $g(x, y) = 6 + 3x^3 + y + 6(x^5 - x)(1 - x^2) + 5(x^2 - x)(x + 1) = y + 6 - 11x + 14x^3 + 6x^5 - 6x^7$. That is, $f(x, y) = g(x, y) + 10v(x, y)$, where $v(x, y)$ is arbitrary polynomial over $\mathbb{Z}$. The both $g(x, y) \bmod 10^n$ and $f(x, y) \bmod 10^n$ are Latin squares modulo 10^n, for every $n = 1, 2, 3, \ldots$.

15.3 The Orthogonality of Latin Squares

Now we explain how the p-adic ergodic theory may be of use to construct mutually orthogonal Latin squares. Recall that two $P \times P$ Latin squares are said to be *orthogonal* if when the squares are superimposed, each of the P^2 ordered pairs of symbols appears exactly once. Here is an example of a pair of orthogonal Latin squares on three symbols: The Latin squares

$$\begin{array}{ccc} 0 & 1 & 2 \\ 1 & 2 & 0 \\ 2 & 0 & 1 \end{array} \qquad \begin{array}{ccc} 0 & 1 & 2 \\ 2 & 0 & 1 \\ 1 & 2 & 0 \end{array}$$

are orthogonal since after we superimpose them, we get a square

$$\begin{array}{ccc} (0,0) & (1,1) & (2,2) \\ (1,2) & (2,0) & (0,1) \\ (2,1) & (0,2) & (1,0) \end{array}$$

where all pairs are different.

Mutually orthogonal Latin squares are used in experiment design to provide consistent testing of samples, as well as in cryptography (e.g., as block mixers for block ciphers, and as cipher combiners), etc.

For instance, consider three programs which must be tested on each of three platforms. To run all these nine tests, we must have a sort of schedule. We can make a schedule using the just mentioned example of orthogonal Latin squares of order 3. Namely, the table of pairs of superimposed squares gives us a schedule: Columns give us days of testing, the first number in a pair is a number of platform, and the second number is a number of program. As the pair $(0,2)$ occurs in the second column, this means that the program No 2 must be tested on the platform No 0 at the second day.

The reasoning to use methods of the p-adic ergodic theory to construct mutually orthogonal Latin squares are similar to that of preceding section: There is no problem to construct a pair of small mutually orthogonal Latin squares; a problem is to create a software that produces pairs of large Latin squares, and that does it in a somewhat "pseudorandom" way. Problems of this kind often arise in genetics, quantitative biology, chemistry, etc.; see [114]. Here we explain the respective method which also exploits Theorem 10.1.

We say that a pair bivariate of 1-Lipschitz functions $f, g\colon \mathbb{Z}_p^2 \to \mathbb{Z}_p$ *is mutually orthogonal Latin squares modulo* p^k whenever the reduced mappings $\bar{f} = f \bmod p^k\colon \mathbb{Z}/p^k\mathbb{Z} \times \mathbb{Z}/p^k\mathbb{Z} \to \mathbb{Z}/p^k\mathbb{Z}$ and $\bar{g} = g \bmod p^k\colon \mathbb{Z}/p^k\mathbb{Z} \times \mathbb{Z}/p^k\mathbb{Z} \to \mathbb{Z}/p^k\mathbb{Z}$ constitute a pair of orthogonal Latin squares on $\mathcal{A} = \mathbb{Z}/p^k\mathbb{Z} = \{0, 1, \dots, p^k - 1\}$.

The base of the method is the following assertion:

Corollary 15.5 of Theorem 10.1 15.5 *Let* $g, f \colon \mathbb{Z}_p^2 \to \mathbb{Z}_p$ *be 1-Lipschitz functions which are uniformly differentiable modulo p, and let f and g be Latin squares modulo* p^k *for all* $k = 1, 2, \ldots$ (cf. Corollary 15.1). *These Latin squares are orthogonal modulo* p^k *for all* $k = 1, 2, \ldots$ *if and only if the function* $F(x, y) = (f(x, y), g(x, y)) \colon \mathbb{Z}_p^2 \to \mathbb{Z}_p^2$ *preserves measure. This holds if and only if f and g are orthogonal modulo* p^k *for some* $k \geqslant \max\{N_1(f), N_1(g)\}$, *and*

$$\det \begin{pmatrix} \frac{\partial_1 f(x,y)}{\partial_1 x} & \frac{\partial_1 g(x,y)}{\partial_1 x} \\ \frac{\partial_1 f(x,y)}{\partial_1 y} & \frac{\partial_1 g(x,y)}{\partial_1 y} \end{pmatrix} \not\equiv 0 \pmod{p}$$

for all $(x, y) \in (\mathbb{Z}/p^{N_1(F)}\mathbb{Z})^2$

Proof From the definition of orthogonal Latin squares, it immediately follows that necessary and sufficient conditions for orthogonality modulo p^k is bijectivity of F modulo p^k; so the Latin squares are orthogonal modulo p^k for all $k = 1, 2, 3, \ldots$ if and only if F is measure-preserving; see Theorem 8.2. Now the conclusion follows from Theorem 10.1. □

Note that *Corollary 15.5 gives no method to construct pairs of orthogonal Latin squares on* 2^k *symbols*: From Corollaries 15.1 and 15.5 it immediately follows that for $p = 2$, no pair of functions f and g satisfy Corollary 15.5. Indeed, from Corollary 15.1 it follows that, as either of functions f and g is a Latin square modulo 2^k, every partial derivative modulo 2 of both f and g must be 1; however, this implies that a determinant from Corollary 15.5 is zero modulo 2.

However, for $p \neq 2$ Corollary 15.5 implies a method to construct large orthogonal Latin squares out of small orthogonal Latin squares. For instance, let $p = 3$, and let

$$f(x, y) \bmod 3 = \begin{pmatrix} 0 & 1 & 2 \\ 1 & 2 & 0 \\ 2 & 0 & 1 \end{pmatrix} \quad g(x, y) \bmod 3 = \begin{pmatrix} 0 & 1 & 2 \\ 2 & 0 & 1 \\ 1 & 2 & 0 \end{pmatrix}$$

be a pair of orthogonal Latin squares of order 3 each. Then, given arbitrary polynomials $v(x, y), w(x, y) \in \mathbb{Z}_3[x, y]$, the functions $f(x, y) = x + y + 3v(x, y)$ and $g(x, y) = 2x + y + 3w(x, y)$ define a pair of orthogonal Latin squares modulo 3^k, for all $k = 1, 2, \ldots$ since

$$\det \begin{pmatrix} 1 & 2 \\ 1 & 1 \end{pmatrix} \equiv 2 \pmod{3}$$

By the same reason, given a set $\mathcal{P}$ of *odd* primes and arbitrary polynomials $v(x, y), w(x, y) \in \mathbb{Z}[x, y]$, the following two Latin squares are orthogonal modulo P for every P such that all prime factors of P are in $\mathcal{P}$:

$$f(x, y) = x + y + \Pi \times v(x, y); \; g(x, y) = -x + y + \Pi \times w(x, y),$$

where $\Pi = \prod_{p \in \mathcal{P}} p$. In the same fashion, Theorem 15.4 can be re-stated for pairs of orthogonal Latin squares; and a method of constructing a pair of orthogonal Latin squares on P symbols for large composite odd P can be derived from this theorem as well. Namely, given N pairs of orthogonal Latin squares on $p_1, \ldots, p_N$ symbols (p_i a prime, $i = 1, 2, \ldots, N$), we construct N pairs of bivariate mappings $f_1(x, y), \ldots, f_N(x, y)$ and $g_1(x, y), \ldots, g_N(x, y)$ modulo $p_1, \ldots, p_N$, respectively, such that every pair $f_i(x, y)$ and $g_i(x, y)$ represents the i-th pair of given orthogonal Latin squares on p_i symbols. For this purpose we apply interpolation formula (1.3). Then, using Chinese Reminder Theorem 1.1, we construct two bivariate polynomials $f(x, y)$ and $g(x, y)$ over $\mathbb{Z}$ such that $f(x, y) \equiv f_{p_i}(x, y) \pmod{p_i}$ and $g(x, y) \equiv g_{p_i}(x, y) \pmod{p_i}$, for all $i = 1, 2, \ldots, N$. After that, with the use of method from Proposition 1.15, we tweak the polynomials $f(x, y)$ and $g(x, y)$ so that that their partial derivatives satisfy conditions of Corollaries 15.1 and 15.5, in a manner we describe in the proof of Theorem 15.4 and in the text thereafter. We leave details to the reader as an exercise.

Concluding the section, we stress that the techniques presented above can be in an obvious way used to construct Latin squares (and mutually orthogonal Latin squares) out of arbitrary uniformly differentiable (modulo p^k) functions, and *not necessarily* out of polynomials, e.g., out of rational functions, analytic functions, etc., if needed.

Chapter 16
Applications to Foundations of Quantum Theory

Abstract In this chapter we explain how the p-adic theory of causality developed in preceding chapters can be applied to the foundations of quantum theory in order to give rigorous mathematical reasoning to the so-called *superdeterministic* interpretations of quantum mechanics. We are aimed to show that superdeterministic models (e.g., the ones which are developing by S. Hossenfelder, T. Palmer, and others) are realistic models in the sense of the so-called model-dependent realism first coined in *The Grand Design* by S. Hawking and L. Mlodinow. We rigorously prove that since observational data (i.e., numerical values of physical quantities) are only rational numbers due to inevitably non-zero measurement errors, a conclusion whether Nature on the smallest of the scales is discrete or continuous, random/chaotic, or strictly deterministic solely depends on *experimenter's free choice of the metrics* (real or p-adic) he chooses to process the observational data; that is, *"free will" is consistent with superdeterminism.* Moreover, the assumption of *causality* at the smallest of the scales inevitably implies superdeterminism.

16.1 G. 't Hooft Causality Postulate; I. Volovich Postulates on Independence of Change of Number Field

A dictum (which is attributed to E. Rutherford) claims that if your experiment needs statistic, you ought to have done a better experiment. However, most (if not all) contemporary quantum experiments need huge statistic and intensive post-processing of experimental data. According to I. Volovich, since the data are rational numbers only, therefore to be consistent with reality (in any reasonable meaning what the "reality" is), physical laws, which are mathematically stated on the base of the data, must "withstand" the change of number field since rational numbers constitute a common subfield of the field of real numbers $\mathbb{R}$ and of all field $\mathbb{Q}_p$ of p-adic numbers, for all prime p. We are going to show in the chapter that if the laws reflect *causality* at the smallest of the scales (in any reasonable meaning what causality at the smallest of the scales is), then inevitably the laws as functions of numerical variables must be continuous with respect to the Archimedean real metric and to

V. Anashin, *Causality: The p-adic Theory*, STEAM-H: Science, Technology, Engineering, Agriculture, Mathematics & Health,
https://doi.org/10.1007/978-3-031-85818-5_16

non-Archimedean p-adic metrics; and moreover, to be causal, the functions must be 1-Lipschitz with respect to p-adic metric; that is, they must be automatic and strictly deterministic. We will derive numerous mathematical consequences which gave an evidence that superdeterministic interpretations of quantum mechanics based on *strict causality* (and thus on the non-Archimedean metric) are as "realistic" as the other interpretations which are based on Archimedean metric, in the sense of *model-dependent realism* [97]. Namely, the models (e.g., the ones being developed by S. Hossenfelder, T. Palmer, and others; cf., e.g., [40, 93, 94, 101, 158] and references therein) are as "realistic" as all other models, in this sense.

Moreover, the results obtained further may be judged as a yet one more contribution to J. A. Wheeler's *it from bit* doctrine (cf. [188]), since the results are information-theoretic by their nature, that is, the post-processing of observational data is information mining, therefore, any physical model constructed this way and expressed in mathematical form is information-theoretic. We mainly follow [22, 23] in this chapter.

The main goal of this chapter is to prove rigorously mathematical statements which show that experimenter's conclusions whether Nature on the smallest of the scales is discrete or continuous [89], random and chaotic, or strictly deterministic [94] solely depend on the experimenter's free choice of the metrics (real or p-adic) he chooses to process the measurement data which basically are rational numbers due to inevitably non-zero measurement errors.

There is a crucial difference between that free choice of the metric and, say, a Conway-Cochen strong free will theorem [42] and the likes. *In order to distinguish between these two faces of experimenter's freedom, within the chapter the two terms "free choice" and "free will" are used; and they are not interchangeable.* There is a resemblance between the meanings of the terms which are used in [94, 101, 158] (where the p-adic metric is briefly mentioned) and in the current chapter; however, here we develop mostly a mathematical model of post-processing of the measurement data rather than broader physical theories though we touch the theories a little bit when discussing possible interpretations of mathematical results of the chapter.

The author of this book, being a mathematician (though, an applied one) rather than a physicist, nevertheless has been motivated in his research of applications of the p-adic dynamics to quantum physics by the ideas of I. Volovich who, in collaboration with V. Vladimirov, in the 1980s has laid a cornerstone of contemporary *p-adic mathematical physics* [182], as well as by the ideas of G. 't Hooft who initiated development of the *cellular automaton interpretation of quantum mechanics* [100] which is based on a suggestion that on some basic level, there is no intrinsic randomness in Nature.

To that end, it is important to mention that the question of "true randomness" is currently not only of theoretical value but quite practical: In many applications there are used generators of "true random numbers," the physical devices which "extract" this "randomness" from physical processes which in modern generators are mostly of quantum nature, e.g., chaotic motion of particles in semiconductor superlattices; see, e.g., [189]. For pseudorandom generators, it is necessary to give mathematical evidence of their "pseudorandomness" before using them for practical purposes, being these computer simulation or password generations (cf. Chap. 14), but what about physical generators of "true random numbers?" As it was mentioned in Chap. 14, when a *program* generator of pseudorandom numbers is used in some sensitive applications, it is absolutely not sufficient just to show by a number of runs that it passes standard statistical tests; a strong mathematical evidence and proofs are needed. No such evidence exists for "physical" generators of "true random numbers," only the references to intrinsic randomness of Nature at quantum level. But this intrinsic randomness is still a belief rather than a fact. This is a yet one more motivator to develop theories which, under assumption of strict determinism, give a possibility to mathematically prove that the mathematical model of respective physical generator of "true random numbers" is so complicated and involved that it is computationally infeasible to exploit this model in order to make predictions on the output sequence produced by the generator.

Results of this chapter may also be regarded as a step toward interpretation of quantum mechanics in terms of *causal functions*, i.e., in terms of the mappings which can be performed by automata; but by automata which are the letter-to-letter transducers (or, sequential Mealy machines whose sets of states are not necessarily finite) (cf. Sect. 1.6), rather than by cellular automata on which G. 't Hooft's interpretation is based. These classes of automata are different from each other both from algorithmic and physical points of view. From the algorithmic point of view, the letter-to-letter transducers can be judged as less powerful computers compared to cellular automata which are the most powerful ones: Any algorithm (i.e., any general recursive function) can be implemented on a suitable cellular automaton since the class of cellular automata is Turing-complete, [43, 162], whereas algorithms which can be implemented by the transducers are necessarily primitive recursive functions and, moreover, constitute a small class of primitive recursive functions.

To begin with, we should state the problems we are addressing in this chapter in formal mathematical language, and we start the formalizing now. First we remark that I. V. Volovich, in his numerous papers, books, talks, etc., many times has stated the following postulates (further referred to as *Volovich postulates*) on which the p-adic mathematical physics is founded (cf., e.g., [51, 54] where the postulates are stated explicitly):

1. Only rational numbers can be observed; irrational numbers cannot.
2. Distances smaller than Planck length cannot be measured.
3. Fundamental physical laws should be invariant with respect to change of number field.

As by Ostrowski's Theorem 2.4, every non-trivial absolute value on the rational numbers $\mathbb{Q}$ is equivalent to either the usual real absolute value or a p-adic absolute value, to ensure the limits of convergent sequences over a field belong to the field; *the mentioned number fields must be the fields* $\mathbb{Q}_p$ *of p-adic numbers or the field* $\mathbb{R}$ *of real numbers* since these fields are the only completions of the field $\mathbb{Q}$ with respect to absolute values on $\mathbb{Q}$. Of course, the fields can be complete extensions of the fields $\mathbb{Q}_p$ and $\mathbb{R}$ like the fields of complex p-adic numbers $\mathbb{C}_p$ or a field of ordinary complex numbers $\mathbb{C}$, but $\mathbb{Q}_p$ and $\mathbb{R}$ are the only "smallest" fields which satisfy the third of Volovich postulates.

Further, G. 't Hooft in his book "The Cellular Automaton Interpretation of Quantum Mechanics" [100] states the following claim (which in what follows is referred to as the *'t Hooft causality postulate*) which is fundamental for the cellular automaton interpretation of quantum mechanics:

> It may well be that, at its most basic level, there is no randomness in Nature, no fundamentally statistical aspect to the laws of evolution. Everything, up to the most minute detail, is controlled by invariable laws. Every significant event in our universe takes place for a reason, it was caused by the action of physical law, not just by chance.

16.2 Causality Over Discrete Time: Non-archimedean 1-Lipschitzness

To be consistent with that causality postulate, a physical system must be *causal*; that is, the "effect," which is the reaction of the system to a "cause," i.e., to the impact the system has been exposed, must be a function of the "cause" and of the "state" of the system. But the very notion of causality is based on the notion of "time" which must be a totally ordered set since the "effect" cannot happen earlier than the "cause" whose function the "effect" is. It is impossible experimentally distinguish rational numbers from real numbers (cf. Volovich first postulate); therefore it is reasonable to assume that "time" is a totally ordered *countable* set. It is well-known that any totally ordered countable set T is order-isomorphic to a subset of $\mathbb{Q}$ (see, e.g., [79]) with respect to the natural order $\leqslant$ on $\mathbb{Q}$. Time T is called "continuous' if the ordering of elements in T is dense; i.e., given $t_i, t_2 \in T$ there exists $t_3 \in T$ such that $t_1 < t_3 < t_2$. Time T is called "discrete" if given any $t_i, t_2 \in T$ there are not more than finitely many $t_3 \in T$ such that $t_1 < t_3 < t_2$.

"Continuous" physical models are based on the assumption that any temporal/spatial interval can be divided into smaller intervals ad infinitum. The "discrete" models assume that spacetime should somehow be "quantized" at the smallest of scales; i.e., that there exist the smallest spatial/temporal intervals which cannot be divided into smaller ones, [89]. In the latter case, it would be reasonable to try to construct a mathematical theory assuming that total amount of that "indivisible" values can be increased ad infinitum. In the both cases, as well as in respective physical theories, the "infinity" just stands for a value which is extremely small (or extremely large) compared to the given value so that calculations involving the notion of infinity result in values which agree with respective measured values up to a small real number, the error. Therefore if theories of either type adequately describe physical reality at respective "ends of scale," the theories must "meet one another somewhere in the middle of the scale." We note that this conclusion is consistent with Hawking-Mlodinow *model-dependent realism* (cf. [97]).

The mathematical notion of causal function over discrete time is given by Definition 1.24. As explained in Sect. 1.6.2, it is reasonable to assume that the both sets $\mathcal{A}$ of "elementary causes" and $\mathcal{B}$ of "elementary effects" from that definition contain at least two elements and, moreover, that the sets are *finite*. Thus we state:

The Finiteness Assumption No physical objects or quantities are known which are proved to be infinite in some natural meaning. Infinity is a mathematical rather than a physical notion which is used in mathematical calculations in order to find good estimates of the values of physical quantities since the values can be measured with a non-zero error only. Therefore physical models should not assume "physical infinity."

From this finiteness assumption, it follows that the *causal functions are exactly the mappings which are produced by letter-to-letter transducers* (or, by *sequential machines*) (cf. Sect. 1.6), which transform input sequences of elementary causes into output sequences of elementary effects so that the second condition of Definition 1.24 (which is just a 1-Lipschitz condition with respect to the natural non-Archimedean metric on sequences) is satisfied.

In what follows, we mostly consider the case when both $\mathcal{A}$ and $\mathcal{B}$ are finite p-element sets $\mathbb{F}_p = \{0, 1, \ldots, p-1\}$ where p is a prime number. The latter restriction is more a technical one imposed for not to overload statements with not too important details which may obscure the very approach. This way we may assume that $\mathbb{F}_p$ is a finite p-element field and that the infinite sequences $(c_i)_{i \in \mathbb{N}_0}$ (where $c_i \in \mathbb{F}_p$) constitute the space $\mathbb{Z}_p$ of p-adic integers, under a natural one-to-one correspondence between the infinite sequences and canonical representations of p-adic integers (cf. Sect. 2.3). In the case when $p = m > 1$ is not a prime number, the sequences also may be put in one-to-one correspondence with the space $\mathbb{Z}_m$ of m-adic integers since the latter spaces are defined for all p, not necessarily for prime p only; see, e.g., [145].

16.3 Formalization of the Postulates

Physical models, speaking loosely, describe functions f which are "physical laws" that express dependencies of physical quantities on other physical quantities; so if time is one of that quantities, it is reasonable to assume causality, i.e., that the functions f are causal. Let us express more formally the conditions the functions f must meet in order to be consistent both with Volovich postulated and 't Hooft causality postulate.

16.3.1 Functions Which Are Consistent with the Postulates

In order to be consistent with Volovich postulates, the following conditions should be satisfied.

- As only rational numbers could be measured, the *functions f, i.e., the closed forms of physical laws which can be experimentally verified, must be mappings of rational numbers to rational numbers; i.e., the functions f must take rational values when values of variables are rational.*
- In order to study functions f when values of variables are "very large" or "very small" with respect to some reasonable metric, one has to expand the laws from the field of rational numbers $\mathbb{Q}$ to a bigger field which is complete with respect to that metric; therefore, that bigger field can only be the field of real numbers $\mathbb{R}$ and/or p-adic fields $\mathbb{Q}_p$ for primes $p = 2, 3, 5, 7, 11, \ldots$; but in order to be invariant with respect to the change of the number field, a *restriction to $\mathbb{Q}$ of any such expansion of f to a bigger field $\mathbb{F} \supset \mathbb{Q}$ must be the same irrespectively of what field $\mathbb{F}$ was used for the expansion, whether $\mathbb{F} = \mathbb{R}$ or $\mathbb{F} = \mathbb{Q}_p$.*

Next, to be consistent also with the 't Hooft causality postulate, the functions f should be causal; however, as it has been argued before, the "time" with respect to that the functions are causal must be order-isomorphic to a subset of $\mathbb{Q}$. But since according to Volovich postulates no temporal interval smaller than Planck's time cannot be measured, the temporal intervals can only be multiples of Planck's time; therefore the "time" over which the functions f are causal must be order-isomorphic to a subset of $\mathbb{Z}$. Thus to up to order isomorphism the timescale is either $\mathbb{N}_0 = \{0, 1, 2, \ldots\}$ or $\mathbb{Z} = \{0, \pm 1, \pm 2, \ldots\}$ depending on whether the "beginning of time" exists or does not exist. According to contemporary physical picture, it is reasonable to assume that the "beginning of time" exists; thus the timescale must be $\mathbb{N}_0$, up to order isomorphism. However, causal functions over the discrete time $\mathbb{N}_0$ can be treated as p-adic 1-Lipschitz functions which are defined on (and valuated in) p-adic integers $\mathbb{Z}_p$ rather than the whole fields $\mathbb{Q}_p$ (see the reasoning which follows Definition 1.24); thus, as the "common part" of $\mathbb{Z}_p$ and $\mathbb{R}$ (which we further denote via $\mathbb{Z}_p \cap \mathbb{Q}$) are p-adic rational numbers, the irreducible fractions whose denominators are co-prime to p, to be consistent with Volovich postulates of the causal functions, must take values from $\mathbb{Z}_p \cap \mathbb{Q}$ rather than from the whole $\mathbb{Q}$, and, moreover, the functions must be expandable to the whole field $\mathbb{R}$ since $\mathbb{Z}_p \cap \mathbb{Q}$ is a dense subset of $\mathbb{R}$. Finally we can specify formal properties the functions f must share in order to be consistent both with Volovich postulates and with the 't Hooft causality postulate, as follows:

Conditions 16.1 (Complete Consistency) *A* (univariate) *continuous real function* $f: \mathbb{R} \to \mathbb{R}$ *which is consistent with both Volovich postulates and the 't Hooft causality postulate must share the properties listed below.*

1. *For every prime p, the restriction $f|_{\mathbb{N}_0}$ must be a causal function over discrete time $\mathbb{N}_0$; i.e., the restriction $f|_{\mathbb{N}_0}$ must satisfy a p-adic Lipschitz condition with a constant 1. That is, for all $m, n \in \mathbb{N}_0$ there must hold the inequality*

$$d_p(f|_{\mathbb{N}_0}(m), f|_{\mathbb{N}_0}(n)) \leqslant d_p(m, n),$$

   where d_p is the p-adic metric (cf., Sect. 2.1).
2. *Since $\mathbb{N}_0$ is a dense subset in $\mathbb{Z}_p$, by (i) for every prime p, there exists a unique extension of $f|_{\mathbb{N}_0}$ to the function $f_p: \mathbb{Z}_p \to \mathbb{Z}_p$ which satisfies a Lipschitz condition with a constant 1 with respect to the p-adic metric d_p. Therefore, to be invariant with respect to the change of the field, the function $f: \mathbb{R} \to \mathbb{R}$ must act on the set $\mathbb{Z}_p \cap \mathbb{Q}$ of all p-adic rational integers exactly as the function f_p does; that is, for every prime p, the restriction $f|_{\mathbb{Z}_p \cap \mathbb{Q}}$ on rational p-adic integers $\mathbb{Z}_p \cap \mathbb{Q}$ must coincide with the restriction $f_p|_{\mathbb{Z}_p \cap \mathbb{Q}}$ on $\mathbb{Z}_p \cap \mathbb{Q}$:*

$$f|_{\mathbb{Z}_p \cap \mathbb{Q}}(r) = f_p|_{\mathbb{Z}_p \cap \mathbb{Q}}(r), \text{ for all } r \in \mathbb{Z}_p \cap \mathbb{Q}.$$

Note 16.2 The second of Conditions 16.1 immediately implies that $f(\mathbb{Z}_p \cap \mathbb{Q}) \subset \mathbb{Z}_p \cap \mathbb{Q}$ for *all* prime p; thus necessarily

$$f(\mathbb{Z}) \subset \bigcap_{p \text{ prime}} (\mathbb{Z}_p \cap \mathbb{Q}) = \mathbb{Z}.$$

The questions which immediately arise are whether there exist functions f which satisfy the conditions and, if such functions do exist, what are these functions. In Sect. 16.3.2 we show that *functions which meet Conditions 16.1 do exist and constitute a class of all polynomials over $\mathbb{Q}$ of a special type* (the class contains, e.g., all polynomials over $\mathbb{Z}$); see Theorem 16.7. Moreover, the functions turn out to be causal with respect to all finite alphabets, and not necessarily with respect to p-symbol alphabets for prime p's. This implies in particular that *the answer to the commonly asked question what p should be chosen by an experimenter in order to make the theory consistent with the observations is: The choice of p is absolutely free if causality, discreteness at Planck's scale, and invariance with respect to the change of the number field are assumed.*

We stress that the functions f which satisfy Conditions 16.1 are causal for all p-symbol alphabets, for all prime p, whence, *for all finite alphabets*. At our view, the latter property looks too restrictive (and somewhat non-physical) since Planck's scale includes a finite number of physical quantities (time, length, etc.) rather than infinite number of ones. Thus, it is reasonable to assume that Conditions 16.1 hold only for a finite set of primes; this will imply that the functions f are causal with respect to finite alphabets the prime power decompositions of the number of elements of whose include only powers of primes from that set. The study of that class of functions can be reduced to the case when the set contains only one prime, p. We show that if in that statement of Conditions 16.1 a prime p is fixed and "for every prime p" is replaced by "for the prime p," then *there exist functions f which satisfy Conditions 16.1 which are continuous real functions on $\mathbb{R}$ but which are not rational functions over $\mathbb{Z}$; i.e., are not of the form $u(x)/v(x)$ where $u(x), v(x) \in \mathbb{Z}[x]$ are polynomials with integer coefficients*; see Theorem 16.19. Note also that in that setting, $f(\mathbb{Z})$ is *not necessarily* a subset of $\mathbb{Z}$ but only a subset of $\mathbb{Z}_p \cap \mathbb{Q}$ (cf. Note 16.2).

16.3.2 Complete Description of the Consistent Functions

Causal functions over discrete time $\mathbb{N}_0 = \{0, 1, 2, \ldots\}$ from Sect. 16.2 are the maps f from the set $\mathcal{W}(\mathcal{A})$ of all one-side infinite words over a finite alphabet $\mathcal{A}$ to the set $\mathcal{W}(\mathcal{B})$ of all one-side infinite words over a finite alphabet $\mathcal{B}$ which are 1-Lipschitz with respect to standard non-Archimedean metric d on the words: $d(f(u), f(v)) \leqslant d(u, v)$; or, which is the same, if and only if words $f(\mathbf{u})$ and $f(v)$ have common prefix of length at least k whenever respective words u, v have common prefix of length k (cf. Sect. 1.6.2).

According to our conventions made above, $\mathcal{A} = \mathcal{B} = \{0, 1, \ldots, p-1\} = \mathbb{F}_p$; therefore the causal functions are automata functions of initial automata whose input/output alphabets are $\mathbb{F}_p$; and the automata functions constitute the class of all 1-Lipschitz maps $\mathbb{Z}_p \to \mathbb{Z}_p$ with respect to the p-adic metric d_p; see Sects. 1.6, 2.3, and Chap. 5.

Example 16.3 (Polynomials Over $\mathbb{Z}$ Are Automata Functions) A polynomial map $f: z \mapsto f(z)$ where $f(x) \in \mathbb{Z}[x]$ is an automaton function; f is never a finite automaton function if $\deg f \geqslant 2$ (see Proposition 13.24 and the text preceding it).

That is, as the set $\mathbb{Z}_p \cap \mathbb{Q}$ of all rational p-adic integers is dense both in $\mathbb{Z}_p$ with respect to the p-adic metric on $\mathbb{Z}_p$ for every p and with respect to usual real metric on $\mathbb{R}$, the map induced by a polynomial $f \in \mathbb{Z}[x]$ is well-defined both on $\mathbb{Z}_p$ for all p and on $\mathbb{R}$; i.e., the map $f: z \mapsto f(z)$, $(z \in \mathbb{Z}_p \cap \mathbb{Q})$, can be uniquely extended both to continuous maps $f: u \mapsto f(u)$, $(u \in \mathbb{Z}_p)$, for all p, and to a continuous map $f: y \mapsto f(y)$, $(y \in \mathbb{R})$: This is because any polynomial map is a composition of additions and multiplications, and these operations are well-defined and continuous both on all $\mathbb{Z}_p$ and on $\mathbb{R}$ with respect to corresponding metrics and agree on $\mathbb{Z}_p \cap \mathbb{Q}$.

16.3.2.1 Universally Causal Functions

The maps $f\colon \mathbb{N}_0 \to \mathbb{Z}$ defined by polynomials over $\mathbb{Z}$ are examples of functions which we call *universally causal*; these are the functions which, speaking loosely, are causal with respect to all finite alphabets $\mathcal{A}$ and $\mathcal{B}$ such that $\#\mathcal{A} = \#\mathcal{B} = r$ whatever $r \in \{2, 3, 4, \dots\}$ is taken. Here is a formal definition.

Definition 16.4 (Universally Causal Functions) A causal function $f\colon (a_i)_{i=0}^{\infty} \mapsto (\varphi_i^f(a_i))_{i=0}^{\infty}$ whose domain are all sequences $\mathbf{a} = (a_i)_{i=0}^{\infty}$ over $\mathcal{A}$ and co-domain is all sequences $\mathbf{b} = (b_i)_{i=0}^{\infty}$ over $\mathcal{B}$ (see Sect. 1.6.2) is called *universally causal* if $\#\mathcal{A} = \#\mathcal{B} = r > 1$ and there exist bijections $\alpha\colon \mathcal{A} \leftrightarrow \{0, 1, \dots, r-1\}$ and $\beta\colon \mathcal{B} \leftrightarrow \{0, 1, \dots, r-1\}$ such that the induced map $\tilde{f}\colon \mathbb{Z}_r \to \mathbb{Z}_r$ defined by $\tilde{f}\colon \sum_{i=0}^{\infty} \alpha(a_i) r^i \mapsto \sum_{i=0}^{\infty} \beta(\varphi_i^f(a_0, \dots, a_i)) r^i$ (see (1.9)), satisfies the following conditions:

1. $\tilde{f}(\mathbb{N}_0) \subset \mathbb{Z}$, where $\mathbb{N}_0$, the rational non-negative integers, are all r-adic integers whose canonical r-adic representations contain only finitely many non-zero terms; and $\mathbb{Z}$, the rational integers, are either non-negative rational integers or negative rational integers; the latter are all r-adic integers whose canonical r-adic representations contain only finitely many terms other than $(r-1)r^i$ (cf. Note 2.9).
2. $\tilde{f}(m) \equiv \tilde{f}(n) \pmod q$ once $m \equiv n \pmod q$, where $m, n, q \in \mathbb{N}_0, q > 1$.

The class of universally causal functions is much wider than functions defined by polynomials over $\mathbb{Z}$: Actually up to the bijections α, β, the universally causal functions constitute a class of the so-called *pseudo-polynomials*, [90], or *universal functions*, [32]; these are maps $g\colon \mathbb{N}_0 \to \mathbb{Z}$ which satisfy second condition from Definition 16.4.

Theorem 16.5 *A map $g\colon \mathbb{N}_0 \to \mathbb{Z}$ is a universal function if and only if g can be represented as*

$$g(z) = c_0 + \sum_{i=1}^{\infty} c_i \operatorname{lcm}\{1, 2, \dots, i\} \binom{z}{i} = c_0 + \sum_{i=1}^{\infty} c_i e^{\psi(i)} \binom{x}{i}, \tag{16.1}$$

where $c_i \in \mathbb{Z}$, $\operatorname{lcm}\{1, 2, \dots, i\}$ is the least common multiple of the numbers $1, 2, \dots, i$ and $\psi(i) = \sum_{q \leqslant i,\ q\ prime} \lfloor \log_q i \rfloor \ln q$ is the second Chebyshev function, $i = 1, 2, \dots$ (recall that $\psi(i) = i + o(i)$).

Proof See [90] or [32] for direct proof; or Corollary 5.13 from Theorem 5.12. □

Note that in literature often only the functions of the form (16.1) which are not polynomials are called pseudo-polynomials; but in the book we call pseudo-polynomials all functions of that form since these functions constitute a wider class than polynomial functions.

Note 16.6 The following is noteworthy.

- Even if all but a finite number of c_i in (16.1) are 0, i.e., if g is a polynomial, then g is not necessarily a polynomial with integer coefficients, though g is polynomial over $\mathbb{Q}$: For instance, put $c_4 = 1$ and put $c_i = 0$ for $i \neq 4$.
- If all but a finite number of c_i are 0, the function g is well-defined on $\mathbb{R}$; i.e., f can be uniquely expanded to a map $\mathbb{R} \to \mathbb{R}$ which is continuous with respect to the real metric.
- For every $p > 1$, the map g can be uniquely expanded to a 1-Lipschitz (thus, automatic) map $\mathbb{Z}_p \to \mathbb{Z}_p$ (cf. (5.25) from Theorem 5.12).

16.3.2.2 The Main Theorem on Complete Consistency

Therefore, polynomials of the form (16.1) satisfy Conditions 16.1 for every prime p. It turns out that the converse statement is also true. Note that a function which satisfies the conditions for all prime p must be universally causal, i.e., must be a pseudo-polynomial; but the only pseudo-polynomials which are well-defined on $\mathbb{R}$ are polynomials since if infinite number of c_i in (16.1) are non-zero, then the series diverges at, e.g., $z = -1$ as the common term at $z = -1$ is $(-1)^i c_i \operatorname{lcm}\{1, 2, \ldots, i\}$ and thus does not go to 0 as $i \to \infty$. However, this argument does not prove the converse claim since for instance if g is a pseudo-polynomial which is not a polynomial then the composition $g(z^2)$ is also a pseudo-polynomial but the map $z \mapsto g(z^2)$ is well-defined on $\mathbb{Z}$. Nonetheless the following theorem holds true:

Theorem 16.7 *A continuous function $f : \mathbb{R} \to \mathbb{R}$ satisfies Conditions 16.1 for all prime p if and only if it is a polynomial of the form* (16.1)*; i.e., when all but a finite number of c_i in* (16.1) *are zero.*

Proof By Theorem 16.5, every polynomial g over $\mathbb{Q}$ of the form (16.1) satisfies the first of Conditions 16.1 (cf. the second condition of Definition 16.4). Therefore g also satisfies the second of Conditions 16.1 since $g(\mathbb{Q}) \subset \mathbb{Q}$ as g is a polynomial over $\mathbb{Q}$.

To prove the converse claim, note that the map $u : \mathbb{Z}_p \to \mathbb{Z}_p$ is 1-Lipschitz if and only if $\Delta^i u(z)/i \in \mathbb{Z}_p$ for all $z \in \mathbb{Z}_p$ and all $i \in \mathbb{N}$ (cf. Proposition 5.4). Here Δ is the (forward) difference operator, i.e., $\Delta^1 u(z) = \Delta u(z) = u(z+1) - u(z)$, $\Delta^{i+1} u(z) = \Delta(\Delta^i u(z))$. Therefore

$$\frac{\Delta^i f(z)}{i} \in \mathbb{Z}, \text{ for all } z \in \mathbb{Z}, i \in \mathbb{N}. \tag{16.2}$$

Further, from the second of Conditions 16.1, it follows that $\frac{1}{h}(f(z+h) - f(z)) = q(z,h) \in \mathbb{Q}$ for all $z, h \in \mathbb{Q}$, $h \neq 0$, since $z, h \in \mathbb{Z}_p \cap \mathbb{Q}$ for all but not more than a finite number of primes p. But $f(z+h) = \sum_{i=0}^{\infty} \binom{h}{i} \Delta^i f(z)$ where the series converges p-adically for all but not more than a finite number of primes p as $\Delta^i f(z)$ tends p-adically to 0 by Theorem 5.12 (cf. the first of Conditions 16.1). Thus the series converges to some $q'(z,h) \in \mathbb{Q}$ by (ii) of Conditions 16.1, and therefore the series converges in $\mathbb{R}$ to that rational number $q'(z, h)$. We have

$$\frac{1}{h}(f(z+h) - f(z)) = \frac{1}{h}\sum_{i=1}^{\infty}\binom{h}{i}\Delta^i f(z) = \sum_{i=1}^{\infty}\binom{h-1}{i-1}\frac{\Delta^i f(z)}{i}, \tag{16.3}$$

where the series in the right hand part converges in $\mathbb{R}$ to a rational number $q(z, h) \in \mathbb{Q}$; therefore the absolute value $|\binom{h-1}{i-1}\frac{\Delta^i f(z)}{i}|$ must tend to 0 in $\mathbb{R}$ as $i \to \infty$. Represent

$$\binom{h-1}{i-1} = \left(\frac{h}{1} - 1\right)\left(\frac{h}{2} - 1\right)\cdots\left(\frac{h}{i-1} - 1\right)$$

From here it follows that

$$\left|\binom{h-1}{i-1}\right| \geqslant \left|1 - \frac{h}{i-1}\right|^{i-1} > 0 \text{ for all } i = 2, 3, \ldots;\ -1 < h < 0$$

As for $-1 < h < 0$ rational it holds

$$\lim_{i\to\infty}\left|1 - \frac{h}{i-1}\right|^{i-1} = \mathrm{e}^{-h} > 0,$$

from the convergency of the series in the right hand part of (16.3) it follows that necessarily $\lim_{i\to\infty}\frac{\Delta^i f(z)}{i} = 0$; therefore by (16.2), given $z \in \mathbb{Z}$ it must hold $\Delta^i f(z) = 0$ for all sufficiently large i. In particular, $\Delta^i f(0) = 0$ for all sufficiently large i. As $f(x) = \sum_{i=0}^{\infty}\binom{x}{i}\Delta^i f(0)$ and in view of the first of Conditions 16.1 the series in right hand part converges p-adically in $\mathbb{Z}_p$, we finally conclude, due to the uniqueness of representation of a continuous function $\mathbb{Z}_p \to \mathbb{Z}_p$ via Mahler series (cf. Sect. 4.2) that f is a polynomial over $\mathbb{Q}$, whence a polynomial of the form (16.1). □

Definition 16.8 (Totally Consistent Functions) Further in the book the functions described by Theorem 16.7 are called *totally consistent*; $C(\mathbb{R})$ denotes the class of all totally consistent functions.

Note 16.9 In view of Theorem 5.12, a similar result holds for multivariate maps $\mathbb{Z}_p^m \to \mathbb{Z}_p^n$ as well.

Note 16.10 From the proof of Theorem 16.7, it follows that relaxation of Conditions 16.1 to functions f whose domain contains a real interval rather than coincides with the whole $\mathbb{R}$ does not widen the class of functions. Moreover, one can relax the conditions assuming that they hold for infinite number of primes, and not necessarily for all primes.

16.4 Totally Consistent Functions: Implications for Quantum Theory

In this section we interpret the results on totally consistent functions and discuss possible implications of these for the foundations of quantum theory.

16.4.1 The Free Choice of Discreteness/Continuity

We stress once again that *in measurement of values of physical quantities the rational p-adic integers $\mathbb{Z}_p \cap \mathbb{Q}$ are indistinguishable from rational numbers $\mathbb{Q}$ since every real number can be approximated by a rational p-adic integer with any desirable accuracy*. Note also that the theorem by M. I. Chlodovsky yields that a continuous real-valued function on a real interval which does not contain integers can be uniformly approximated by polynomials over $\mathbb{Z}$ (cf. [39, 66]). Therefore, by Theorem 16.7, *any continuous real function on the real interval* $[\alpha, \beta]$ *where* $0 < \alpha < \beta < 1$ *can be uniformly approximated* (with respect to real metric) *by completely consistent functions*.

On the other hand, Theorems 5.12 and 16.7 imply that *any p-adic 1-Lipschitz function* $f : \mathbb{Z}_p \to \mathbb{Z}_p$ *can be uniformly approximated* (with respect to the p-adic metric) *by completely causal functions, whatever prime p is taken*. Indeed, by (5.25), the function f can be represented by Mahler expansion $f(z) = \sum_{i=0}^{\infty} b_i p^{\lfloor \log_p i \rfloor} \binom{z}{i}$ where $b_i \in \mathbb{Z}_p$. By Theorem 16.7, given $n \in \mathbb{N}$ we must find a polynomial $g(x) = c_0 + \sum_{i=1}^{\infty} c_i \mathrm{lcm}\{1, 2, \ldots, i\}\binom{x}{i}$ where all $c_i \in \mathbb{Z}$ such that $f(z) \equiv g(z) \pmod{p^n}$ for all $z \in \mathbb{Z}_p$. As $\mathrm{lcm}\{1, 2, \ldots, i\} = \prod_{(q)} q^{m_{i,q}}$ where $q^{m_{i,q}}$ is the largest power of a prime q which does not exceed i, then $m_{i,q} = \lfloor \log_q i \rfloor$ and therefore $c_i \mathrm{lcm}\{1, 2, \ldots, i\} = c_i p^{\lfloor \log_p i \rfloor} a_i$ where $a_i = \mathrm{lcm}\{1, 2, \ldots, i\}/p^{\lfloor \log_p i \rfloor}$ is in $\mathbb{Z}$ and is coprime to p. Hence, given $b_i \in \mathbb{Z}_p$, a congruence $b_i \equiv c_i a_i \pmod{p^n}$ has an integer solution $c_i \in \mathbb{Z}$. Put $c_i = 0$ for all i such that $\lfloor \log_p i \rfloor \geqslant n$, let $c_0 \in \mathbb{Z}$ be the least non-negative residue of $b_0 \in \mathbb{Z}_p$ modulo p^n. Then the so-defined polynomial g is the one we need.

All the preceding considerations may be judged as an evidence in favor of the following plausible statement which answers the question which entitles monograph [89]:

Interpretation 16.11 (Observer's Free Choice of Discreteness/Continuity) *Due to inevitable non-zero error in measurements of values of physical quantities, an observer's conclusion whether Nature on the smallest of the scales is discrete or continuous completely depends on the observer's free*

(continued)

choice of metric with respect to which the observer processes the measured numerical data. Moreover, the very "degree of the discreteness," the number p, is a subject of observer's free choice.

16.4.2 The Free Choice of Chaoticity/Predictability

The next important question which should be addressed is related to 't Hooft causality postulate and can be stated in the following form: Can an observer determine by numerical observational data whether Nature on the smallest of the scales is random or absolutely predictable? In what follows, the second term is understood as causality, i.e., that if an observer probes a system by exposing it to some impacts, reactions of the system coincide whenever impacts coincide, up to a precision of measurement equipment; that is, *same causes imply same effects*, so the behavior of the system is completely predictable since a cause results in a unique effect, within the measurement precision. The *randomness means that "same" causes may result in different effects*: Namely, causes whose numerical values are indistinguishable by measurement since the values coincide up to the precision of measurement equipment may result in effects which are distinguishable by measurement, i.e., differences of numerical values of respective effects exceed the measurement error. That is why we treat in what follows the randomness as chaos in a broad meaning since the definitive feature of chaos is its extreme sensitivity to negligible distortions/perturbations.

Recall that there are many non-equivalent mathematical notions of chaos; see, e.g., expository paper [127]. One of the most common of these definitions is due to R. L. Devaney [48, Definition 8.5] which reads:

Definition 16.12 (Devaney's Chaos on Metric Spaces) Let $F\colon \mathbb{X} \to \mathbb{X}$ be a continuous function on a metric space $\mathbb{X}$ equipped with a metric d. The function F is said to be *chaotic* if it satisfies the following three conditions:

1. *Sensitive dependence on initial conditions:* There is $\delta > 0$ such that, for any $x \in \mathbb{X}$ and any neighborhood $A \subset \mathbb{X}$ of x there exists $y \in A$ and $n \in \mathbb{N}_0$ such that $d(F^n(x), F^n(y)) > \delta$.
2. *Topological transitivity:* Given any pair of open subsets $U, V \subset \mathbb{X}$ there exists $k \in \mathbb{N}$ such that $F^k(U) \cap V \neq \emptyset$.
3. *Density of periodic points:* The set of all periodic points of F is dense in $\mathbb{X}$ (a point $x \in \mathbb{X}$ is called periodic if $F^k(x) = x$ for some $k \in \mathbb{N}$).

It is known that Conditions 1–3 of the definition are *not independent*: In [30] it is proved that sensitive dependence on the initial conditions is a redundant element in Devaney's definition, because it follows from topological transitivity and denseness of the periodic points; in [28] it is shown by construction of counter examples that neither topological transitivity nor denseness of the periodic points follow from the remaining two properties. In [122] it is proved that chaos according to Devaney's definition may exist on bounded but non-compact spaces without any non-periodic orbits. For bounded metric spaces, however, the following theorem is true:

Theorem 16.13 (C. Knudsen, [122]) *Let F, $\mathbb{X}$, d be as in Definition 16.12; let $\mathbb{X}$ be bounded; let $f = F|_{\mathbb{Y}}$ be a restriction of F to a dense subset $\mathbb{Y}$ of $\mathbb{X}$. Then*

- *$F\colon \mathbb{X} \to \mathbb{X}$ is topologically transitive if and only if $f\colon \mathbb{Y} \to \mathbb{Y}$ is topologically transitive;*
- *$F\colon \mathbb{X} \to \mathbb{X}$ exhibits sensitive dependence on the initial conditions if and only if $f\colon \mathbb{Y} \to \mathbb{Y}$ exhibits sensitive dependence on the initial conditions.*

The following definition of chaos on a bounded metric space is due to Knudsen.

Definition 16.14 (Knudsen's Chaos on Bounded Metric Spaces, [122]) Let F be a continuous transformation of a bounded metric space $\mathbb{X}$. If F has a dense orbit in $\mathbb{X}$ and F exhibits sensitive dependence on the initial conditions, then F is said to be *chaotic*.

We stress that to the best of our knowledge *all definitions of chaos on metric spaces contain sensitive dependence on initial conditions as an inherent property*; other conditions vary but that one is always present, [125]. For other various types of chaos on compact metric spaces $\mathbb{X}$, see [127]. We only mention that a continuous map $F\colon \mathbb{X} \to \mathbb{X}$ is called *topologically chaotic* if topological entropy of F is positive. The topological chaos implies *Li-Yorke chaos*, yet one more widely known type of chaos for the definition of which the reader s referred to [127]. Also, positive topological entropy implies *distributional chaos* of type DC2, [53]. Chaos can also be defined in terms of measure-preserving transformations of measure spaces rather than of metric spaces; see [53].

The "chaos-like" behavior may also be expressed in terms of "blending capability" which we illustrate by an example taken from [91]. If in cocktail shaker of volume 1, there are 10 shares of gin and 90 shares of vermouth, then after *ergodic* shaking in every volume V of the shaker, there will be 10 shares of gin and 90 shares of vermouth *on average*, whereas after *strong mixing* shaking in every V, there will be approximately 10 shares of gin and 90 shares of vermouth, whereas after *weak mixing* shaking *in every V with the exception of some rare instants* there will be 10 shares of gin and 90 shares of vermouth. Formally, a measure-preserving transformation F is by the definition *strongly mixing* if $\lim_{n\to\infty} \mu(F^{-n}(A) \cap B)) = \mu(A)\mu(B)$ for every μ-measurable subsets A, B. Thus, if μ is a probability measure, the strongly mixing transformation after being applied sufficiently large number of times makes any two "events" A, B "independent" in the probabilistic meaning. The map F is called *weakly mixing* if for any two measurable sets A, B there exists a sequence $n_k \to \infty$ over $\mathbb{N}_0$ such that $\mu(F^{-n_k}(A) \cap B) \to \mu(A)\mu(B)$ as $k \to \infty$. If $n_k = k$, the weak mixing is called (strong) mixing. Weak mixing imply ergodicity but is a stronger condition than ergodicity: The map F is weakly mixing if and only if the map $(x, y) \mapsto (F(x), F(y))$ of $\mathbb{X}\times\mathbb{X}$ into $\mathbb{X}\times\mathbb{X}$ is ergodic.

For the convenience of referencing, we state some of important results of Chap. 8 as the following proposition:

Proposition 16.15 *A function $F\colon \mathbb{Z}_p^n \to \mathbb{Z}_p^n$ is measure-preserving and 1-Lipschitz if and only if it is an isometry of $\mathbb{Z}_p^n$ onto itself. A measure-preserving 1-Lipschitz function F is ergodic if and only if it has a dense orbit; moreover, all orbits of ergodic 1-Lipschitz function $F\colon \mathbb{Z}_p^n \to \mathbb{Z}_p^n$ are dense.*

As the space $\mathbb{Z}_p^n$ is a probability space and a metric (thus, topological) space, one can define for a continuous function $\mathbb{Z}_p^n \to \mathbb{Z}_p^n$ a *metric entropy* (related to the probability) and a *topological entropy* (related to the topology). In general, given F, these entropies may differ. But *for 1-Lipschitz functions F the both entropies coincide and are 0*. Indeed, it is known that if $G\colon \mathbb{X} \to \mathbb{X}$ is a 1-Lipschitz map of a compact metric space $\mathbb{X}$ into itself, then topological entropy of G is 0 (cf., e.g., [52, Exercise 6.3]). Yet variational principle for the topological entropy yields that the topological entropy of a continuous transformation G of a compact metric space $\mathbb{X}$ is a supremum of all metric entropies of G with respect to G-invariant measures on $\mathbb{X}$ (cf. [52, Theorem 6.8.1]); thus we are done. Also, from Proposition 16.15 it follows that given a 1-Lipschitz ergodic map $F\colon \mathbb{Z}_p \to \mathbb{Z}_p$, the map $F \times F\colon (x, y) \mapsto (F(x), F(y))$ of $\mathbb{Z}_p^2$ to $\mathbb{Z}_p^2$ is never ergodic since an orbit which starts from $(z, z) \in \mathbb{Z}_p^2$ is never dense in $\mathbb{Z}_p^2$ (cf. Theorem 8.2), so F *is never weakly mixing*.

We summarize:

- A function $F\colon \mathbb{Z}_p^n \to \mathbb{Z}_p^n$ is measure-preserving and 1-Lipschitz if and only if it is an isometry.
- A 1-Lipschitz function $F\colon \mathbb{Z}_p^n \to \mathbb{Z}_p^n$ is an isometry if and only if it is bijective; i.e., if and only if the respective automaton is *reversible* (cf. Sect. 13.1.2). Recall that an automaton $\mathfrak{A}$ whose automaton functions is F is called *reversible* if there exists an automaton $\mathfrak{B}$ whose automaton function is G and such that $G = F^{-1}$, i.e., the composition $G(F)$ is an identity map $\mathbb{Z}_p^n \to \mathbb{Z}_p^n$. Therefore that automata reversibility may be treated as *time-reversibility* of respective causal function F since G may be treated as a causal function which acts in the opposite direction of the discrete time (cf. Sect. 1.6.2)
- All 1-Lipschitz functions $\mathbb{Z}_p^n \to \mathbb{Z}_p^n$ have zero topological entropy; thus, zero metric entropy.
- All 1-Lipschitz ergodic maps $F\colon \mathbb{Z}_p^n \to \mathbb{Z}_p^n$ are uniquely ergodic.
- None of the 1-Lipschitz ergodic maps $F\colon \mathbb{Z}_p^n \to \mathbb{Z}_p^n$ is weakly mixing.
- Every orbit of every 1-Lipschitz ergodic map $F\colon \mathbb{Z}_p^n \to \mathbb{Z}_p^n$ is dense.

1-Lipschitz measure-preserving map can be neither strongly nor weakly mixing; only the ergodicity is possible.

Finalizing the considerations of chaos, we claim:

1-Lipschitz functions $F: \mathbb{Z}_p^n \to \mathbb{Z}_p^n$ are deterministic and non-chaotic with respect to chaos of any type.

Indeed, due to the 1-Lipschizness, these functions exhibit no sensitive dependence on initial conditions; the topological entropy of these functions is zero; whence any metric entropy of the functions is zero. Moreover, as measure-theoretical chaos is defined only for measure-preserving maps, and as a 1-Lipschitz map $F: \mathbb{Z}_p^n \to \mathbb{Z}_p^n$ preserves the Haar probability measure μ_p if and only if it is an isometry, it can be easily shown that F is chaotic with respect to no type of measure-theoretic chaos defined in [53]. One may say therefore that totally consistent functions (see Definition 16.4) are the best candidates to be called *superdeterministic*. The latter term may have not exactly the same meaning in which it is used in superdeterministic interpretations of quantum mechanics; see, e.g., [101], but as a mathematical notion, it stresses the "extreme non-chaoticity" of the functions.

However, on the other hand, the totally consistent functions are like *Ianus Bifrons*:

Being strictly deterministic with respect to a p-adic metric for every p, *the totally consistent functions can nevertheless be chaotic when treated as real functions on a real interval.*

Let us consider an illustrative example. A well-known "canonical" example of real chaotic maps, the logistic map $L(x) = 4x(1 - x)$ maps real closed interval $[0, 1]$ to $[0, 1]$; the map L has positive topological entropy $\log 2$; see, e.g., [50].

On the other hand, L is a polynomial with integer coefficients whence it is a totally consistent function; thus its entropy as a p-adic 1-Lipschitz map $z \mapsto 4z(1 - z)$ (both topological and metric with respect to Haar probability measure) is 0, and $L: \mathbb{Z}_p \to \mathbb{Z}_p$ is not sensitive to initial conditions. The map L on $\mathbb{Z}_2$ is not measure-preserving with respect to the Haar probability measure on $\mathbb{Z}_2$; it has the only point of attraction (namely, 0) to which all orbits converge; thus L is not topologically transitive on $\mathbb{Z}_2$.

However, the map L is *ergodic on the 5-adic sphere* $\mathbf{S}_{1/25}(7)$ *of radius 1/25 centered at 7 since 7 is a fixed point of* L *and the derivative* $L'(7) \equiv 23 \pmod{25}$ *is a generator of the group of units modulo 25*; see Theorem 11.10. Namely, by (11.10), the sphere $\mathbf{S}_{1/25}(7)$ is a disjoint union of four 5-adic balls:

$$\mathbf{S}_{1/25}(7) = \mathbf{B}_{1/125}(32) \cup \mathbf{B}_{1/125}(57) \cup \mathbf{B}_{1/125}(82) \cup \mathbf{B}_{1/125}(107)$$

The sphere is invariant under action of L on $\mathbb{Z}_5$ and is measurable with respect to the Haar probability measure on $\mathbb{Z}_5$. Thus, the probability measure on $\mathbb{Z}_5$ induces a probability measure on $\mathbf{S}_{1/25}(7)$ with respect to which the action of L on that sphere is measure-preserving and ergodic. The set of all rational 5-adic numbers from $\mathbf{S}_{1/25}(7)$ which lie in the real closed interval $[0, 1]$ is dense in $[0, 1]$ with respect to real metric. Therefore, *as 5-adic rational integers are indistinguishable from real numbers by measurement due to inevitable non-zero error, the map L should be judged as measure-preserving and ergodic.*

Now consider the map L on the 5-adic sphere $\mathbf{S}_{1/25}(0)$. The sphere is a disjoint union of four balls:

$$\mathbf{S}_{1/25}(0) = \mathbf{B}_{1/125}(25) \cup \mathbf{B}_{1/125}(50) \cup \mathbf{B}_{1/125}(75) \cup \mathbf{B}_{1/125}(100)$$

The sphere is invariant under action of L on $\mathbb{Z}_5$ and is measurable with respect to the probability measure on $\mathbb{Z}_5$, *the map L on $\mathbf{S}_{1/25}(0)$ is measure-preserving with respect to the induced probability measure, but is not ergodic* by Theorem 11.10 since $L'(0) = 4$ is not a generator of the group of units modulo 25. The set $\mathbf{S}_{1/25}(0) \cap [0, 1]$ is dense in $[0, 1]$ with respect to real metric; therefore by the reasoning similar to that as above, *the map L should be judged as measure-preserving but not ergodic.*

Finally, *the map $L\colon \mathbb{Z}_p \to \mathbb{Z}_p$ is measure-preserving for no p* as L is not bijective modulo p; cf. Corollary 10.4 and the texts which follows it. But the set $\mathbb{Z}_p \cap \mathbb{Q} \cap [0, 1]$ is also dense both in $\mathbb{Z}_p$ and in $[0, 1]$ with respect to the p-adic and to real metrics accordingly, so the values the map L takes on $\mathbb{Z}_p \cap \mathbb{Q} \cap [0, 1]$ define a unique map both on $[0, 1]$ and on $\mathbb{Z}_p$ for every p. However, an observer's measurement data may only be rational numbers due to the inevitable non-zero measurement error.

To put it in other words, numbers from $\mathbb{Z}_p \cap \mathbb{Q} \cap [0, 1]$ (as well as from $\mathbf{S}_{1/25}(7) \cap \mathbb{Q} \cap [0, 1]$, or from $\mathbf{S}_{1/25}(0) \cap \mathbb{Q} \cap [0, 1]$) are indistinguishable from numbers in $\mathbb{Q} \cap [0, 1]$ and from numbers in $[0, 1]$ by measurements due to non-zero measurement error; but *the choice of metric (and of dense subset) with respect to which the measurement data are processed is crucial for the observer's conclusion whether the obtained data are completely random or satisfy a strictly deterministic law.*

All these facts can be judged as an evidence in favor of the following assertion.

Interpretation 16.16 (Observer's Free Choice of Determinism/Randomness) *Due to the inevitable non-zero error in measurements of values of physical quantities, an observer's conclusion whether Nature on the smallest of the scales is superdeterministic or random completely depends on the observer's free choice of metric with respect to which the observer processes the measured numerical data.*

16.5 p-Consistency: Main Properties and Their Interpretations

In view of the *finiteness assumption* (cf., Sect. 16.2), Conditions 16.1 may look too restrictive since by physical reasons the number of "elementary causes" and "elementary effects" cannot be arbitrarily large; so this number does not exceed some p. This is a motivation to introduce the following class of causal functions:

Definition 16.17 Given a prime p, denote via $C_p(\mathbb{R})$ the class of all continuous (with respect to usual real metric on $\mathbb{R}$) functions $\breve{f}\colon \mathbb{R} \to \mathbb{R}$ such that:

1. $\breve{f}(\mathbb{Z}_p \cap \mathbb{Q}) \subset \mathbb{Z}_p \cap \mathbb{Q}$;
2. There exists a p-adic 1-Lipschitz function $f\colon \mathbb{Z}_p \to \mathbb{Z}_p$ such that
   - $f(\mathbb{Z}_p \cap \mathbb{Q}) \subset \mathbb{Z}_p \cap \mathbb{Q}$
   - $f(z) = \breve{f}(z)$ for every $z \in \mathbb{Z}_p \cap \mathbb{Q}$

We call the functions from $C_p(\mathbb{R})$ the (univariate) *p-consistent functions*

The multivariate p-consistent functions $\mathbb{R}^m \to \mathbb{R}^n$ can be defined similarly. In what follows, we are taking liberty to use notations f and $\breve{f}$ for $C_p(\mathbb{R})$-functions interchangeably since this does not cause ambiguities.

The functions from $C_p(\mathbb{R})$ "reside in two worlds simultaneously," in the Archimedean world and in the non-Archimedean world.

Namely, any $\breve{f} \in C_p(\mathbb{R})$ defines a unique 1-Lipschitz (i.e., automaton) function $f\colon \mathbb{Z}_p \to \mathbb{Z}_p$ since $\mathbb{Z}_p \cap \mathbb{Q}$ is dense in $\mathbb{Z}_p$ with respect to p-adic metric, and vice versa, any that f defines a unique continuous real function $\breve{f}\colon \mathbb{R} \to \mathbb{R}$ since $\mathbb{Z}_p \cap \mathbb{Q}$ is dense in $\mathbb{R}$ with respect to real metric (that is why in what follows, we use the same symbol f for $\breve{f}$ as well).

The functions from $C_p(\mathbb{R})$ may suit the best for physical modeling of causal dependencies both at macro- and micro-scales since *values of* $f \in C_p(\mathbb{R})$ *on, e.g.,* $\mathbb{N}_0$ *completely define the function* f *on* $\mathbb{R}$.

From the definition it immediately follows that *any function from* $C_p(\mathbb{R})$ *can be represented via Mahler series* (5.25) *where all* c_i *are in* $\mathbb{Z}_p \cap \mathbb{Q}$; *the series converges both on* $\mathbb{R}$ *and on* $\mathbb{Z}_p$ *with respect to real and, accordingly, to the p-adic metric*. It would be interesting to find necessary and sufficient conditions on the coefficients c_i when the series (5.25) defines a $C_p(\mathbb{R})$-function. The general conditions are not known yet, but nevertheless it is clear that the class $C_p(\mathbb{R})$ is rich: For instance, it contains not only polynomials over $\mathbb{Z}_p \cap \mathbb{Q}$ but also some rational functions:

Example 16.18 Given polynomials $u, v \in \mathbb{Z}[x]$ such that $v(z) \not\equiv 0 \pmod p$ for all $z \in \mathbb{Z}_p$ and $v(x) \neq 0$ for all $x \in \mathbb{R}$, the rational function $f(x) = u(x)/v(x)$

is in $C_p(\mathbb{R})$. The rational functions f *are differentiable with respect to both p-adic metric and real metric; moreover,* $\check{f}' = f'$ *everywhere on* $\mathbb{Z}_p \cap \mathbb{Q}$ *and* $f' \in C_p(\mathbb{R})$; see Proposition 6.14.

For $k \in \mathbb{N}$ denote via $C_p^k(\mathbb{R})$ (resp., via $C_p^\infty(\mathbb{R})$) the subclass of all functions which are k-times (resp., infinitely many times) differentiable with respect to both p-adic and real metric whose derivatives are also in $C_p(\mathbb{R})$; put $C_p^0(\mathbb{R}) = C_p(\mathbb{R})$. *It is natural to ask therefore whether there exist functions in* $C_p(\mathbb{R})$ *which are not rational functions.* The answer is affirmative:

Theorem 16.19 *There exist functions in* $C_p^\infty(\mathbb{R})$ *which are not rational functions.*

Proof The theorem can be proved by employing ideas from [80, 146].

The set $\mathbb{Z}_p \cap \mathbb{Q}$ is countable; enumerate its elements by $z_1, z_2, \ldots$. Define by simultaneous induction a sequence of functions $g_0, g_1, g_2, \ldots$ and integers $m_0 < m_1 < m_2 < \cdots$ as follows: Put $g_0(x) = 0$, $m_0 = 1$. For $n \geqslant 1$ consider the following polynomial over the ring $\mathbb{Z}_p \cap \mathbb{Q}$:

$$h_n(x) = \prod_{i=1}^{n}(r_i - x) = a_{n,0} + a_{n,1}x + \cdots + a_{n,n-1}x^{n-1} + (-1)^n x^n.$$

Put

$$g_n(x) = \frac{p^n h_n(x)}{(p^{2n}+1)\lceil |a_{n,0}| + |a_{n,1}|m_{n-1} + |a_{n,n-1}|m_{n-1}^{n-1} + m_{n-1}^n \rceil_p},$$

where $|\cdot|$ is real absolute value and $\lceil r \rceil_p$ for $r \in \mathbb{R}$ is the smallest $\ell \in \mathbb{N}$ such that $r \leqslant \ell$ if $\ell \not\equiv 0 \pmod p$, or $\lceil r \rceil_p = \ell + 1$ if $\ell \equiv 0 \pmod p$. Then $g_n(x)$ is a polynomial over $\mathbb{Z}_p \cap \mathbb{Q}$ and $\|g_n(c)\| < p^{-n}$ for every $c \in \mathbb{C}$ whose complex absolute value $\|c\| \leqslant m_{n-1}$.

Now if n is even, let m_n be first integer larger than m_{n-1} such that $\sum_{i=0}^{n} g_i(m_n) \geqslant 2$. If n is odd, let m_n be the first integer larger than m_{n-1} such that $\sum_{i=0}^{n} g_i(m_n) \leqslant -2$. Since the leading coefficient of $h_n(x)$ is $(-1)^n$, these conditions are always true for large enough m_n. After defining all $g_n(x)$, put $g(x) = \sum_{i=1}^{\infty} g_i(x)$. Then

- The sum $g(x) = \sum_{i=1}^{\infty} g_i(x)$ converges uniformly in the open complex disk $D_{m_n}(0)$ of radius m_n centered at 0 for all $n \in \mathbb{N}$, because except for the first n terms, every term $g_i(c)$ is bounded absolutely by p^{-i}, and the sum of these converges.
- Because uniform convergence in an open subset of $\mathbb{C}$ preserves analyticity, the function g is analytic on every $D_{m_n}(0)$ and so on the whole $\mathbb{C}$.
- $g(\mathbb{Z}_p \cap \mathbb{Q}) \subset \mathbb{Z}_p \cap \mathbb{Q}$ since for every $z = z_k \in \mathbb{Z}_p \cap \mathbb{Q}$ all $g_j(z) = 0$ for $j \geqslant k$.
- In the sequence $(g(m_k))_{k=1}^{\infty}$ the terms having odd indices k are less than -1 whereas the terms having even indices k are greater than 1 since the sum

$\sum_{i=0}^{n} g_i(m_n) \geqslant 2$ for even n, the sum $\sum_{i=0}^{n} g_i(m_n) \leqslant -2$ for odd n, and remaining terms in $g(m_n)$ cannot change the whole sum for more than 1.

Thus the function g is well-defined on the whole $\mathbb{R}$; g is a continuous function with respect to real metric, and by the intermediate value theorem, the function g has a zero between m_n and m_{n+1}, for all sufficiently large $n \in \mathbb{N}$. Therefore g has infinitely many zeroes in $\mathbb{R}$ and whence cannot be of the form $u(x)/v(x)$, where $u(x), v(x) \in \mathbb{Z}[x]$. The function g by the construction is a complex analytic function which is analytic on the whole $\mathbb{C}$; thus the restriction of g on $\mathbb{R}$ is a function $\mathbb{R} \to \mathbb{R}$ which is infinitely many times differentiable everywhere in $\mathbb{R}$; each derivative is continuous with respect to real metric and thus is uniquely defined by its values on $\mathbb{Z}_p \cap \mathbb{Q}$ as $\mathbb{Z}_p \cap \mathbb{Q}$ is dense in $\mathbb{R}$ with respect to real metric.

On the other hand, given $a, b \in \mathbb{Z}_p \cap \mathbb{Q}$, $a \neq b$, there are unique $k, n \in \mathbb{N}$ such that $a = z_k$, $b = z_n$ with respect to the numeration of numbers in $\mathbb{Z}_p \cap \mathbb{Q}$. If $n > k$ then $g(a) = \sum_{i=0}^{n-1} g_i(a)$, $g(b) = \sum_{i=0}^{n-1} g_i(b)$; therefore $|g(a) - g(b)|_p \leqslant |a - b|_p$ since $\sum_{i=0}^{n-1} g_i(x)$ is a polynomial over $\mathbb{Z}_p \cap \mathbb{Q}$; thus, a unique continuation of g to the whole $\mathbb{Z}_p$ is a p-adic 1-Lipschitz function. Let $\bar{g}_k = g \bmod p^k$ be a polynomial over $\mathbb{N}_0$ obtained by reduction modulo p^k of the function g (it is clear that then $\deg \bar{g}_k \leqslant k$ by the construction of g). Then the function $g\colon \mathbb{Z}_p \to \mathbb{Z}_p$ can be uniformly approximated by the polynomials $\bar{g}_k$ with respect to the p-adic sup-norm which is defined as follows: Given a p-adic 1-Lipschitz function $u\colon \mathbb{Z}_p \to \mathbb{Z}_p$, the p-adic sup-norm is $\max\{|u(z)|_p \colon z \in \mathbb{Z}_p\}$. Therefore the function $g\colon \mathbb{Z}_p \to \mathbb{Z}_p$ is a $\mathcal{B}$-function, the Stone-Weierstrass completion of the polynomials over $\mathbb{N}_0$ with respect to the said p-adic sup-norm; thus g is infinitely many times differentiable with respect to the p-adic metric; all derivatives are $\mathcal{B}$-functions; thus the derivatives are uniquely defined by their values on $\mathbb{Z}_p \cap \mathbb{Q}$ as $\mathbb{Z}_p \cap \mathbb{Q}$ is dense in $\mathbb{Z}_p$ with respect to the p-adic metric; see Sect. 6.3.

Therefore g is infinitely many times differentiable both on $\mathbb{R}$ and on $\mathbb{Z}_p$, and values of derivatives both with respect to real and to the p-adic metric coincide on $\mathbb{Z}_p \cap \mathbb{Q}$. This finally proves that g is a $C_p^{\infty}(\mathbb{R})$-function. □

16.5.1 The Hologram-Likeness of $C_p(R)$-Functions

The $C_p(\mathbb{R})$-functions exhibit a sort of "hologram-likeness": Recall that a complete hologram can be restored from a small piece of a holography plate.

> Any two $C_p(\mathbb{R})$-functions coincide everywhere on $\mathbb{Z}_p$ (equivalently, everywhere on $\mathbb{R}$) if and only if they coincide on an open subset of $\mathbb{Z}_p$, no matter how small, equivalently, if they coincide on a real interval, no matter how small.

Here is the theorem.

Theorem 16.20 *Let $f, g \in C_p(\mathbb{R})$.*

1. *Let $(\alpha, \beta) \subset \mathbb{R}$ be any open interval; then $f = g$ if and only if $f(x) = g(x)$ for all $x \in (\alpha, \beta) \subset \mathbb{R}$ (equivalently, for all $x \in (\alpha, \beta) \cap \mathbb{Z}_p \cap \mathbb{Q}$).*
2. *Let $\mathbf{U} \subset Z_p$ be any non-empty open subset; then $f = g$ if and only if $f(z) = g(z)$ for all $x \in \mathbf{U} \subset \mathbb{Z}_p$ (equivalently, for all $z \in \mathbf{U} \cap \mathbb{Z}_p \cap \mathbb{Q}$).*

Proof For $n \in \mathbb{N}$, $d \in \{1, \ldots, p^n - 2\}$ put $z_{n,d} = \frac{d}{1-p^n}$; then $z_{n,d} \in (-1, 0) \subset \mathbb{R}$, $d \in \mathbb{Z}_p \cap \mathbb{Q}$. If $d = \xi_0 + \xi_1 p + \cdots \xi_{n-1} p^{n-1}$ is the base-p expansion of z then

$$z_{n,d} = \xi_0 + \xi_1 p + \cdots \xi_{n-1} p^{n-1} + \xi_0 p^n + \xi_1 p^{n+1} + \cdots \xi_{n-1} p^{2n-1} + \cdots \qquad (16.4)$$

is the p-adic canonical form of $z_{n,d}$ as $(1-p^n)^{-1} = 1 + p^n + p^{2n} + p^{3n} + \cdots \in \mathbb{Z}_p$; cf. Sect. 2.3.2. From 16.4 it immediately follows that the set $\mathcal{Z}$ of all these $z_{n,d}$ is dense in $\mathbb{Z}_p$. Therefore, $f = g$ on $\mathbb{Z}_p$ on if and only if $f = g$ on $\mathcal{Z}$; but $f = g$ on $\mathbb{Z}_p$ if and only if $f = g$ on $\mathbb{R}$.

Let now $\alpha, \beta \in \mathbb{Z}_p \cap \mathbb{Q}$, $\alpha < \beta$. Put $\gamma = -1$ if $\alpha - \beta \leqslant -1$; let $\gamma = \frac{1}{1-p^t}$ be such that $0 > \gamma > \alpha - \beta$ if $\alpha - \beta > -1$, for a suitable $t \in \mathbb{N}$. It is clear from what we have already proved that $f(x) = g(x)$ for all $x \in \mathbb{R}$ if and only if $f(x) = g(x)$ for all $x \in (\gamma^{-1}(\beta - \alpha), 0)$ as $(\gamma^{-1}(\beta - \alpha), 0) \supset (-1, 0)$. This proves claim 1 of the theorem.

Claim 2 is nearly obvious: Every open set $\mathbf{U} \subset \mathbb{Z}_p$ is a countable union of a clopen balls $\mathbf{B}_{p^{-r}}(c)$ of non-zero radii (cf. Sect. 2.2.1); but $\mathbf{B}_{p^{-r}}(c) \cap \mathbb{Q}$ is dense in $\mathbb{R}$ by Theorem 2.19; thus if $f(z) = g(z)$ for all $z \in \mathbf{B}_{p^{-r}}(c) \cap \mathbb{Q}$ then $f(z) = g(z)$ for all $z \in \mathbb{R}$ as the both functions are continuous real functions and $\mathbf{B}_{p^{-r}}(c) \cap \mathbb{Q}$ is dense in $\mathbb{R}$; whence, $f(z) = g(z)$ for all $z \in \mathbb{Z}_p \cap \mathbb{Q}$, thus $f(z) = g(z)$ for all $z \in \mathbb{Z}_p$ as the both functions f and g are continuous p-adic functions and $\mathbb{Z}_p \cap \mathbb{Q}$ is dense in $\mathbb{Z}_p$. □

Interpretation 16.21 ("Causality" vs "Locality") *"Local" behavior of $C_p(\mathbb{R})$-functions completely defines their "global" behavior: Given values a $C_p(\mathbb{R})$-function takes on arbitrarily small neighborhood of arbitrary point, the values the function takes at all other points can be "restored uniquely." If the points of $\mathbb{R}$ are treated as "positions" and values of the function as "measurement data" of a physical system to which the function is ascribed, then the data an observer get by probing a system in a given position let him completely predict values of physical quantities obtained by measurements at all other positions.*

Note that "position" here should not be understood literally, as a position in physical space necessarily, though it *may* be treated in this way also, depending on interpretations. The "position" *may* also be considered as "temporal," i.e., as a position at the arrow of time when the point of $\mathbb{R}$ is treated as time instant. From this latter view, the "hologram-likeness" property of $C_p(\mathbb{R})$-functions resembles the property of wave function whose time evolution is given by Schrödinger equation which is deterministic with respect to time (cf. (16.14) further).

The "hologram-likeness" property of $C_p(\mathbb{R})$-functions is especially important since various classes of real functions can be approximated by $C_p(\mathbb{R})$-functions.

Theorem 16.22 *The following is true:*

1. *Any continuous function* $g\colon [a,b] \to \mathbb{R}$ *can be uniformly approximated on* $[a,b] \subset \mathbb{R}$ *by* $C_p^\infty(\mathbb{R})$*-functions.*
2. *Any continuous function* $g\colon [a,b] \to \mathbb{R}$ *can be uniformly approximated on* $[a,b] \subset \mathbb{R}$ *by* $C_p^\infty(\mathbb{R})$*-functions which are automata functions of reversible automata.*
3. *Any continuous function* $g\colon [a,b] \to \mathbb{R}$ *can be uniformly approximated by ergodic automata functions from* $C_p^\infty(\mathbb{R})$.
4. *Any continuous function* $g\colon \mathbb{R} \to \mathbb{R}$ *which vanishes at infinity can be uniformly approximated on* $\mathbb{R}$ *by* $C_p^\infty(\mathbb{R})$*-functions.* (Recall that a continuous function $g\colon \mathbb{R} \to \mathbb{R}$ vanishes at infinity, if, for every $\varepsilon > 0$, there exists a compact set $K \subset \mathbb{R}$ such that $|g(x)| < \varepsilon$ for all $x \in \mathbb{R} \setminus K$.)
5. *Any continuous function* $g\colon \mathbb{R} \to \mathbb{R}$ *which vanishes at infinity can be uniformly approximated on* $\mathbb{R}$ *by* $C_p^\infty(\mathbb{R})$*-functions which are automata functions of time-reversible automata.*
6. *Any square-integrable function* (and moreover, any function which is integrable with its n-th power, for some $n \in \mathbb{N}$) $g\colon \mathbb{R} \to \mathbb{R}$ *can be uniformly approximated by* $C_p^\infty(\mathbb{R})$*-functions which are automata functions of reversible automata.*

Proof The class $C_p^\infty(\mathbb{R})$ contains all polynomial functions over $\mathbb{Z}$. The Chlodovsky's theorem reads that a continuous real-valued function on a real interval which does not contain integers can be uniformly approximated by polynomials over $\mathbb{Z}$, [39, 66]. If the interval $[a,b]$ contains integers, take $\alpha, \beta \in \mathbb{Z}_p \cap \mathbb{Q}$ such that the interval $[a',b'] = [\alpha a + \beta, \alpha b + \beta]$ contains no integers (e.g., take $m \in \mathbb{N}$ such that $|b-a| < p^m - 1$ and put $\alpha = 1/(p^m - 1)$). Given a continuous function $g\colon [a,b] \to \mathbb{R}$, the function $g(\alpha^{-1}(x-\beta))$ can be uniformly approximated by polynomials $u_i(x) \in \mathbb{Z}[x]$ on $[a',b']$ by Chlodovsky's theorem; thus the function g can be uniformly approximated by polynomials $u_i(\alpha x + \beta))$ on $[a,b]$. But $\hat{u}_i(x) = u_i(\alpha x + \beta)$ is a polynomial over $\mathbb{Z}_p \cap \mathbb{Q}$ in variable x since $\alpha, \beta \in \mathbb{Z}_p \cap \mathbb{Q}$; thus $\hat{u}_i \in C_p^\infty(\mathbb{R})$. This proves claim 1.

To prove claim 2, consider the function $\tilde{g}(x) = \frac{g(x)-x}{p}$. In view of claim 1, since $\tilde{g}(x)$ is continuous on $[a,b]$, $\tilde{g}$ can be uniformly approximated by $C_p^\infty(\mathbb{R})$-functions u_i; thus g can be uniformly approximated by functions $x + pu_i(x)$ which are also in $C_p^\infty(\mathbb{R})$. But given any 1-Lipschitz function $u\colon \mathbb{Z}_p \to \mathbb{Z}_p$, the function $z + pu(z)$ is 1-Lipschitz measure-preserving, by Lemma 9.6. Thus the $C_p^\infty(\mathbb{R})$-functions $x + pu_i(x)$ are automata functions of reversible automata.

To prove claim 3, note that the function g can be uniformly approximated by polynomials $w_j(x)$ over $\mathbb{Z}_p \cap \mathbb{Q}$; see the proof of claim 1. Then the difference equation $\frac{w_j(x)-x-1}{p} = \Delta\tilde{w}_j(x)$ has a solution $\tilde{w}_j(x)$ which is a polynomial since $\frac{w_j(x)-x-1}{p}$ is. These $\tilde{w}_j(x)$ can be uniformly approximated by polynomials $u_{ji}(x)$ over $\mathbb{Z}_p \cap \mathbb{Q}$, see, the proof of claim 1. Therefore g can be uniformly approximated by polynomials over $\mathbb{Z}_p \cap \mathbb{Q}$ of the form $1 + x + p\Delta u(x)$ which are all ergodic by Lemma 9.6.

To prove claim 4, consider functions of the form $\frac{ru(x)}{1+pv(x)^2}$ where $r \in \mathbb{Z}_p \cap \mathbb{Q}$, $u(x), v(x) \in \mathbb{Z}[x]$, and $1 \leqslant \deg u(x) \leqslant \deg v(x)$. Then all these functions vanish at infinity, are $C_p^\infty(\mathbb{R})$-functions (see Example 16.18), and separate points. Therefore the $\mathbb{R}$-algebra $\mathbf{A}$ generated by the set A of all these functions satisfies conditions of Stone-Weierstrass theorem for locally compact spaces, i.e., the algebra is dense with respect to the topology of uniform convergence in the Banach algebra of all real-valued continuous functions on $\mathbb{R}$ which vanish at infinity. But the set A is dense in $\mathbf{A}$.

In order to prove claim 5 note that in view of the proof of claim 4 it suffices to approximate uniformly on $\mathbb{R}$ the functions of the form $h(x) = \frac{ru(x)}{1+pv(x)^2}$ where $r \in \mathbb{Z}_p \cap \mathbb{Q}$, $u(x), v(x) \in \mathbb{Z}[x]$, $1 \leqslant \deg u(x) \leqslant \deg v(x)$, by $C_p^\infty(\mathbb{R})$-functions which are automata functions of reversible automata. Represent $h(x) = \frac{x}{1+pv(x)^2} + \frac{p\tilde{u}(x)}{p(1+pv(x)^2)}$, then $\tilde{u}(x) \in \mathbb{Z}_p[x]$, $\deg \tilde{u}(x) \leqslant \deg v(x)$. Given $c_i \in \mathbb{Z}_p \cap \mathbb{Q}$, the function $\frac{x}{1+pv(x)^2} + c_i \frac{p\tilde{u}(x)}{1+pv(x)^2}$ vanishes at infinity; moreover, this function is a $C_p^\infty(\mathbb{R})$-function and is a measure-preserving 1-Lipschitz function $\mathbb{Z}_p \to \mathbb{Z}_p$ since the function is bijective modulo p and its derivative vanishes modulo p nowhere; see Corollary 10.4 and the text thereafter. Taking a sequence $(c_i)_{i=0}^\infty$ over $\mathbb{Z}_p \cap \mathbb{Q}$ which converges to $1/p$ in $\mathbb{R}$, we conclude that the function $h(x)$ can be uniformly approximated on $\mathbb{R}$ by $C_p^\infty(\mathbb{R})$-functions which are measure-preserving 1-Lipschitz functions $\mathbb{Z}_p \to \mathbb{Z}_p$; that is, automata functions of reversible automata.

It is well-known that functions which are integrable with their n-th powers, for some $n \in \mathbb{N}$, can be uniformly approximated by Schwartz functions which by the definition are infinitely many times differentiable functions $f : \mathbb{R} \to \mathbb{R}$ such that $\lim_{t\to\pm\infty} f^{(j)}(x)/x^n = 0$ for any $n \in \mathbb{N}$, $j \in \mathbb{N}_0$, where $f^{(j)}$ is j-th derivative of f. The Schwartz functions are smooth and vanish at infinity. By claim 5, this proves claim 6 and the theorem. □

Interpretation 16.23 (Free Choice of the Direction of Arrow of Time)
Due to the inevitable non-zero error in measurements of values of physical quantities, an observer's conclusion on the direction of "arrow of time" completely depends on the observer's free choice of metric with respect to which the observer processes the measured numerical data: By claims

(continued)

2, 5–6 of Theorem 16.22, "causes" can be recovered from "effects," with any desirable accuracy provided the measurements are precise enough. The "entropic arrow of time" also depends on the choice of metric since the value of entropy does; see Sect. 16.4.2.

A *finite* automaton $C_p(\mathbb{R})$-function is a function $\check{f} \in C_p(\mathbb{R})$ such that $f : \mathbb{Z}_p \to \mathbb{Z}_p$ is an automaton function of a finite automaton. As said, in what follows we use the same notation f for $\check{f}$ as well (cf. Definition 16.17). The finite automata $C_p(\mathbb{R})$-functions are important further in the book.

Theorem 16.24 *Let a finite automaton function $f \in C^1_p(\mathbb{R})$; i.e., let f be differentiable both over $\mathbb{R}$ and over $\mathbb{Z}_p$; let $f' \in C_p(\mathbb{R})$. Then f is an affine function over $\mathbb{Z}_p \cap \mathbb{Q}$; i.e., $f(x) = ax + b$ for suitable $a, b \in \mathbb{Z}_p \cap \mathbb{Q}$. Vice versa, all these affine functions are finite automata functions from $C^\infty_p(\mathbb{R})$.*

Proof Given a 1-Lipschitz function f, for $n \in \mathbb{N}_0$, $k \geqslant \lfloor \log_p n \rfloor + 1$, consider functions $f_{n,k} : \mathbb{Z}_p \to \mathbb{Z}_p$ defined as follows:

$$f_{n,k}(z) = \frac{1}{p^k}\Big(f(n + p^k z) - (f(n) \bmod p^k)\Big) =$$

$$\frac{f(n + p^k z) - f(n)}{p^k} - \frac{f(n) - f(n) \bmod p^k}{p^k} = \frac{f(n + p^k z) - f(n)}{p^k} - f_{n,k}(0), \tag{16.5}$$

for all $z \in \mathbb{Z}_p$. The function f is an automaton function of a finite automaton if and only if the collection $\mathcal{F}$ of function $f_{n,k}$ (where $n \in \mathbb{N}_0$, $k \in \mathbb{N} = \{1, 2, 3, \ldots\}$, $k \geqslant \lfloor \log_p n \rfloor + 1$) contains only finitely many pairwise distinct functions: Note that $f_{n,k}$ is the automaton function that corresponds to the automaton $\mathfrak{A}(s(n_k)) = \langle \mathbb{F}_p, \mathcal{S}, \mathbb{F}_p, S, O, s(n_k) \rangle$, where $s(n_k) \in \mathcal{S}$ is the state the automaton $\mathfrak{A} = \mathfrak{A}_f = \langle \mathbb{F}_p, \mathcal{S}, \mathbb{F}_p, S, O, s_0 \rangle$ reaches after it has been fed by the input word n_k (of length p^k) that corresponds to the base-p expansion of n (so the word n_k may contain some leading zeros that correspond to higher order digits of the expansion); cf., the beginning of the proof of Theorem 13.22. So, there are $N, K \in \mathbb{N}$ such that for every $n \in \mathbb{N}_0$, $k \in \mathbb{N}$, one finds $\check{n} \leqslant N$, $\check{k} \leqslant K$ such that $f_{n,k}(z) = f_{\check{n},\check{k}}(z)$ for all $z \in \mathbb{Z}_p$.

Take $z_k = \frac{h}{p^{2k} - 1}$ where $h \in \mathbb{N}$. Note that $z_k \in \mathbb{Z}_p \cap \mathbb{Q}$ and that $\lim_{k\to\infty} p^k z_k = 0$ both with respect to real metric and to the p-adic metric. Then

$$\frac{f_{n,k}(z_k)}{z_k} = \frac{f(n + p^k z_k) - f(n)}{p^k z_k} - \frac{f(n) - f(n) \bmod p^k}{p^k z_k}, \tag{16.6}$$

where $\lim_{k\to\infty} \frac{f(n+p^k z_k)-f(n)}{p^k z_k} = f'(n)$ both with respect to real metric and to the p-adic metric. Thus from (16.5) it follows that $\lim^p_{k\to\infty}(f_{n,k}(u)-f_{n,k}(0)) = uf'(n)$ for every $u \in \mathbb{Z}_p, n \in \mathbb{N}_0$. However, $f'(n)$ is a derivative at $n \in \mathbb{N}_0$ both with respect to real metric and to the p-adic metric; but $f'(n)$ may take only finitely many values due to the finiteness of the number of pairs n, k which enumerate pairwise distinct $f_{n,k}$. Therefore $f'(z)$ may take not more than finitely many values on $\mathbb{Z}_p$ since any $z \in \mathbb{Z}_p$ is a p-adic limit of some sequence over $\mathbb{N}_0$ and f' is a continuous function $\mathbb{Z}_p \to \mathbb{Z}_p$ by the conditions of the theorem. Hence, f' may take not more than finitely many values on $\mathbb{Z}_p \cap \mathbb{Q}$ and thus on $\mathbb{R}$ since $\mathbb{Z}_p \cap \mathbb{Q}$ is dense in $\mathbb{R}$ and f' is a continuous real function by the conditions of the theorem. Therefore, the derivative f' is a constant function over $\mathbb{R}$, thus over $\mathbb{Z}_p \cap \mathbb{Q}$, thus over $\mathbb{Z}_p$. That is, $f(x) = ax + b$ for some $a, b \in \mathbb{Z}_p \cap \mathbb{Q}$. Proposition 13.24 proves converse claim of the theorem. □

Note 16.25 The theorem remains true for multivariate $C_p(\mathbb{R})$-maps $F : \mathbb{Z}_p^m \to \mathbb{Z}_p^n$ as well: The only these maps are affine maps over $\mathbb{Z}_p \cap \mathbb{Q}$.

Interpretation 16.26 (Finiteness Implies Linearity) *Theorem 16.24 and Note 16.25 may serve as a sort of a hint why mathematical formalism of quantum mechanics is the theory of linear operators over Hilbert space: As all "real-world" systems have a finite number of states, then when duration of temporal interval measured in the smallest (say, Planck) time units becomes comparable to the number of the states, the finiteness reveals itself as the linearity.*

16.6 Observations at the "Both Ends of the Scale"

In Sect. 16.2 we conjectured that if both "continuous" and "discrete" models adequately describe physical reality at respective "ends of the scale," the models must "meet one another somewhere in the middle of the scale." *In this Section we argue that "in the middle of the scale" there is wave function* where the models meet each other. To do this, we first need to formalize the notion of an observer; actually we will consider observers of two kinds, each for respective end of the scale.

16.6.1 Little-Endian and Big-Endian Observers

To begin with, let us introduce two types of observers, the *Big-endian* and the *Little-endian*. The names of the two observers are more related to big-end and little-end orders the bytes of representation of a number are read in computer science and less to the hero of the novel *Gulliver's Travels* by Jonathan Swift: in computer science, the end of the registry which contains lest lower-order (i.e., less significant) digits of the base-2 expansion of a number is called a little end; the opposite end of the registry is called the big end. Since the content of the registry can be read from the left to right or from the right to left, for the most significant for each way of reading are the digits which are read the first. So in what follows the Little-endian may be associated with a micro-observer whereas Big-endian to a macro-observer. Of course, the representation of numbers are assumed to be in the base p, where p is not necessarily 2.

To put it in other words, given a large non-negative number having a very long base-p expansion, the Big-endian is capable to observe only highest-order digits of the expansion, i.e., he knows the order of magnitude of the number and (up to a non-zero error) a mantissa, since the Big-endian is not able to see the rightmost digits of the numbers. On the contrary, the Little-endian sees the rightmost digits of the number, starting with the smallest order digit, but has no idea what are the rightmost digits and the order of magnitude of the number (though he assumes that the order is finite but very large). One may call the Big-endian a macro-observer whereas Little-endian a micro-observer. But the both observers measure values of observable which are rational p-adic integers: As it was already mentioned, real numbers are indistinguishable during measurements from rational p-adic integers $\mathbb{Z}_p \cap \mathbb{Q}$, due to inevitable non-zero measurement error with respect to real metrics. That is why we assume that numerical values of observable are in $\mathbb{Z}_p \cap \mathbb{Q}$; and the Little-endian sees the first terms of the canonical p-adic expansion of the value of the observable whereas Big-endian sees the highest-order digits of the base-p expansion of the same value as of a real number. We explain this more formally.

We already know from Sect. 2.3.2 that $z \in \mathbb{Z}_p \cap \mathbb{Q}$ if and only if z can be represented as $z = c + \frac{d}{p^t - 1}$ for some $t \in \mathbb{N}$ and $d \in \{1, 2, \ldots, p^t - 1\}$, $c \in \mathbb{Z}$; or, if and only if the p-adic canonical representation of z is eventually periodic

$$z = \alpha_0 + \alpha_1 p + \cdots + \alpha_{r-1} p^{r-1} + (\beta_0 + \beta_1 p + \cdots + \beta_{t-1} p^{t-1}) p^r +$$
$$(\beta_0 + \beta_1 p + \cdots + \beta_{t-1} p^{t-1}) p^{r+t} + (\beta_0 + \beta_1 p + \cdots + \beta_{t-1} p^{t-1}) p^{r+2t} + \cdots \tag{16.7}$$

for suitable $\alpha_j, \beta_i \in \{0, 1, \ldots, p-1\}$, $r \in \mathbb{N}_0$, $t \in \mathbb{N}$ (the sum $\alpha_0 + \alpha_1 p + \cdots + \alpha_{r-1} p^{r-1}$ is absent in the above expression once $r = 0$). In that case the base-p representation of fractional part of z as of a real number is

$$z \bmod 1 = 0.(\hat{\beta}_{t-1-\bar{r}} \hat{\beta}_{t-2-\bar{r}} \ldots \hat{\beta}_0 \hat{\beta}_{t-1} \hat{\beta}_{t-2} \ldots \hat{\beta}_{t-\bar{r}})^{\infty} \bmod 1, \tag{16.8}$$

where $\hat{\beta} = p - 1 - \beta$ for $\beta \in \{0, 1, \ldots, p-1\}$ and $\bar{r}$ is the least non-negative residue of r modulo t if $t > 1$ or $\bar{r} = 0$ otherwise.

To illustrate what are Big-endian observations and Little-endian observations let $r = 1$, $t \gg 1$, $\alpha_0 = 1$, $\beta_{t-2} = \beta_{t-1} = 0$; so the both "-endians" measure physical quantity z which takes values in $[0, 1]$. Then, as none of the observers is able to measure the value z with a non-zero error, the Big-endian will obtain only the digits $\hat{\beta}_{t-2}, \hat{\beta}_{t-3}, \ldots, \hat{\beta}_{t-n}$ for some $n < t$, whereas the Little-endian will get $\beta_0, \beta_1, \ldots, \beta_m$ for some $m < t - 1$. Thus, the only information about z which possibly may share the both observers is the values $\beta_{t-\ell}, \beta_{t-\ell-1}, \ldots, \beta_k$ for some $k > 0$, $\ell > 1$. The two observers may communicate with each other and thus make only common guesses about what z is; however, both do not know what t is; therefore as $t \gg 1$ the only thing the both observers may know for sure is that $1 \geqslant z \geqslant 1 - 1/p$. Note that t is large but finite by the finiteness assumption (cf. Sect. 16.2); so the Big-endian sees the leftmost digits in (16.8) and assumes that t is infinite to the right while Little-endian sees the rightmost digits and assumes that t is infinite to the left (cf. (16.7)).

In connection with these two kinds of observers, it is worth mentioning late V. S. Vladimirov's model of measurement which justifies the p-adic approach to quantum phenomena (cf. [116, Chapter 1, Remark 6.2]): For a distant observer somewhere far in Universe, our human scale may look like a quantum one since Earth distances are too small for the observer to be observed directly. On the other hand, for Earth astronomers only much larger distances like parsecs, light years, etc. matter. A similar model connected with a time measurement was proposed by M. Sernetz and F. Jestczemski of the Institute of Endocrinology of the University of Giessen. If we try to study the evolution of a biological organism which lived millions years ago, then these millions years are not so important with respect to the period of the life of this organism. These ideas which go back to V. S. Vladimirov are developed in [116].

16.6.2 Real Time, p-Adic Time, and "Common" Time

In this subsection we introduce a p-adic model of the instrument which measure and indicates time, the *p-adic clock*; then we prove that there exist the only clock which is the same for all Little-endians and Big-endian, the *universal clock*.

A timekeeping element of contemporary physical clock is harmonic oscillator of particular frequency, which is assumed to be a positive integer showing the number of periods per unit interval; so the shortest time interval which can be measured is a reciprocal of the frequency. In order to measure the value of time elapsed, one just counts the number of periods since one time moment till another one and represent this non-negative integer in some base, say, p where p is the frequency of the oscillator. In what follows, we assume that p is a prime just for not to overload the exposition with unimportant technical details. Thus, a model of such clock can be represented by the *p-adic odometer*, a dynamical system $f = \tau_p \colon z \mapsto z + 1$ on the space of p-adic integers $\mathbb{Z}_p$. If initial point is $x_0 \in \mathbb{Z}_p$, e.g., $x_0 = 0$, put $f^0(x_0) = x_0$, $f^1(x_0) = f(x_0) = x_1, \ldots, f^i(x_0) = f(f^{i-1}(x_0)) = x_i$, then the

Fig. 16.1 The p-adic clock

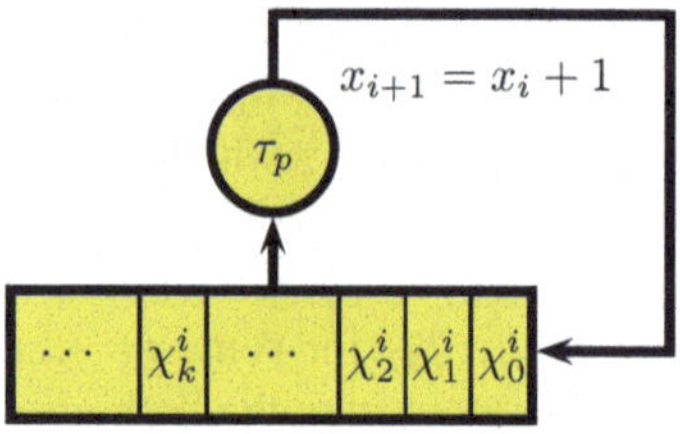

base-p expansion of x_i represents the time elapsed, $i = \sum_{j=0}^{\lfloor \log_p i \rfloor + 1} \chi^i_j p^j$, where $\chi^i_j = \delta_j(i) \in \{0, 1, \dots, p-1\}$ is the j-th digit of the base-p expansion of i. In loose terms, the p-adic clock is just a counter whose face consists of windows, at each time moment i each j-th window shows $\delta_j(i)$. It is convenient to assume that the number of windows is infinite to have the time elapsed unrestricted; thus we get the dynamical system τ_p on $\mathbb{Z}_p$. Note that initial state x_0 may be taken arbitrary, and not necessarily $x_0 = 0$; then to get the base-p representation of time elapsed since the initial moment on has to perform subtraction $x_i - x_0$ in $\mathbb{Z}_p$. The p-adic clock is depicted by Fig. 16.1. To the right, the content of the registry is like a standard representation of time in decimal (rather than p-ary) fractions of a second (millisecond, microsecond, nanosecond, ...) with Planck time the rightmost position, whereas to the left are decimal multiples of a second (petasecond, exasecond, etc.).

Speaking loosely, the registry in Fig. 16.1 is like a face of a mechanical counter consisting of cogwheels: The period of the sequence of states of the rightmost cell of the registry (which can be judged as the rightmost cogwheel) is p, the period of the sequence of states of the second rightmost cell is p^2 since the figure in that cell changes once in a period of the rightmost cell, etc. The latter property is a definitive property of an ergodic transformation on $\mathbb{Z}_p$ (cf. Theorem 8.2); therefore, all ergodic 1-Lipschitz transformations on $\mathbb{Z}_p$ should be considered as clocks (cf. claim 2 of Theorem 8.13) as they can be "adjusted" one to another since they all are conjugate to the p-adic odometer.

16.6.2.1 Continuous p-Adic Time

The p-adic time has been considered in a number of works on p-adic and adelic mathematical physics; see, e.g., [51, 54, 194].

We are going to define the p-adic time in another way, via the odometer which, as we have already seen, is merely a "digital clock."

If the initial state of the odometer is taken to be 0 (i.e., each cell of the registry depicted by Fig. 16.1 is 0), then after $n \in \mathbb{N}$ time units elapsed, the registry will contain the base-p expansion of the number n since $\tau_p^n(0) = n$. Let us now take any ergodic 1-Lipschitz map $f : \mathbb{Z}_p \to \mathbb{Z}_p$ and take any $t \in \mathbb{Z}_p$ and any sequence $(n_i)_{i=0}^{\infty}$ over $\mathbb{N}_0$ which converges p-adically to t (such a sequence exists as $\mathbb{N}_0$ is dense in $\mathbb{Z}_p$). It turns out that then for any $z \in \mathbb{Z}_p$, the p-adic limit $\lim^p_{i\to\infty} f^{n_i}(z)$

exists; denoting this limit via $f^t(z)$, then $(z; t) \mapsto f^t(z)$ is a 1-Lipschitz map $\mathbb{Z}_p^2 \to \mathbb{Z}_p$ which is measure-preserving with respect to t; therefore, the *p-adic time t is well-defined.* This is what we are going to prove now.

Speaking more exactly, we are going to show that *every discrete 1-Lipschitz ergodic dynamical system f on $\mathbb{Z}_p$ can be extended to a dynamical system with a continuous p-adic time.* In other words, whenever $f: \mathbb{Z}_p \to \mathbb{Z}_p$ is 1-Lipschitz and ergodic, the function $f^n(x)$, $n \in \mathbb{N}_0$, can be expanded to the function $f^t(x)$, $(t, x) \in \mathbb{Z}_p^2$, which is continuous as a 2-variate function (actually, is a 1-Lipschitz). Moreover, in the case $p = 2$ we show that given an arbitrary 1-Lipschitz measure-preserving function $f: \mathbb{Z}_2 \to \mathbb{Z}_2$, which is *not* necessarily ergodic, the function $f^n(x)$, $n \in \mathbb{N}_0$, can be expanded to the function $f^t(x)$, $(t, x) \in \mathbb{Z}_2^2$, which is continuous as a two-variate function. Thus, we stress that the p-adic time arises very naturally in the p-adic ergodic theory.

Let $f: \mathbb{Z}_p \to \mathbb{Z}_p$ be a 1-Lipschitz ergodic transformation on $\mathbb{Z}_p$. For every $n \in \mathbb{N}_0$, the n-th iteration $f^n(x)$ is well-defined. We assert that given $t \in \mathbb{Z}_p$, there exists a limit (with respect to the p-adic metric)

$$\overset{p}{\lim_{n_j \to t}} f^t(x),$$

where $(n_j \in \mathbb{N}_0)_{j=0}^{\infty}$ is an arbitrary sequence that tends p-adically to $t \in \mathbb{Z}_p$.

Indeed, let $n = m + p^k\ell$, $m, n, \ell \in \mathbb{N}_0$, $k \in \mathbb{N}$. Then $f^n(x) \equiv f^m(x) \pmod{p^k}$ as $f(x)$ is transitive modulo p^k for all $k \in \mathbb{N}$, by Theorem 8.2. That is, $|f^n(x) - f^m(x)|_p \leqslant p^{-k}$ whenever $|n - m|_p \leqslant p^{-k}$. This proves our assertion as $\mathbb{N}_0$ is dense in $\mathbb{Z}_p$.

Thus, the two-variate function $f^t(x): \mathbb{Z}_p^2 \to \mathbb{Z}_p$, $t, p \in \mathbb{Z}_p$, is well-defined. Note that the argument above implies that $f^t(x)$ is 1-Lipschitz with respect to the variable $t \in \mathbb{Z}_p$. We claim that the function $f^t(x)$ is continuous as a two-variate function of $t, x \in \mathbb{Z}_p$. Indeed, given $x, x', t, t' \in \mathbb{Z}_p$ such that $|x - x'|_p \leqslant p^{-n}$, $|t - t'|_p \leqslant p^{-m}$ we see that $|f^t(x) - f^{t'}(x')|_p \leqslant p^{-k}$ for $k = \min\{n, m\}$ since in this case $t \equiv t' \equiv r \pmod{p^k}$ and $x \equiv x' \equiv z \pmod{p^k}$ for suitable $r, z \in \{0, 1, \ldots, p^k - 1\}$; so $f^t(x) \equiv f^{t'}(x') \equiv f^r(z) \pmod{p^k}$ as f is transitive modulo p^k by Theorem 8.2. Thus we have proved the following proposition.

Proposition 16.27 *Given a 1-Lipschitz ergodic transformation f on $\mathbb{Z}_p$, the function $f^t(x)$ is a 1-Lipschitz function defined for all $(t, z) \in \mathbb{Z}_p^2$ and valuated in $\mathbb{Z}_p$.*

Foremost, for every $x \in \mathbb{Z}_p$ *the function $f^t(x)$ is measure-preserving as a function of variable $t \in \mathbb{Z}_p$.* Indeed, given $n, m \in \mathbb{N}_0$, $n \neq m$, take $k \in \mathbb{N}$ such that $p^k > n, m$. Then $f^n(x) \not\equiv f^m(x) \pmod{p^k}$ for every $x \in \mathbb{Z}_p$ since f is transitive modulo p^k. This proves that given $x \in \mathbb{Z}_p$, the function $f^t(x)$ of variable $t \in \mathbb{Z}_p$ is bijective modulo p^k for all $k \in \mathbb{N}$; thus this function $f^t(x)$ is measure-preserving by Theorem 8.2. Thus the following proposition is true:

Proposition 16.28 *The two-variate function $f^t(x)$ from Proposition 16.27 is measure-preserving with respect to variable $t \in \mathbb{Z}_p$.*

Example 16.29 Given an ergodic affine transformation $f(x) = ax + b$ on $\mathbb{Z}_p$, the two-variate function from Proposition 16.27 is of the form $f^t(x) = bt + x$ if $a = 1$, and

$$f^t(x) = b\frac{a^t - 1}{a - 1} + a^t x,$$

if $a \neq 1$. Note that by Theorem 9.1, $p \nmid b$ and $a \equiv 1 \pmod{p}$.

Indeed, if $a = 1$ then $f^n(x) = bn + x$ for all $n \in \mathbb{N}_0$; so given $t \in \mathbb{Z}_p$, we have that $\overset{p}{\lim_{n\to t}} f^n(x) = bt + x$.

Let now $a \neq 1$. Then by Theorem 9.1, $a - 1 = pz$ if $p \neq 2$, and $a - 1 = 4z$ if $p = 2$, for a suitable $z \in \mathbb{Z}_p$. Given $n \in \mathbb{N}$, we have then that $f^n(x) = b(a^{n-1} + a^{n-2} + \cdots + 1) + a^n x$. Let $k = \operatorname{ord}_p z$, i.e., $z = p^k \hat{z}$ where $p \nmid \hat{z}$, then

$$a^{n-1} + a^{n-2} + \cdots + 1 = \frac{a^n - 1}{a - 1} = \begin{cases} \sum_{i=1}^{n} p^{(k+1)(i-1)} \hat{z}^{i-1} \binom{n}{i}, & \text{if } p > 2; \\ \sum_{i=1}^{n} 2^{(k+2)(i-1)} \hat{z}^{i-1} \binom{n}{i}, & \text{if } p = 2, \end{cases}$$

is a p-adic integer. Under the above restrictions on a, the function a^t is analytic on $\mathbb{Z}_p$ (cf. Example 4.5); so we see that $\overset{p}{\lim_{n\to t}} a^n = a^t$, and the conclusion follows.

Now we consider the 2-adic case. Let $f\colon \mathbb{Z}_2 \to \mathbb{Z}_2$ be a 1-Lipschitz measure-preserving function. Thus, by Theorem 8.2, f is bijective modulo 2^n, for all $n \in \mathbb{N}$; so every map $f \bmod 2^n : x \mapsto f(x) \bmod 2^n$ of the residue ring $\mathbb{Z}/2^n\mathbb{Z}$ into itself is a permutation on $\{0, 1, \ldots, 2^n - 1\}$. We claim that *every cycle of this permutation has the length 2^ℓ, for a suitable $\ell \in \mathbb{N}_0$*. We proceed by induction on n.

For $n = 1$ the claim is obvious since $f \bmod 2$ is either the identity map (whose cycles are all of length $2^0 = 1$) or the map $x \mapsto x + 1$, which consist of the only cycle, of length 2.

Let the claim be true for all $1 \leqslant n < k$; let us prove it for $n = k$. Given $x \in \mathbb{Z}/2^k\mathbb{Z}$, denote $\chi_i^j = \delta_i(f^j(x))$, the i-th digit in the base-2 expansion of the j-th iterate of f. Note that $\chi_i^0 = \delta_i(x)$, $i = 0, 1, 2, \ldots$. From Theorem 9.4 it follows that

$$\chi_i^1 = \chi_i^0 \oplus \varphi_i(\chi_0^0, \ldots, \chi_{i-1}^0),$$

where ψ_i is a Boolean function in i Boolean variables and $\oplus$ stands for the addition modulo 2. Iterating this equality, we obtain that

$$\chi_i^j = \chi_i^0 \oplus \sum_{\ell=0}^{j-1} \varphi_i(\chi_0^\ell, \ldots, \chi_{i-1}^\ell), \tag{16.9}$$

for all $i = 0, 1, 2, \ldots$, where the sum $\sigma_i^j = \sum_{\ell=0}^{j-1} \varphi_i(\chi_0^\ell, \ldots, \chi_{i-1}^\ell)$ in the right-hand side is taken modulo 2; so $\sigma_i^j \in \{0, 1, \}$. If $f^r(x) \equiv x \pmod{2^k}$, then $f^r(x) \equiv x \pmod{2^{k-1}}$, so by induction hypothesis, the smallest $r \in \mathbb{N}$ that satisfies the latter congruence is a power of 2: $r = 2^s$, for a suitable $s \in \mathbb{N}_0$. Hence, either $f^{2^s}(x) \equiv x \pmod{2^k}$, or $f^{2^s}(x) \not\equiv x \pmod{2^k}$, and in the latter case $\chi_i^{2^s} = \chi_i^0$, $\chi_{k-1}^{2^s} \equiv \chi_{k-1}^0 + 1 \pmod 2$, $i = 0, 1, 2, \ldots, k-2$. Thus, in the latter case $\sigma_{k-1}^{2^s} \equiv 1 \pmod 2$ in view of congruence (16.9). But then necessarily $\chi_{k-1}^{2^{s+1}} \equiv \chi_{k-1}^0 + \sigma_{k-1}^{2^{s+1}} \equiv \chi_{k-1}^{2^{s+1}} \pmod 2$ since $\sigma_{k-1}^{2^{s+1}} = 2\sigma_{k-1}^{2^s} \equiv 0 \pmod 2$; we just note that $\sigma_{k-1}^{2^s}$ is a sum modulo 2 of all values of the Boolean function φ_{k-1} when the number $\chi_0 + \chi_1 2 + \cdots \chi_{k-2} 2^{k-2}$ runs through the cycle of the permutation $f \bmod 2^{k-1}$ that contains the residue $x \bmod 2^{k-1}$. So in this case the length of the cycle of the permutation $f \bmod 2^k$ that contains the residue $x \bmod 2^k$ is 2^{s+1}.

Now everything is ready to prove the following proposition:

Proposition 16.30 *For every 1-Lipschitz measure-preserving function $f : \mathbb{Z}_2 \to \mathbb{Z}_2$, the function $f^n(x)$, $n \in \mathbb{N}_0$, can be expanded to the function $f^t(x)$, $(t; x) \in \mathbb{Z}_2^2$, which is a 1-Lipschitz* (thus, continuous) *two-variate function defined for every $(t, x) \in \mathbb{Z}_2^2$ and valuated in $\mathbb{Z}_2$.*

Proof We mimic the proof of Proposition 16.27. Let $n = m + 2^k\ell$, $m, n, \ell \in \mathbb{N}_0$, $k \in \mathbb{N}$. Then $f^n(x) \equiv f^m(x) \pmod{2^k}$ as the residue $x \bmod 2^k$ lies on some cycle of length 2^ℓ, $\ell \leqslant k$. That is, $|f^n(x) - f^m(x)|_2 \leqslant 2^{-k}$ whenever $|n - m|_2 \leqslant 2^{-k}$. This proves that given $t \in \mathbb{Z}_2$, the limit $\lim^2_{n \to t} f^n(x)$ exists.

Now, given $x, x', t, t' \in \mathbb{Z}_2$ such that $|x - x'|_2 \leqslant 2^{-n}$, $|t - t'|_2 \leqslant 2^{-m}$ we see that $|f^t(x) - f^{t'}(x')|_2 \leqslant 2^{-k}$ for $k = \min\{n, m\}$ since in this case $t \equiv t' \equiv r \pmod{2^k}$ and $x \equiv x' \equiv z \pmod{2^k}$ for suitable $r, z \in \{0, 1, \ldots, 2^k - 1\}$; so $f^t(x) \equiv f^{t'}(x') \equiv f^r(z) \pmod{2^k}$ as z lies on a cycle of length 2^ℓ, $\ell \leqslant k$, of the permutation $f \bmod 2^k$. □

16.6.2.2 The Universal Clock

The problem which immediately arises is that the p-adic time t is well-defined for every $t \in \mathbb{Z}_p$, but if q is a prime number distinct from p, the p-adic time t may be meaningless for a q-adic observer, the q-adic Little-endian, not speaking of the Big-endian. But fortunately *there is a clock (and therefore time) which is common both for all Little-endians and for the Big-endian. This clock/time is unique up to direction of time arrow.* It is clear that the clock which is common for all p-adic Little-endian and for the Big-endian must be a totally consistent function. The following theorem holds:

Theorem 16.31 *The totally consistent functions which are measure-preserving for all prime p are exactly the functions $x \mapsto \pm x + c$, where $c \in \mathbb{Z}$; only the functions $\tau_{\pm}(x) = x \pm 1$ are ergodic for all prime p.*

This means that "universal clock" is a standard odometer which runs forward ($\tau_{+}(x) = x + 1$) or backward ($\tau_{-}(x) = x - 1$).

Proof of Theorem 16.31 By Theorem 16.7, any totally consistent g function is a polynomial; therefore to be measure-preserving on $\mathbb{Z}_p$, g must be (1) bijective modulo p, and (2) its derivative $g'(x)$ must vanish modulo p nowhere, for all prime p; see Corollary 10.4. As g is 1-Lipschitz on $\mathbb{Z}_p$ for all prime p, and $g'(x)$ is a polynomial, the derivative exists and takes values from $\mathbb{Z}_p$ for all prime p; hence $g'(\mathbb{Z}) \subset \mathbb{Z}$. Therefore (as $g'(x)$ is a polynomial), condition (2) implies that $g'(\mathbb{Z}) \in \{1, -1\}$ which means that g' is a constant, ± 1. This means that g is affine function, namely, either $g(x) = -x + c$ or $g(x) = x + c$ for some $c \in \mathbb{Z}$ (since $g(0)$ must be an integer as $g(\mathbb{Z}) \subset \mathbb{Z}$ due to total consistency). This proves the claim about measure-preservation.

The ergodicity claim follows from the ergodicity criterion for affine maps $z \mapsto az + b$ which implies that if the map is ergodic on $\mathbb{Z}_p$, then $a \equiv 1 \pmod{p}$ and $b \not\equiv 0 \pmod{p}$; see Theorem 9.1. As these conditions must hold for all prime p, we conclude that $a = 1$ and $b \in \{1, -1\}$. □

Interpretation 16.32 (No Ordering of Events at the Smallest of Scales) *The only clock which is common for the "both ends of scale" is standard odometer $\tau(t) = t_0 + t$ which shows the time $t - t_0 \in \mathbb{R}$ elapsed since the moment $t_0 \in \mathbb{R}$. All observers acquire the value of time elapsed up to a non-zero error with respect to corresponding metrics. Therefore in a contrast to a real observer (the Big-endian), the p-adic observers (the Little-endians) generally cannot determine by "time stamps" of events which one of two events happened earlier and which one later since there is no order on the field of p-adic numbers which agrees with field operations.*

Note 16.33 It is known that generally there is no ordering of events in quantum mechanics; see, e.g., [157].

16.7 Wave Functions Produced by Automata

Within our approach, the initial automaton is a model of a (generally, open) physical system prepared in some fixed state; when experimenter probes the system (actually, a large collection of identical systems prepared in the same state), he exposes the system to time series of "elementary impacts," the system responds in time series of "elementary reactions": The impacts/reactions occurs at discrete instants of time since time is assumed to be discrete; e.g., at Planck's scale, the smallest time interval is Planck time $5.391247(60) \times 10^{-44}$ seconds. Concrete values of that smallest time interval depends on the process which is modeled (e.g., in smart contracts of digital economy that smallest interval is usually assumed to be 24 hours) and are not specified; the definitive feature of the model is that "time flow" consists of that "indivisible time intervals."

16.7.1 The Model of Measurement

Experimenter prepares a number of identical systems in the same state and probe them by exposing to different impacts, observing reactions and thus obtaining a number of experimental points ($\langle$impact$\rangle$; $\langle$reaction$\rangle$); in angle brackets there are measured numerical values of impacts and reactions. The Big-endian experimenter then treats any measured value as a real number up to a non-zero real error; i.e., the least significant digits of numerical values of $\langle$impact$\rangle$ and $\langle$reaction$\rangle$ remain unobserved for him. This is a typical "black-box" approach when the "internal structure" of the black box is absolutely unaccessible for the observer which may only make guesses/developing theories about this internal structure judging *only* by its reactions to impacts. *If causality is assumed, then the automaton model of the black box is looking a consistent one.*

For not to overload the exposition, in what follows we consider a one-dimensional case mostly when the values $\langle$impact$\rangle$ and $\langle$reaction$\rangle$ are numbers rather than vectors. Up to normalization, we may assume that the measured numerical values are all in the unit real interval $[0, 1]$; thus the Big-endian experimenter obtains a number of experimental points in the real unit square $[0, 1] \times [0, 1] = \mathbb{I}^2 \subset \mathbb{R}^2$. Namely, let an automaton $\mathfrak{A}$ be such a model; let $f = f_{\mathfrak{A}} \colon \mathbb{Z}_p \to \mathbb{Z}_p$ be the automaton function (i.e., a 1-Lipschitz map) of $\mathfrak{A}$. Consider a subset $\mathcal{E}(f)$ of all the following points of the Euclidean unit square $\mathbb{I}^2 = [0, 1] \times [0, 1] \subset \mathbb{R}^2$:

$$e_k^f(z) = \left(\frac{z \bmod p^k}{p^k}; \frac{f(z) \bmod p^k}{p^k}\right) \in \mathbb{I}^2,$$

$z \in \mathbb{Z}_p$, $k = 1, 2, \ldots$. Here $z \bmod p^k = \sum_{i=0}^{k-1} \chi_i p^i$ if $z \in \mathbb{Z}_p$ is represented by its canonical form $z = \sum_{i=0}^{\infty} \chi_i p^i$, ($\chi_i \in \mathbb{F}_p$; $i = 0, 1, 2, \ldots$). Note that $f(x) \bmod p^k$ corresponds to a k-letter output word $\xi_{k-1} \cdots\cdots\cdots \xi_1\xi_0$ of the automaton feeded

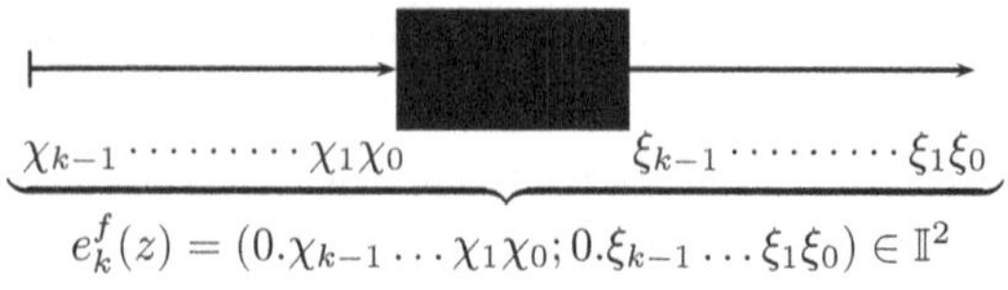

Fig. 16.2 A point in the unit square $\mathbb{I}^2 \subset \mathbb{R}^2$ which the Big-endian obtains by probing a black-box model of open system

by the k-letter input word $\chi_{k-1}\cdots\cdots\cdots\chi_1\chi_0$ which corresponds to $x \bmod p^k$ (cf. Fig. 16.2).

This way the experimenter in series of experiments obtains a number of experimental points in $\mathbb{I}^2$, for different $z \in \mathbb{Z}_p$, $k \in \mathbb{N}$; then the experimenter somehow interpolate the cluster points of the so-obtained point set in $\mathbb{I}^2$ and treats the cluster points as output of the experiment/observation. Mathematically, this process is completely described in Sect. 12.1; the so-obtained cluster points constitute the *limit plot* $\mathcal{P}'(\mathfrak{A})$ of the automaton $\mathfrak{A}$; see Definition 12.1.

A crucial feature of the *model of measurements* is that *due to the non-zero real error, the Big-endian experimenter (i.e., the macro-observer) is able neither to measure nor to control the low-order digits of the both values* $\langle$impact$\rangle$ *and* $\langle$reaction$\rangle$ represented by the base-p expansion of $z \bmod p^k/p^k$ whatever reasonable p is chosen. By measurements in $\mathbb{R}$ with accuracy ε worse than Planck's scale one cannot find prefix of input word since if $z \in (0, 1) \cap \mathbb{Z}_p \cap \mathbb{Q}$ then $z = d/(p^T - 1)$; thus higher-order digits of the base-p expansion of z as of a real number are determined by higher-order digits of $d \in \{0, 1, \ldots, p^T - 2\}$ while prefix of the word which corresponds to z is determined by low-order digits of the base-p expansion of d (cf. Sect. 2.3.2).

Speaking loosely, the plot is a picture the experimenter obtains as an output of the experiment which consists of a number of individual probes of a physical system which is prepared in the same state before each probe.

Note that *none of the two experimenters can be sure that the systems their probe* ***are prepared in the same state*** since this preparation of the system means that it is fed by a (specially chosen) long finite input word which switches the automaton into some fixed state which is then taken as initial; but the specially chosen word is so long that Little-endian can control only the prefix of the word which corresponds to the base p-expansion of the number $z = d/(p^T - 1)$, i.e., the low-order digits only, whereas the Big-endian can control the higher-order digits only. However, to control the state in which the automaton should be prepared, it is necessary to control the whole long word.

Note also that *the set of cluster points of the pictures of the both experimenters, the Little-endian and the Big-endian, obtain as result of the experiment look very similar for the both since Little-endian makes the word lengths as long as possible to construct the cluster points while the Big-endian is capable to get only the points which correspond to sufficiently long words, i.e., the points which are close to the cluster points. This fact is crucial for the future construction of the* ***wave function*** *by the both experimenters as well as for the* ***uncertainty relation*** *on which the both agree.*

Therefore, as both the preparation procedure and the measurement procedure are crucially dependent on the settings of the experimenter's equipment, the model of measurement introduced in the current subsection agrees on the violation of statistical independence with superdeterministic model (cf., e.g., [93]):

> The key property of a superdeterministic model is that it is a locally causal hidden variables model which violates statistical independence: that is, the probability distribution of the hidden variables is correlated with the detector settings at the time of measurement.

Also it is worth noting that the model of measurement we consider here agrees with the superdeterministic model of quantum mechanics regarding fine-tuning; see, e.g., [101]. We quote from that paper:

> ...let us merely note that two points that are close according to the Euclidean metric may be far away according to the p-adic metric.
> This means from the perspective of the p-adic metric, the distance between the actual world where the parity of the millionth digit of the input to Bell's pseudorandom number generator was, say, 0, and the counterfactual world where the parity was a 1 could be very large, even though it is small using an Euclidean measure of distance. A theory that seems fine-tuned with respect to the latter metric would not be fine-tuned with respect to the former metric. Like with Penrose's triangle, the seemingly impossible becomes understandable if we are prepared to modify our intuition about distance.

We will return to the comparison of superdeterministic and "classical" models of quantum mechanics once again in Sect. 16.9 after introducing wave functions of two different kinds which correspond to respective types of models.

16.7.2 "Sharp" and "Fuzzy" Wave Functions

Recall that in Sect. 12.1 for $s = \sum_{j=-k}^{\infty} \zeta_j p^j \in \mathbb{Q}_p$, ($\zeta_j \in \{0, 1, \dots, p-1\}$, $j \in \mathbb{Z}$), we denote via $[s]_p = \zeta_0 + \zeta_1 p + \zeta_2 p^2 + \cdots \in \mathbb{Z}_p$ and $]s[_p = \zeta_{-k} p^{-k} + \cdots + \zeta_{-1} p^{-1}$ the integral and the fractional parts of s, respectively. Recall that any complex character of additive group $\mathbb{Q}_p^+$ of the field $\mathbb{Q}_p$ of p-adic numbers is of the form $\chi_r(s) = \mathrm{e}^{2\pi \mathrm{i}]sr[_p}$, where $r \in \mathbb{Q}_p$; χ_r is a continuous group epimorphism into the group of complex roots of unity (which is isomorphic to the group $\mathbb{Q}^+/\mathbb{Z}^+$).

Take $r = 1$, denote $\chi_1 = \chi$; Then, given a 1-Lipschitz map $f: \mathbb{Z}_p \to \mathbb{Z}_p$, consider mappings

$$\check{f}_k: \mathrm{e}^{2\pi \mathrm{i}]p^{-k}z[_p} \mapsto \mathrm{e}^{2\pi \mathrm{i}]p^{-k}f(z)[_p}, \quad (z \in \mathbb{Z}_p),$$

for all $k \in \mathbb{N}_0$. As every $\check{f}_k$ maps points of the unit circle $\mathbb{S}$ into points of $\mathbb{S}$, the pairs $(\mathrm{e}^{2\pi \mathrm{i}]p^{-k}z[_p}; \mathrm{e}^{2\pi \mathrm{i}]p^{-k}f(z)[_p})$ constitute a set of points on the unit torus $\mathbb{T}^2 = \mathbb{S} \times \mathbb{S}$. The unit square $\mathbb{I}^2$ is a universal cover of the torus $\mathbb{T}^2$; this way the points $e_k^f(z) \in \mathbb{I}^2$ are identified with the points $(\mathrm{e}^{2\pi \mathrm{i}]p^{-k}z[_p}; \mathrm{e}^{2\pi \mathrm{i}]p^{-k}f(z)[_p}) \in \mathbb{T}^2$, and in what follows, we do not differ the both point sets and speak either of the points on the surface of the torus $\mathbb{T}^2$ or on the square $\mathbb{I}^2$ when and what is more convenient. The closure $\mathcal{P}(f) = \mathcal{P}(\mathfrak{A})$ of all the points

$$e_k^f = \left(\frac{z \bmod p^k}{p^k}; \frac{f(z) \bmod p^k}{p^k} \right)$$

in the square $\mathbb{I}^2$ (or, of all the points $(\mathrm{e}^{2\pi \mathrm{i}]p^{-k}z[_p}; \mathrm{e}^{2\pi \mathrm{i}]p^{-k}f(z)[_p})$ in the torus $\mathbb{T}^2$), where $k \in \mathbb{N}$, $z \in \mathbb{Z}_p$, we call a (one-dimensional) plot of the automaton $\mathfrak{A}$, or, which is the same, of the automaton function $f = f_{\mathfrak{A}}$. The set $\mathcal{P}'(f) = \mathcal{P}'(\mathfrak{A})$ of all the limit points of the plot, the derived set of the set $\mathcal{P}(f) = \mathcal{P}(\mathfrak{A})$, we call the limit plot of the automaton $\mathfrak{A}$ (of the automaton function $f_{\mathfrak{A}}$).

The Lebesgue measure of the plot $\mathcal{P}(\mathfrak{A})$ can be only 0 or 1, by the automata 0-1 law; see Theorem 13.8. This fact may be interpreted as manifestation of the two ways along which chaos emerges.

Interpretation 16.34 (Emergence of Chaos/Randomness: The Two Ways)

- ***First way***: *Chaos emerges from infinite "chaotic sequences" like randomly chosen real numbers from the unit interval* $[0, 1] \subset \mathbb{R}$ *by iterating the numbers via Bernoulli-shift-like mappings, logistic mappings, etc.: Thus, it is assumed a priory that "chaos/randomness do exist immanently" since the digits of the base-*p *expansion of real number from the unit interval* $[0, 1]$ *form a random string with probability 1 since the Martin-Löf random numbers constitute a set of Lebesgue measure 1 in* $[0, 1]$.

The plots of automata whose function is a polynomial over $\mathbb{Z}$ of degree greater than 1 not only are of measure 1 but the points $e_k^f(z)$ are uniformly distributed in $\mathbb{I}^2$ (and in $\mathbb{I}^n$ for all dimensions n), by E. Lerner's Theorem 13.17. This fact can be interpreted as the second way of how chaos emerges:

Interpretation 16.35 (Emergence of Chaos/Randomness: The Two Ways)

- ***Second way***: *Chaos emerges from the "lack of knowledge what elementary causes happened at the very beginning"; i.e., if the Big-endian observer is incapable to determine what are the digits* $\xi_0, \xi_1, \ldots, \xi_{k-1}, \ldots$ *of the input* $z = \sum_{j=0}^{\infty} \xi_j p^j \in \mathbb{Z}_p$ *of the causal function* f *if* k *is small enough and the number of iterations of* f *is large enough* (for instance, the function f may be a polynomial over $\mathbb{Z}$ whose degree is greater than 1; cf. Theorem 13.17)

Note that according to the *finiteness assumption* from Sect. 16.2, the chaos of the first type should be judged as "unphysical." In Sect. 16.4.2 we have already shown that chaos of the first type should not be judged as ontic due to the dependence on the metric the observer chooses to decide whether the phenomenon is chaotic or not.

Note also that polynomials over $\mathbb{Z}$ whose degrees are greater than 1 are automata functions of *infinite* automata (cf. Example 16.3). However, *any automaton function* $f: \mathbb{Z}_p \to \mathbb{Z}_p$ *of infinite automaton can be uniformly approximated on* $\mathbb{Z}_p$ *by automata functions of finite automata*; for instance, by the functions $f_n: z \mapsto f(z) \bmod p^n$. This fact, together with the finiteness assumption from Sect. 16.2, emphasizes the distinguished role which finite automata play in further considerations; so we now pay special attention to finite automata.

Theorems 13.8 imply that plots of finite automata are all of measure 0; therefore the limit plots of finite systems cannot contain "figures" but may contain "lines." *These lines are of the most importance in further considerations since they may naturally be treated as "experimental curves" obtained by probing a physical system* both by Little-endian and Big-endian observers since the smooth lines from limit plots of finite systems are *windings of torus*; therefore the lines may be treated as *sine waves*, so *the smooth lines in the limit plot of a finite automaton constitute a collection of sine waves*. Moreover, *the waves are limit plots of finite affine automata*. The waves should be judged as ontic since they are observed in quantum mechanical experiments. That is why we are going to derive wave function from the automata models of open systems.

From Sect. 13.6 we already know that C^2-smooth curves (i.e., those which are twice differentiable and have continuous second derivatives) in $\mathcal{P}'(f_{\mathfrak{A}})$ are windings of the torus $\mathbb{T}^2$ provided the automaton $\mathfrak{A}$ is finite (cf. Figs. 13.19 and 13.20).

All the necessary notions about the torus windings, torus knots, cables of torus, torus links, etc. are given in Sect. 13.6.2. In particular, from Theorem 13.70, it follows that the links which belong to limit plots of finite automata are exactly graphs of the maps $x \mapsto (ax + b) \bmod 1$, where $a, b \in \mathbb{Z}_p \cap \mathbb{Q}$. In cylindrical coordinates, every torus winding of a torus which is obtained by revolving around Z-axis a circle which is coplanar with the axis and has radius r and center at the distance R from the origin can be represented by the following parametric equations:

$$\begin{bmatrix} r_0 \\ \theta \\ z \end{bmatrix} = \begin{bmatrix} R + r\cos(ax+b) \\ x \\ r\sin(ax+b) \end{bmatrix}, \quad x \in \mathbb{R}.$$

If $a \in \mathbb{Z}_p \cap \mathbb{Q}$ then a is an irreducible fraction α/β where $\alpha, \beta \in \mathbb{Z}$ and $p \nmid \beta$; then corresponding winding winds β times around Z-axis and $|\alpha|$ times around a circle in the interior of the torus, the sign of α determines whether the rotation is clockwise or counterclockwise. Hence "physical meaning" that can be ascribed to the coefficient $a = \frac{\alpha}{\beta}$ of the affine map $z \mapsto az + b$, $(z \in \mathbb{Z}_p)$, which is a finite automaton function of affine automaton if and only if $a, b \in \mathbb{Z}_p \cap \mathbb{Q}$, is *frequency* (or, as a *wave number*, under proper choice of units). The choice of sign $+$ or $-$ depends only on what direction of rotation is assumed to be "positive" or "negative"; so *polarization*, *spin* can be ascribed to the sign of a in relevant models. In view of the above representation of the winding in cylindrical coordinates, Corollary 13.50 implies that limit plot of a finite automaton whose function is $z \mapsto az + b$, where $a, b \in \mathbb{Z}_p \cap \mathbb{Q}$, z runs over $\mathbb{Z}_p$, is in a one-to-one correspondence to a complex-valued function $\psi : \mathbb{R} \times \mathbb{N}_0 \to \mathbb{C}$:

$$\psi(x, k) = \mathrm{e}^{\mathrm{i}(ax - 2\pi p^k b)}, \quad \text{where } x \in \mathbb{R}, k \in \mathbb{N}_0 \tag{16.10}$$

The function $\psi(x, k)$ is well-defined for all $k \in \mathbb{Z}$ since p is invertible modulo β'/d and thus $\mathrm{e}^{-2\pi \mathrm{i} p^k b}$ is well-defined for every $k \in \mathbb{Z}$; see Theorem 13.46.

Note 16.36 By Theorem 13.46, different affine functions $z \mapsto az + b$ may have identical limit plots. For instance, all the functions $f(z) = z + c$ where $c \in \mathbb{Z}_p \cap \mathbb{Q}$ have identical limit plots which correspond to the function $\psi(x) = \mathrm{e}^{\mathrm{i}x}$. Note also that whenever a limit plot of a finite automaton $\mathfrak{A}$ is the same as of finite automaton whose automaton function f is affine, $f(z) = az + b$, there exist a *minimal sub-automaton* of $\mathfrak{A}$ (i.e., the one having no sub-automata other than itself) which has exactly the same limit plot, see Figs. 16.3 and 16.4.

Recall that finite automaton is minimal if and only if its reduced state transition diagram is totally connected: Given two states $s, t \in \mathcal{S}$, there is finite word w such that when the automaton in state s accepts the word w, the automaton changes

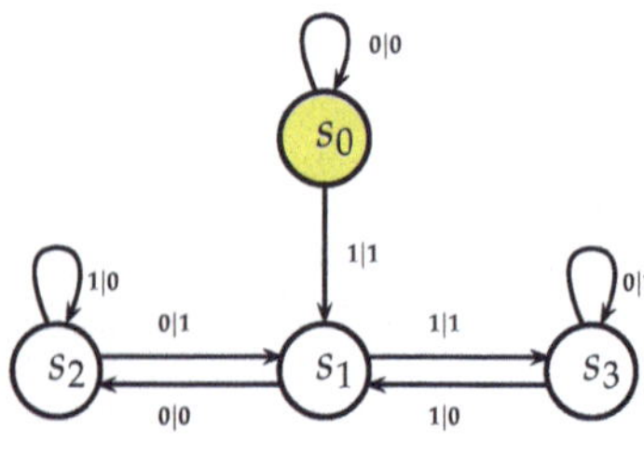

Fig. 16.3 Automaton function is $z \mapsto -\frac{1}{3}z$; minimal sub-automaton function is $z \mapsto -\frac{1}{3}z - \frac{2}{3}$, $(z \in \mathbb{Z}_2)$, s_0, s_1 are initial

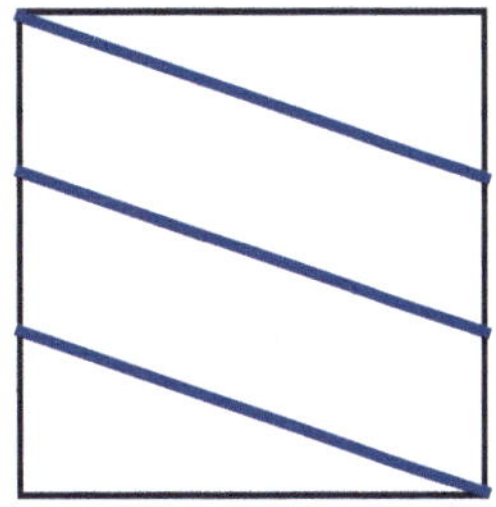

Fig. 16.4 Limit plots of the automaton and of its minimal sub-automaton coincide

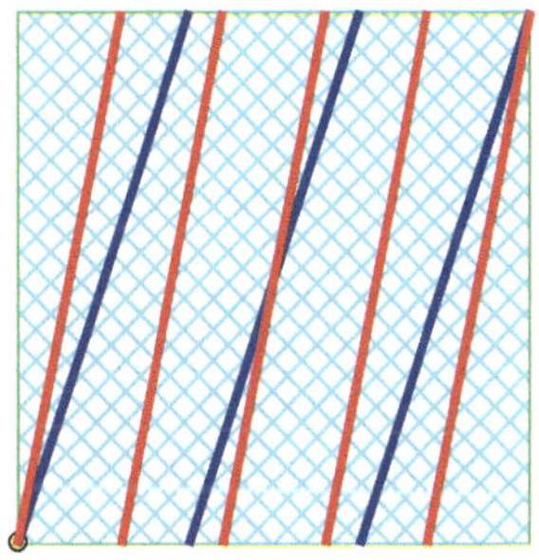

Fig. 16.5 Limit plot of the automaton having two sub-automata whose functions are $z \mapsto 3z$ and $z \mapsto 5z$, $(z \in \mathbb{Z}_2)$

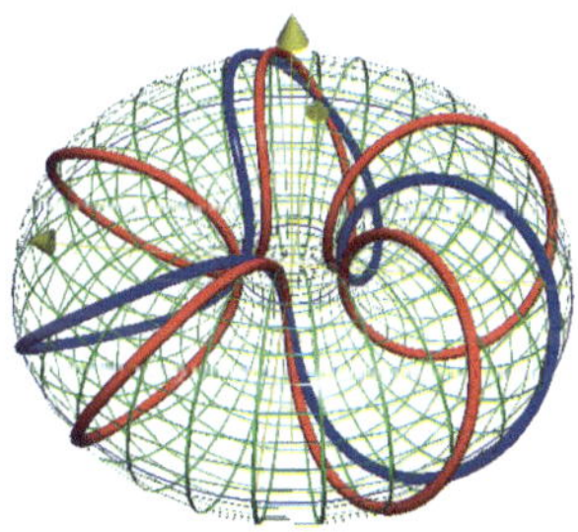

Fig. 16.6 Limit plot of the same automaton on the torus $\mathbb{T}^2 \subset \mathbb{R}^3$. The surface of the torus is made visible by cross-hatching

its state to t. If an automaton reaches a state which belongs to its (minimal) sub-automaton, the automaton will never reach a state which does not belong to the sub-automaton; cf. Sect. 1.6.

More examples of limit plots of automata follow. Figures 13.14 and 13.15 show limit plot of a constant function which is an automaton function of finite *autonomous* automaton; autonomous automata may be judged as models of either *isolated* or *closed* physical systems. Parallel lines shown by Fig. 13.14 may be ascribed to energy levels.

The rest examples are *non-autonomous* automata; these can serve as models of *open* physical systems. Figures 13.14 and 13.15 depict limit plots produced of an autonomous automaton whose state transition diagram depicts Fig. 13.16. Figures 16.5 and 16.6 show limit plot of an automaton having two minimal sub-automata; the state transition diagram of the automaton is shown by Fig. 16.7.

Figure 16.8 represents a plot of a finite automaton which approximates a measure-1 (whence, infinite) automaton whose automaton function is $z \mapsto 1 +$

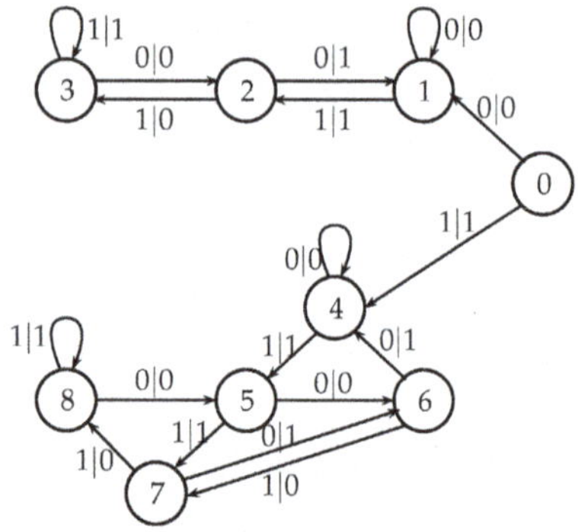

Fig. 16.7 State transition diagram of the automaton having two minimal sub-automata whose automata functions are $z \mapsto 3z$ and $z \mapsto 5z$, $z \in \mathbb{Z}_2$ Initial state is 0.

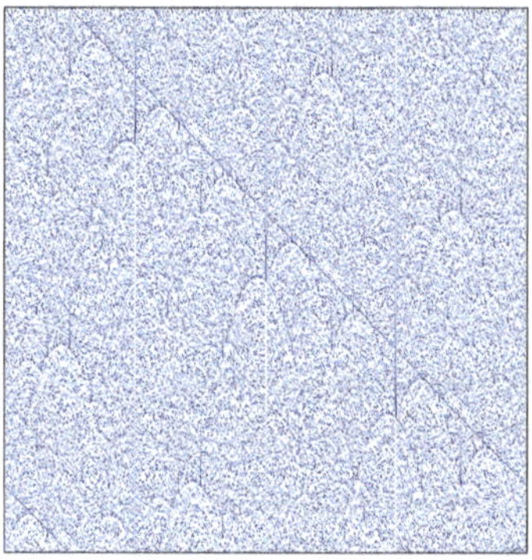

Fig. 16.8 Plot of finite automaton which approximates infinite automaton whose function is $z \mapsto 1 + 3z + 2z^2$, $(z \in \mathbb{Z}_2)$

$3z + 2z^2$, $(z \in \mathbb{Z}_2)$. Note straight lines in the plot; these lines constitute limit plot of a minimal sub-automaton.

Figure 16.9 depicts a plot of a measure-0 (but infinite) automaton which has the only minimal finite affine sub-automaton; the automaton function of the latter sub-automaton is $z \mapsto 5z$, $(z \in \mathbb{Z}_2)$. The limit plot of the latter automaton is red lines (cf. Fig. 16.5); the state transition diagram is the lower part of the diagram shown by Fig. 16.7.

Basically limit plot of a finite automaton whose minimal sub-automata are affine consists of families of parallel straight lines in the unit square or, respectively, of links of torus windings whose slopes are in $\mathbb{Z}_p \cap \mathbb{Q}$ (cf. Figs. 13.19, 13.20 and 16.5, 16.6). The minimal sub-automata from the first example "exhibit non-zero phase shifts," while the ones from the second example "phase shifts" are 0. The both examples are automata having two minimal affine sub-automata. The minimal sub-automata from the first example (Figs. 13.19 and 13.20) have limit plots defined by the functions $f_1(z) = -2z + \frac{1}{3}$ (red and green windings) and $f_2(z) = \frac{3}{5}z + \frac{2}{7}$, (yellow, brown, and blue windings), respectively, $z \in \mathbb{Z}_2$. The minimal sub-automata from the second example (Figs. 16.5 and 16.6) have limit plots defined by the respective functions $z \mapsto 3z$ (blue lines) and $z \mapsto 5z$ (red lines), $z \in \mathbb{Z}_2$.

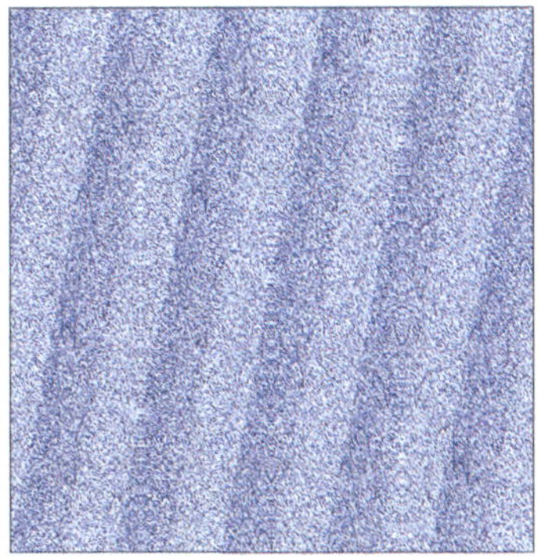

Fig. 16.9 Plot of a measure-0 automaton having the only minimal sub-automaton whose automaton function is $z \to 5z$, $(z \in \mathbb{Z}_2)$

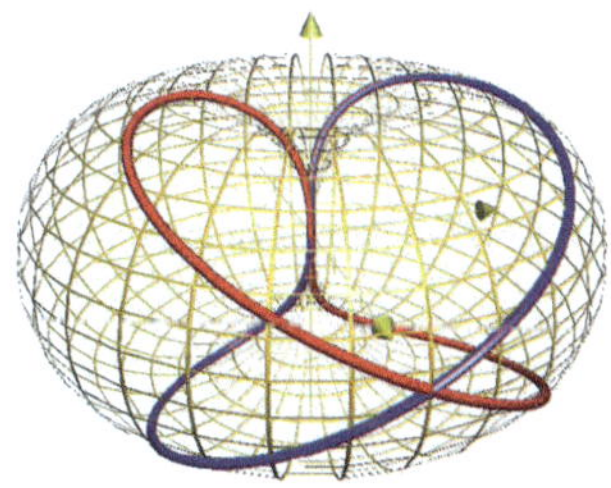

Fig. 16.10 Limit plot of a finite automaton whose automaton function is $z \mapsto ((z \text{ AND } 1) - ((\text{NOT}(z)) \text{ AND } 1))z$, $(z \in \mathbb{Z}_2)$ on the (horn) torus

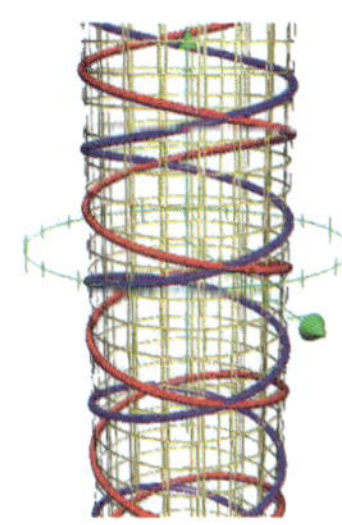

Fig. 16.11 Solenoid which is a limit plot of the automaton having the same automaton function $f(z) = ((z \text{ AND } 1) - ((\text{NOT } (z)) \text{ AND } 1))z$

Limit plot of a finite affine automaton whose automaton function is $z \mapsto az + b$ in the unit square $\mathbb{I}^2$ consists of parallel straight lines with slope $a = \alpha/\beta \in \mathbb{Z}_p \cap \mathbb{Q}$; so the plot may be considered not only on the torus obtained by "gluing together" opposite sides of the square but also on a cylinder obtained by "gluing together" only a pair of opposite sides of the square. This way one obtains *solenoid* rather than torus link. That representation of limit plot is also convenient in some cases. For instance, Figs. 16.10 and 16.11 depict limit plot of the automaton whose automaton function is $f(z) = ((z\text{AND}1) - ((\text{NOT}(z))\text{AND}1)) \times z$, where AND and NOT are, respectively, bitwise logical "and" and bitwise logical "not" operations on base-2 expansions of numbers with no carries; see Sect. 9.5.1; "$\times$" and "$-$" are usual multiplication and subtraction of numbers (with carries).

Figure 16.12 represents state transition diagram of a general automaton all whose minimal automata are finite and affine.

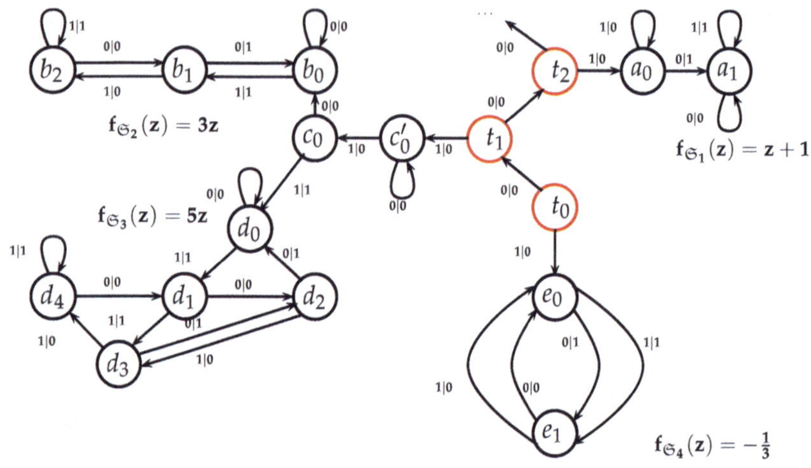

Fig. 16.12 General automaton all whose minimal sub-automata are finite and affine

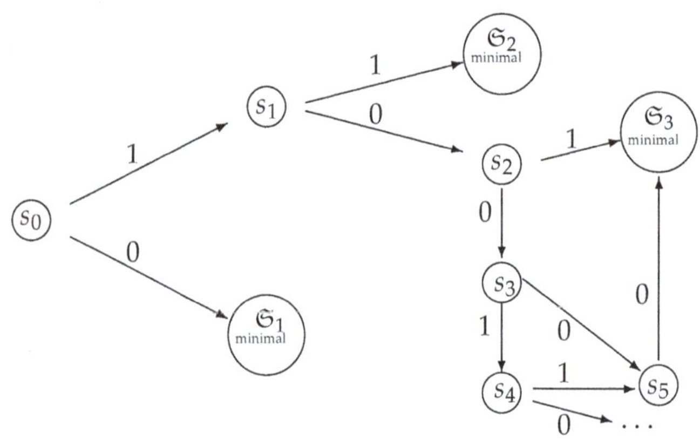

Fig. 16.13 Example state transition diagram of 2-adic automaton having minimal sub-systems (output symbols in the labels of arrows are omitted); s_0 is the initial state. Respective probabilities of reaching subsystems $\mathfrak{S}_1$, $\mathfrak{S}_2$, $\mathfrak{S}_3$ are $1/2$; $1/4$; $11/64 = 1/8 + 1/32 + 1/64$

As said, our goal is derive wave functions from causal functions, that is, from automata. Functions (16.10) are building blocks of the construction of the wave function on the base of causal maps. To start with, we briefly outline general idea of the construction.

Recall that reduced state transition diagram of a finite automaton is a digraph of each path in which it ultimately reaches a minimal sub-automaton. There are no outgoing paths from sub-automata. By feeding the automaton with random long words, to each minimal sub-automaton we assign a probability when the automaton reaches states which belong to the sub-automaton (cf. Fig. 16.13). Let automaton $\mathfrak{A}$ be such that, being fed by random long words, with probability 1 the automaton reaches (at a finite step) a state which belongs to a minimal sub-automaton which is finite and affine. Limit plot of every such automaton is described by a complex-valued function of the form (16.10). To every minimal sub-automaton which is finite and affine, it is ascribed a limit plot. There are only countably many such limit plots since there only countable many affine functions $\mathbb{Z}_p \to \mathbb{Z}_p$ which are automata functions of these sub-automata: Due to the finiteness, coefficients of these affine function must belong to a countable set $\mathbb{Z}_p \cap \mathbb{Q}$. As every two minimal sub-automata have no common states due to the minimality, and as to every minimal sub-automaton, it is assigned a probability of reaching the sub-automaton, to every limit plot one assigns a probability to "observe" that limit plot in experiment, i.e., to obtain accumulation points in the unit square which constitute that limit plot. The probability is equal to the sum of all probabilities to reach minimal sub-automata having that plot. Therefore these probabilities constitute a distribution assigned to the automaton, the *characteristic function* of that distribution is a (generally infinite) series whose terms are functions $\psi(x, k)$ (cf., (16.10)), multiplied by values of respective probabilities. We are going to argue that this characteristic function of the distribution may be treated as a wave function since the characteristic function of a distribution is a Fourier transform of the probability measure (cf. [68, Chapter 10]). We recall the definition (in terms of probability theory) of characteristic functions of distributions and its basic properties following [179].

Let X be a real-valued random variable with distribution function $F(x)$. The characteristic function $f(t)$ of the distribution function $F(x)$ (or of the random variable X) is defined as

$$f(t) = \int_{-\infty}^{+\infty} e^{itx} dF(x) = \mathsf{E}e^{itX}, \ -\infty < t < +\infty \tag{16.11}$$

Here E stands for expectation. If $F(x)$ is absolutely continuous with density $p(x)$, then

$$f(t) = \int_{-\infty}^{+\infty} p(x)e^{itx} dx \tag{16.12}$$

If X is discrete with $\text{Prob}(X = x_n) = p_n$, $n = 1, 2, \ldots$, $\sum_{n=1}^{\infty} p_n = 1$, then

$$f(t) = \sum_{n=1}^{\infty} p_n e^{ix_n t} \tag{16.13}$$

Two distribution functions are identical if and only if their characteristic functions are identical.

Now recall that the wave function $\Psi(x, t)$ of real variables x (for the position of a particle) and t (for time) must be a solution of Schrödinger equation; therefore the wave function (under the assumption that spatial and temporal variables are separated) is

$$\Psi(x,t)=\sum_{n=1}^{\infty}c_n\psi_n(x)\mathrm{e}^{-\mathrm{i}E_nt/\hbar}, \tag{16.14}$$

since every solution to the (time-dependent) Schrödinger equation can be written in this form; see [81, Section 2.1]. In what follows *we argue that the similarity between the expressions* (16.13) *and* (16.14) *is not occasional but reflects intrinsic ties between quantum theory and the theory of causal functions over discrete time* (i.e., of automata functions, the 1-Lipschitz p-adic functions $\mathbb{Z}_p \to \mathbb{Z}_p$) which is the leading theme of the book.

We are going to construct automata of special kind which in some precise meaning "produce" wave functions. First we briefly outline the idea of the construction.

Recall from Sect. 1.6 that the reduced state transition diagram of a finite automaton is a digraph of each path in which it ultimately reaches a minimal sub-automaton. There are no outgoing paths from sub-automata. By feeding the automaton with random long words, to each minimal sub-automaton, we assign a probability when the automaton reaches states which belong to the sub-automaton (cf. Fig. 16.13).

Let the automaton $\mathfrak{A}$ be such that being fed by random long words, with probability one the automaton reaches at some finite step some state which belongs to some minimal automaton which is finite and affine. Limit plot of every such automaton is described by a complex-valued function of the form (16.10). To every minimal sub-automaton which is finite and affine, it is ascribed a limit plot. There are only countably many such limit plots since there only countable many affine functions $\mathbb{Z}_p \to \mathbb{Z}_p$ which are automata functions of these sub-automata: Due to the finiteness, coefficients of these affine function must belong to a countable set $\mathbb{Z}_p \cap \mathbb{Q}$. As every two minimal sub-automata have no common states due to the minimality, and as to every minimal sub-automaton, it is assigned a probability of reaching the sub-automaton, to every limit plot one assigns a probability to "observe" that limit plot in experiment, i.e., to obtain accumulation points in the unit square which constitute that limit plot. The probability is equal to a sum of all probabilities to reach minimal sub-automata having that plot. Therefore these probabilities constitute a distribution assigned to the automaton, the characteristic function of that distribution which is a (generally infinite) series whose terms are functions $\psi(x,k)$ from (16.10) multiplied by values of respective probabilities. We argue that this characteristic function of the distribution may be treated as a wave function.

Going to a formal rigorous construction, let us make some preliminary conventions. In what follows:

- we do not distinguish affine automata whose limit plots coincide, so actually the probability distribution related to automaton is taken over classes of finite affine sub-automata having coinciding limit plots;

- we use terms "p-adic integer," "infinite word over p-symbol alphabet," and "infinite path in a state transition diagram" as synonyms; see Sects. 1.4.1, 1.6.1, and 2.3.3.

A word of caution: There is a one-to-one correspondence between all paths of length k in the state transition diagram and all numbers from $\{0, 1, \ldots, p^k - 1\}$; to every number from $\mathbb{N}_0 = \{0, 1, 2, \ldots\}$ it corresponds infinitely many infinite paths. Every such path has a prefix which is just the base-p expansion of a number, and a suffix which consists of zeros only (cf. Sect. 1.4.1).

Given an automaton $\mathfrak{A}$, let $\mathfrak{S}$ be its sub-automaton. Let $W(\mathfrak{S})$ be the set of all infinite paths starting from the initial state of $\mathfrak{A}$ in the state transition diagram (the p-adic tree) of $\mathfrak{A}$ which reach states of $\mathfrak{S}$ at finite steps. Note that if a path w reaches $\mathfrak{S}$ at k-th step, then all paths which correspond to infinite words having the same prefix of length k reach $\mathfrak{S}$ at the k-th step; therefore, the p-adic integers which correspond to these paths constitute a p-adic ball of radius p^{-k}. Therefore, *all p-adic integers that correspond to infinite paths which reach the sub-automaton $\mathfrak{S}$ at finite steps constitute a disjoint union $\mathbf{B}(\mathfrak{S})$ of balls of non-zero radii; whence, $\mathbf{B}(\mathfrak{S})$ is a μ_p-measurable subset of $\mathbb{Z}_p$ with respect to the probability measure μ_p on $\mathbb{Z}_p$ (cf. Sect. 2.4). This way to $\mathfrak{S}$ we assign a probability $\mu(\mathfrak{S}) = \mu_p(\mathbf{B}(\mathfrak{S}))$.*

Note that the set $W(\mathfrak{S})$ does not depend on concrete state transition diagram of the automaton $\mathfrak{A}$ but to be more definite one may assume that the state transition diagram of the automaton is reduced, thus, given the automaton, a unique (cf. Sect. 1.6.1). In that case some care should be taken speaking of paths since some arrows in the reduced state transition diagram may actually be loops; see, e.g., Fig. 16.12: The paths (which are written from left to right) that begin at the initial state t_0 and have prefixes 0111, 01011, 010011, 0100011, ... all reach the sub-automaton $\mathfrak{S}_3$ on 4-th, 5-th, 6-th, 7-th,.. steps, respectively, so the probability to reach the sub-automaton $\mathfrak{S}_3$ is $1/16+1/32+1/64+1/128+\cdots = 1/8$ and $\mathbf{B}(\mathfrak{S}_3)$ is a disjoint union of balls $\mathbf{B}_{1/16}(14)$, $\mathbf{B}_{1/32}(26)$, $\mathbf{B}_{1/64}(50)$, ..., $\mathbf{B}_{1/2^k}(2+3\times 2^{k-2})$,... where $k = 4, 5, 6, \ldots$.

Given two minimal sub-automata $\mathfrak{S}$ and $\mathfrak{T}$ of the automaton $\mathfrak{A}$ which are finite and affine, by the minimality one has $\mathbf{B}(\mathfrak{S}) \cap \mathbf{B}(\mathfrak{T}) = \emptyset$; so the probability that a random infinite path starting from the initial state reaches at a finite step some minimal sub-automaton of the automaton $\mathfrak{A}$ is the sum $\sum \mu_p(\mathbf{B}(\mathfrak{S}))$ taken over all minimal sub-automata $\mathfrak{S}$ which are finite and affine. Call an automaton $\mathfrak{A}$ *ultimately affine* if the probability is 1. Note that by König's lemma (see Proposition 1.21), there are infinite paths which never reach states belonging to these minimal sub-automata. These paths constitute a μ_p-measurable subset in $\mathbb{Z}_p$, but the measure of the subset is 0 since the subset is a complement to a countable union of balls whose measure is 1. For instance, the path $111\ldots$ in the reduced state transition diagram of the 2-adic odometer depicted by Fig. 13.11 (whose non-reduced diagram is the 2-adic tree depicted by Fig. 1.2) never reaches minimal sub-automaton (which has the only state s_1) but all other paths reach the sub-automaton at finite steps; so the probability to reach that minimal sub-automaton is 1.

Definition 16.37 (Plot Equivalence of Automata) Call finite affine automata $\mathfrak{S}$ and $\mathfrak{T}$ *plot equivalent*, $\mathfrak{S} \equiv_{\mathcal{P}} \mathfrak{T}$, if their respective functions $\psi\colon \mathbb{R}\times\mathbb{Z} \to \mathbb{C}$ defined by (16.10) coincide; that is, if their limit plots coincide, $\mathcal{P}'(\mathfrak{S}) = \mathcal{P}'(\mathfrak{T})$; i.e., if the limit plots are links of the same number of torus windings having common slope (cf., Theorem 13.46).

Given $a, b \in \mathbb{Z}_p \cap \mathbb{Q}$ denote $\mathfrak{S}_{a,b}$ an automaton whose automaton function is $z \mapsto az + b$. Let $[\mathfrak{S}_{a,b}]$ be the set of all minimal sub-automata of $\mathfrak{A}$ which are plot-equivalent to $\mathfrak{S}_{a,b}$. By the minimality, given $\mathfrak{S}, \mathfrak{T} \in [\mathfrak{S}_{a,b}]$ the sub-automata $\mathfrak{S}$ and $\mathfrak{T}$ have no common states; therefore $\mathbf{B}(\mathfrak{S}) \cap \mathbf{B}(\mathfrak{T}) = \emptyset$. That is, the probability

$$q_{[\mathfrak{S}_{a,b}]} = \sum_{\mathfrak{S}\in[\mathfrak{S}_{a,b}]} \mu_p(\mathbf{B}(\mathfrak{S}))$$

is well-defined. Given $a, b \in \mathbb{Z}_p \cap Q$, the equivalence relation $\equiv_{\mathcal{P}}$ induces an equivalence relation on the set of all pairs $(a; b) \in (\mathbb{Z}_p \cap \mathbb{Q}) \times (\mathbb{Z}_p \cap \mathbb{Q})$ which we denote by the same symbol, i.e., $(a; b) \equiv_{\mathcal{P}} (c; d)$ if and only if $\mathfrak{S}_{a,b} \equiv_{\mathcal{P}} \mathfrak{S}_{c,d}$.

Let $\operatorname{Spec}(\mathfrak{A})$ be the set of all equivalence classes defined by minimal sub-automata of $\mathfrak{A}$ which are finite and affine. Then the following series:

$$\Psi_{\mathfrak{A}}(\rho, k) = \sum_{[\mathfrak{S}_{a,b}]\in\operatorname{Spec}(\mathfrak{A})} q_{[\mathfrak{S}_{a,b}]} \mathrm{e}^{\mathrm{i}(a\rho - 2\pi p^k b)} \tag{16.15}$$

converges absolutely for all $\rho \in \mathbb{R}$, $k \in \mathbb{Z}$ and therefore defines a complex-valued function $\Psi_{\mathfrak{A}}(\rho, k)$. Call the function $\Psi_{\mathfrak{A}}$ the *sharp wave function* assigned to the automaton $\mathfrak{A}$.

Theorem 16.38 *Given non-negative real numbers $q_1, q_2, \ldots$ such that*

$$\sum_{j=1}^{\infty} q_j = 1$$

and finite affine automata $\mathfrak{S}_j = \mathfrak{S}_{a_j,b_j}$, $(a_j, b_j \in \mathbb{Z}_p \cap \mathbb{Q}$, $j = 1, 2, \ldots)$ which are pairwise plot-nonequivalent, there exists an ultimate affine automaton $\mathfrak{A}$ such that $\operatorname{Spec}(\mathfrak{A}) = \{[\mathfrak{S}_j]\colon j = 1, 2, \ldots\}$, $q_j = q_{[\mathfrak{S}_{a_j,b_j}]}$, *and* $\Psi_{\mathfrak{A}}(\rho, k) = \sum_{j=1}^{\infty} q_j \mathrm{e}^{\mathrm{i}(a_j\rho - 2\pi p^k b_j)}$.

To prove the theorem, we need a lemma. Actually *the lemma proves that any discrete random variable can be "modeled" on $\mathbb{Z}_p$*. This result is a sort of the one stated by Theorem 7.14 but concerns only discrete distributions on $\mathbb{Z}_p$.

Lemma 16.39 *Given convergent series $\sum_{j=0}^{\infty} q_j = 1$ of positive real numbers $q_j \in \mathbb{R}_{\geqslant 0}$, there exist pairwise disjoint open sets $W_j \subset \mathbb{Z}_p$ such that the normalized Haar measure μ of W_j is q_j, $j = 0, 1, 2, \ldots$.*

Proof of Lemma 16.39 Consider the Monna map $\mathsf{Mon}(z) = \sum_{i=0}^{\infty} \alpha_i p^{-i-1} = 0.\alpha_0\alpha_1\alpha_2 \ldots \in [0, 1] \subset \mathbb{R}$ where $z = \sum_{i=0}^{\infty} \alpha_i p^i$ is a p-adic canonical expansion of $z \in \mathbb{Z}_p$; that is, $\mathsf{Mon}(\mathbf{B}_{1/p^k}(a)) = [0.\alpha_0\alpha_1 \ldots \alpha_{k-1}, 0.\alpha_0\alpha_1 \ldots \alpha_{k-1} + p^{-k}] \subset [0, 1]$, where $a = \sum_{i=0}^{k-1} \alpha_i p^i \in \mathbb{Z}_p$. The Monna map maps p-adic balls $\mathbf{B}_{1/p^k}(a) \subset \mathbb{Z}_p$ of radii $1/p^k$ centered at $a \in \mathbb{Z}_p$ onto closed subintervals of length $1/p^k$ of the unite interval $[0, 1]$; note that $\lambda(\mathsf{Mon}(\mathbf{B}_{1/p^k}(a))) = \mu_p(\mathbf{B}_{1/p^k}(a))$ where λ is Lebesgue measure on the unit real interval $[0, 1]$, i.e., the length of the closed interval (cf. Sect. 7.2.1).

Without loss of generality assume that all $q_j > 0$. Split the unit interval $[0, 1]$ into pairwise disjoint open intervals Q_j such that the length of the j-th interval Q_j is q_j: Namely, let $Q_1 = (0, q_1)$, $Q_2 = (q_1, q_1 + q_2)$, $Q_3 = (q_1 + q_2, q_1 + q_2 + q_3)$, ...; then $Q = \bigcup_{j=0}^{\infty} Q_j$ is λ-measurable and $\lambda(Q) = 1$.

For each Q_j let B_j be the set of all balls of non-zero radii such that $\mathsf{Mon}(\mathbf{B}) \subset Q_j$ for every $\mathbf{B} \in B_j$. As any two p-adic balls either disjoint or one is a subset of another one, the set B_j is a countable disjoint union of balls of non-zero radii. Thus, B_j is open as each p-adic ball of non-zero radius is clopen; whence B_j is μ_p-measurable. As every point from Q_j lies in Mon-image of some ball from B_j, we conclude that $\mu_p(B_j) = q_j$ and $\mu_p(\bigcup_{j=1}^{\infty} B_j) = \sum_{j=1}^{\infty} \mu_p(B_j) = 1$ as $B_j \cap B_k = \emptyset$ when $j \neq k$, by the construction. □

Proof of Theorem 16.38 Follows immediately from the proof of Lemma 16.39. Every B_j, $j = 1, 2, \ldots$, is a countable disjoint union of balls $\mathbf{B}_{1/p^{r_{jm}}}(a_{jm})$, $m = 1, 2, 3, \ldots$, centered at $a_{jm} = \sum_{k=0}^{r_{jm}-1} \alpha_{j,m,k} p^k \in \mathbb{Z}_p$. Let branches of a p-adic tree be $\alpha_{j,m,0}\alpha_{j,m,1} \cdots \alpha_{j,m,r_{jm}-1}$; let leafs be $\mathbf{B}_{1/p^{r_{jm}}}(a_{jm})$, $j, m = 1, 2, \ldots$. In this digraph replace all leafs $\mathbf{B}_{1/p^{r_{jm}}}(a_{jm})$ by state transition diagrams of automata $\mathfrak{S}_m \in [\mathfrak{S}_j]$. The so constructed digraph is a state transition diagram of the automaton $\mathfrak{A}$ which is ultimate affine and such that $\Psi_{\mathfrak{A}}(\rho, k) = \sum_{j=1}^{\infty} q_j e^{\mathrm{i}(a_j\rho - 2\pi p^k b_j)}$. □

Note 16.40 From the proof of Theorem 16.38, it follows that ultimate affine automaton may be either measure-0 or measure-1. The first case occurs when, e.g., in the series $\sum_{j=0}^{\infty} q_j$ is finite as then $\mathfrak{A}$ is finite, whence, measure-0; the measure-1 case occurs when, e.g., all coefficients $a_j \in \mathbb{Z}_p \cap \mathbb{Q}$ constitute a dense subset in $\mathbb{R}$ and all $b_j = 0$.

In what follows, we will need a slightly generalized version of Lemma 16.39.

Corollary 16.41 (Generalized Lemma 16.39) *Given convergent series $\sum_{j=0}^{\infty} q_j = q \leqslant 1$ of positive real numbers $q_j \in \mathbb{R}_{\geqslant 0}$, there exist pairwise disjoint open sets $W_j \subset \mathbb{Z}_p$ such that the measure μ_p of W_j is q_j, $j = 0, 1, 2, \ldots$.*

Proof of Corollary 16.41 Take $[0, q]$ instead of $[0, 1]$ in the proof of Lemma 16.39 and modify the argument in obvious way. □

Sharp wave functions can be considered as wave functions with respect to *discrete time* since the map $\mathrm{e}^{2\pi i b} \mapsto \mathrm{e}^{2\pi i p^k b}$ is equivalent to a k-digit *shift* of the base-p representation of b and reduction of the result modulo 1.

As k is the order of time elapsed (and measured by p-adic clock; see Sect. 16.6.2 and Fig. 16.1) since the moment the automaton reaches a state from its minimal affine sub-automaton whose automaton function is $z \mapsto az + b$, the sharp wave function may be judged as the one the Little-endian can construct by observing reactions of a physical system at the smallest of scales.

We argue that a wave function with respect to *continuous time* can also be constructed by using ultimate affine automata. The core idea of the construct is using of beta-representations of numbers rather than the base-p expansions. The beta-representations of real numbers were first introduced by A. Rényi in 1957 and since that time attracted substantial attention in ergodic theory and symbolic dynamics; see, e.g., monograph [143].

Recall that given real $\beta > 1$, a *β-representation* of real $b \geqslant 0$ is infinite word $\chi_0\chi_1\cdots$ over the alphabet $\mathcal{B} = \{0, 1 \ldots, \lfloor\beta\rfloor\}$ such that $b = \sum_{j=-k}^{\infty} \chi_{k+j}\beta^{-k-j}$. Note that we consider β-representations of real $b \geqslant 0$, and not only of real $b \in [0, 1]$ as in [143]. Of course, in (16.15) we always may assume that $b \in [0, 1]$; but to assign real numbers to paths in state transition diagrams of automata, we need beta-representations of numbers from $\mathbb{N}_0$ which then are converted into real numbers in a way similar to that we use in Sect. 16.7.1 by exploiting p-adic representations.

Namely, first we use β instead of p: Thus, each arrow in a state transition diagram of the automaton whose input and output alphabets are $\mathcal{B}$ is labeled by the pair $\chi|\xi$, where $\chi, \xi \in \mathcal{B}$; to infinite path which starts from the initial state it corresponds infinite word $w = \chi_0\chi_1\cdots$ over the alphabet $\mathcal{B}$, to w we put into the correspondence the $(\lfloor\beta\rfloor + 1)$-adic integer $\sum_{j=0}^{\infty} \chi_j(\lfloor\beta\rfloor + 1)^j$. To construct a plot, we convert these $(\lfloor\beta\rfloor + 1)$-adic integers into sequences of real numbers $\chi_0\beta^{-1}$, $\chi_1\beta^{-1} + \chi_0\beta^{-2}$, $\chi_2\beta^{-1} + \chi_1\beta^{-2} + \chi_0\beta^{-3}$,... thus obtaining points $(\chi_{k-1}\beta^{-1} + \cdots + \chi_0\beta^{k-2}; \xi_{k-1}\beta^{-1} + \cdots + \xi_0\beta^{k-2}) \in \mathbb{R}^2$. To put it in other words, we just use β-representations for input/output words of the automaton $\mathfrak{A}$ when constructing a plot of the automaton, but the automaton function is still a 1-Lipschitz map from $(\lfloor\beta\rfloor + 1)$-adic integers to $(\lfloor\beta\rfloor + 1)$-adic integers. This way we have constructed the sharp wave function $\Psi_{\mathfrak{A}}(\rho, k) = \sum_{[\mathfrak{S}_{a,b}]\in\mathrm{Spec}(\mathfrak{A})} q_{[\mathfrak{S}_{a,b}]}\mathrm{e}^{\mathrm{i}(a\rho - 2\pi(\lfloor\beta\rfloor+1)^k b)}$ (cf. (16.15)), which is a well-defined complex valued-function of $\rho \in \mathbb{R}$ and $k \in \mathbb{Z}$; then we replace $(\lfloor\beta\rfloor + 1)$ by β in the formula thus resulting in another complex-valued function of $\rho \in \mathbb{R}$ and $k \in \mathbb{Z}$. The crucial point is that if $1 < \beta \lll 2$; i.e., if $\beta = 1 + \tau$ where $0 < \tau \lll 1$, then $\beta^k = (1 + \tau)^k \approx 1 + k\tau$. When τ is small (e.g., if $\tau = 5.391247(60) \times 10^{-44}$ s, the Planck time) then for the Big-endian observer who is incapable to perform measurements with that accuracy (which is currently only about 10^{-20} sec) $k\tau \in R$ is indistinguishable from continuous time. This way we have constructed the *fuzzy wave function*:

$$\tilde{\Psi}_{\mathfrak{A}}(\rho, t) = \sum_{[\mathfrak{S}_{a,b}] \in \mathrm{Spec}(\mathfrak{A})} q_{[\mathfrak{S}_{a,b}]} \mathrm{e}^{\mathrm{i}(a\rho - 2\pi t b)} \tag{16.16}$$

which is ascribed to the automaton $\mathfrak{A}$: The function is well-defined for all $\rho, t \in \mathbb{R}$ since the series converges absolutely.

It is also worth noting that *since* $\lfloor\beta\rfloor = \lfloor 1+\tau \rfloor = 1$, *i.e.,* $\mathcal{B}$ *is a two-letter alphabet, then necessarily* $p = 2$; *see sharp wave function formula* (16.15). This observation can be judged as a yet one more contribution in support of J. A Wheeler's "it from bit" doctrine [188].

By the construction, the *sharp wave function* (which is a discrete time function) *can be viewed as approximation of a fuzzy wave function which is a continuous time function: Actually, the fuzzy wave function is a "usual" wave function given by the expression* (16.14).

The term "approximation" here is not rigorous: To prove that statement with a full rigor is a separate problem which will be addressed in the future. Now we only are going to find the exact representation for $\beta = 1 + \tau$ under the *finiteness assumption* from Sect. 16.2; but before doing this, we illustrate the usage of that β-representation by the analogy with movies: Each frame of a movie contains a number of details, but to cause an illusion of motion to a viewer, only a little share of the whole number of details are changed from frame to the next frame; the smaller is the share, the slower the motion is, from the viewer's standpoint. For a Little-endian viewer that share is $p - 1$ since he uses the base-p representation of numbers; in the case when that share is τ one has the $(1 + \tau)$-representation. If $0 < \tau \lll 1$ we are at a Big-endian viewer's standpoint.

16.7.3 Consequences of the Finiteness Assumption: The "Size" of Automata

To start with, it is important to stress that *to represent numbers from* $\mathbb{N}_0$ *in the base* β, *we use only non-negative powers of* β *in order to guarantee the uniqueness of* β*-representation for each number from* $\mathbb{N}_0$ since if negative powers of $\beta = 1 + \tau$ when $\tau \lll 1$ are allowed in β-representations, then every number from $(0; \tau^{-1})$ has a continuum of distinct β-representations provided $\tau < \frac{\sqrt{5}-1}{2}$, [61]. But in that case the very problem of assigning a number to a finite path in state transition diagram becomes ill-posed. Under the said convention, we assert:

The finiteness assumption implies $\beta = \sqrt[N]{2}$

That is, the following theorem is true.

Theorem 16.42 *Let $1 < \beta < 2$. If an automaton which performs addition of β-representations of numbers from $\mathbb{N}_0$ is finite then necessarily $\beta = \sqrt[N]{2}$ for some $N \in \mathbb{N}$. For each $N \in \mathbb{N}$, addition of numbers from $\mathbb{N}_0$ which are represented by $\sqrt[N]{2}$-representations can be performed by a finite automaton.*

Proof The number 1 admits the only β-representation $1 = 1+0\times\beta+0\times\beta^2+\cdots$ in non-negative powers of β as $\beta > 1$. A finite automaton maps eventually periodic sequences to eventually periodic sequences; therefore if a finite automaton which maps pairs of infinite words into infinite words over the alphabet $\mathcal{B} = \{0, 1\}$ and performs $1 + 1 = 2$, then necessarily

$$\begin{aligned}
&2 = \alpha_0 + \alpha_1\beta + \cdots + \alpha_{n-1}\beta^{n-1} + \\
&+ (\gamma_0 + \gamma_1\beta + \cdots + \gamma_{s-1}\beta^{s-1})\beta^n + (\gamma_0 + \gamma_1\beta + \cdots + \gamma_{s-1}\beta^{s-1})\beta^{2n} + \cdots = \\
&= \alpha_0+\alpha_1\beta+\cdots+\alpha_{n-1}\beta^{n-1}+(\gamma_0+\gamma_1\beta+\cdots+\gamma_{s-1}\beta^{s-1})\beta^n(1+\beta^n+\beta^{2n}+\cdots),
\end{aligned}$$

where $\alpha_i, \gamma_j \in \{0, 1\}$. As the series $1 + \beta^n + \beta^{2n} + \cdots$ diverges, then all $\gamma_j = 0$; whence

$$2 = \alpha_0 + \alpha_1\beta + \cdots + \alpha_{k-1}\beta^{k-1} + \beta^k, \tag{16.17}$$

for suitable $k \leqslant n - 1$, $\alpha_i \in \{0, 1\}$. If $\alpha_0 = 1$ then the right-hand side of (16.17) is not equal to the left-hand side; so $\alpha_0 = 0$ and therefore substituting $\beta = 1 + \tau$ and collecting terms of positive degrees in τ we get (by binomial theorem)

$$2 = \alpha_1\beta + \cdots + \alpha_{k-1}\beta^{k-1} + \beta^k = (\alpha_1 + \cdots + \alpha_{k-1} + 1) + \tau u(\tau),$$

where $u(x)$ is a polynomial of variable x whose coefficients are in $\mathbb{N}_0$. Hence, $1 = \alpha_1 + \cdots + \alpha_{k-1} + \tau u(\tau)$, where $\alpha_j \in \{0, 1\}$, $j = 1, 2, \ldots, k - 1$.

If $u(x)$ is a non-zero polynomial, then $\tau u(\tau) > 0$; so as $\alpha_1 + \cdots + \alpha_{k-1} \in \mathbb{N}_0$ we conclude that $\alpha_1 + \cdots + \alpha_{k-1} = 0$ since otherwise right-had side in $1 = \alpha_1 + \cdots + \alpha_{k-1} + \tau u(\tau)$ is strictly greater than left-hand side. Therefore all $\alpha_j = 0$ and thus $2 = \beta^k$, i.e., $\beta = \sqrt[k]{2}$.

If $u(x)$ is a zero polynomial then necessarily $\alpha_1 + \cdots + \alpha_{k-1} = 1$. Therefore there must be exactly one non-zero α_j; hence $2 = \beta^j + \beta^k$, where $0 < j < k$. However, $2 \neq \beta^j + \beta^k$ since $\beta > 1$, a contradiction.

The converse assertion of the theorem is obvious since addition of numbers represented by $\sqrt[N]{2}$ expansions is an "addition with carry to the N-th digit"; i.e., for example when $N = 2$ one has

$$
\begin{array}{cccccccccccc}
 & \ldots 1 & 1 & 1 & 1 & 1 & 1 & 1 & = & - & \sqrt{2} & - 1 \\
+ & & & & & & & & & & & \\
 & \ldots 0 & 0 & 0 & 0 & 0 & 0 & 1 & = & 1 & & \\
\hline
 & \ldots 0 & 1 & 0 & 1 & 0 & 1 & 0 & = & - & \sqrt{2} &
\end{array}
$$

□

It is worth warning the reader that Theorem 16.42 is *not* about calculation of Planck time whose value depends on the choice of units. In words, *Theorem 16.42 is about how much information one needs to agree the both worldviews, the Little-endian's and the Big-endian's.* Namely, Theorem 16.42 implies that the fuzzy wave function is the one which corresponds to an automaton over a 2^N-symbol alphabet, that is, to the automaton whose function is $f: \mathbb{Z}_2^N \to \mathbb{Z}_2^N$, i.e., a N-variate 2-adic 1-Lipschitz map which correspond to the automaton with N input binary channel and N output binary channels; see Sects. 1.6.1, 13.1.2, and Fig. 13.2. Also, f may be treated as a 1-Lipschitz map $\mathbb{Z}_2(\sqrt[N]{2}) \to \mathbb{Z}_2(\sqrt[N]{2})$, where $\mathbb{Z}_2(\sqrt[N]{2})$ is the ring of integers of the field $\mathbb{Q}_2(\sqrt[N]{2})$. Note that $x^N - 2$ is irreducible over $\mathbb{Q}_2$ by the Eisenstein's criterion (one may take $p = 2$ to fulfil conditions of the criterion since $a_0 = 2$, $a_N = 1$).

Theorem 13.74 implies that in the multivariate case the sharp wave function is of the form

$$\Psi_{\mathfrak{A}}(\vec{\mathbf{x}}, r) = \sum_{\left[\mathfrak{S}_{\mathbf{A},\vec{\mathbf{b}}}\right] \in \mathrm{Spec}(\mathfrak{A})} q_{\left[\mathfrak{S}_{\mathbf{A},\vec{\mathbf{b}}}\right]} e^{i(\vec{\mathbf{x}}\mathbf{A} - 2\pi p^r \vec{\mathbf{b}})}; \ (\vec{\mathbf{x}} \in \mathbb{R}^m; \ \vec{\mathbf{b}} \in \mathbb{R}^n; r \in \mathbb{Z}).$$

Therefore, *Theorem 16.42 implies that a univariate fuzzy wave function is actually a multivariate sharp wave function*; however, for a large number of dimensions: For instance, if $\sqrt[N]{2} = 1 + \tau$ where τ is of order of Planck time then $N \approx \frac{\ln 2}{\tau} \approx 10^{43}$; i.e., the automaton function of respective automaton is a 1-Lipschitz map $\mathbb{Z}_2^{10^{43}} \to \mathbb{Z}_2^{10^{43}}$. This means that the matrices $\mathbf{A}$ (see the formula above) in the sharp wave function $\Psi_{\mathfrak{A}}(\vec{\mathbf{x}}, r)$ are $10^{43} \times 10^{43}$; i.e., each of the matrices contains more entries than the number of atoms in our Universe.

An infinite-dimensional space is an adequate model for a 10^{43}-dimensional space; that is why both the Big-endian and Little-endian would agree on wave functions which "live" in Hilbert spaces.

The functions may be judged as "physical" since thy are elements of Hilbert space $\ell^2(\mathrm{Spec}(\mathfrak{A}))$ of square-summable complex sequences whose terms are indexed by elements of the set $\mathrm{Spec}(\mathfrak{A})$ (which is countable) since a "physical" wave function must be square-summable and the sum of squares of probability amplitudes must be 1. Recall that any separable Hilbert space is metrically isomorphic to ℓ^2 and that Fourier transform on the circle is such isomorphism between Hilbert space of square-integrable functions on $[0, 1] = \mathbb{I}$ and the space $\ell^2(\mathbb{Z})$ of square-summable complex sequences whose terms are enumerated by integers. It is not difficult to construct sharp wave functions which can be judged as "physical" in that meaning. Indeed, take any sequence $q_1, q_2, \ldots$ of positive real numbers such that $\sum_{j=1}^{\infty} q_j = 1$ and the series $\sum_{j=1}^{\infty} \sqrt{q}_j$ of positive square roots converges; construct by using Theorem 16.38 the automaton $\mathfrak{A}$. Then the function $\sum_{j=1}^{\infty} \sqrt{q}_j e^{i(a_j\rho - 2\pi p^k b_j)}$ is the one we are seeking for.

We finalize the subsection with the following interpretation:

Interpretation 16.43 (Discrete Spectrum; Continuous Spectrum) *The measure-0 ultimate affine automata may be treated as models of physical systems having discrete (energy, frequency, ...) spectra while measure-1 ultimate affine automata may be treated as models of physical systems having continuous spectra.*

16.8 The Entropy of Automata and Uncertainty Relation

In this section we formally derive an uncertainty relation which holds for automata wave functions. We stress that *despite Little-endian is capable to perform observation at the smallest of the scales while Big-endian is not, the uncertainty relation which can be treated as time-energy uncertainty holds for the both observers, i.e., for the Little-endian as well as for the Big-endian.*

The uncertainty relation we are going to deduce is an *entropic* one. To entropic uncertainty relations a number of research papers is devoted; see, e.g., expository paper [41] and references therein. The entropic uncertainty relation which is derived below is of a novel type since it relates time during which a system reaches a "pure state" which can be ascribed to a minimal affine sub-automaton, and the state (i.e., an element of $\mathrm{Spec}(\mathfrak{A})$) itself. Note that as Little-endian is capable to perform measurements at the smallest of the scales, the time the system reaches a state which belongs to some minimal sub-automaton is not 0, i.e., the "wave function collapse" is not momentary, it takes some minimal time intervals (e.g., some Planck times). Note that collapse of wave functions as a finite-time process is discussed in literature; see, e.g., [102].

To start with, we need to re-state the some results of Sect. 16.7.2 in terms of *prefix codes* since in what follows we use some basic properties of the codes; see Sect. 1.4.2 about these.

Note 16.44 From the proof of Theorem 16.38 it follows that *branches of the state transition diagram constitute a prefix code since each word which corresponds to a branch of length k reaches some minimal affine sub-automaton exactly at k-th step; thus the word cannot be a prefix of any other word which corresponds to another branch*: Note that words begin from the root of the tree; the root is initial state in the state-transition diagram. From the construction, it follows that the Kraft-MacMillan inequality (1.5) for that code is equality. However, by using Corollary 16.41 rather than Lemma 16.39 in the proof, one constructs a prefix code such that $\sum_{i=1}^{\infty} p^{-\ell_i} = q \leqslant 1$, for any given $0 < q \leqslant 1$. In that case the rest infinite paths of complete p-adic tree which lead to no minimal finite affine sub-automaton constitute a set of measure $1 - q$. The automaton having such state transition diagram will reach minimal sub-automata which are finite and affine with probability $0 < q \leqslant 1$ rather than exactly 1. To that automaton $\mathfrak{A}$ it corresponds a sharp wave function of the form (16.15) such that $\sum_{[\mathfrak{S}_{a,b}]\in \mathrm{Spec}(\mathfrak{A})} q_{[\mathfrak{S}_{a,b}]} = q$.

Let X be a random variable on the prefix code $\mathbf{G} = (w_i)_{i=1}^{\infty}$; denote $q_i = \mathrm{Prob}(X = w_j)$, the probability that X is equal to the word w_i. By the definition, the *entropy* $\mathbf{H}(X)$ *of the random variable* X is $\mathbf{H}(X) = -\sum_{i=1}^{\infty} q_i \log_p q_i$, whereas the *mean* length of the codeword is $\mathbf{E}(X) = \sum_{i=0}^{\infty} q_i \ell_i$. If $\mathbf{H}(X), \mathbf{E}(X) < +\infty$, then

$$\mathbf{H}(X) \leqslant \mathbf{E}(X) < \mathbf{H}(X) + 1, \tag{16.18}$$

with left-hand side inequality becoming an equality if and only if $q_i = p^{-\ell_i}$; there exists a prefix code such that $\ell_i = \lceil \log_p(1/q_i) \rceil$ for which the right-hand side inequality (which is not true in general) holds; see Sect. 1.4.2 and (1.6) there.

The time which a (both sharp and fuzzy) wave function takes to collapse can be expressed via the length of a word w which reaches a state from some minimal affine sub-automaton since the length of the word is the order of time t expressed in the respective base. That is why *in what follows we deal with lengths of the words rather than with time itself.* Note that when dealing with the lengths of words, we may restrict our considerations to the words over the alphabet $\{0, 1, \ldots, p-1\}$ where p is a prime since fuzzy wave functions are constructed by using words over the alphabet $\{0, 1\}$; see Sect. 16.7.2. The only difference between sharp and fuzzy wave functions constructions for $p = 2$ are numerical values which are assigned to words by the both observers: The Little-endian assigns numbers to words by reading the words as the base-2 expansions of numbers, whereas the Big-endian reads these words as $(1+\tau)$-representations of numbers for $0 < \tau \lll 1$. We stress that in what follows "mean time of collapse" stands for "mean word length" though the *actual* mean time of collapse *measured by the both observers* is different due to the inevitable non-zero measurement error: For instance, to the word of length

k whose prefix is $k-1$ zeros and suffix is 1, the Little-endian assigns the value 2^k, whereas the Big-endian assigns the value $(1+\tau)^k$ which for small τ and not too large k is indistinguishable for that observer from 1 due to the measurement error. To put it in other words, the Little-endian's measurements of time elapsed is much more accurate than the Big-endian's; the time within the wave function collapses is large for Little-endian whereas that time is zero for Big-endian up to the measurement accuracy of his equipment: Although the both clocks the observers use are 2-adic, however, by Theorem 16.42 the Big-endian can observe digits in the windows which are to the left of the $(N-1)$-th window at the face of the clock for N large whereas Little-endian observes digits to the left of the lowest-order position, i.e., from the rightmost window (cf. Fig. 16.1). Nevertheless, we are going to show that *"time-energy" uncertainty in terms of length of words in state transition diagrams of automata holds for the both observers.*

Let $\mathfrak{A}$ be an ultimately affine automaton; cf. Sect. 16.7.2. Define the *automaton entropy* as

$$H_{\mathfrak{A}} = -\sum_{[\mathfrak{S}_{a,b}]\in\mathrm{Spec}(\mathfrak{A})} q_{[\mathfrak{S}_{a,b}]}\log_p q_{[\mathfrak{S}_{a,b}]}.$$

For every $[\mathfrak{S}_{a,b}] \in \mathrm{Spec}(\mathfrak{A})$ the probability $q_{[\mathfrak{S}_{a,b}]}$ is equal to the sum of all $p^{-\Lambda(w)}$, where $\Lambda(w)$ is the length of a finite word w which reaches some state which belongs to some sub-automaton from $[\mathfrak{S}_{a,b}]$ exactly at the $\Lambda(w)$-th step; the sum is taken over all these words. Let $C[\mathfrak{S}_{a,b}]$ be a code whose codewords are all these words w; then $\sum_{(w\in[\mathfrak{S}_{a,b}])} p^{-\Lambda(w)} = q_{[\mathfrak{S}_{a,b}]}$. Note that $\sum_{[\mathfrak{S}_{a,b}]\in\mathrm{Spec}(\mathfrak{A})}\sum_{(w\in[\mathfrak{S}_{a,b}])} p^{-\Lambda(w)} = 1$, the codes $C[\mathfrak{S}_{a,b}]$ are disjoint for different $[\mathfrak{S}_{a,b}] \in \mathrm{Spec}(\mathfrak{A})$, the union of all these codes for all $[\mathfrak{S}_{a,b}] \in \mathrm{Spec}(\mathfrak{A})$ is a prefix code $C(\mathfrak{A})$ such that $\sum_{w\in C(\mathfrak{A})} p^{-\Lambda(w)} = 1$. According to the above convention, the mean time $T(\mathfrak{A})$ of wave function collapse is the mean length of a codeword of the code $C(\mathfrak{A})$: $T(\mathfrak{A}) = \sum_{w\in C(\mathfrak{A})} \Lambda(w)p^{-\Lambda(w)}$. The inequality (16.18) implies that

$$T(\mathfrak{A}) \geqslant H_T = -\sum_{w\in C(\mathfrak{A})} p^{-\Lambda(w)}\log_p(p^{-\Lambda(w)}) = \sum_{w\in C(\mathfrak{A})} \Lambda(w)p^{-\Lambda(w)} = T(\mathfrak{A}),$$

where H_T is the entropy of the code $C(\mathfrak{A})$. Therefore *the mean time of collapse of automaton wave function is equal to the entropy of the code $C(\mathfrak{A})$.*

For every $n \in \mathbb{N}$, let T_n be the set of all codewords of length n from the code $C(\mathfrak{A})$. If $T_n \neq \emptyset$ then T_n is a prefix code. All these codes are disjoint; their union is $C(\mathfrak{A})$. Therefore, $P_N = \sum_{n=1}^{N}\sum_{w\in T_n} p^{-n}$ is a probability the wave function collapses for a time not greater than N. Denote $H_{\mathrm{time}\leqslant\mathrm{N}}(\mathfrak{A})$ the entropy of the prefix code $T(N) = \bigcup_{n=1}^{N} T_n$; i.e., $H_{\mathrm{time}\leqslant\mathrm{N}}(\mathfrak{A}) = -\sum_{w\in T(N)} p^{-\Lambda(w)}\log_p(p^{-\Lambda(w)}) = \sum_{w\in T(N)} \Lambda(w)p^{-\Lambda(w)}$. As the probability assigned to $w \in C(\mathfrak{A})$ is $p^{-\Lambda(w)}$ and as $\sum_{w\in C(\mathfrak{A})} p^{-\Lambda(w)} = 1$, then, for N not less than the length of the shortest word from $C(\mathfrak{A})$) it holds $H_{\mathrm{time}\leqslant\mathrm{N}}(\mathfrak{A}) + H_{\mathfrak{A}} > 0$. Moreover, as $H_{\mathrm{time}\leqslant+\infty}(\mathfrak{A}) =$

$\sum_{w\in C(\mathfrak{A})} \Lambda(w)p^{-\Lambda(w)} \geqslant \sum_{w\in C(\mathfrak{A})} p^{-\Lambda(w)} = 1$, then

$$H_{\text{time}\leqslant+\infty}(\mathfrak{A}) + H_{\mathfrak{A}} \geqslant 1.$$

If to minimal affine sub-automata are ascribed "energy levels" (e.g., if in the sub-automata functions $z \to az + b$ coefficients a are different and $b = 0$),s these inequalities may be judged as time-energy uncertainty since *if an observer measures the time which takes wave function to collapse he does not know for sure to which of the states the wave function has collapsed; and if he knows to which of the states the wave function has collapsed, he does not know for sure how much time the collapse has taken.*

In general case, these inequalities cannot be sharpen: Since $\bigcup_{n=1}^{\infty} T_n = C(\mathfrak{A})$, then $C(\mathfrak{A})$ can be split arbitrarily into the disjoint union of sets $D_1, D_2, \ldots$ and, as each of D_j is itself a prefix code, there is an automaton $\mathfrak{D}$ such that $C[\mathfrak{S}_{a,b}] = D_j$, $(j = 1, 2, \ldots)$. Indeed, the entropy $H_{\text{time}\leqslant+\infty}(\mathfrak{A})$ is determined by the code $C(\mathfrak{A})$ only, whereas $H_{\mathfrak{A}}$ is determined completely by the partition of the code $C(\mathfrak{A})$ into arbitrary non-empty subsets and by the "assignment of limit plots" to each of the subsets.

The codeword lengths in $C(\mathfrak{A})$ can be arbitrary also by the following theorem on maximal prefix codes (see [174]):

Theorem 16.45 *For every non-decreasing function $\ell\colon \mathbb{N} \to \mathbb{N}$ such that $\sum_{n=1}^{\infty} p^{\,\ell(n)} = d \leqslant 1$ there exists a (maximal) prefix code $C = \{w_n : n \in \mathbb{N}\}$ such that $\Lambda(w_n) = \ell(n)$, for all $n \in \mathbb{N}$.*

Though it is not important at the moment we note that without the condition of maximality (which is also not needed at the moment), the statement of Theorem 16.45 remains true for every partial recursive function f, and respective inequality (the Kraft-McMillan inequality (1.5)) is known as Kraft-Chaitin inequality in this case, [35].

By Theorem 16.45, one can take any prefix code C for $d = 1$, split all its codewords into the partition $\mathbf{P}(C)$ of non-empty subsets, assign to every subset $S \in \mathbf{P}(C)$ a limit plot of a finite automaton $\mathfrak{S}_{a,b}(S)$ arbitrarily, and then construct respective automaton $\mathfrak{A}$ so that $C(\mathfrak{A}) = C$ and all finite paths from every S lead to the $\mathfrak{S}_{a,b}(S)$.

We have that $H_{\mathfrak{A}} \in [0, +\infty]$, $H_{\text{time}\leqslant+\infty}(\mathfrak{A}) \in [1, +\infty]$, $T(\mathfrak{A}) \in [1, +\infty]$ (as $T(\mathfrak{A}) = H_T = H_{\text{time}\leqslant+\infty}(\mathfrak{A})$), and nothing more definite can be said in general case.

It is possible that $H_{\text{time}\leqslant+\infty}(\mathfrak{A}) = 1$: For instance, take $p = 2$, $C(\mathfrak{A}) = \{1, 01, 001, 0001, \ldots\}$. Then the entropy $H_{\mathfrak{A}}$ may be equal to 1 if different limit plots are assigned to different balls $\mathbf{B}_{2^{-n-1}}(2^n)$. The entropy $H_{\mathfrak{A}}$ may be zero if limit plots which are assigned to all these balls are equal one to another. One may split the set of all these balls into a partition of pairwise disjoint non-empty subsets and assign to each ball a limit plot so that to all ball from a subset the same limit plot is assigned but to balls which belongs to different subsets one assigns different limit plots. In all these cases $H_{\text{time}\leqslant+\infty}(\mathfrak{A}) = 1$ (as the entropy is equal to $T(\mathfrak{A})$), but the entropies $H_{\mathfrak{A}}$ are different.

Finally, consider generating series $f_{\mathfrak{A}}(x) = \sum_{w\in C(\mathfrak{A})} x^{\Lambda(w)} = \sum_{n=1}^{\infty} t_n x^n$, where t_n is the number of all words of length n in the prefix code $C(\mathfrak{A})$. As $f_{\mathfrak{A}}(1/p) = 1$ then for the radius of convergence of the series it holds $R_{\mathfrak{A}} \geqslant 1/p$, with $f_{\mathfrak{A}}(1/p) = 1$. Hence, the function $f_{\mathfrak{A}}(x)$ is differentiable at all points from $(-R_{\mathfrak{A}}, R_{\mathfrak{A}})$ but if $R_{\mathfrak{A}} = 1/p$ then the derivative $f'_{\mathfrak{A}}(x)$ may not exist at $x = 1/p$ or may go to $+\infty$. Note that $f'_{\mathfrak{A}}(1/p) = pT(\mathfrak{A})$, i.e., the derivative $f'_{\mathfrak{A}}(1/p)$ determines the entropy $H_{\text{time}\leqslant+\infty}(\mathfrak{A})$.

Interpretation 16.46 *The uncertainty relation which is proved in this subsection may be treated as establishing some limitations of an experiment with a large number of identical systems prepared in identical states. All the systems are measured simultaneously, thus reaching some state whose observed value (e.g., frequency) is E_n, where n is the number of a system. The "time of relaxation" when all the systems reach the states may be very short, but never zero. The experimenter assumes that all the systems (or, nearly all) will relax in some definite state E and wants to determine the value of E which belongs to the set $\mathbb{E}(t)$, where t is time of relaxation, i.e., the time elapsed since the beginning of measurement until the numerical result of the experiment, the value of E, is supposed to be determined with a sufficient degree of reliability. Then the uncertainty relation which is proved above just means that the higher this degree of reliability of determining E, i.e., the smaller is the collection $\mathbb{E}(t)$, the bigger is t and vice versa; and it is impossible to achieve simultaneously the highest reliability within the shortest time of relaxation.*

16.9 Superdeterminism vs Classical Interpretations: Never the Twain Shall Meet?

The first remark is about very concrete applied problem, the "true randomness." Random numbers are needed in a number of applications, from computer simulations to data protection; that is why on the market there are two types of generators of random numbers. The first type are pseudorandom number generators (PRNG), the software/hardware implementations of algorithms which produce "random looking" sequences of numbers which by their algorithmic nature are, of course, *not* random, but are *mathematically proved* to pass a lot of statistical tests (see Chap. 14) or are shown that to distinguish the sequences from a "truly random" ones is an infeasible task since the problem is *mathematically proved* to be polynomially equivalent to some "hard" mathematical problem for which no algorithms solving the problem in polynomial time of the volume of input data are known.

The second type of generators constitute the ones which derive random numbers from some physical processes which are *believed to be "chaotic" and "unpredictable" by physical reasons*. The latter generators are usually called true random number generators (TRNG) and currently are mostly based on quantum processes, e.g., a chaotic motion in semiconductor superlattices, [189]. The *only* reason why the sequences of numbers produced by the generators of that kind are judged as "random" is that they pass standard batteries of statistical tests, e.g., NIST battery of tests or DIEHARD battery. But numerous PRNG have been found to produce sequences of unsatisfactory statistical quality for sensitive applications (in, e.g., cryptography) and rejected by that reason; but *all* these PRNG also pass all standard batteries of statistical tests. Thus the very practical question arises: How to guarantee that quantum TRNG produce "truly random" sequences of numbers which suit for application to sensitive areas? Statistical tests which the sequences pass as well as reference to "immanent randomness of quantum world" are way not enough to persuade a designer that a concrete TRNG is suitable for some concrete application. A real TRNG is a complex device which digitalize output of some physical process which is (or assumed to be) a "continuous" one, e.g., a current in electric circuit. The result of that digitalizing is what must be estimated. Therefore, we have an isolated physical system (or, which is assumed to be isolated) with which the digitalizing equipment interacts. To put it in other words, we have a dynamical system, the quantum one, and an observer, the digitalizing equipment; i.e., basically the model should be the same as the one depicted by Fig. 14.1, where the state update circuit is the *isolated* quantum system and output function is digitalizing equipment. This way we come to an *automaton model of isolated physical system which is an autonomous automaton whose state update and output functions are causal*. Moreover, semiconductor superlattice-based devices which supposedly perform "truly random mappings" of binary words to binary words are already known to exist; the idea of these devices originates from the paper [193]. Therefore, the *respective physical system*, the device, *is not isolated; it is open*, and the mapping of words the system performs is supposedly "truly random," i.e., is a result of random choice of 2^{2N} mappings, where N is bitlength of the words. *The system is causal since identical input words result in identical output words*. The system as a whole consists of a digit-to-analog converter, the superlattice, and the analog-to digit converter; i.e., is even more complex than a TRNG.

The question which immediately arises is as follows: In order to guarantee sufficient statistical quality of the output of TRNG or of the mapping performed by the device, one must at least *estimate the number of states of the automaton* since if the number is too small, then the statistical quality cannot be satisfactory. Estimates which follow Theorem 16.42 imply that the number of states is of order 10^{80}. Is this enough or not? A physicist may say "yes" since this is of order of number of all atoms in our Universe, but a mathematician may say "no" since this number is of order 2^{256}, i.e., ensures the so-called 256-bit security and no more. However, in the next generation of cryptography applications, the 512-bit security will be a must and maybe even a 1024-bit security. So what is better for such sensitive applications as cryptographic ones, quantum TRNG or software PRNG with 512-bit (or 1024-bit) seed (cf., Sect. 14.1) ? Moreover, in connection with the supposed randomness of the mapping from N-bit words to N-bit words which is performed by the superlattice-based device, if $N = 256$, then the number of states of the respective automaton must be not less than $2^{512} \approx 1.3 \times 10^{154}$. Is there enough particles in our Universe to ensure the "true randomness" of the mapping?

The practical importance of the "true randomness vs pseudorandomness" issue is a justification for the need to develop superdeterministic models of quantum mechanics.

The next remark is about the notion of wave function which is understood (or, better to say, interpreted) in quantum theory in very diverse meanings; see, e.g., [92]. To that end, the interpretation of automata wave functions is quite close to the *ψ-ensemble interpretation* from the just mentioned paper.

The automata which are ascribed to quantum systems should be understood as supposedly identical (i.e., have the identical automata functions), supposedly prepared in identical states (i.e., the initial states of all the identical automata are identical), and are probed in supposedly identical experiments by the Big-endian, the macro-observer (i.e., both the sets of finite input/output words of the identical automata are identical, and respective numerical data are identically read out by the equipment). The states of the state transition diagrams of the automata have nothing to do with the states of physical systems; these states are "epistemic"; only the minimal finite affine sub-automata (to be more exact, their limit plots) may be ascribed to pure states of a physical system.

Note that mixed states can also be interpreted in that manner: For instance the collection $\{c_0, c_0'\}$ of states of the automaton depicted by Fig. 16.12 may be treated as a mixed state of the whole system depicted by this figure.

The following remark is of a general character; it concerns the question which of the two kinds of interpretations of quantum mechanics, superdeterministic or the rest ones, better captures the most important physical features of Nature. We already have touched this question; see the end of Sect. 16.7.1. To continue, we note:

By its very construction, the sharp wave function (which is a discrete time function) can be viewed as an approximation of a fuzzy wave function which is a continuous time function. However, this may be stated in a completely opposite way: A "usual" (also, a fuzzy) wave function is an *interpolation* of a sharp wave function since there is no way to determine which of the functions is a "genuine" one as the very "continuity" (i.e., the Archimedean property of time) is just an assumption, a postulate which emerged from the everyday human experience at macro-level.

From that view, in a contrast to the main claim of [45], namely "standard" models of quantum mechanics may be judged as "overfitted" compared to the superdeterministic models since numerical data of measured physical quantities are always "discrete" rather than "continuous", the "experimental curves" are interpolations of the discrete set of points in Euclidean space, the numerical output of the experiment: It is impossible to determine experimentally for sure what sort of distribution represents statistics of the experiment, whether the distribution is absolutely continuous, discrete, or singular or the mixture of all three.

The very existence of the two kinds of models is a confirmation of *model-dependent realism* concept in the sense of Hawking-Mlodinow, [97]; so basically the superdeterministic models should not be disfavored against standard quantum mechanical models on the ground of "overfitting" (cf., [45]), only experimentally verified predictions which implies a particular model are a reason to accept or reject the model.

The final remark is also a general comment.

Causality By Its Very Nature Is Essentially *non-Archimedean*: The very notion of causality, being stated in rigorous mathematical terms, assumes that natural metric on the space of causes/effects is an ultrametric; see Sect. 16.2. However, the most of contemporary interpretations of quantum mechanics are based on *Archimedean* metric; therefore the conflict on basic concepts between superdeterminism and other interpretations is, in some sense, inevitable.

This is why the both types of interpretations may never agree on whether Nature is "discrete" or "continuous" since the "continuity" in its standard meaning merely assumes that any segment can be divided into smaller segments "ad infinitum"; i.e., the *Archimedean Axiom is an unspeakable assumption* in that case. By the same reason, both may never agree whether randomness is intrinsic feature of Nature or not (see Interpretations 16.34 and 16.35); or what is observer's "free will" (see Interpretations 16.11–16.16). ***A conflict between "locality" and "causality" seems***

therefore possible and likely since the "locality" is commonly understood as a spatial phenomenon in Archimedean metric space whereas "causality" resides in a non-Archimedean metric space; and never the twain shall meet. Nonetheless, the "two strong men," the Big-endian and the Little-endian, will "meet each other in the middle of the scales": The wave function is the meeting point.

References

1. Ahmed, E., Savchuk, D.: Endomorphisms of regular rooted trees induced by the action of polynomials on the ring F_d of d -adic integers. J. Algebra Appl. **19**(8), 2050154 (2020)
2. Aliprantis, C.D., Burkinshaw, O.: Principles of Real Analysis, 3rd edn. Academic, Cambridge (1998)
3. Allouche, J.P., Shallit, J.: Automatic Sequences. Theory, Applications, Generalizations. Cambridge University Press, Cambridge (2003)
4. Amice, Y.: Interpolation p-adique. Bull. Soc. Math. France **92**, 117–180 (1964)
5. Anashin, V.: Uniformly distributed sequences over p-adic integers. In: van der Poorten, I.S.A.J., Zimmer, H.G. (eds.) Number Theoretic and Algebraic Methods in Computer Science. Proceedings of the Int'l Conference (Moscow, June–July, 1993), pp. 1–18. World Scientific, Singapore (1995)
6. Anashin, V.S.: Mixed identities in groups. Mat. Zametki **24**(1), 19–30 (1978)
7. Anashin, V.S.: Solvable groups with operators and commutative rings having transitive polynomials. Algebra Logic **21**(6), 419–432 (1982). Translated from Russian
8. Anashin, V.S.: Mixed identities and mixed varieties of groups. Mat. Sb. (N. S.) **129**(2), 163–174 (1986)
9. Anashin, V.S.: Uniformly distributed sequences of p-adic integers. Math. Notes **55**(2), 109–133 (1994)
10. Anashin, V.S.: Uniformly distributed sequences in computer algebra, or how to construct program generators of random numbers. J. Math. Sci. **89**(4), 1355–1390 (1998)
11. Anashin, V.S.: Uniformly distributed sequences of p-adic integers, II. Discrete Math. Appl. **12**(6), 527–590 (2002)
12. Anashin, V.: Ergodic transformations in the space of p-adic integers. In: Khrennikov, A.Y., Rakić, Z., Volovich, I.V. (eds.) p-Adic Mathematical Physics. 2-nd Int'l Conference (Belgrade, Serbia and Montenegro 15–21 September 2005), AIP Conference Proceedings, vol. 826, pp. 3–24. American Institute of Physics, Melville (2006)
13. Anashin, V.: Noncommutative algebraic dynamics: Ergodic theory for profinite groups. Proc. Steklov Instit. Math. **265**, 30–58 (2009)
14. Anashin, V.: Non-Archimedean ergodic theory and pseudorandom generators. Comput. J. **53**(4), 370–392 (2010). https://doi.org/10.1093/comjnl/bxm101
15. Anashin, V.: Automata finiteness criterion in terms of van der Put series of automata functions. p-Adic Num. Ultramet. Analy. Appl. **4**(2), 151–160 (2012)
16. Anashin, V.: The non-Archimedean theory of discrete systems. Math. Comp. Sci. **6**(4), 375–393 (2012)

V. Anashin, *Causality: The* p*-adic Theory*, STEAM-H: Science, Technology, Engineering, Agriculture, Mathematics & Health,
https://doi.org/10.1007/978-3-031-85818-5

17. Anashin, V.S.: Quantization causes waves: Smooth finitely computable functions are affine. p-Adic Num. Ultramet. Analy. Appl. **7**(3), 169–227 (2015)
18. Anashin, V.S.: Smooth finitely computable functions are affine, or why quantum systems are wave systems. Doklady Math. **92**(3), 165–167 (2015)
19. Anashin, V.: Discreteness causes waves. Facta Univ. **14**(6), 143–196 (2016)
20. Anashin, V.S.: On automata models of blockchain. Inf. Appl. **13**(2), 29–36 (2019). In Russian, English Summary
21. Anashin, V.: The p-adic theory of automata functions. In: Advances in Non-Archimedean Analysis and Applications, STEAM-H: Science, Technology, Engineering, Agriculture, Mathematics & Health, pp. 9–113. Springer Nature, Berlin (2021)
22. Anashin, V.: Toward the (non-cellular) automata interpretation of quantum mechanics: Volovich postulates as a roadmap. Int. J. Modern Phys. A **37**(20–21), 2243003 (2022)
23. Anashin, V.: Free choice in quantum theory: A p-adic view. Entropy **25**(5), 830 (2023)
24. Anashin, V., Khrennikov, A.: Applied Algebraic Dynamics. de Gruyter Expositions in Mathematics, vol. 49. Walter de Gruyter GmbH, Berlin (2009)
25. Anashin, V., Khrennikov, A., Yurova, E.: Using van der Put basis to determine if a 2-adic function is measure-preserving or ergodic w.r.t. Haar measure. In: Advances in Non-Archimedean Analysis. Contemporary Mathematics, vol. 551, pp. 33–38. American Mathematical Society, Providence (2011)
26. Anashin, V., Khrennikov, A., Yurova, E.: Ergodicity criteria for non-expanding transformations of 2-adic spheres. Discrete Contin. Dyn. Syst. **34**(2), 367–377 (2014)
27. Anashin, V., Khrennikov, A., Yurova, E.: T-functions revisited: new criteria for bijectivity/transitivity. Designs Codes Cryptogr. **71**(3), 383–407 (2014)
28. Assaf IV, D., Gadbois, S.: Definition of chaos. Am. Math. Month **99**(9), 865–865 (1992)
29. Axelsson, E.Y., Khrennikov, A.: Generation of genetic codes with 2-adic codon algebra and adaptive dynamics. Biosystems **240**, 105230 (2024)
30. Banks, J., Brooks, J., Cairns, G., Davis, G., Stacy, P.: On Devaney's definition of chaos. Am. Math. Month. **99**, 332–334 (1992)
31. Brauer, W.: Automatentheorie. B. G. Teubner, Stuttgart (1984)
32. Bruijn, N.G.D.: Some classes of integer-valued functions. Nederl. Akad. Wetensch. Proc, Ser. A **58**, 363–367 (1955)
33. Bryk, J., Silva, C.E.: Measurable dynamics of simple p-adic polynomials. Am. Math. Monthly **112**(3), 212–232 (2005)
34. Cahen, P.J., Chabert, J.L.: Integer-Valued Polynomials. Mathematical Surveys and Monographs, vol. 48. American Mathematical Society, Providence (1997)
35. Calude, C., Grozea, C.: Kraft-Chaitin inequality revisited. J. Univ. Comput. Sci. **2**(5), 306–310 (1996)
36. Chaudhuri, S., Sankaranarayanan, S., Vardi, M.Y.: Regular real analysis. In: 28th Annual ACM/IEEE Symposium on Logic in Computer Science (LICS 2013), pp. 509–518. IEEE Computer Society, Los Alamitos (2013)
37. Cherepov, A.N.: On approximation of continuous functions by determinate functions with delay. Discrete Math. Appl. **22**(1), 1–24 (2010)
38. Cherepov, A.N.: Approximation of continuous functions by finite automata. Discrete Math. Appl. **22**(4), 445–453 (2012)
39. Chlodovsky, M.I.: Une remarque sur la représantation des focntions continues par des polynomes á coefficients entiers. Math. Sb., pp. 472–474 (1925)
40. Ciepielewski, G.S., Okon, E., Sudarsky, D.: On superdeterministic rejections of settings independence. Br. J. Philos. Sci. **74**(2), 435–467 (2023)
41. Coles, P.J., et al.: Entropic uncertainty relations and their applications. Rev. Mod. Phys. **89**(1), 015002–1–015002–58 (2017)
42. Conway, J.H., Kochen, S.: The strong free will theorem. Notices Am. Math. Soc. **56**(2), 226–232 (2009)
43. Cook, M.: Universality in elementary cellular automata. Compl. Syst. **15**(1), 1–40 (2004)
44. Crowell, R., Fox, R.: Introduction to the Knot Theory. Ginu and Co., Boston (1963)

45. Daley, P.J., Resch, K.J., Spekkens, R.W.: Experimentally adjudicating between different causal accounts of Bell-inequality violations via statistical model selection. Phys. Rev. A **105**, 042220 (2022)
46. Delaygue, E., Rivoal, T.: On primary pseudo-polynomials (around Ruzsa's conjecture). Int. J. Num. Theory **18**, 1613–1636 (2022)
47. Desjardins, D.L., Zieve, M.E.: On the structure of polynomial mappings modulo an odd prime power (2001). Available at http://arXiv.org/math.NT/0103046
48. Devaney, R.L.: An Introduction to Chaotic Dynamical Systems, 2nd edn. Addison-Wesley, Boston (1989)
49. Diestel, J., Spalsbury, A.: The Joys of Haar Measure. American Mathematical Society, Providence (2014)
50. Dilao, R., Amigó, J.: Computing the topological entropy of unimodal maps. Int. J. Bifurcat. Chaos **22**(6), 1250152 (2012)
51. Djordjievich, G.S., Dragovich, B., Neshich, L.D., Volovich, I.V.: p-adic and adelic minisuperspace quantum cosmology. Int. J. Modern Phys. A **17**(10), 1413–1433 (2002)
52. Downarowicz, T.: Entropy in Dynamical Systems. Cambridge University Press, Cambridge (2011)
53. Downarowicz, T., Lacroix, Y.: Measure-theoretic chaos. Ergod. Th. Dynam. Sys. **34**, 110–131 (2012)
54. Dragovich, B.: p-adic and adelic cosmology: p-adic origin of dark energy and dark matter. AIP Conf. Proc. **826**(25), 25–42 (2006)
55. Dubrovin, B.A., Fomenko, A.T., Novikov, S.P.: Modern Geometry–Methods and Applications, vol. II. Springer, Berlin (1985)
56. Durand, F., Paccaut, F.: Minimal polynomial dynamics on the set of 3-adic integers. Bull. London Math. Soc. **41**(2), 302–314 (2009)
57. Eichenauer, J., Lehn, J., Topuzoglu, A.: A nonlinear congruential pseudorandom number generator with power of two modulus. Math. Comp. **51**, 757–759 (1988)
58. Eichenauer-Herrmann, J., Grothe, H.: A new inversive congruential pseudorandom number generator with power of two modulus. ACM Trans. Modell. Comput. Simul. **2**, 1–11 (1992)
59. Eichenauer-Herrmann, J., Herrmann, E., Wegenkittl, S.: A survey of quadratic and inversive congruential pseudorandom numbers. In: Hellekalek, P., Larcher, G., Niederreiter, H., Zinterhof, P. (eds.) Monte Carlo and Quasi-Monte Carlo Methods 1996. Lecture Notes in Statistics, vol. 127, pp. 66–97. Springer, Berlin (1998)
60. Eilenberg, S.: Automata, Languages, and Machines, vol. A. Academic, Cambridge (1974)
61. Erdös, P., Joó, I., Komornik, V.: Characterization of the unique expansions $1 = \sum_{i=1}^{\infty} q^{-n_i}$ and related problems. Bull. Soc. Math. Fr. **118**, 377–390 (1990)
62. Everest, G., van der Poorten, A., Shparlinsky, I., Ward, T.: Recurrence Sequences. American Mathematical Society Surveys, vol. 104. American Mathematical Society, Providence (2003)
63. Fan, A.H., Liao, L.M.: On minimal decomposition of p-adic polynomial dynamical systems. Adv. Math. **228**, 2116–2144 (2011)
64. Fan, A.H., Wang, Y.F.: On p-adic dynamical systems. In: Proceedings of the International Congress of Chinese Mathematicians, vol. 2, pp. 773–799 (2007)
65. Fan, A., Fan, S., Liao, L., Wang, Y.: On minimal decomposition of p-adic homographic dynamical systems. Adv. Math. **257**, 92–135 (2014)
66. Ferguson, L.B.O.: Approximation by Polynomials with Integral Coefficients. American Mathematical Society, Providence (1980)
67. Flood, M.D., Goodenough, O.R.: Contract as automaton: the computational representation of financial agreements. SSRN Electr. J. (2015). https://doi.org/10.2139/ssrn.2538224
68. Folland, G.B.: Real Analysis: Modern Techniques and their Applications, 2nd edn. Wiley, Hoboken (1999)
69. Furno, J.: Orbit equivalence of p-adic transformations and their iterates. Monatsh. Math. **175**, 249–276 (2014)
70. Furno, J.: Singular p-adic transformations for Bernoulli product measures. N. Y. J. Math. **20**, 799–812 (2014)

71. Furno, J.: Natural extensions for p-adic β-shifts and other scaling maps. Indag. Math. **30**, 1099–1108 (2019)
72. Gégalkine, J.J.: Sur le calcul des propositions dans la logique symbolique. Matematicheskii Sbornik **34**(1), 9–28 (1927). Russian and French
73. Gill, A.: Introduction to the Theory of Finite-State Machines. McGraw-Hill, New York (1963)
74. Goldin, D., Smolka, S.A., Wegner, P. (eds.): Interactive Computation, Chap. A Theory of System Interaction: Components, Interfaces, and Services, pp. 41–96. Springer, Berlin (2006)
75. Golomb, S.W., Peile, R.E., Scholtz, R.A.: Basic Concepts in Information Theory and Coding. Springer, Berlin (1994)
76. Gorman, A.B., et al.: Continuous regular functions. Logical Methods Comput. Sci. **16**(1), 17:1–17:24 (2020)
77. Gouvêa, F.Q.: p-adic Numbers. An Introduction, 2nd edn. Springer, Berlin (1997)
78. Graham, R.L., Knuth, D.E., Patashnik, O.: Concrete Mathematics: A Foundation for Computer Science, 2nd edn. Addison–Wesley, Reading (1998)
79. Grätzer, G.: Lattice Theory: Foundation. Birkhäuser, Basel (2011)
80. Green, J.W.: Functions which assume rational values at rational points. Duke Math. J. **5**(1), 164–171 (1939)
81. Griffith, D.J.: Introduction to Quantum Mechanics, 2nd edn. Pearsom Prentice Hall, Upper Saddle River (2005)
82. Grigorchuk, R.I.: Some topics in the dynamics of group actions on rooted trees. Proc. Steklov Inst. Math. **273**, 64–175 (2011)
83. Grigorchuk, R., Savchuk, D.: Solenoid maps, automatic sequences, van der Put series, and Mealy-Moore automata. J. Austral. Math. Soc. **114**(1), 78–109 (2023). https://doi.org/10.1017/S1446788722000027
84. Grigorchuk, R.I., Nekrashevich, V.V., Sushchanskii, V.I.: Automata, dynamical systems, and groups. Proc. Steklov Instit. Math. **231**, 128–203 (2000)
85. Gross, J.L., Yellen, J., Anderson, M.: Graph Theory and Its Applications, 3rd edn. CRC Press, Boca Raton (2019)
86. Gundlach, M., Khrennikov, A.Y., Lindahl, K.O.: Topological transitivity for p-adic dynamical systems. In: Katsaras, A.K., Schikhof, W., van Hamme, L. (eds.) P-adic Functional Analysis. Lecture Notes in Pure and Applied Mathematics, vol. 222, pp. 127–132. Dekker, New York (2000)
87. Gundlach, V.M., Khrennikov, A.Y., Lindahl, K.O.: Ergodicity on p-adic sphere. In: German Open Conference on Probability and Statistics, pp. 15–21. University of Hamburg Press, Hamburg (2000)
88. Gundlach, M., Khrennikov, A.Y., Lindahl, K.O.: On ergodic behavior of p-adic dynamical systems. Inf. Dim. An., Quantum Prob. Related Fields **4**(4), 569–577 (2001)
89. Hagar, A.: Discrete or Continuous? Cambridge University Press, Cambridge (2014)
90. Hall, R.R.: On pseudo-polynomials. Arch. Math. **18**, 71–77 (1971)
91. Halmos, P.R.: Lectures on Ergodic Theory. Publications of the Mathematical Society of Japan, Kenkyusha (1956)
92. Hance, J.R., Hossenfelder, S.: The wave function as a true ensemble. Proc. R. Soc. A **478**, 20210705 (2022)
93. Hance, J., Hossenfelder, S.: Comment on "experimentally adjudicating between different causal accounts of bell-inequality violations via statistical model selection". Phys. Rev. A **109**(2), 26201 (2024)
94. Hance, J.R., Hossenfelder, S., Palmer, T.N.: Supermeasured: Violating Bell statistical independence. Found. Phys. **52**, 81 (2022)
95. Hasselblatt, B., Katok, A. (eds.): Handbook of Dynamical Systems, vol. 1A. Elsevier Science B. V., Amsterdam (2002)
96. Hasselblatt, B., Katok, A.: A First Course in Dynamics. Cambridge University Press, Cambridge (2003)
97. Hawking, S., Mlodinow, L.: The Grand Design. Bantam Books, New York (2010)

98. Hieronymi, P., Walsberg, E.: A tetrachotomy for expansions of the real ordered additive group. Selecta Math. **27**, 54 (2021)
99. Hofmann, K.H., Morris, S.A.: The Structure of Compact Groups, 3rd edn. Walter de Gruyter GmbH, Berlin (2013)
100. 't Hooft, G.: The Cellular Automaton Interpretation of Quantum Mechanics, Fundamental Theories of Physics, vol. 185. Springer, Berlin (2016)
101. Hossenfelder, S., Palmer, T.: Rethinking superdeterminism. Front. Phys. **8**, Article 139 (2020)
102. Ignatiev, A.Y.: How fast is the wave function collapse? J. Phys. Conf. Ser. **410**, 012153 (2013)
103. Jang, Y., Jeong, S., Li, C.: Criteria of measure-preservation for 1-Lipschitz functions on $\mathbb{F}_q[[T]]$ in terms of the van der Put and its applications. Finite Fields Appl. **37**, 131–157 (2016)
104. Jang, Y., Jeong, S., Li, C.: Measure-preservation criteria for 1-Lipschitz functions on $\mathbf{F}_q[[T]]$ in terms of the three bases of Carlitz polynomials, digit derivatives, and digit shifts. Finite Fields Appl. **46**, 304–325 (2017)
105. Jeong, S.: Characterization of the ergodicity of 1-Lipschitz functions on $\mathbb{Z}_2$ using the q-Mahler basis. J. Num. Theory **151**, 116–128 (2015)
106. Jeong, S.: Measure-preservation and the existence of a root of p-adic 1-Lipschitz functions in Mahler's expansion. p-Adic Num. Ultrametr. Analy. Appl. **10**(3), 192–208 (2018)
107. Jeong, S., Li, C.: Measure-preservation criteria for a certain class of 1-Lipschitz functions on $\mathbb{Z}_p$ Mahler's expansion. Discrete Contin. Dyn. Syst. **37**(7), 3787–3804 (2017)
108. Kaiser, H.K., Nöbauer, W.: Permutation polynomials in several variables over residue class rings. J. Austral. Math. Soc. **A43**, 171–175 (1987)
109. Kalman, R.E., Falb, P.L., Arbib, M.A.: Topics in Mathematical System Theory. McGraw-Hill, New York (1969)
110. Kato, T., Wu, L.M., Yanagihara, N.: On a nonlinear congruential pseudorandom number generator. Math. Comp. **65**, 227–233 (1996)
111. Katok, A., Hasselblatt, B.: Introduction to the Modern Theory of Dynamical Systems. Cambridge University Press, Cambridge (1998)
112. Katok, S.: p-adic Analysis in Comparison with Real. MASS Selecta. American Mathematical Society, Providence (2003)
113. Kechris, A.S.: Classical Descriptive Set Theory. Springer, Berlin (1995)
114. Keedwel, D., Dénes, J.: Latin squares and their applications, 2nd edn. North-Holland, Amsterdam (2015)
115. Kemeny, J.G., Snell, J.L.: Finite Markov Chains. Springer, Berlin (1976)
116. Khrennikov, A.Y.: Non-Archimedean Analysis: Quantum Paradoxes, Dynamical Systems and Biological Models. Kluwer Academic Publishers, Dordrecht (1997)
117. Khrennikov, A., Yurova, E.: Criteria of measure-preserving for p-adic dynamical systems in terms of the van der Put basis. J. Num. Theory **133**, 484–491 (2013)
118. Khrennikov, A.Y., Lindahl, K.O., Gundlach, M.: Ergodicity in the p-adic framework. In: Operator Methods in Ordinary and Partial Differential Equations. Operator Methods: Advances and Applications, vol. 132. Birkhauser, Basel (2002)
119. Kingsbery, J., Levin, A., Preygel, A., Silva, C.E.: On measure-preserving c1 transformations of compact-open subsets of non-archimedean local fields. Trans. Am. Math. Soc. **361**(1), 61–85 (2009)
120. Kingsbery, J., Levin, A., Preygel, A., Silva, C.E.: Dynamics of the p-adic shift and applications. Discrete Contin. Dyn. Syst. **30**(1), 209–218 (2011)
121. Klimov, A., Shamir, A.: A new class of invertible mappings. In: Kaliski, B.S. (ed.) Cryptographic Hardware and Embedded Systems 2002. Lecture Notes in Computer Science, vol. 2523, pp. 470–483. Springer, Berlin (2003)
122. Knudsen, C.: Chaos without nonperiodicity. Amer. Math. Month. **101**(6), 563–565 (1994)
123. Knuth, D.: The Art of Computer Programming, vol. 2:Seminumerical Algorithms, 3rd edn. Addison-Wesley, Boston (1997)
124. Koblitz, N.: p-Adic Numbers, p-Adic Analysis, and Zeta-Functions. Graduate Texts in Mathematics, vol. 58, 2nd edn. Springer, Berlin (1984)

125. Kolesov, A.Y., Rozov, N.K.: On the definition of "chaos". Russ. Math. Surv. **64**(4), 701–744 (2009)
126. Kolmogorov, A.N., Fomin, S.V.: Elements of the Theory of Functions and Functional Analysis, vol. 1. Graylock Press, Rochester (1957)
127. Kolyada, S.F.: Li-Yorke sensitivity and other concepts of chaos. Ukrainian Math. J. **56**(8), 1242–1257 (2004)
128. Konečný, M.: Real functions computable by finite automata using affine representations. Theor. Comput. Sci. **284**, 373–396 (2002)
129. Kozyrev, S.V.: Wavelet theory as p-adic spectral analysis. Izvestiya: Math. **66**(2), 367–376 (2002)
130. Kuipers, L., Niederreiter, H.: Uniform Distribution of Sequences. Wiley, New York (1974)
131. Lane, S.M.: Categories for the Working Mathematician, 2nd edn. Springer, Berlin (1998)
132. Larin, M.V.: Transitive polynomial transformations of residue class rings. Disc. Math. Appl. **12**(2), 141–154 (2002)
133. Lausch, H., Nöbauer, W.: Algebra of Polynomials. North-Holl, American Elsevier, Cambridge (1973)
134. Laywine, C.F., Mullen, G.L.: Discrete Mathematics Using Latin Squares. Wiley, New York (1998)
135. Lerner, E.: On synchronizing automata and uniform distribution. In: Han, Y.S., Salomaa, K. (eds.) Implementation and Application of Automata. Lecture Notes in Computer Science, vol. 9705, pp. 202–212. Springer, Berlin (2016)
136. Lerner, E.E.: Uniform distribution of sequences generated by iterated polynomials. Doklady Math. **92**(3), 704–706 (2015)
137. Lerner, E.: The uniform distribution of sequences generated by iterated polynomials. p-Adic Num. Ultramet. Analy. Appl. **11**(4), 280–298 (2019)
138. Lidl, R., Niederreiter, H.: Finite Fields. Addison-Wesley, Boston (1983)
139. Lin, D., Shi, T., Yang, Z.: Ergodic theory over $\mathbb{F}_2[[T]]$. Finite Fields Appl. **18**, 473–491 (2012)
140. Lindahl, K.O.: Dynamical systems in p-adic geometry. Licentiate Thesis. School of Mathematics and Systems Engineering, Växjö University (2001)
141. Lisovik, L.P., Shkaravskaya, O.Y.: Real functions defined by transducers. Cyber. Syst. Analy. **34**(1), 69–76 (1998)
142. Loève, M.: Probability Theory I, 4th edn. Springer, Berlin (1977)
143. Lothaire, M.: Algebraic Combinatorics on Words. Cambridge University Press, Cambridge (2002)
144. Lunts, A.G.: The p-adic apparatus in the theory of finite automata. Problemy Kibernetiki **14**, 17–30 (1965). In Russian
145. Mahler, K.: p-adic Numbers and Their Functions, 2nd edn. Cambridge University Press, Cambridge (1981)
146. Makholm, H.: Functions that take rationals to rationals. http://math.stackexchange.com/questions/167620/functions-that-take-rationals-to-rationals (2012)
147. Mansurov, V.: Knot Theory. Chapman & Hall/CRC, Boca Raton (2004)
148. Marsaglia, G.: Random numbers fall mainly in the planes. Proc. Natl. Acad. Sci. **61**(1), 25–28 (1968)
149. McDonald, B.R.: Finite Rings with Identity. Dekker, New York (1974)
150. Memić, N.: Characterization of ergodic rational functions on the set of 2-adic units. Int. J. Num. Theory **13**(05), 1119–1128 (2017)
151. Memić, N.: Ergodic polynomials on 2-adic spheres. Bull. Polish Acad. Sci. Math. **65**, 35–44 (2017)
152. Memić, N.: Mahler coefficients of 1-Lipschitz measure-preserving functions on $\mathbb{Z}_p$. Int. J. Num. Theory **16**(6), 1247–1261 (2020)
153. Mishchenko, A., Fomenko, A.: A course of differential geometry and topology. Mir, Moscow (1988)

154. Monna, A.F.: Sur une transformation simple des nombres p-adiques en nombres réels. Indag. Math. **14**, 1–9 (1952)
155. Mukhamedov, F.M., Khakimov, O.N.: On metric properties of unconventional limit sets of contractive non-Archimedean dynamical systems. Dyn. Syst. **31**(4), 506–524 (2016)
156. Niederreiter, H.: Random number generation and quasi-Monte Carlo methods. Society for Industrial and Applied Mathematics (SIAM), Philadelphia (1992)
157. Oppenheim, J., Reznik, B., Unruh, W.G.: Temporal ordering in quantum mechanics. J. Phys. A Math. Gen. **35**, 7641–7652 (2002)
158. Palmer., T.N.: Discretization of the Bloch sphere, fractal invariant sets and Bell's theorem. Proc. R. Soc. A **406**, 20190350 (2020)
159. Pettigrew, J., Roberts, J.A.G., Vivaldi, F.: Complexity of regular invertible p-adic motions. Chaos **11**, 849–857 (2001)
160. Pin, J.E.: Profinite methods in automata theory. In: Symposium on Theoretical Aspects of Computer Science — STACS 2009, pp. 31–50. Freiburg (2009)
161. Qu, C., Zhu, Z., Zhou, Z.: A note on the perturbed monomial mapping. Appl. Math. J. Chinese Univ. **34**(1), 76–81 (2019)
162. Rendell, P.: Turing Machine Universality of the Game of Life. Emergence, Complexity and Computation, vol. 18. Springer, Berlin (2016)
163. Rivest, R.: Permutation polynomials modulo 2^w. Finite Fields Appl. **7**(2), 287–292 (2001)
164. Robert, A.M.: A Course in p-adic Analysis. Springer, Berlin (2000)
165. Rogers, C.A.: Hausdorff Measures. Cambridge Universy Press, Cambridge (1970)
166. Rozikov, U.A., Sattarov, I.A.: Dynamical systems of the p-adic (2, 2)-rational functions with two fixed points. Results Math. **75**, 100 (2020)
167. Salomaa, A.: Theory of Automata. Pergamon Press, Oxford (1969)
168. Schikhof, W.H.: Ultrametric Calculus: An Introduction to p-adic Analysis. Cambridge University Press, Cambridge (1984)
169. Shi, T., Anashin, V., Lin, D.: Linear weaknesses in T-functions. In: Helleseth, T., Jedwab, J. (eds.) SETA 2012. Lecture Notes in Computer Science, vol. 7280, pp. 279–290. Springer, Berlin (2012)
170. Shkaravskaya, O.Y.: Affine mappings defined by finite transducers. Cyber. Syst. Analy. **34**(5), 781–783 (1998)
171. Shor, O., Benninger, F., Khrennikov, A.: Quantization of events in the event-universe and the emergence of quantum mechanics. Sci. Rep. **13**(17865) (2023)
172. Silva, C.E.: Invitation to Ergodic Theory. American Mathematical Society, Providence (2008)
173. Smyshlyaeva, T.I.: A criterion for functions defined by automata to be bounded-determinate. Diskret. Mat. **25**(2), 121–134 (2013)
174. Staiger, L.: On maximal prefix codes. Bull. EATCS **91**, 205–207 (2007)
175. Tao, R.: Finite Automata and Application to Cryptography. Tsinghua University Press, Springer, Berlin (2008)
176. Tyapaev, L.B.: Measure-preserving and ergodic asynchronous automata mappings. In: Kasim-Zade, O.M. (ed.) Proceedings of the 12th International Workshop on Discrete Mathematics and its Applications (June 20-26, 2016, Moscow), pp. 398–400. Lomonosov Moscow State University, Moscow (2016). In Russian
177. Tyapaev, L.B.: Transitive families and measure-preserving an n-unit delay mappings. In: Proceedings of the International Conference on Computer Science and Information Technologies (June 30-July 2, 2016, Saratov), pp. 425–429. Publishing Center Nauka, Saratov (2016)
178. Tyapaev, L.B.: Ergodic automata mappings with delay. In: Zhuravlev, Y.I. (ed.) Proceedings of the International Conference on Problems of Theoretical cybernetics (June 19-23, 2017, Penza), pp. 242–244. Maks Press, Moscow (2017). In Russian
179. Ushakov, N.G.: Selected Topics in Characteristic Functions. DeGruyter, Berlin (1999)
180. van Rooij, A.: Non-Archimedean Functional Analysis. Dekker, New York (1978)
181. Vanchurin, V.: The World as a Neural Network. Entropy **20**, 1210 (2020)
182. Vladimirov, V.S., Volovich, I.V., Zelenov, E.I.: p-adic Analysis and Mathematical Physics. World Scientific, Singapore (1994)

183. Vuillemin, J.: On circuits and numbers. IEEE Trans. Comput. **43**(8), 868–879 (1994)
184. Vuillemin, J.: Finite digital synchronous circuits are characterized by 2-algebraic truth tables. In: Advances in Computing Science - ASIAN 2000. Lecture Notes in Computer Science, vol. 1961, pp. 1–7 (2000)
185. Vuillemin, J.: Digital algebra and circuits. In: Verification: Theory and Practice. Lecture Notes in Computer Science, vol. 2772, pp. 733–746 (2003)
186. Washington, L.: On the self-duality of $\mathbf{Q}_p$. Am. Math. Month. **81**(4), 369–371 (1974)
187. Weaver, G.: König's infinity lemma and Beth's tree theorem. History Philos. Logic **38**(1), 48–56 (2017)
188. Wheeler, J.A.: Information, physics, quantum: The search for links. In: Zurek, W.H. (ed.) Complexity, Entropy, and the Physics of Information, pp. 309–336. Addison-Wesley, Redwood City (1990)
189. Wu, H., et al.: Design and implementation of true random number generators based on semiconductor superlattice chaos. Microelectron. J. **114**, 105119 (2021)
190. Yablonsky, S.V.: Introduction to Discrete Mathematics. Mir, Moscow (1989)
191. Yang, Z.: Ergodic functions over $\mathbb{F}_q[[T]]$. Finite Fields Appl. **53**, 189–204 (2018)
192. Yurov, I.A.: On p-adic functions preserving the Haar measure. Math. Notes **63**(5–6), 823–836 (1998)
193. Zhang, Y., Kastrup, J., Klann, R., Ploog, K.H.: Synchronization and chaos induced by resonant tunneling in GaAsyAlAs superlattices. Phys. Rev. Lett. **77**(14), 3001–3004 (1996)
194. Zuniga-Galindo, W.A.: Non-archimedean quantum mechanics via quantum groups. Nuclear Phys. B **985**, 116021 (2022)
195. Zuniga-Galindo, W.A., He, C., Zambrano-Luna, B.A.: p-adic statistical field theory and convolutional deep. Prog. Theor. Exp. Phys. **2023**, 063A01 (2023)

Index

V. Anashin, *Causality: The* p*-adic Theory*, STEAM-H: Science, Technology, Engineering, Agriculture, Mathematics & Health,
https://doi.org/10.1007/978-3-031-85818-5

The manufacturer's authorised representative in the EU is Springer Nature Customer Service Centre GmbH, Europaplatz 3, 69115 Heidelberg, Germany. If you have any concerns regarding our products, please contact ProductSafety@springernature.com

Printed and bound by CPI Group (UK) Ltd, Croydon, CR0 4YY
30/09/2026
02225568-0017